W9-CUI-024

Greenhouse Gas Sinks

To John and Joan

Greenhouse Gas Sinks

Edited by

David S. Reay

School of GeoSciences, University of Edinburgh, UK

C. Nick Hewitt

Lancaster Environment Centre, University of Lancaster, UK

Keith A. Smith

School of GeoSciences, University of Edinburgh, UK

John Grace

School of GeoSciences, University of Edinburgh, UK

CABI is a trading name of CAB International

CABI Head Office
Nosworthy Way
Wallingford
Oxfordshire OX10 8DE
UK
Tel: +44 (0)1491 832111
Fax: +44 (0)1491 833508
E-mail: cabi@cabi.org
Website: www.cabi.org

CABI North American Office
875 Massachusetts Avenue
7th Floor
Cambridge, MA 02139
USA
Tel: +1 617 395 4056
Fax: +1 617 354 6875
E-mail: cabi-nao@cabi.org

A catalogue record for this book is available from the British Library, London, UK.

A catalogue record for this book is available from the Library of Congress, Washington, DC.

ISBN: 978 1 84593 189 6

Typeset by SPi, Pondicherry, India.
Printed and bound in the UK by Athenaeum Press Ltd, Gateshead.

Contents

Contributors

Michele Aresta, *Department of Chemistry and CIRCC, University of Bari, Via Celso Ulpiani 27, Bari, Italy, E-mail: m.aresta@chimica.uniba.it*

Malcolm Asadoorian, *Joint Program on the Science and Policy of Global Change, Massachusetts Institute of Technology, 77 Massachusetts Avenue, Cambridge, MA 02139-4307, USA*

Klaus Butterbach-Bahl, *Institute for Meteorology and Climate Research, Atmospheric Environmental Research (IMK-IFU), Kreuzeckbahnstrasse 19, D-82467 Garmisch-Partenkirchen, Germany*

Pascal Boeckx, *Laboratory of Applied Physical Chemistry (ISOFYS), Ghent University, Coupure 653, 9000 Ghent, Belgium*

Lex Bouwman, *Netherlands Environmental Assessment Agency, RIVM, Bilthoven, The Netherlands, E-mail: Lex.Bouwman@mnp.nl*

Christopher L. Butenhoff, *Department of Physics, Portland State University, PO Box 720, Portland, OR 97202-0751, USA, E-mail: christop@global.phy.pdx.edu*

Angela Dibenedetto, *Department of Chemistry and CIRCC, University of Bari, Via Celso Ulpiani 27, Bari, Italy*

Peter F. Dunfield, *Institute of Geological and Nuclear Sciences, Wairakei Research Station, Private Bag 2000, Taupo, New Zealand, E-mail: p.dunfield@gns.cri.nz*

Alex De Visscher, *Department of Chemical and Petroleum Engineering, Schulich School of Engineering, University of Calgary, 2500 University Drive N.W., Calgary, Alberta T2N 1N4, Canada*

Wim de Vries, *Wageningen University and Research Centre, Alterra, Droevendaalse steeg 3, 6700 AA Wageningen, The Netherlands, E-mail: Wim.devries@wur.nl*

Ursula Edwards, *Occidental Oil and Gas Corporation, Houston, TX 77056, USA, E-mail: andy@seao2.org, Ursula.Edwards@BHPBilliton.com*

Richard A. Feely, *NOAA Pacific Marine Environmental Laboratory, 7600 Sand Point Way NE, Seattle, WA 98115, USA, E-mail: richard.a.feely@noaa.gov*

Benjamin Felzer, *The Ecosystems Center, Marine Biological Laboratory, 7 MBL Street, Woods Hole, MA 02543, USA*

John Grace, *School of GeoSciences, University of Edinburgh, Crew Building, West Mains Road, Edinburgh EH9 3JN, UK, E-mail: jgrace@ed.ac.uk*

C. Nick Hewitt, *Lancaster Environment Centre, University of Lancaster, Lancaster LA1 4YQ, UK, E-mail: N.Hewitt@lancaster.ac.uk*

Graham Hymus, *Department of Biological Sciences, Northern Arizona University, Building 21, South Beaver St, Flagstaff, AZ 86011-5604, USA, E-mail: Graham.Hymus@NAU.EDU*

Phil Ineson, *Department of Biology (SEIY), University of York, YO10 5YW, UK, E-mail: pi2@york.ac.uk*

H. Henry Janzen, *Agriculture and AgriFood Canada, Lethbridge, Alberta, Canada, E-mail: Janzen@AGR.GC.CA*

M. Aslam K. Khalil, *Department of Physics, Portland State University, PO Box 720, Portland, OR 97202-0751, USA*

David Kicklighter, *The Ecosystems Center, Marine Biological Laboratory, 7 MBL Street, Woods Hole, MA 02543, USA*

Carolien Kroeze, *Environmental Sciences, Environmental Systems Analysis Group, Wageningen University, PO Box 47, 6700 AA Wageningen, The Netherlands, E-mail: Carolien.kroeze@wur.nl*

Reynald L. Lemke, *Agriculture and AgriFood Canada, Swift Current, Saskatchewan, Canada, E-mail: Lemker@AGR.GC.CA*

Jerry Melillo, *The Ecosystems Center, Marine Biological Laboratory, 7 MBL Street, Woods Hole, MA 02543, USA*

Oene Oenema, *Wageningen University and Research Centre, Alterra, Droevendaalse steeg 3, 6700 AA Wageningen, The Netherlands*

David S. Reay, *School of GeoSciences, University of Edinburgh, Crew Building, West Mains Road, Edinburgh EH9 3JN, UK, E-mail: David.Reay@ed.ac.uk*

John Reilly, *Joint Program on the Science and Policy of Global Change, Massachusetts Institute of Technology, 77 Massachusetts Avenue, Cambridge, MA 02139-4307, USA*

Andy Ridgwell, *School of Geographical Sciences, University of Bristol, Bristol, UK*

Christopher L. Sabine, *NOAA Pacific Marine Environmental Laboratory, 7600 Sand Point Way NE, Seattle, WA 98115, USA, E-mail: chris.sabine@noaa.gov*

Dudley E. Shallcross, *Biogeochemistry Research Centre, School of Chemistry, University of Bristol, Bristol BS8 1TS, UK, E-mail: d.e.shallcross@bris.ac.uk*

Caroline P. Slomp, *Department of Earth Sciences – Geochemistry, Faculty of Geosciences, Utrecht University, PO Box 80021, 3508 TA Utrecht, The Netherlands, E-mail: slomp@geo.uu.nl*

Keith A. Smith, *School of GeoSciences, University of Edinburgh, Crew Building, West Mains Road, Edinburgh EH9 3JN, UK, E-mail: Keith.Smith@ed.ac.uk*

Pete Smith, *School of Biological Sciences, University of Aberdeen, Cruickshank Building, St Machar Drive, Aberdeen AB24 3UU, UK, E-mail: pete.smith@abdn.ac.uk*

Hanqin Tian, *School of Forestry and Wildlife Sciences, Auburn University, SFWS Building, 602 Duncan Drive, Auburn, AL 36849-5418, USA, E-mail: jreilly@mit.edu*

Riccardo Valentini, *Universitá della Tuscia, DISAFRI, Via S. Camillo de Lellis, Viterbo 01100, Italy*

Oswald Van Cleemput, *Laboratory of Applied Physical Chemistry (ISOFYS), Ghent University, Coupure 653, 9000 Ghent, Belgium, E-mail: Oswald.VanCleemput@UGent.be*

Hugo Denier van der Gon, *TNO Environment and Geosciences, Laan van Westenenk 501, 7300 AH Apeldoorn, The Netherlands*

Preface

The debate about human-induced climate change may still be raging in certain sections of the media, but here we start from a position of acceptance: acceptance that increased greenhouse gas (GHG) concentrations in the atmosphere are leading to increased global temperatures, and that human activities are playing an ever-growing role in this increase.

To briefly set the scene, let us look at the evidence for human influence on global climate. The 'greenhouse effect' is essentially the trapping of infrared radiation from the earth's surface by 'greenhouse gases' (though greenhouses warm up primarily through the glass keeping heat in, rather than through GHGs trapping the heat). We actually need GHGs in our atmosphere; without them the average temperature on earth would be around −18°C. Unfortunately, in the last 200 years or so the activities of humans have pumped a lot of extra GHG into the atmosphere and this has led to an 'enhanced greenhouse effect' and thereby to the extra global warming we are now experiencing.

By increasing the atmosphere's ability to absorb infrared energy, our greenhouse emissions are disturbing the way the climate maintains the balance between incoming and outgoing energy. A doubling of the concentration of GHGs (predicted in the next 100 years) would, if nothing else changed, reduce the rate at which the planet can shed energy into space by about 2%. The climate will somehow have to adjust – by heating up – and while 2% does not seem much, across the entire earth it is equivalent to trapping the energy content of about 3 million tonnes of oil every 10 min.

Long-term data-sets for global temperature indicate a clear and consistent increase in global temperatures, particularly since the industrial revolution. Much of the argument surrounding climate change hinges not on whether global warming is occurring, but rather whether this warming is a result of human activity. Natural drivers of global climate, such as solar and volcanic activity, have been invoked by some to explain the accelerated rate of global warming observed in recent decades. However, such natural drivers of global temperatures cannot explain the warming that has occurred since 1840. The prime culprit for the observed increase: elevated GHG concentrations.

As concentrations of the main GHGs have risen in our atmosphere, global temperatures have also increased. A rapid increase in atmospheric concentrations of the three main anthropogenic (human-made) GHGs – carbon dioxide (CO_2), methane (CH_4) and nitrous oxide (N_2O) – is clear from measurements over the past few decades. Ice-core records for these gases show that their concentrations in the atmosphere are now higher than at any time in the last 650,000 years, and probably even in the last 20 million years.

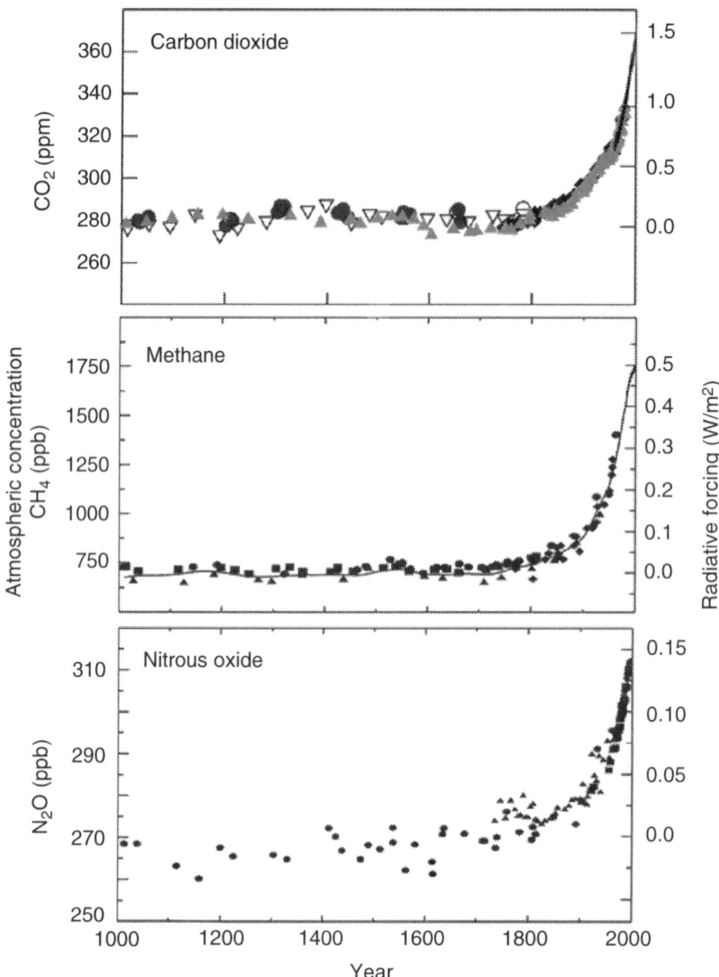

Fig. P.1. Global atmospheric concentrations of three well-mixed greenhouse gases. (From the Intergovernmental Panel on Climate Change (IPCC). Reproduced with permission.)

Since around the time of the industrial revolution, levels of CO_2, CH_4 and N_2O have all risen dramatically. Fossil fuel combustion, increasingly intensive agriculture and an expanding global human population have been the primary causes for this rapid increase.

Sulphate aerosols, though not GHGs, are none the less very important to global climate. Sulphate in our atmosphere has a net cooling effect and therefore reduces the warming effect of the GHGs to a certain extent. The same increases in fossil fuel burning that have led to elevated GHG concentrations in the last 200 years have also led to an increase in sulphate emissions. Cleaner fuel technologies are now leading towards a reduction in sulphate emissions and their incidental cooling effect (commonly known as global dimming) on our climate. If GHG emissions continue to increase, their overall warming effect may become even greater.

The complex interaction of positive and negative influences (the feedbacks) on global climate, together with the uncertainty over how anthropogenic GHG emissions will change in coming decades makes predicting future warming difficult. The problem is exacerbated by our poor level of understanding of exactly how some key factors, such as albedo (the reflectance of the earth's surface) and cloud formation, operate and interact with a changing climate. What we do know is that increases in concentrations of GHGs in our atmosphere risk rapid increases in global temperature. This in turn is likely to lead to rapid changes in climate and sea level that may threaten civilization itself. Indeed, the publication of the Stern Report in 2006 highlighted the grave economic consequences of our failing to tackle human-induced global warming immediately.

It is against this backdrop that *Greenhouse Gas Sinks* has been written. This book brings together chapters written by leading scientists from across Europe, the USA, Canada and Australia. It represents an expert synthesis of GHG sink science and how this applies to the past, current and future changes in climate. The 'sinks' discussed here are the planet's storage areas, where GHGs are locked away from the atmosphere and thus prevented from contributing to global warming.

As will become apparent, the sinks for CO_2, CH_4 and N_2O play a vital role in determining the concentrations of these GHGs in our atmosphere, but all are vulnerable to human activities.

The aim of this book is to provide readers with in-depth, authoritative information on these sinks. We will explore how the sinks may respond to increased GHG emissions and global temperatures, and whether we can protect and even enhance them to help mitigate climate change. These are urgent questions that must be answered as we face this greatest of challenges to humankind in the 21st century.

David S. Reay
Edinburgh, UK
November 2006

Acknowledgements

Any book with the scope of *Greenhouse Gas Sinks* relies a great deal on the time, effort and expertise of numerous contributors. The editors would like to acknowledge their world-class input and their patient assistance with peer review of each of the chapters. David Reay would also like to acknowledge the assistance of the UK's Natural Environment Research Council in helping to support this and ongoing work relating to greenhouse gas research.

1 Carbon Dioxide: Importance, Sources and Sinks

David S. Reay and John Grace
School of GeoSciences, University of Edinburgh, Edinburgh, UK

1.1 Introduction

Carbon dioxide (CO_2) is without doubt the best-known anthropogenic greenhouse gas. As long ago as 1895, the Swedish Nobel laureate Svante Arrhenius saw that the increased emissions of CO_2 resulting from a rapid rise in fossil fuel burning had the potential to affect global temperatures. In his landmark paper 'On the Influence of Carbonic Acid (Carbon Dioxide) in the Air upon the Temperature of the Ground' he considered the radiative effects of CO_2 and water vapour on the surface temperature of the earth. Arrhenius calculated that if the concentrations of CO_2 increased by 250–300% compared to 1895 levels, temperatures in the Arctic could rise by 8–9°C. At the time his paper was published, such an increase in atmospheric CO_2 concentrations remained theoretical. Even if global CO_2 concentrations were increasing, there was no way to reliably measure such increases.

In 1958 the late Charles Keeling and his colleagues began measurements of CO_2 at Mauna Loa in Hawaii. Mauna Loa is a huge mid-ocean volcano, rising to over 4000 m above sea level. The height and location of Mauna Loa make it an ideal place to monitor changes in the composition of the atmosphere. It is away from any major local sources of CO_2, and its great height means that CO_2 concentrations measured there are representative of those across much of the northern hemisphere.

The classic data-set of climate changes obtained at Mauna Loa – the 'Keeling record' (Fig. 1.1) – has shown a rapid increase in the mixing ratio of CO_2 in the atmosphere, rising from 315 ppm in 1960 to over 380 ppm in 2005. The Mauna Loa record is not just notable for the rapid increase in CO_2 that it shows, but also for the sink signal it contains. Rather than increasing as a steady rate, the line zigzags up and down over each year, being highest in the northern hemisphere winter and lowest in the summer. This is the atmospheric signal of a sink – the massive uptake of CO_2 by terrestrial and oceanic plants during summer in the northern hemisphere.

This zigzag provides clear evidence of the importance of such sinks in controlling the concentrations of greenhouse gases in our atmosphere, but it also shows that this control is limited. The zigzag may be obvious, but even more so is the underlying upward trend – the sinks are no longer balancing the sources.

Since the industrial revolution, concentrations of CO_2 in our atmosphere have increased at an ever-faster rate, and are now 30% greater. The rapid increase in CO_2 emissions observed during the last 250 years is expected to continue for several decades to

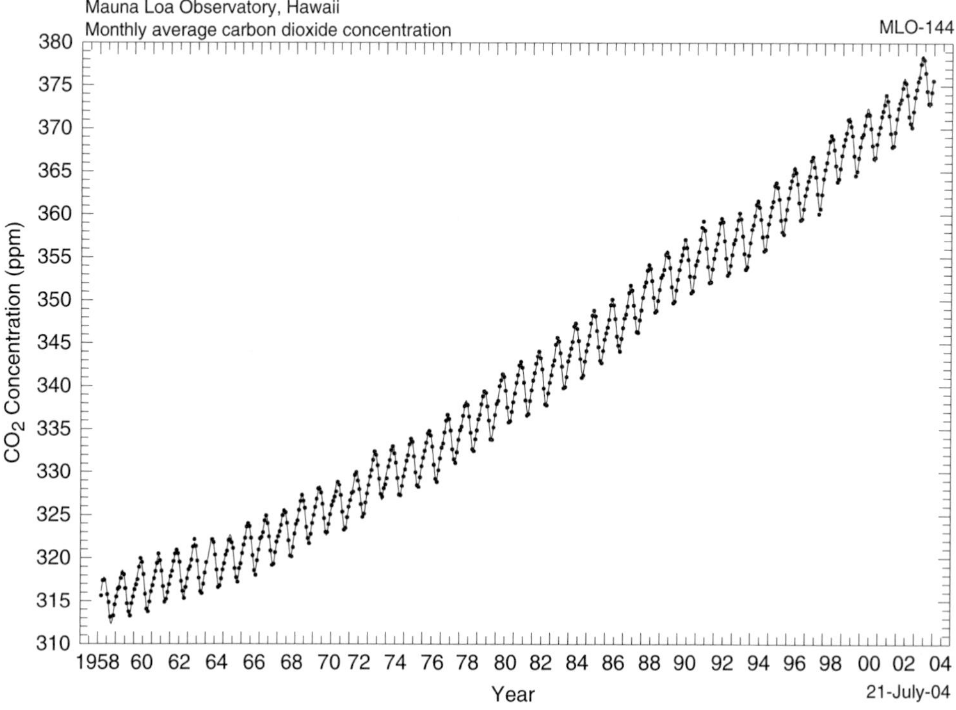

Fig. 1.1. Monthly averaged carbon dioxide (CO_2) concentration measured at Mauna Loa from 1958 to 2004. (From Keeling, C.D. and Whorf, T.P. (2004) Atmospheric CO_2 concentrations derived from flask air samples at sites in the SIO network. In: *Trends: A Compendium of Data on Global Change*. Carbon Dioxide Information Analysis Center, Oak Ridge National Laboratory, US Department of Energy, Oak Ridge, Tennessee.)

come. Various scenarios have been examined, depending on factors like fossil fuel use and efficiency. Even the best case scenario predicts further increases in CO_2 emissions until at least 2040.

Many of the scenarios indicate that by the mid-21st century emissions of CO_2 should at least start to level off, though some predict increases in emissions throughout this century. Although the different scenarios predict a wide range of trends in emissions, the predicted net effect on atmospheric CO_2 concentrations in the future is fairly consistent. All predict a further increase in CO_2 concentrations by the end of this century, with some of the scenarios predicting a doubling or even trebling of current levels. If the predicted increases in CO_2 and other greenhouse gas concentrations are translated into temperature changes, a global temperature increase of 1.4–5.8°C is predicted for 2010. This compares to an increase of ~0.6°C over the last 100 years.

The large variation between predictions of the different scenarios underlines the complexity involved in making such predictions and the great uncertainty inherent in climate change models. Key to these predictions is a host of assumptions about what effect the varied feedbacks to the global carbon cycle will have. For instance, there is the possibility of a reversal from CO_2 sink to source in Amazonia, due to increased drying caused by climate change. There is also the so-called CO_2 fertilization effect, where increased concentrations of CO_2 in the atmosphere may promote faster plant growth and result in CO_2 removal from the atmosphere (see Hymus and Valentini, Chapter 2, this volume). In soils there is the positive feedback of increased temperatures resulting in higher decomposition and respiration rates, leading to greater CO_2 emissions (Smith and Ineson, Chapter 4, this volume). In particular, the thawing of permafrost soils may result in greatly increased emissions of both CO_2 and methane.

Even in the oceans, where the buffering capacity for increased global temperatures is much larger than on land, increased stratification arising from warming of surface waters may reduce rates of CO_2 uptake by phytoplankton due to nutrient limitation (Sabine and Feely, Chapter 3, this volume).

In the following chapters we will see how the various carbon sinks and sources interact, and how the myriad of positive and negative feedbacks within the global climate system affect them now, and could affect them in the future.

1.2 Global Carbon Cycle

Figure 1.2 shows the main components of the natural carbon cycle with the various storages represented in terms of Pg (billions of tonnes) of carbon (C) and fluxes (Pg/year) estimated for the 1980s. The thick black lines represent the most important fluxes, while the thinner lines denote smaller yet significant transfers of carbon. These natural fluxes into and out of the atmosphere are approximately balanced over the course of a year. Note that the CO_2 emissions from fossil fuel burning and cement production are estimated to have increased from 5.4 Pg C/year in the 1980s to 6.3 Pg C/year in the 1990s. As we will see, this human perturbation to the global carbon cycle is still relatively small compared to the huge natural fluxes, but its cumulative effect on concentrations of CO_2 in the atmosphere, and therefore on global climate, is substantial.

In considering the global balance, or imbalance, of carbon we should first examine the natural sources of CO_2.

1.3 Natural Sources

1.3.1 Respiration

Respiration, both on land and in the sea, is a key component of the global carbon cycle. On land, an estimated 60 Pg C/year is emitted to the atmosphere by autotrophic respiration. A similar amount, ~55 Pg C, is emitted as a result of heterotrophic respiration.

In the sea, autotrophic respiration is thought to account for ~58 Pg C/year of the dissolved inorganic carbon in surface waters, with the contribution of heterotrophic respiration being 34 Pg C. Most of the input of dissolved organic carbon (DOC) to oceans from rivers (~0.8 Pg C/year) is respired and released to the atmosphere. Physical air–sea exchange of this dissolved CO_2 leads to an emission to the atmosphere of ~88 Pg C/year.

Although the worldwide human population has now grown to more than 6 billion,

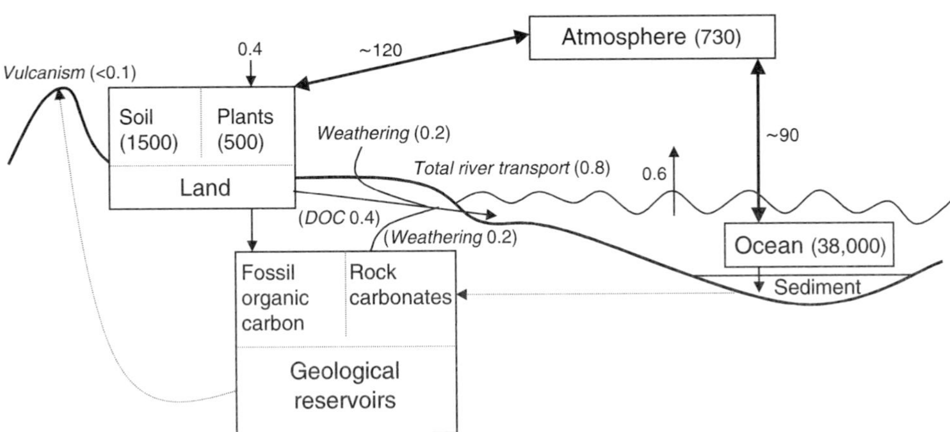

Fig. 1.2. Main components of the natural carbon cycle. (Reproduced with permission from the Intergovernmental Panel on Climate Change (IPCC).)

our direct contribution to atmospheric CO_2 concentrations via respiration is relatively insignificant. Of greater concern is the impact human-induced global warming could have on global respiration rates. As temperatures increase, rates of respiration also rise in many organisms; most microbes, for instance, double their rate of respiration with every 10°C increase in temperature. There exists therefore a danger of warming, inducing further increases in atmospheric CO_2 concentrations, and thus more warming – an example of a positive feedback.

1.3.2 Vulcanism and biomass burning

Emissions of CO_2 due to volcanic activity, though sometimes large on a local scale, are relatively minor on a global scale, accounting for 0.02–0.05 Pg C/year. Of more importance, climate-wise, is the release of aerosols from volcanic eruptions, as these can cause a net cooling of the planet's climate.

Although fires caused by lightning strikes have accounted, and still do account, for some large biomass burning events – and thus emissions of CO_2 – the impact of this source on atmospheric CO_2 concentrations is relatively short term due to rapid uptake of CO_2 during regrowth of vegetation. Globally, direct and indirect biomass burning activities of humans in the past few centuries have dwarfed emissions from such natural biomass burning. Let us now consider this and the numerous other anthropogenic sources of CO_2.

1.4 Anthropogenic Sources

1.4.1 Land-use change

Land-use change is estimated to contribute 10–30% of all current anthropogenic CO_2 emissions. It is estimated that human-made changes in land use have, until now, produced a cumulative global loss from the land of ~200 Pg C. Widespread deforestation has been the main source of this loss, estimated to be responsible for nearly 90% of losses since the mid-19th century. These losses primarily occur due to the relatively long-term carbon stocks of forests being replaced by agricultural land.

The conversion of land from forested to agricultural land can have a wide range of effects as far as CO_2 emissions are concerned. Soil disturbance and increased rates of decomposition in converted soils can both lead to emission of CO_2 to the atmosphere, with increased soil erosion and leaching of soil nutrients further reducing the potential for the area to act as a sink for atmospheric carbon. Similarly, land reclamation and changes in land use management can effect an increase in terrestrial carbon uptake. Current estimates suggest that such land-use changes lead to the emission of 1.7 Pg C/year in the tropics, mainly as a result of deforestation, and to a small amount of uptake (~0.1 Pg C) in temperate and boreal areas, thus producing a net source of ~1.6 Pg C/year.

This balance between land-use change being a source and being a sink of CO_2 is highly vulnerable to social and economic pressures. Humankind's need for wood – for fuel and construction – and our ever-increasing need for agricultural land have led to systematic clearances of forests across the planet in the past few hundred years. Today the pressure on forested areas is huge, with a rapidly growing human population requiring food and the land necessary for its cultivation. Increased awareness of the most sensitive way to manage land, tied to better agricultural practice and combined with political agreement on food trade and avoidance of deforestation, is required if land-use change is not to become an even greater net source of carbon to the atmosphere in years to come.

1.4.2 Energy-related emission

Strictly speaking, energy-related CO_2 emissions should also include transport, but to simplify things this section will deal only with 'stationary sources', such as power stations. Of the emissions arising from fossil fuel burning – a total of ~6.5 Pg C/year – nearly one-half is a result of electricity generation, either by electric utilities or through domestic energy use.

The amount of CO_2 emitted as a result of the generation of a given unit of electricity varies greatly depending on the fuel used and the level of efficiency at which the power plant operates. Generally speaking, coal-fuelled power generation is the most carbon-intensive, with the emission of up to 1 kg of CO_2 for every kilowatt-hour (kWh) of electricity that it provides. Oil- and gas-fuelled electricity generation tends to have a lower CO_2 emission cost. Even those energy-generation strategies without apparent use of fossil fuels generally have some associated CO_2 emissions. Nuclear power, for instance, relies on large amounts of energy use for fuel extraction and processing, and so indirectly results in CO_2 emissions. The construction of any power station, wind turbine or other power-generation facility carries with it an emission cost through the embodied energy of the materials used in its construction. This cost must be included if the full climate benefits of any one type of energy generation are to be accurately assessed and compared with others.

Humankind's demand for, and generation of, electricity for heating, cooking, light and so on has grown at an astonishing rate during the last century. Now, as in the past, there are huge geographic disparities in energy consumption. Individuals in developed countries such as the UK and USA use much more electricity per person than in most developing countries.

The efficiency of electricity generation has increased greatly and continues to do so, although there is a large amount of energy wastage in the form of waste heat in most fossil fuel-burning power stations. The introduction of combined heat and power (CHP) systems, in which waste heat is captured and used for residential or business purposes, can raise the efficiency of coal-burning power stations from below 40% to 60%.

Such improvements in efficiency, however, have been outstripped by energy demand, and therefore energy-related emissions of CO_2 continue to rise rapidly. In the longer term, some developed countries may have to accept that their level of energy consumption is unsustainable. In the shorter term, greater use of 'low carbon' energy-generation technologies, such as geothermal, wind, wave, solar and tidal power, may provide significant reductions in emissions.

1.4.3 Transport

Globally, transport-related emissions of CO_2 are growing rapidly. The use of petroleum as a fossil fuel for transportation dominates CO_2 emissions from this source. In 1999, in the USA, more than 30% of fossil fuel-related CO_2 emissions were a direct result of transportation, with about two-thirds of this coming from petrol consumption by motor vehicles and the remainder coming from diesel and jet fuel use in lorries and aircraft, respectively.

In addition to a rapid increase in worldwide motor vehicle use, the widespread use of large-engined private cars has led to increased CO_2 emissions in many countries, despite substantial increases in engine efficiencies. The falling price of national and international flights has also led to an explosion in air traffic in recent years, with air travel being one of the most greenhouse gas-intensive forms of transport. Technological advances are continuing to increase efficiencies, but at the same time car, lorry and airplane use is escalating. The emission of greenhouse gas from transport is likely to be one of the most fiercely fought battlegrounds of environmental reform during this century.

1.4.4 Industry

CO_2 emissions arising from industrial processes are substantial. In addition to the emissions associated with electricity generation, as discussed previously, processes such as cement and lime production also represent significant sources of anthropogenic CO_2.

CO_2 is produced in lime and cement manufacture as a result of the heating of limestone.

The final amount of CO_2 produced varies depending on the type of cement being made. Globally, this source of emission is estimated to amount to 0.2 Pg C/year to the atmosphere.

It is hoped (as discussed by Reilly *et al.*, Chapter 8, this volume) that future trading of 'carbon credits' and the likely increase in financial cost of greenhouse gas emissions to individual companies will lead to substantial cuts in CO_2 emissions from industry through improved efficiency and greenhouse gas capture.

1.4.5 Biomass burning

Exact CO_2 emissions from biomass burning are difficult to quantify due to a dearth of information on fire-carbon fluxes, and the longer-term balance of carbon emissions with carbon uptake by regrowth of vegetation. In savannah regions of the world, burning is often carried out every few years to promote regeneration of the vegetation. Large amounts of CO_2 are therefore produced, but in many cases the subsequent regrowth and CO_2 uptake of savannah areas mean that the net carbon emission is much reduced or completely negated. Biomass burning is therefore an important short-term source of CO_2. In the longer term it can actually contribute to a net carbon sink in the form of relatively stable 'black carbon' deposited to the soils where burning has taken place (as discussed in Section 1.5.3).

Huge areas of woodland and grassland are periodically burned for land clearance, with this change in land use often reducing the size of the soil and vegetation carbon sinks, thus having the net effect of increasing concentrations of CO_2 in the atmosphere. Additionally, wood burning as a domestic fuel source and for charcoal production releases significant amounts of CO_2 on a global scale. Accidental fire and arson account for further large-scale biomass burning events each year. The uncontrolled way in which most biomass burning occurs means that the only real route to reducing emissions from this source is to lessen the amount of burning itself. Some biomass burning is required if environments such as the savannah are to be retained, but it is the large-scale destruction of forest areas for cash crop agriculture and urban spread that stands out as an area to be tackled.

Controlled biomass burning is being developed as an alternative to traditional fossil fuel-based energy production, with power stations fuelled by forest residues already a reality. By making use of a renewable resource like pinewood chips, these biomass power stations are able to have a greatly reduced net greenhouse gas impact per kilowatt-hour of energy output. The CO_2 they release is effectively in a closed loop, where it is incorporated into more trees, which, when large enough, are harvested and used to fuel more energy output from the biofuel power plant.

1.5 Sinks

The sources of CO_2 to the atmosphere are diverse. In particular, the wide range of sources of anthropogenic CO_2 emissions means that reducing these emissions is far from straightforward. In the debate over how best to tackle human-induced climate change, a parallel is often drawn between this problem and that of ozone depletion by chlorofluorocarbons (CFCs). In the case of CFCs, efforts to cut emissions were targeted at a small number of sources (e.g. refrigerants and aerosol propellants) and, through implementation of the Montreal Protocol, large emission reductions were achieved in a relatively short period of time. However, attempts to cut anthropogenic greenhouse gas emissions must address emissions from myriad sources, rather than from just a few.

Even more problematic is the lack of straightforward alternatives. Where CFCs were replaced by hydrofluorocarbons (HFCs), and then by hydrocarbons, without great difficulty and cost, alternatives to key sources of anthropogenic greenhouse gas emissions, such as fossil fuel burning, are much harder to implement.

In addressing rising greenhouse gas concentrations in the atmosphere, cutting anthropogenic emissions is only part of the equation. Protecting, and potentially increasing, the sinks for CO_2, CH_4 and N_2O may play a crucial role in determining how the climate of the 21st century unfolds.

1.5.1 Terrestrial vegetation

Current estimates of the amount of carbon stored in terrestrial vegetation range from 450 to 650 Pg C, with forests comprising 80% of this carbon sink. Plants utilize CO_2 during photosynthesis, but also produce it during respiration. The net effect is an uptake of CO_2 from the atmosphere equivalent to ~60 Pg C/year. As such, terrestrial vegetation plays a key role in the global carbon cycle. Hymus and Valentini (Chapter 2, this volume) examine the role plants play in controlling atmospheric CO_2 concentrations and the potential changes in this large sink for CO_2 in the face of changing land use and climate.

They discuss the CO_2 fertilization effect, whereby plants respond to elevated CO_2 concentrations in the atmosphere by increasing CO_2 uptake, and the uncertainties that revolve around this potentially very important negative feedback to anthropogenic climate change.

The determinants of terrestrial primary production, past and present, are also discussed, as is the potential for increased CO_2 sequestration in the 21st century through both natural and managed changes in the carbon sink strength of terrestrial vegetation. Our primary impact on the terrestrial vegetation CO_2 sink has been through our alterations in land use. A switch from forested land to agricultural crops means that carbon incorporated into plant tissues is taken out of the atmosphere for a much shorter time, and so the effectiveness of the plants as CO_2 sinks is much reduced.

Globally, through a combination of afforestation and reforestation, it has been estimated that between 0.2 and 0.6 Pg C/year could be sequestered in above- and below-ground biomass by 2010, with improved management practices and land-use change providing an additional 0.57 and 0.44 Pg C/year, respectively.

Hymus and Valentini conclude that, though terrestrial vegetation has played a vital role in buffering anthropogenic CO_2 emissions in the past and has the potential to do so in the future, its continuation as a leading sink for our own CO_2 emissions cannot be taken for granted given the continuing uncertainty in its drivers.

1.5.2 Oceans

Across the world's oceans there is a continual cycle of equilibration of dissolved CO_2 in water with CO_2 in the atmosphere. About 88 Pg C/year is released from the surface of the world's oceans, with an annual uptake by the oceans of 90 Pg C. Consequently, the net uptake by oceans is estimated to be ~2 Pg C/year.

The carbon that dissolves in our oceans occurs in three main forms. In addition to CO_2, it is also found as bicarbonate and carbonate ions. About 90% exists as bicarbonate, and about 8% as carbonate.

Like terrestrial vegetation, the world's oceans have provided a substantial buffer to increases in atmospheric CO_2 emissions arising from human activities. Sabine and Feely (Chapter 3, this volume) examine the various oceanic sources and sinks for CO_2, the net air–sea flux of CO_2 over time and the potential changes in the strength of the oceans as a sink for CO_2 in response to increasing atmospheric CO_2 concentrations and changes in climate in the 21st century.

The oceans contain the bulk of the world's natural carbon, far more than land or the atmosphere. They also have a huge capacity to store CO_2 released by human activity. It is estimated that ~118 Pg anthropogenic CO_2-C is accumulated in the world's oceans between 1800 and 1994 – equivalent to almost half of all emissions related to fossil fuel and cement manufacturing over that time. However, this sink capacity is controlled by a complex mix of factors and, over the next few decades to centuries, the maintenance of this capacity is far from guaranteed. For instance, the efficiency with which the oceans can take up CO_2 at the surface – known as the Revelle factor – has decreased since the pre-industrial era in response to reductions in pH, themselves caused by elevated atmospheric CO_2 concentrations. As mentioned earlier, rising sea surface temperatures may also have significant effects on the oceanic CO_2 sink. As water temperatures increase, the solubility of CO_2 is reduced and the likelihood of stratification (and so nutrient limitation of phytoplankton) is increased, both leading to an overall reduction in oceanic CO_2 uptake.

Such impacts of global warming and elevated atmospheric CO_2 concentrations represent a positive feedback effect that, other things being equal, will result in a reduction in the carbon sink strength of the oceans.

1.5.3 Soils

Containing ~2000 Pg C at any one time, soils play a key role in the global carbon cycle. About 300 Pg C can be found as detritus in the topsoil, with this carbon-rich material decomposing at varying rates depending on factors such as temperature and soil conditions. During this decomposition some of the carbon in soil detritus is respired by the decomposing organisms (often fungi and bacteria), and the carbon returned to the atmosphere as CO_2. The rest of the detritus carbon can be converted into 'modified soil carbon', which decomposes at a slower pace and so keeps the carbon from the atmosphere for a longer time. A small amount of this carbon is further decomposed to 'inert' carbon and this can remain locked away from the atmosphere for over a thousand years. Biomass burning may be particularly important here, as a small percentage of the material burnt can be left behind in the soil as inert 'black carbon'. These recalcitrant burning residues can constitute small (<0.1 Pg C/year) but long-term sinks for atmospheric CO_2.

Smith and Ineson (Chapter 4, this volume) review our current understanding of CO_2 fluxes in soils, the sensitivity of these fluxes to land use and the potential impacts of increasing global temperatures on these fluxes as the 21st century progresses.

They go on to discuss the potential of soils to help mitigate anthropogenic climate change through land management aimed at protecting the existing sink and, where possible, increasing the soil carbon sink strength. With anthropogenic CO_2-C emissions already standing at ~6.3 Pg/year, and set to go on increasing, achieving a stable CO_2 concentration in the atmosphere that avoids potentially dangerous climate change is a huge challenge. However, the use of carbon sequestration in soils in the short term (next 20 years) as part of a wider suite of measures to offset increasing emissions may allow stabilization at reasonable levels (450–600 ppm) by 2010.

1.5.4 Increasing the soil carbon sink

Human conversion of soils from natural to agricultural use has led to substantial reductions in the soil carbon sink. Greater soil disturbance, such as that caused by ploughing, can induce rapid respiration and loss of large amounts of soil carbon which would otherwise decompose more slowly. Sensitive land use practice is key to better balancing of the soil carbon sink, and perhaps to reversing recent trends of loss of carbon from soils. Farming practices such as 'no-till' – whereby agricultural land is used without the soil disturbance and carbon loss that comes with ploughing – are becoming more widespread and land use remains a key area of research in studies of anthropogenic greenhouse gas emissions and strategies to reduce them. Lemke and Janzen (Chapter 5, this volume) discuss the impact of no-till farming on emissions or removal of CO_2, CH_4 and N_2O, considering the overall net effect on the earth's climate in both the short and long terms. By doing so, they demonstrate the importance of considering all of these climate-forcing emissions together when assessing the value of no-till agriculture as a climate change mitigation strategy.

1.5.5 The geological sink

The calcium carbonate cliffs of Dover and the petrol at motorway service stations both represent large geological reservoirs of carbon and, as such, potentially very long-term carbon sinks. Ridgwell and Edwards (Chapter 6, this volume) examine the key determinants of these sinks and their role in the global carbon budget.

Marine sediments provide the ultimate long-term 'geologic' sink for CO_2 emitted to the atmosphere. For instance, carbon extracted from the surface ocean and transformed into organic matter by photosynthesizing organisms can be 'locked away' by burial in accumulating sediments. Sediments, mainly those of the oceans, can provide long-term sinks for carbon. The remains of oceanic plants and animals sink to the sea bed as 'marine snow', with productive areas of the world's oceans producing huge amounts of such particulate organic carbon. In coastal areas, dissolved and particulate carbon carried by rivers can also provide a significant source of carbon to marine and estuarine sediments. Through the sedimentation of dissolved and particulate organic carbon, sediments provide a global carbon sink of ~10 million tonnes per year. We are familiar with the end result of this – fossil fuel deposits such as natural gas and oil.

Carbon is also taken up by plants on land: the anoxic conditions prevailing in swamps and peatlands allow for efficient preservation of this plant material, ultimately forming coal measures. However, the rate of formation of new fossil fuel deposits is more than 100 times slower than the rate at which we are burning them (~6 Pg C/year). It is also unlikely that there would be any significant increase in their rate of formation in the future. This means that the geologic organic carbon sink may not be of any appreciable help in removing the greenhouse gases we are adding to the atmosphere.

Carbon is also buried as an inorganic solid – carbonate, the common reservoirs of which include chalks and limestones. Precipitation of calcium carbonate and its deposition in sediments lead to a further 200 Tg C, ending up in sediments annually. This deposition of carbon is made even more important by the very long periods (thousands of years) during which such carbon is generally taken out of the atmosphere. The global burial rate of carbonate carbon is several times faster than for organic carbon but, like fossil fuel deposits, formation of rock carbonate is unlikely to match the current rate of use of this carbon sink by humans in industries such as cement production. At present, 100–200 Tg C/year is released to the atmosphere from use of this carbonate carbon reservoir.

More importantly, accumulation of carbonate in the deep ocean is very sensitive to environmental conditions, particularly ocean pH. Geochemical interactions between dissolved CO_2 and marine sediments will act to enhance the capacity of the ocean to sequester CO_2 from the atmosphere on a timescale of thousands of years.

Ultimately, virtually all the fossil fuel carbon that humankind can find to burn is likely to become locked up in geological formations as carbonates. Unfortunately, the timescale for this slowest removal process is hundreds of thousands of years. In contrast, rising greenhouse concentrations in the atmosphere and associated climate change are occurring at a much quicker, century-scale pace.

1.5.6 Artificial sinks

Aresta and Dibenedetto (Chapter 7, this volume) explore the mechanisms by which CO_2 may be utilized in industrial chemical synthesis or for technological applications. Although the sink strength represented by the direct incorporation of CO_2 into long-lived (>10 years) compounds such as polymers appears rather limited, the indirect reduction in atmospheric CO_2 concentrations through reduced industrial energy use makes these 'artificial sinks' a significant component of overall anthropogenic climate forcing.

Currently, industrial use of CO_2 is equivalent to ~110 million tonnes CO_2/year, which is dominated by urea and carbonate production. Aresta and Dibenedetto identify significant potential for greater use of CO_2 in these and other processes, particularly in the production of polymers, methanol and carboxylates. The life cycle analysis (LCA) approach is discussed as a means of assessing the net CO_2 reduction achieved by industrial CO_2 utilization, with inclusion of both the direct (CO_2 incorporation) and indirect (reduced energy use) components highlighted.

Examples with significant potential include the addition of CO_2 to CH_4 in the production of methanol, an approach that can lead to a net reduction of 200 kg CO_2 for each tonne of methanol produced. Likewise, the synthesis of organic carbonates using CO_2 can result in both energy savings and the avoidance of toxic compounds like phosgene. Given increased exploitation of these and other processes incorporating CO_2, the net CO_2 avoidance through such 'artificial sinks' may be increased to between 250 and 300 Tg/year (this is equivalent to offsetting all CO_2 emissions arising from aviation in developed countries during 2002).

The wider use of such carbon capture and sequestration (CCS), especially that making use of geological formations to store captured CO_2 for hundreds or even thousands of years, has huge potential as a strategy to mitigate anthropogenic climate change (see Reay, Chapter 16, this volume).

1.5.7 Carbon sinks and emissions trading

Much of the current interest in carbon sequestration by vegetation relates to those parts of the Kyoto Protocol that deal with forests and land-use change (Articles 3.3, 3.4 and 12). The interpretation and implementation of these sections of the Protocol has proved to be challenging and has certainly occupied much of the delegates' time at successive conferences of the parties. Reilly *et al.* (Chapter 8, this volume) provide insight into some of the issues about the use of sinks in carbon trading. To include biological sinks to 'offset' carbon emissions is inherently controversial, as these sinks change with environmental conditions, are not permanent and can be difficult to measure. Moreover, different countries have different perspectives on biological sinks, because of their different circumstances. Reilly *et al.* work through this issue using an ingenious approach, comparing hypothetical countries with different circumstances. Finally, they propose a solution, the 'cap and trade' system in which most of the perceived difficulties are minimized. Such creative thinking is much needed for the second commitment period of the Kyoto Protocol.

2 Terrestrial Vegetation as a Carbon Dioxide Sink

Graham Hymus[1] and Riccardo Valentini[2]
[1]*Department of Biological Sciences, Northern Arizona University, Flagstaff, Arizona, USA;* [2]*Universitá della Tuscia, DISAFRI, Viterbo, Italy*

2.1 Introduction

In the past few decades a global excess of photosynthesis over respiration has resulted in an annual accumulation of 0.2–1.4 Pg C in the terrestrial biosphere (Sabine *et al.*, 2004), and slowed down the rate of rise in atmospheric CO_2 (C_a). However, the processes driving this increase in productivity, with its spatial distribution and magnitude, remain uncertain. Specifically, it is not clear whether this fertilization effect has been driven primarily by global changes in climate and atmospheric composition, or is largely the consequence of historic changes in land use. This uncertainty has important implications for future carbon sequestration by terrestrial vegetation. If global change has been largely responsible for increased carbon sequestration, continued climatic and atmospheric changes could continue to increase carbon sequestration in terrestrial vegetation. However, published predictions of the effect of future global changes on carbon sequestration in terrestrial ecosystems vary enormously, ranging from ~150 to ~530 Pg C by 2100 (Cramer *et al.*, 2001; Gruber *et al.*, 2004). Conversely, if past land-use change has driven recent increases in carbon sequestration, future sequestration may be limited. Managed carbon sequestration strategies, such as afforestation and reforestation projects permitted by the Kyoto Protocol, could provide for significant carbon sequestration in terrestrial biomass in addition to that which may occur due to global change. However, these projects are few in number and predictions of the extent to which they can be established globally are extremely uncertain. Estimates of potential global carbon storage range from a 'theoretical maximum' of 5.0 Pg C/year to an 'actually achievable' minimum of 0.2 Pg C/year (Cannell, 2003). If carbon sequestration in terrestrial biomass is to be used effectively as one mechanism among many by which the rise in C_a can be attenuated in the future, these uncertainties need resolving. Here we will provide an assessment of our current understanding of: (i) the drivers of past productivity of terrestrial vegetation; (ii) current terrestrial productivity; and (iii) the potential for natural and managed increases in carbon sequestration in terrestrial vegetation in the coming century.

2.2 Quantifying Carbon Sequestration by Terrestrial Vegetation

Photosynthetic carbon reduction is the basis of productivity, providing photosynthate for biomass accumulation from the level of an individual leaf up to an ecosystem and, further, the globe. However, because productivity is variously defined, a short summary of the derivations of the different definitions is provided as a foundation to this chapter (Fig. 2.1).

Rates of gross photosynthesis summed for individual plants, entire canopies, regions or the globe constitute gross primary productivity (GPP). However, direct measurements of gross photosynthesis are seldom made. The problem is that leaf dark respiration (R_{leaf}) occurs simultaneously with photosynthesis. Consequently net rates of leaf photosynthesis (A_{net}) are typically measured. Therefore,

$$\mathrm{GPP} = A_{net} + R_{leaf} \quad (2.1)$$

These limitations aside, GPP provides a good starting point from which to develop a structural framework for assessment of productivity of terrestrial vegetation. When moving up in scale from leaves to an individual plant to a forest stand, productivity estimates must account for losses of CO_2 from autotrophic respiration by all respiring components of the plant (R_a), including the shoots (R_{shoot}) and

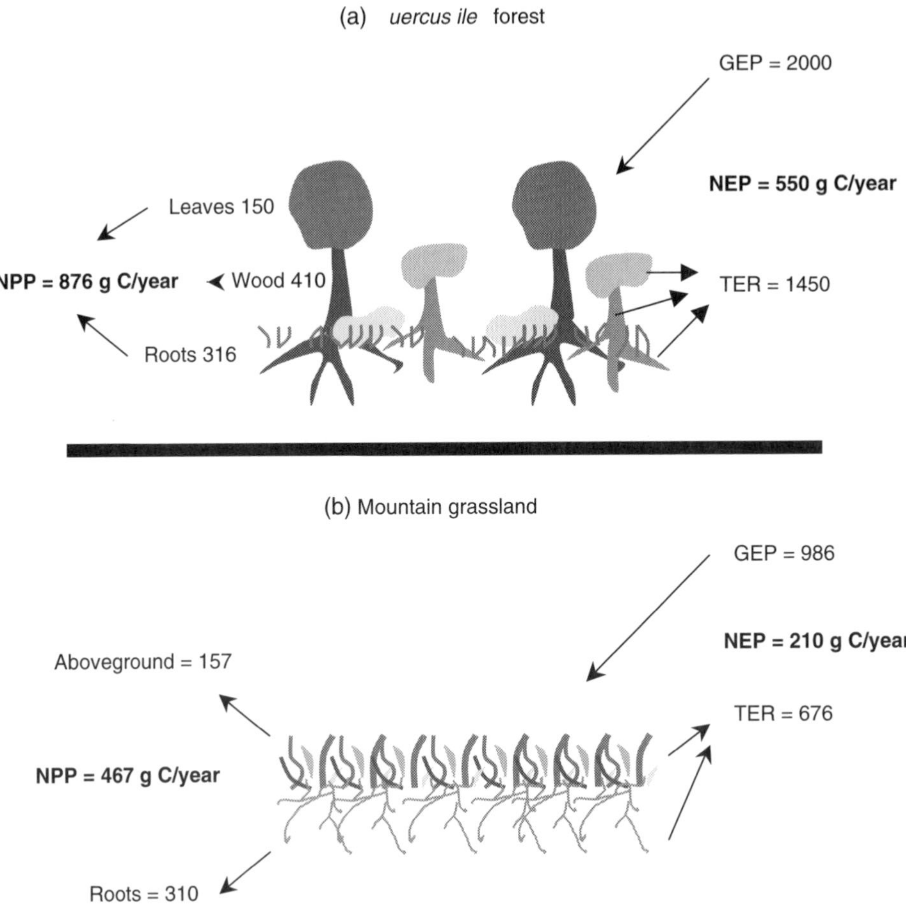

Fig. 2.1. Representative measurements of annual ecosystem productivity are shown for two Italian ecosystems: (a) an evergreen *Quercus ilex* forest and (b) a mountain grassland. The two ecosystems display similar GEP/NEP ratios of 0.27 (forest) and 0.22 (grassland), respectively, evidence of similar respiratory costs per unit of carbon fixed. However, productivity in the forest ecosystem is double that of the grassland and from the perspective of long-term carbon sequestration in biomass, the accumulation of carbon in woody biomass is much more significant. (From unpublished data provided by Sabina Dore.)

roots (R_{root}) in addition to R_{leaf}. The result is an estimate of net primary productivity (NPP):

$$NPP = GPP - R_a \quad (2.2)$$

$$\text{where } R_a = R_{leaf} + R_{shoot} + R_{root} \quad (2.3)$$

Unlike GPP, NPP can in theory be directly measured. In practice, however, it is a time-consuming labour-intensive process, in which accumulation and turnover of biomass, not just in leaves, shoots and roots but in all carbon pools such as exudates and volatile organic compounds, should in theory be assessed over a given period of time. In reality they seldom are, and the accuracy of NPP assessments have long been questioned (Long *et al.*, 1989; Geider *et al.*, 2001). Of course, vegetation forms but one component of ecosystems and consequently assessments of whole ecosystem productivity need to account for further respiratory losses of CO_2 from heterotrophic organisms within the soil (R_h), as well as those from vegetation. The result is an estimate of net ecosystem production (NEP):

$$NEP = GPP - R_a - R_h \quad (2.4)$$

or

$$NEP = NPP - R_h \quad (2.5)$$

Recent application of the eddy covariance technique to terrestrial ecosystems has provided a method to continually measure instantaneous net CO_2 exchange between the biosphere and the atmosphere (NEE) and, when these measurements are integrated over time, NEP (recent review Baldocchi, 2003). Arguably the principle methodological uncertainty surrounding estimates of NEP made using eddy covariance centre on night time measurements, and the fact that the atmospheric conditions required for the technique to work are often not met (Baldocchi, 2003). However, eddy covariance measurements of just CO_2 will not measure non-CO_2 losses of carbon from an ecosystem such as other carbon trace gases (Crutzen *et al.*, 1999; Walter *et al.*, 2001) or dissolved organic or inorganic carbon (Ludwig *et al.*, 1998), which in situations where they are substantially present would result in overestimation of NEP. The addition of total ecosystem respiration (TER = $R_a + R_h$) during the photoperiod to NEP measured during the photoperiod only provides an estimate of gross ecosystem productivity (GEP). Although GPP and GEP are often used interchangeably, they do differ to a small extent, with GEP containing a photorespiratory component that GPP does not (Waring and Running, 1998).

Over longer periods of time (decades and longer) ecosystem productivity is often subject to large-scale disturbance that results in large carbon losses (R_d). Disturbance from fire, pest outbreaks or logging are examples. The subtraction of these terms from NEP results in an estimate of net biome productivity (NBP) (Schulze *et al.*, 2000):

$$NBP = NEP - R_d$$

Changes in productivity provide the basis for changes in the size of the terrestrial carbon sink. However, as detailed above, productivity is defined in many ways and because there is no guarantee that photosynthesis and respiration are tightly coupled over a range of timescales from days to decades, these definitions can be contradictory. For example, GPP could in theory increase at the same time as NPP decreases, a consequence of a more than proportionate increase in R_a over GPP, as may happen during a period of active growth. Alternatively increases in GPP and NPP can mask decreases in NEP, as a result of large increases in respiratory losses from heterotrophic soil organisms, and may occur after a short-term rain event or a longer-term disturbance event. Determining the extent to which GPP/NPP or GEP/NEP ratios are conserved across ecosystems and timescales remains a major research priority. These complications need to be borne in mind as we make both general and specific references to productivity throughout this chapter.

Ultimately, partitioning of photosynthate into plant biomass components, all with different turnover rates and resistance to decomposition, will have as significant an impact on carbon sequestration as rates of productivity. While some plant compounds such as carbohydrates and proteins are important energy sources for soil organisms and respire quickly, others such as lipids, lignin and cellulose are considered more resistant to breakdown (Gleixner, 2005). From the perspective of carbon sequestration in terrestrial vegetation, the accumulation of woody biomass and production of long-lived wood products are the keys to increasing carbon sequestration and are the focus of this chapter.

2.3 Recent Changes in the Terrestrial Carbon Sink

Over the last 200 years the terrestrial carbon cycle has undergone profound direct and indirect changes. Direct changes have been driven by the interaction between an increasing global population and agrarian as well as industrial changes in land use. Indirect changes have been consequences of these economic and social changes. In particular, the rise in C_a from 280 to ~380 ppm has been caused by the addition of ~400 Pg C to the atmosphere, primarily through the burning of fossil fuels and land-use change (Sabine *et al.*, 2004). The rate of input of nitrogen into the terrestrial nitrogen cycle has doubled due to humans (Vitousek *et al.*, 1997a), with extreme inputs of up to 60 kg/ha/year recorded in some European ecosystems (MacDonald *et al.*, 2002). Our understanding of the relative roles that direct land-use change and indirect climate changes have played in changing terrestrial productivity in the recent past has important implications for our ability to predict the future, and is the focus of this section.

2.3.1 Land-use change

It has been estimated that up to 50% of the earth's land surface has been transformed by human action (Vitousek *et al.*, 1997b). Key amongst many changes have been: (i) conversion of forest to agricultural land; (ii) the subsequent abandonment of agricultural land and the natural or managed re-establishment of woody vegetation; (iii) fire suppression; and (iv) expansion of woody species. All these land-use change have altered ecosystem productivity through changes in species composition, above- and belowground allocation of carbon, rooting depth and soil faunal communities (Nepstad *et al.*, 1994; Jackson *et al.*, 2000) and consequent changes in nutrient cycling and carbon storage (Trumbore, 1997; Jackson *et al.*, 2002). The net consequence of the sum of land-use changes is thought to be a loss of carbon from terrestrial ecosystems to the atmosphere of 100 Pg C since 1850 (Houghton, 1995). The following subsections discuss in more detail the key land-use changes highlighted earlier. The specific changes discussed are not an exhaustive list but instead reflect our understanding of those changes currently thought to have been most significant from the perspective of recent changes in carbon storage in terrestrial vegetation.

Agricultural expansion

The clearing of forest to make way for agriculture has been the most significant historical human-induced land-use change. Currently ~1800 million hectares are cultivated globally (Turner *et al.*, 1993). In the USA cropland has increased from 50 million hectares to 200 million hectares in the last 150 years (Richards, 1990), with 140 million hectares of forest cleared during this time (Ramankutty and Foley, 1999). In Europe ~50% of the land surface is currently used for agriculture, with ~30% devoted to the production of crops; in addition, grasslands cover 55 million hectares (CarboEurope GHG, 2004a). Prior to 1950, land-use change was predominantly in the temperate zone; however, since then and arguably for the last century, forested regions in the tropics have seen the highest and most uncertain rates of conversion (Houghton, 2003). In general, conversion of forest to agricultural land removes a large stable aboveground and belowground carbon pool, both of which will likely be accumulating carbon. Depending on the harvest technique, and the use to which the harvested wood is put, forest clearance will result in the oxidation of a major proportion of the aboveground biomass and belowground soil carbon pools (Guo and Gifford, 2002). Once established, croplands and grasslands exhibit idiosyncratic carbon sequestration patterns. These patterns are strongly dependent on the exact use to which they are put, the intensity of their management and their geographical location (Grünzweig *et al.*, 2004). Janssens *et al.* (2003) estimated that croplands in Europe lose 300 million tonnes of carbon per year, and are consequently the largest annual biospheric source of CO_2 to the atmosphere. European grasslands were, conversely, thought to be a sink for atmospheric CO_2 of 101 million tonnes of carbon per year (CarboEurope GHG, 2004b). However, it must be acknowledged that both these estimates were bounded by very large uncertainties.

Abandonment of agricultural lands

From the perspective of carbon accumulation in biomass, abandonment of agricultural land that results in managed or natural regeneration of secondary forest is of great consequence. Nabuurs *et al.* (2003) explained that the current age structure of European forests is the consequence of large-scale afforestation of lands cleared for grazing or other agricultural practices in the early 20th century, and past management practices that centred on replanting after clear-cutting, followed by regular thinning of mono-specific even-aged stands. The result is that previous agricultural land is now covered by young forest, which is around 57 years, with an average aboveground carbon stock in stemwood of 40–50 t C/ha and an age structure in which there is an almost complete absence of forest cover that is more than 150 years. The age structure of these young European forests is important, because it is considered as the main reason that they are still actively sequestering carbon at a high rate. Valentini *et al.* (2000) highlighted how 25 of 27 forest study sites located throughout Europe were accumulating carbon, with a mean annual NEP for the 27 sites of 3 t C/ha/year. Within the Valentini study the oldest forest site was 110 years, while the mean age was 60 years. Significantly, the accumulation of stemwood in European forests has been predicted to continue until 2025 when age-related declines in productivity will become important (Nabuurs, 2004). In comparison with Europe, the average age of forests in the USA and European Russia is older at 76 and 80 years, respectively (Nabuurs, 2004). In the USA this is because there has not been such a pronounced abandonment of agriculture, with only 6% of the forestland lost to agriculture being reforested since 1920 (Houghton and Hackler, 2000). However, although managed afforestation of abandoned agricultural land in the USA has been small, the natural re-establishment of woody species has been significant. Casperson *et al.* (2000) calculated that the encroachment of native vegetation on abandoned agrcultural land in eastern USA constituted almost the entire regional sink for carbon.

While not the focus of this chapter, changes in vegetation biomass induced by land-use change cannot be considered independently of changes in soil carbon stocks. However, there is much specificity in soil carbon responses to land-use change. For example, the conversion of pastureland to pine plantation forestry resulted in losses of soil carbon sufficient to offset increases in vegetation; but conversion to broad-leaved or naturally regenerated secondary forest had no effect on soil carbon (Guo and Gifford, 2002). Conversely, the conversion of cropland to forest or plantation typically increases soil carbon in addition to the expected increases in aboveground biomass (Guo and Gifford, 2002).

Fire and fire suppression

Fire is a critical component of many ecosystems and has profound instant and long-term implications for carbon cycling and storage. Both historical reconstructions and current understanding of global fire frequency and extent are limited. However, in what the authors describe as a ‘first approximation’, Mouillot and Field (2005) estimated that 608 million hectares burned per year at the end of the 20th century, 86% of which were in tropical savannahs. Fires in forests have more impact on global carbon stocks, and Mouillot and Field (2005) estimated that 70.7 million hectares of forest burned annually at the beginning of the century, the majority being in the boreal and temperate forests of the northern hemisphere. As a consequence of fire suppression policies the area of temperate and boreal forest subject to fires decreased to 15.2 million hectares per year in the 1960s and a current figure of 11.2 million hectares per year. Over the same time period fires in tropical forests have increased exponentially to 54 million hectares per year. Although it is thought that fire suppression has certainly resulted in significant carbon accumulation in terrestrial ecosystems over the last century (Houghton *et al.*, 2000), the quantitative consequences of fire, and thus fire suppression, on the terrestrial carbon cycle are hard to know. This is because difficult-to-determine factors, such as intensity of burn, vegetation type and standing biomass, control the amount of carbon emitted during a fire. However, experiments have demonstrated the impact of fire on ecosystem carbon pools. In a Minnesota oak savannah, fire suppression led to an average increase of 1.8t C/ha/year, with most of the carbon stored in woody biomass and on the forest floor (Tilman *et al.*, 2000).

Expansion of woody vegetation

Unmanaged encroachment of woody species on primarily herbaceous ecosystems, either as a consequence of climate change, human activity or the interaction of both, typically increases carbon sequestration in aboveground biomass. The literature increasingly documents carbon sequestration in aboveground biomass due to woody encroachment in diverse locations. These include the Arctic, where Sturm *et al.* (2001) document a doubling of woody vegetation in some Alaskan Arctic tundra study plots over the last 50 years. Increasing size of individual trees, filling in of gaps and increased density of forests at tree line sites were all observed. Historical photos were used to document increases in woody vegetation cover of up to 500% over a 63-year period in unmanaged Texas rangelands (Asner *et al.*, 2003). Knapp and Soulé (1998) observed a 59% increase in the cover of western juniper in central Oregon over a 23-year period. Similar trends have been documented in savannah ecosystems (Scholes and Archer, 1997) and grasslands (Jackson *et al.*, 2002). However, the study of Jackson *et al.* (2002) was unique in that it clarified the need for consideration of soil carbon changes when assessing carbon sequestration due to woody encroachment, as at wetter study sites losses of soil carbon completely offset carbon accumulation in woody biomass.

2.3.2 Enhanced growth

The role land-use change has played in increasing terrestrial productivity over the last 200 years, relative to climate and atmospheric changes, is controversial. If changing atmospheric chemical composition is increasing productivity, increasing C_a is without question regarded as the key driver, stimulating photosynthesis and widely documented as being able to provide varying degrees of 'protection' against many plant stresses and the deleterious effects of other important atmospheric changes, such as the increase in tropospheric ozone concentrations (Drake *et al.*, 1997; Hymus *et al.*, 2001; Ashmore, 2005). However, the role that increasing C_a has played in recent global productivity remains the subject of much debate. Recently Casperson *et al.* (2000) concluded that increases in C_a over the last century were not a factor that could explain increases in US forest productivity over the same time period. This finding has been challenged by Joos *et al.* (2002) who, using the same data-set, concluded that the same increases in productivity could indeed have been the result of CO_2 fertilization. Some studies have made more specific conclusions. In a modelling study, Osborne *et al.* (2000) reconciled a 25% increase in NPP in the Mediterranean in the last 100 years primarily with a 70ppm increase in C_a and consequent increases in water use efficiency. Thornton *et al.* (2002) modelled carbon accumulation after disturbance for coniferous forests and showed that increases in C_a reduced the period in which the forest was a source of CO_2 to the atmosphere following disturbance, the total carbon lost during this period, the time to reach peak sink strength and increased total carbon storage. Nitrogen deposition was an important additive factor in this assessment, with the effects described earlier increasing with nitrogen deposition.

What is less contentious than attributing increases in terrestrial productivity with increases in C_a specifically is that a combination of changes in climate and atmospheric composition has had significant effects on terrestrial productivity over the last few decades. The last two decades have seen rapid changes in key atmospheric and climate variables: C_a increased by 9%, they were two of the warmest decades in the instrumental record, they experienced three persistent El Niño events and a human population increase of 37% (Nemani *et al.*, 2003). These changes combined with nitrogen deposition and forest regrowth are well documented to have contributed to regional increases in NPP, particularly in the mid to high latitudes (Myneni *et al.*, 1997; Houghton, 1999). However, Nemani *et al.* (2003) reconciled 18 years of satellite data with climate records collected between 1982 and 1999 to conclude that a broad suite of climatic changes over that time period had eased climatic restrictions on NPP and increased NPP by 6% or 3.4 PgC. Importantly, this study highlighted much regional variability, with the greatest increase in NPP, of 42%, observed in the tropics primarily due to decreases in cloud cover. This observation of an increase in productivity in the tropics is in keeping with the recent findings of Lewis *et al.* (2004) who concluded that there had been significant increases in productivity over the last three decades in South American tropical forests which would have resulted in the region being an abundant carbon sink. This was also explained in terms of a combination of increases in key resources, including C_a and radiation.

In addition to easing of limitations to productivity across the globe, global change has also caused specific changes in vegetation phenology that have implications for productivity. We highlight the example of advancing spring bud break and the greening up of forest canopies in mid- and high-latitude forests at a rate of several days per decade over the last 50 years (Badeck *et al.*, 2004). This effect lengthens the growing season, which is in turn well correlated with increase in NEP (Baldocchi and Wilson, 2001).

2.4 The Current Global Carbon Cycle and the Terrestrial Carbon Sink

Terrestrial biomass represents a dynamic and therefore important carbon pool, turning over relatively quickly compared with other carbon pools with an average residence time of 10 years (Gruber *et al.*, 2004). Estimates of the amount of carbon stored globally in terrestrial vegetation and in specific biomes vary significantly (Table 2.1). Two recent estimates are 466 Pg C (Watson *et al.*, 2000) and 652 Pg C (Saugier *et al.*, 2001), amounts broadly similar to that found in the atmosphere, but only 25% of that stored in soils. Regardless of the estimate, ~80% of the carbon is stored in the world's forests, with 45–50% in tropical forests alone. Set against these carbon stocks the biosphere and the atmosphere exchange ~120 Pg C annually. Sabine *et al.* (2004) estimated annual global NPP of 57 Pg C averaged throughout the 1990s (~40% of GPP). This figure was almost balanced by 55.5 Pg C/year lost from terrestrial ecosystems through respiration, fire and land-use change. The small difference between these figures is the net land–atmosphere flux, an annual estimate of NEP which was estimated at 1.4 Pg C/year by Prentice *et al.* (2001) throughout the 1990s (Table 2.2). Carbon sequestration of this magnitude represented ~20% of the carbon being emitted from fossil fuel burning and cement production annually through the 1990s (Table 2.2), and is a significant increase on the NEP estimate of 0.2 Pg C/year during the 1980s (Table 2.2). While informative, this global estimate of NEP is continually being revised with other recent studies providing estimates of annual global NEP between 1.2 ± 0.8 (Le Quéré *et al.*, 2003) and 3.4, ranging from 1.8 to 5.0 Pg C/year (Houghton, 2003).

Table 2.1. Net primary productivity (Pg C/year) and carbon stored in plant biomass (Pg C) for the world's major biomes.

	NPP[a]	Plant C[b]	Plant C[c]
Tropical forests	20.1	340	212
Temperate forests	7.4	139	59
Boreal forests	2.4	57	88
Arctic tundra	0.5	2	6
Mediterranean shrub	1.3	17	–
Crops	3.8	4	3
Tropical savannahs and grasslands	13.7	79	66
Temperate grasslands	5.1	6	9
Deserts	3.2	10	8
Wetlands	–	–	15
Total	57	652	466

[a]From Sabine *et al.* (2004).
[b]From Saugier *et al.* (2001).
[c]From Watson *et al.* (2000).

Estimates of the terrestrial carbon sink are made by subtracting the measured increase in C_a from total anthropogenic emissions. The difference is the combined uptake of the oceans and terrestrial biosphere. Partitioning carbon sequestration between the ocean and land is subsequently achieved through simultaneous high-precision measurements of atmospheric CO_2, $^{13}CO_2$, O_2 and $^{18}O_2$ (recently reviewed in Ciais *et al.*, 2005). Once determined, the terrestrial sink can be partitioned between that associated with changes in land-use and that associated with increasing carbon sequestration in existing ecosystems. Throughout the 1980s and 1990s land-use change resulted in significant fluxes of carbon to the atmosphere; consequently NEP of terrestrial ecosystems not undergoing change was significantly higher, estimated at 1.9Pg C throughout the 1980s (Prentice *et al.* 2001; Table 2.2). Of these key steps to determining global productivity there are two notable areas of uncertainty. The first is the need for improved understanding of the processes controlling the changing ratios of $^{12}C/^{13}C$ and $^{16}O/^{18}O$ in the atmosphere (Ciais *et al.*, 2005; Randerson, 2005). The second is the need for improved estimates of land-use change and the type of land-use change. Houghton (2003) recently suggested that the proportion of the carbon sink in North America that is due to regrowth following land-use change could be as high as 98% (Casperson *et al.*, 2000) or as low as 40% (Pacala *et al.*, 2001) depending on land use figures used for the calculation.

2.4.1 Regional

Estimating carbon balances of large regions is no trivial matter and is increasingly done by trying to reconcile results gained at two levels of enquiry: 'top-down' atmospheric

Table 2.2. The global carbon budget (Pg C/year).

	1980s	1990s
Atmospheric increase	+3.3 ± 0.1	+3.2 ± 0.1
Emissions (fossil fuels, cement)	+5.4 ± 0.3	+6.3 ± 0.4
Ocean–atmosphere flux	−1.9 ± 0.6	−1.7 ± 0.5
Land		
Carbon uptake (NPP)		57
Carbon loss (TER + fire)		55.5
Net land–atmosphere flux (NEP)	−0.2 ± 0.7	−1.4 ± 0.7
Land-use change	+1.7 ± (+0.6 to + 2.5)	–
Residual terrestrial sink	−1.9 ± (−3.8 to + 0.3)	–

All values are taken from Prentice *et al.* (2001), except those for carbon uptake and carbon loss from the land, which are from Sabine *et al.* (2004). Positive numbers represent an atmospheric increase in carbon (source), and negative numbers an atmospheric decrease in carbon (sink).

inversions, and 'bottom-up' scaling up of direct NEE measurements or inventory measurements that track changes in biomass (typically wood) through time (Pacala *et al.*, 2001; Janssens *et al.*, 2003).

Of the world's major biomes, tropical forest is thought to be the most productive area of the earth, followed by tropical grasslands and savannah (Saugier *et al.*, 2001; Table 2.1). However, NPP does not necessarily provide a good measure for the amount of carbon that is sequestered within a biome over a given period of time, as it excludes all losses of CO_2 from the ecosystem via processes other than leaf respiration. Although tropical forests unequivocally have the highest NPP of the earth's biomes, estimates of NEP are more equivocal. Inverse models suggest that the tropics appear to be somewhere between a small net source of carbon to the atmosphere and carbon neutral (Gurney *et al.*, 2002). However, there are still challenges to reconcile this finding with ground-based data. For example, lowland humid tropical forests appear to be accumulating woody biomass at a mean rate of 3.1t C/ha/year (Malhi *et al.*, 2004). Although direct measurements of NEP at some study sites suggest a strong sink (Grace *et al.*, 1995; Malhi *et al.*, 1998), the same measurements at others do not (Saleska *et al.*, 2003). Saleska *et al.* (2003) highlighted significant losses of carbon associated with disturbance, specifically from fallen trees during the wet season, as an important factor that reduced NEP. Seemingly there are clear limitations to the extent to which tropical forest can be thought of as a homogeneous biome for the purposes of scaling up. Uncertainties associated with estimates of deforestation within the tropics (Houghton, 2003; Grace, 2004), losses of CO_2 from study areas, e.g. in solution in rivers, and different temporal durations of field studies, some of which may be made during unusual climatic conditions or different times after disturbance, all complicate reconciling ground-based measurements with estimates from atmospheric inversions. In light of these uncertainties, Houghton (2003) concluded that the small sink for carbon in the tropics was the result of a large release of carbon from land-use change that is being partially offset by a large sink for carbon in undisturbed forests. A second, mutually exclusive, possibility is that the source of CO_2 from deforestation is smaller and the net accumulation of carbon in undisturbed forests is essentially zero.

In contrast to the tropics, there is compelling evidence that the northern temperate regions of the earth are sequestering carbon. Houghton (2003) describes how a northern mid-latitude carbon sink of 2Pg C/year appears robust. For the northern USA alone Pacala *et al.* (2001) reconciled top-down estimates of the carbon sink of the continental USA with bottom-up estimates that included land-use change data and forest inventories. The study concluded that the continental USA was a substantial sink for carbon of 0.15–0.35Pg C/year, of which accumulation of wood and other biomass was the largest component of the sink. Similar studies have been conducted for Europe. Janssens *et al.* (2003) calculated a European carbon sink of 0.11Pg C/year from inventory studies, and a mean atmospheric inversion estimate of 0.29Pg C/year, thereby offsetting 7–12% of Europe's CO_2 emissions. Intercontinental trade in organic products was thought to largely account for the differences in the two estimates. Other estimates are available. Papale and Valentini (2003) used an artificial neural network to estimate that European forests were sequestering 0.47Pg C/year.

2.4.2 Biome and ecosystem

The integrated nature of global and regional studies means that they are of limited use in advancing specific understanding of controls and drivers of ecosystem productivity. For this smaller-scale studies are required. The application of eddy covariance systems, capable of continuous measurements of NEE and integrated measurements of NEP, to terrestrial ecosystems, has enabled insights into carbon cycling through terrestrial ecosystems at temporal and spatial scales, which was previously impossible.

The number of sites and years of data collected is enabling syntheses that are assessing the extent to which carbon cycling through different ecosystems responds similarly to biotic and abiotic drivers. Among these studies several stand out as providing genuinely new insights into the factors controlling ecosystem productivity.

As already detailed in the preceding section, Valentini *et al.* (2000) highlighted how the relatively young, largely managed forests of Europe were significant sinks for carbon. Importantly, Valentini *et al.* (2000) concluded that it was variation in respiration rather than GEP that was the key control on the net size of the carbon sink of European forest systems, with respiration rates of European forests well correlated with latitude, and, surprisingly, the control exerted by respiration on NEP being greater in colder northern latitudes.

It has long been thought that productivity of trees and forest stands declines with age. The theory proposed by Odum (1969) is one of several commonly cited as a reason for this. It proposes that as trees or forests age, leaf area decreases relative to woody biomass; the consequence is that NPP slowly declines, tending towards zero as ultimately photosynthesis is balanced by autotrophic respiration (Odum, 1969). As a result, mature forest stands have not been thought to be significant sinks for carbon. Recent studies have begun to challenge this idea. Although Pregitzer and Euskirchen (2004) showed clear declines in NPP of tropical, temperate and boreal forests with age, forests older than 120 years still exhibited significant NPP. A study by Carey *et al.* (2001) discovered that productivity of subalpine forest stands only began to plateau after 500 years and that these 500-year-old stands had an NPP of close to 6t C/ha. A study by Malhi *et al.* (2004) showed that the lowland new world humid tropical forests are accumulating carbon in woody biomass at an average rate of 3.1t C/ha/year. One characteristic of mature unmanaged forest stands is a large dead wood pool. Carbon losses from this pool can be significant (Saleska *et al.*, 2003) and will reduce NEP in relation to NPP. However, studies that have measured NEP in the Amazon have shown significant NEP (Grace *et al.*, 1995; Malhi *et al.*, 1998). Similarly Knohl *et al.* (2003) measured NEP of ~500g C/year in a 250-year-old beech forest in Germany.

Major disturbances, either managed or natural, as a consequence of fire, wind, pest or disease are a central feature of forest ecosystems. New quantitative insights into the effect of these disturbances on productivity immediately after and during recovery are being achieved by the use of chronosequences of forest stands at different ages, thereby replacing space for time. As expected, disturbance decreases photosynthetic capacity, completely in the case of logging, and typically converts a forest carbon sink into a carbon source (Knohl *et al.*, 2002; Kowalski *et al.*, 2004). Recovery is characterized by increases in NPP over time. However, of key importance from the perspective of carbon sequestration and its management is the time required for a stand to become carbon-neutral and ultimately sequester carbon after the initial disturbance. Studies are beginning to provide insight into this. Thornton *et al.* (2002) combined measurements with an ecosystem process model and found this time to be highly variable, depending on type of disturbance, intensity and postdisturbance management for evergreen forests. Howard *et al.* (2004) found it to be 10 years for a jack pine stand in Canada, Litvak *et al.* (2003) recorded 11 years for a black spruce stand in Manitoba, while NEP of a Siberian Scots pine forest recovered to zero 12 and 24 years after a stand replacing fire in forests with a *Vaccinium* sp. and lichen understorey, respectively (Knohl *et al.*, 2002).

Important new environmental drivers of productivity have also been uncovered. Instantaneous measurements of NEE of forest ecosystems at a given PPFD have been shown to be dependent on the ratio of diffuse to direct PPFD, with NEE increasing as the diffuse component increases and the effect becoming greater as PPFD increases (Gu *et al.*, 2002; Law *et al.*, 2002). While diffuse light will penetrate a forest canopy to a greater extent than direct light, increasing photosynthesis in shade leaves, it will also change the canopy microclimate, with likely decreases in both air and leaf temperature, and changes in leaf air vapour pressure deficit. All are key drivers of either photosynthesis or respiration, or both. The net effect of all these changes is increased net photosynthesis (Law *et al.*, 2002).

These important advances aside, there is still much uncertainty in our understanding of the biosphere–atmosphere feedbacks that drive spatial and temporal variability in carbon cycling through specific ecosystems and biomes. We highlight the need to better understand: (i) how representative current study sites are of larger biomes, which has important consequences for scaling up using biogeochemical models and 'ground truthing' the algorithms used for satellite-based global productivity estimates; (ii) the role that large-scale disturbance and extreme climatic events play in driving ecosystem productivity in the short and long term; and (iii) the extent to which an understanding of the drivers of current carbon cycling through terrestrial ecosystems can inform models predicting carbon cycling into the next century.

2.5 Assessing the Potential for Future Managed and Unmanaged Carbon Sequestration

It is expected that over the next 100 years the productivity of terrestrial ecosystems will be increasingly driven by frequent extreme weather events, set against unprecedented increases in air temperature and the concentration of key trace gases in the atmosphere. These changes will drive a myriad of feedbacks between the biosphere and the atmosphere that are expected to alter the ecosystem functions that control productivity and carbon sequestration, resulting in unmanaged changes in carbon cycling through terrestrial ecosystems. However, management of terrestrial ecosystems is being undertaken with the aim to slow the rate of rise in C_a. The extent to which these unmanaged and managed changes in carbon sequestration in terrestrial vegetation will mediate against rising C_a is highly uncertain with estimates ranging from 10% to 60% over the next 100 years (Fig. 2.2). In Section 2.6 we will assess the potential for both unmanaged, as a consequence of global change, and managed carbon sequestration in terrestrial ecosystems.

(a)

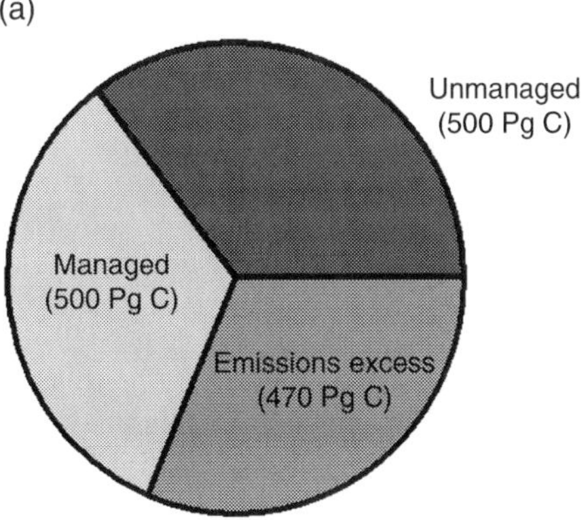

(b)

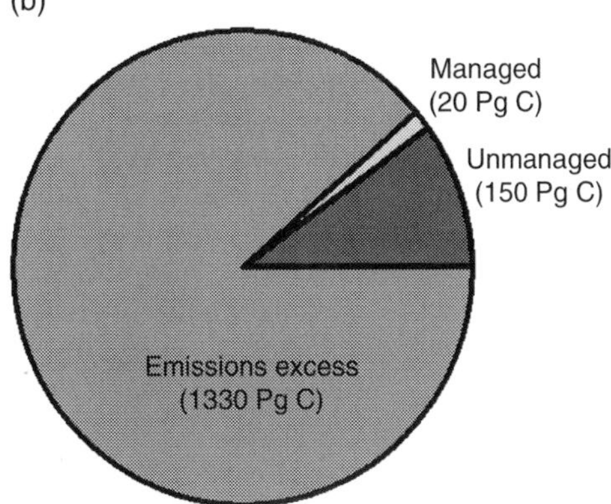

Fig. 2.2. The (a) upper and (b) lower limits of both unmanaged (global change-driven) and managed changes in the carbon sequestration capacity of the terrestrial biosphere summed for the next 100 years, relative to predicted CO_2 emissions over the same period. Predicted CO_2 emissions of 1500 Pg C over the next 100 years are from IPCC scenario IS92a, a mid-range scenario of rising C_a. The upper limit of unmanaged carbon sequestration is predicted from both CO_2 and climate change effects on terrestrial carbon cycling by the TRIFFID model of Cramer *et al.* (2001), and the lower limit is from Gruber *et al.* (2004). The upper (maximum theoretical potential) and lower (actually achievable) limits of managed carbon sequestration are from Cannell (2003).

2.5.1 Unmanaged increases in carbon sequestration

Tremendous uncertainty surrounds the extent to which global changes over the next century will change carbon sequestration by terrestrial ecosystems. A recent cross-comparison of the output of dynamic global vegetation models (DGVM), which incorporate understanding of physiology, phenology and elemental cycles with predicted changes in atmospheric composition and temperature, predicted a widely distributed terrestrial carbon sink sequestering 260–530 Pg C by 2100 (Cramer *et al.*, 2001), which translates to 16–34% of expected anthropogenic emission over the same time period. However, a more recent assessment has concluded that terrestrial ecosystems will provide a much smaller sink of up to 20 and 150 Pg C over the next 20 and 100 years, respectively (Gruber *et al.*, 2004). One key to this revised estimate was the incorporation of constraints to CO_2 fertilization of terrestrial ecosystems by nitrogen availability into the DGVM runs (Hungate *et al.*, 2003).

These findings are informative. In addition to highlighting the massive uncertainty associated with how carbon cycling through terrestrial ecosystems will change in the coming century, they highlight: (i) the need for experimental results to continue to inform model prediction; and (ii) for models to be used to highlight the key process that will control carbon sequestration by terrestrial ecosystems in the future. It is a certainty that new experimental insights into carbon cycling through ecosystems will result in important revisions to model predictions in the coming years.

Uncertainty in predicting the magnitude of carbon sequestration in terrestrial ecosystems over the next century should not detract from the fact that impressive advances in our understanding of how key global changes affecting biological processes have been made in the past decades. A detailed review of these advances is beyond the scope of this chapter; however, it is necessary to say that they encompass new understanding from the gene to the ecosystem, and have been facilitated by research performed at scales from highly controlled growth chambers to large-scale open-air manipulations. What is now required is an effort to understand the extent to which our extensive understanding of the action of single-factor manipulations on biological processes can be used to predict the effects of multifactor global change on ecosystem carbon cycling in the coming century. Specifically, we urgently need to know whether the optimism of Norby and Luo (2004) is appropriate when they suggest that a thorough understanding of the modes of action of single-factor manipulations, when factored into models, should provide the appropriate multifactor outputs required for predicting into the next century, or whether the caution of Osmond *et al.* (2004) is more appropriate when they highlight how ‘our mechanistic understanding of feedbacks between the biosphere and atmosphere is vague at best, often provided by models, which themselves lack mechanistic understanding gained from experiments made at appropriate scale’. Clearly the only way to determine the more valid point of view is the urgent development of targeted, appropriately scaled multifactor experiments, to assess whether multifactor model predictions can be derived from single-factor inputs (Beier, 2004). If it turns out that our largely single-factor understanding is sufficient for predicting into the multifactor future, well and good. If not, a more general requirement for large-scale, multifactor experiments in key ecosystems must be undertaken. Ideally these studies will run in tandem with the sort of controlled large-scale, multifactor experiment described by Osmond *et al.* (2004). The rational behind this approach is to identify generally applicable mechanisms that will drive future carbon cycling within and between biomes, while minimizing the possibility that findings from individual studies are useful only as case studies.

Of course it must be acknowledged that numerous multifactor manipulations, typically elevated C_a and a second factor (e.g. ozone or temperature) have been performed at scales of enquiry from growth chambers to in-field open-topped chambers and increasingly within free-air CO_2 enrichment (FACE) studies. However, care must be taken to make sure that results coming from such studies are either directly relevant to what is likely to happen in the coming century or can be weighted in some way to take account of potential artefacts. These artefacts could be due to experiments being performed at small scale, or vegetation at a specific growth stage. As an example, single-factor CO_2 manipulation experiments have shown that effects of increases in CO_2 on key physiological and whole-plant processes of numerous species do differ depending on the experimental conditions (Norby *et al.*, 1999; Ainsworth and Long, 2004). In addition, care must be taken that treatments are applied appropriately, particularly heating treatments. It is highly likely that commonly used heating techniques such as soil heating, or infrared (IR) lamps that directly heat a vegetation canopy, will uncouple the normal controls on carbon flows through the plant into the soil. Lastly, experiments must be conscious of the inherent difficulties associated with scaling results onwards and upwards in time and space. Given these limitations to experimentation to date, it is unfortunate that funding constraints continue to prevent troubleshooting and ultimately setting up of a single, large-scale CO_2 and temperature, FACE-type manipulation in a forest ecosystem. Such an experiment is essential for providing the first opportunities for validating our current understanding of CO_2 and temperature interactions on carbon cycling through forest ecosystems.

2.5.2 Managed carbon sequestration in terrestrial biomass

There is an acceptance that, if used as part of a portfolio of carbon management options, managed sequestration of carbon in terrestrial biomass could have significant contribution to attempts to slow the rise of C_a in the coming decades (Caldiera *et al.*, 2004). The 'coming decades' time frame is significant as models suggest that the future trajectory of C_a rise during the rest of the century will be set in these coming decades. Watson *et al.* (2000) recognized this potential and concluded that projects that both conserve and increase the size of existing carbon pools should be encouraged. Prospects for significant increases in carbon pools were recently improved by the ratification of the Kyoto Protocol, which provides incentives for Annex 1 (developed countries) to offset domestic carbon emissions by setting up afforestation and reforestation projects in Non-Annex 1 (developing countries). The protocol provides not only incentives for such projects but also protocols for their implementation as part of the Clean Development Mechanism. In Section 2.6 we will assess the potential for carbon sequestration and problems that may arise as a consequence of attempts to manage carbon cycling through the terrestrial biosphere.

Conservation of existing carbon pools

It was recently assessed that 7.7% (50 Pg) and 31% (200 Pg) of the carbon stock of terrestrial biomass would be vulnerable to oxidation in the next 20 and 100 years, respectively (Gruber *et al.*, 2004). Land-use changes would account for almost the complete short-term threat, rendering 40 and 100 Pg C vulnerable to oxidation in the next 20 and 100 years, respectively. Of the multitude of potential land-use changes deforestation constitutes the single biggest threat. Given that deforestation over the last 200 years is thought to have constituted 30% of the total anthropogenic increase in C_a over that time frame (Raupach *et al.*, 2004), decreasing the rates of deforestation has considerable potential to slow C_a rise. Indeed it has been calculated that a reduction in deforestation of 3–10% in Non-Annex 1 countries by 2010 would offset 0.053–0.177 Pg C/year, or 1–3.5 % of Annex 1 countries' base year (for Kyoto commitments) emissions (1990) (Watson *et al.*, 2000). Unfortunately deforestation is being driven by social and political factors, not by the needs to protect carbon stocks. These underlying causes will need addressing before deforestation rates can fall. The remaining 10 and 100 Pg C stored in terrestrial biomass is considered vulnerable to oxidation in the coming 20 and 100 years, respectively, due to climate changes and particular effects on species' ability to adapt (Gruber *et al.*, 2004).

Fire suppression is a management strategy that could in theory protect carbon stocks in the coming decades. Schlesinger (1997) calculated that a 20% decrease in fire in tropical savannah and woodlands would protect 1.4 t C/ha/year from oxidation. Many studies have come to similar conclusions. Dixon *et al.* (1994) concluded that fire management in Russia could lead to long-term carbon storage of 0.6 Pg/year, while Tilman *et al.* (2000) calculated that fire suppression in the Siberian boreal forests, tropical savannah and woodlands would significantly decrease the rate of increase in C_a by 1.3 Pg C/year. There are, however, several significant problems associated with fire suppression. Tilman *et al.* (2000) describe how: (i) carbon storage following fire occurs during the period of rapid tree biomass accumulation, e.g. much of the potential for carbon storage in the USA via fire suppression may have already occurred; (ii) some of the increased carbon stores associated with fire suppression represent an accumulation of fuel that might lead to catastrophic, stand-destroying fires; (iii) it is unclear that global climate change might interact with fire frequency to impact carbon storage; and (iv) fire suppression can have negative impacts via its effects on the diversity, composition and functioning of savannah and forested ecosystems.

Increasing the size of carbon pools

The greatest potential for increased sequestration of carbon in terrestrial biomass lies in managing forests to increase the carbon stores that living wood and wood products represent. Deriving estimates of the potential for carbon sequestration in terrestrial vegetation by specific management activities at a range of scales has been, and continues to be, the focus of much research.

Nabuurs *et al.* (2000) concluded that, for the USA, Canada, Australia, Iceland, Japan and the old EU15, alternative forest management practices to improve carbon sequestration were feasible on 10% of the forest area and that in total there was a potential for increased carbon sequestration that averaged 0.6 Pg C/year for the six regions. The potential was variable between regions, with Canada being the highest, and between management activity, with the effectiveness of specific management practices ranging from 0.02 t C ha/year for forest fertilization to 1.2 t C ha/year for a targeted combination of measures in a loblolly pine stand. Nabuurs *et al.* (2000) also assessed the utility of different management strategies in different regions, and again there was much variability; for example, in the USA, Canada and Australia, fire management had the potential to contribute ~40% of carbon sequestration potential. Restoration of degraded lands accounted for another 40% of the carbon sequestration potential in Australia. Neither strategy was considered useful for carbon sequestration for the old EU15.

The potential for carbon sequestration in Non-Annex 1 countries is especially relevant, as Article 3 of the Kyoto Protocol makes provision for Annex 1 countries to offset domestic CO_2 emissions by implementing afforestation and reforestation projects in Non-Annex 1 countries. In specific Non-Annex 1 countries considerable potential for carbon sequestration in forests has been assessed. Sathaye and Ravindranath (1998) suggest that in 10 tropical and temperate countries in Asia 300 million hectares may be available for diverse forest management strategies and that afforestation or plantation projects could increase carbon stocks by 70–100 t C/ha in many places. Specifically commercial timber forestry in Indonesia and India could increase carbon stocks by 165 and 120 t C/ha, respectively.

In 2001 a special report of the IPCC was prepared to provide a state-of-the-art assessment of the global potential for carbon sequestration strategies relevant to land use, land-use change and forestry (Watson *et al.*, 2000). This assessment concluded that global carbon sequestration in terrestrial ecosystems of 1.2–1.5 Pg C/year was possible by 2010. Carbon sequestration in above- and belowground biomass as a consequence of afforestation and/or reforestation projects accounted for 0.2–0.6 Pg C/year; of this estimate, improved management practices within a given land use resulted in an additional 0.57 Pg C/year and land-use change of 0.44 Pg C/year. By 2050 a cumulative increase in carbon stocks in terrestrial ecosystems of up to 100 Pg C was assessed as possible – a sequestration equivalent to ~10–20% of projected fossil fuel emissions during that period. A recent study by Cannell (2003) concluded that increasing terrestrial carbon sinks and biomass energy substitution could sequester 0.2–5 Pg C/year, with 5 Pg C/year considered the theoretical maximum achievable and 0.2 Pg C/year the most conservative actually achievable estimate.

Problems associated with managed carbon sequestration

The expansion of forestlands and decreases in rates of deforestation are considered win–win options, in that they not only increase carbon sequestration but should result in a number of key ecological improvements, including preservation of biodiversity, protection of top soils and watersheds (Schulze *et al.*, 2002). However, there is concern that particular management options may ultimately prove to be less carbon-sound than others; indeed there is the danger that some may not result in a net increase in carbon sequestration. Most, if not all, these concerns have been central to the drafting of the protocols for the inclusion of biological sinks into the Kyoto Protocol. Sanz *et al.* (2004) describe both the 'loophole' argument, where biological sinks are seen as a way of doing little to reduce industrial emissions, and the 'floodgates' argument, whereby scientific and technical limitations to assessing carbon sequestration could allow for inflated claims of carbon sequestration. Fundamental to the success of carbon sequestration projects is ensuring the 'permanence' of carbon sinks, the 'additionality' of carbon sequestration over and above what would have occurred naturally and the protection against 'leakage', whereby a project to increase carbon storage in one location increases carbon release in another, while at the same time removing apparent 'perverse incentives' to clear off old-growth forests and replace them with a fast-growing tree species. This latter concern arises from the lack of protection towards old-growth forests in the Kyoto Protocol (Schulze *et al.*, 2003).

A final concern is the need for complete carbon accounting over the complete lifetime of a project as well as knowledge of the ultimate use of the carbon pools, not just increases in aboveground biomass. The dangers of a narrow focus on wood production rather than complete carbon accounting are clearly illustrated in a recent study by Deckmyn *et al.* (2004), who modelled carbon sequestration for two different afforestation projects, initiated on agricultural land. They demonstrated that while NPP and wood production of a short-rotation poplar coppice (SRC) was much higher than that for an oak–beech forest (OBF), complete accounting after 150 years revealed that carbon stored in all carbon pools in the OBF was double that of the SRC. However, even though wood produced in the OBF was used for much longer-lived wood products, that from the poplar plantation substituted for fossil fuels, which ultimately resulted in a fourfold higher mitigation potential for the SRC of ~26 t CO_2 ha/year. Complete carbon flux accounting over the life of a project also requires appreciation of the carbon costs of all components of the project. It has been argued that the carbon costs of applications of water or fertilizer on forests to increase productivity could offset the increase in carbon sequestration that they yield, due to the carbon costs of production and application (Schlesinger, 1997). Care must also be taken that application of fertilizer or irrigation does not trigger oxidation of soil carbon stores by microbes whose activity is also water- or nitrogen-limited.

2.6 Conclusions

The terrestrial biosphere is a dynamic component of the global carbon cycle. This fact is clearly evident from the seasonal and annual fluctuations in C_a that are caused by changes in the balance between photosynthesis and respiration on a global scale. That much of the terrestrial biosphere has been sequestering carbon in living biomass during the recent past, thereby mitigating against the rise in C_a, suggests that terrestrial ecosystems could buffer us from the worst effects of global change. However, as highlighted in this chapter, there remain major uncertainties in our understanding of the factors that have driven historical, are driving current, and will drive future, changes in carbon sequestration by terrestrial vegetation. The imperative for the development of new and existing policy initiatives to stabilize C_a requires continued experimentation across the broad spectrum of terrestrial carbon cycle research to reduce these uncertainties.

References

Ainsworth, E.A. and Long, S.P. (2004) What have we learned from 15 years of free-air CO_2 enrichment (FACE)? A meta-analytic review of the responses of photosynthesis, canopy properties and plant production to rising CO_2. *New Phytologist* 165, 351–372.

Ashmore, M.R. (2005) Assessing the future global impacts of ozone on vegetation. *Plant, Cell and Environment* 28, 949–964.

Asner, G.P., Archer, S., Flint Hughes, R., Ansley, R.J. and Wessman, C.A. (2003) Net changes in regional woody vegetation cover and carbon storage in Texas Drylands, 1937–1999. *Global Change Biology* 9, 316–335.

Badeck, F.W., Bondeau, A., Böttcher, K., Doktor, D., Lucht, W., Schaber, J. and Sitch, S. (2004) Responses of spring phenology to climate change. *New Phytologist* 162, 295–309.

Baldocchi, D.D. (2003) Assessing the eddy covariance technique for evaluating carbon dioxide exchange rates of ecosystems: past, present and future. *Global Change Biology* 9, 479–492.

Baldocchi, D.D. and Wilson, K.B. (2001) Modelling CO_2 and water vapour exchange over a temperate broad-leafed forest across hourly to decadal time scales. *Ecological Modelling* 142, 155–184.

Beier, C. (2004) Climate change and ecosystem function: full-scale manipulations of CO_2 and temperature. *New Phytologist* 162, 243–245.

Caldiera, K., Granger Morgan, M., Baldocchi, D., Brewer, P.G., Chen, C.T.A., Nabuurs, G.J., Nakicenoviv, N. and Robertson, G.P. (2004) A portfolio of carbon management options. In: Field, C.B. and Raupach, M.R. (eds) *The Global Carbon Cycle*. Island Press, Washington, DC, pp. 103–129.

Cannell, M.G.R. (2003) Carbon sequestration and biomass energy offset: theoretical, potential and achievable capacities globally, in Europe and the UK. *Biomass and Bioenergy* 24, 97–116.

CarboEurope GHG (2004a) Greenhouse gas emissions from European croplands. *Report 3, Specific Study 2,* European Commission.

CarboEurope GHG (2004b) Greenhouse gas emissions from European grasslands. *Report 4, Specific Study 3.* European Commision.

Carey, E.V., Sala, A., Keane, R. and Callaway, R.M. (2001) Are old forests underestimated as global carbon sinks? *Global Change Biology* 7, 339–344.

Casperson, J.P., Pacala, S.W., Jenkins, J.C., Hurtt, G.C., Moorcroft, P.R. and Birdsey, R.A. (2000) Contributions of land-use history to carbon accumulation in U.S. forests. *Science* 290, 1148–1151.

Ciais, P., Cuntz, M., Scholze, M., Mouillot, F., Peylin, P. and Gitz, V. (2005) Remarks on the use of ^{13}C and ^{18}O isotopes in atmospheric CO_2 to quantify biospheric carbon fluxes. In: Flanagan, L.B., Ehleringer, R. and Pataki, D.E. (eds) *Stable Isotopes and Biosphere–Atmosphere Interactions: Processes and Biological Controls*. Academic Press, London, pp. 235–267.

Cramer, W., Bondeau, A., Woodward, F.I., Prentice, I.C., Betts, R.A., Brovkin, V., Cox, P.M., Fisher, V., Foley, J.A., Friend, A.D., Kucharik, C., Lomas, M.R., Ramankutty, N., Sitch, S., Smith, B., White, A. and Young-Molling, C. (2001) Global response of terrestrial ecosystem structure and function to CO_2 and climate change: results from six dynamic global vegetation models. *Global Change Biology* 7, 357–373.

Crutzen, P.J., Fall, R., Galbally, I. and Lindinger, W. (1999) Parameters for global ecosystem models. *Nature* 399, 535.

Deckmyn, G., Muys, B., Garcia Quijano, J. and Ceulemans, R. (2004) Carbon sequestration following afforestation of agricultural soils: comparing oak/beech forest to short-rotation poplar coppice combining a process and a carbon accounting model. *Global Change Biology* 10, 1482–1491.

Dixon, R.K., Brown, S., Houghton, R.A., Solomon, A.M., Trexler, M.C. and Wisniewski, J. (1994) Carbon pools and flux of global forest ecosystems. *Science* 263, 185–190.

Drake, B.G., González-Meler, M.A. and Long, S.P. (1997) More efficient plants: a consequence of rising atmospheric CO_2? *Annual Review of Plant Physiology and Plant Molecular Biology* 48, 609–639.

Geider, R.J., Delucia, E.H., Falkowski, P.G., Finzi, A.C., Grime, J.P., Grace, J., Kana, T.M., La Roche, J., Long, S.P., Osborne, B.A., Platt, T., Prentice, I.C., Raven, J.A., Schlesinger, W.H., Smetacek, V., Stuart, V., Sathyendranath, S., Thomas, R.B., Vogelmann,T.C., Williams, P. and Woodward, F.I. (2001) Primary productivity of planet earth: biological determinants and physical constraints in terrestrial and aquatic habitats. *Global Change Biology* 7, 849–882.

Gleixner, G. (2005) Stable isotope composition of soil organic matter. In: Flanagan, L.B., Ehleringer, J.R. and Pataki, D.E. (eds) *Stable Isotopes and Biosphere–Atmosphere Interactions: Processes and Biological Controls*. Elsevier Academic Press, San Diego, California, pp. 29–46.

Grace, J. (2004) Understanding and managing the global carbon cycle. *Journal of Ecology* 92, 189–202.

Grace, J., Lloyd, J., McIntyre, J., Miranda, A.C., Meir, P., Miranda, H., Nobre, C., Moncrieff, J.M., Massheder, J., Malhi, Y., Wright, I.R. and Gash, J. (1995) Carbon dioxide uptake by an undisturbed tropical rain forest in South-West Amazonia 1992–1993. *Science* 270, 778–780.

Gruber, N., Friedlingstein, P., Field, C.B., Valentini, R., Heimann, M., Richey, J.E., Lankao, P.R., Schulze, E.D. and Chen, C.T.A. (2004) The vulnerability of the carbon cycle in the 21st century: an assessment of carbon–climate–human interactions. In: Field, C.B. and Raupach, M.R. (eds) *The Global Carbon Cycle*. Island Press, Washington, DC, pp. 45–76.

Grünzweig, J.M., Sparrow, S.D., Yakir, D. and Chapin III, F.S. (2004) Impact of agricultural land-use change on carbon storage in boreal Alaska. *Global Change Biology* 10, 452–472.

Gu, L., Baldocchi, D., Verma, S.B., Black, T.A., Vesala, T., Falge, E.M. and Dowty, P.R. (2002) Advantages of diffuse radiation for terrestrial ecosystem productivity. *Journal of Geophysical Research* 107(D6), 4050; doi: 10.1029/2001JD001242.

Guo, L.B. and Gifford, R.M. (2002) Soil carbon stocks and land use change: a meta analysis. *Global Change Biology* 8, 345–360.

Gurney, K.R., Law, R.M., Denning, A.S., Rayner, P.J., Baker, D., Bousquet, P., Bruhwiler, L., Chen, Y-H., Ciais, P., Fan, S., Fung, I.Y., Gloor, M., Heimann, M., Higuchi, K., John, J., Maki, T., Maksyutov, S., Masarie, K., Peylin, P., Prather, M., Pak, B.C., Randerson, J., Sarmiento, J., Taguchi, S., Takahashi, T. and Yuen, C.-W. (2002) Towards robust regional estimates of CO_2 sources and sinks using atmospheric transport models. *Nature* 415, 626–630.

Houghton, R.A. (1995) Land-use change and the carbon cycle. *Global Change Biology* 1, 275–287.

Houghton, R.A. (1999) The annual net flux of carbon to the atmosphere from changes in land use 1850–1990. *Tellus* 51B, 298–313.

Houghton, R.A. (2003) Why are estimates of the terrestrial carbon balance so different? *Global Change Biology* 9, 500–509.

Houghton, R.A. and Hackler, J.L. (2000) Changes in terrestrial carbon storage in the United States 1: the roles of agriculture and forestry. *Global Ecology and Biogeography* 9, 125–144

Houghton, R.A., Hackler, J.L. and Lawrence, K.T. (2000) Changes in terrestrial carbon storage in the United States 2: The role of fire and fire management. *Global Ecology and Biogeography* 9, 145–170.

Howard, E.A., Gower, S.T., Foley, J.A. and Kuchark, C.J. (2004) Effects of logging on carbon dynamics of a jack pine forest in Saskatchewan, Canada. *Global Change Biology* 10, 1267–1284.

Hungate, B.A., Dukes, J.A., Shaw, M.R., Luo, Y. and Field, C.B. (2003) Nitrogen and climate change. *Science* 302, 1512–1513.

Hymus, G.J., Baker, N.R. and Long, S.P. (2001) Growth in elevated CO_2 can both increase and decrease photochemistry and photoinhibiton of photosynthesis in a predictable manner: *Dactylis glomerata* grown at two levels of nitrogen nutrition. *Plant Physiology* 127, 1204–1211.

Jackson, R.B., Schenk, H.J., Jobbágy, E.G., Canadell, J., Colello, G.D., Dickinson, R.E., Field, C.B., Friedlingstein, P., Heimann, M., Hibbard, K., Kicklighter, D.W., Kleidon, A., Neilson, R.P., Parton, W.J., Sala, O.E. and Sykes, M.T. (2000) Belowground consequences of vegetation change and their treatment in models. *Ecological Applications* 10, 470–483.

Jackson, R.B., Banner, J.L., Jobbágy, E.G., Pockman, W.T. and Wall, D.H. (2002) Ecosystem carbon loss with woody plant invasion of grasslands. *Nature* 418, 623–626.

Janssens, I.A., Freibauer, A., Ciais, P., Smith, P., Nabuurs, G.-J., Folberth, G., Schlamadinger, B., Hutjes, R.W.A., Ceulemans, R., Schulze, E.D., Valentini, R., Dolman, A.J. (2003) Europe's terrestrial biosphere absorbs 7 to 12% of European anthropogenic CO_2 emissions. *Science* 300, 1538–1542.

Joos, F., Prentice, I.C. and House, J.I. (2002) Growth enhancement due to global atmospheric change as predicted by terrestrial ecosystem models: consistent with US forest inventory data. *Global Change Biology* 8, 299–303.

Knapp, P.A. and Soulé, P.T. (1998) Recent *Juniperus occidentalis* (western juniper) expansion on a protected site in central Oregon. *Global Change Biology* 4, 347–357.

Knohl, A., Kolle, O., Minayeva, T.Y., Milyukova, I.M., Vygodskaya, N.N., Foken, T. and Schulze, E.D. (2002) Carbon dioxide exchange of a Russian boreal forest after disturbance by wind throw. *Global Change Biology* 8, 231–246.

Knohl, A., Schulze, E.D., Kolle, O. and Buchmann, N. (2003) Large carbon uptake by an unmanaged 250-year-old deciduous forest in Central Germany. *Agricultural and Forest Meteorology* 118, 151–167.

Kowalski, A.S., Loustau, D., Berbigier, P., Manca, G., Tedeschi, V., Borghetti, M., Valentini, R., Kolari, P., Berninger, F., Rannik, U., Hari, P., Rayment, M., Mencuccini, M., Moncrief, J. and Grace, J. (2004) Paired comparisons of carbon exchange between undisturbed and regenerating stands in four managed forests in Europe. *Global Change Biology* 10, 1707–1723.

Law, B.E., Falge, E., Gu, L., Baldocchi, D.D., Bakwin, P., Berbigier, P., Davis, K., Dolman, A.J., Falk, M., Fuentes, J.D., Goldstein, A., Granier, A., Grelle, A., Hollinger, D., Janssens, I.A., Jarvis, P., Jensen, N.O., Katul, G., Mahli, Y., Matteucci, G., Meyers, T., Monson, R., Munger, W., Oechel, W., Olson, R., Pilegaard, K., Paw, U.K.T., Thorgeirsson, H., Valentini, R., Verma, S., Vesala, T., Wilson, K. and Wofsy, S. (2002) Environmental controls over carbon dioxide and water vapor exchange of terrestrial vegetation. *Agricultural and Forest Meteorology* 113, 97–120.

Le Quéré, C., Aumont, O., Bopp, L., Bousquet, P., Ciais, P., Francey, R., Heimann, M., Keeling, C.D., Keeling, R.F., Kheshgi, H., Peylin, P., Piper, S.C., Prentice, I.C. and Rayner, P.J. (2003) Two decades of ocean CO_2 sink and variability. *Tellus* 55B, 649–656.

Lewis, S.L., Phillips, O.L., Baker, T.R., Lloyd, J., Malhi, Y., Almeida, S., Higuchi, N., Laurance, W.F., Neill, D.A., Silva, J.N.M., Terborgh, J., Torres Lezama, A., Vásquez Martínez, R., Brown, S., Chave, J., Kuebler, C., Núñez Vargas, P. and Vinceti, B. (2004) Concerted changes in tropical forest structure and dynamics: evidence from 50 South American long-term plots. *Philosophical Transactions of the Royal Society of London Series B* 359, 421–436.

Litvak, M.E., Miller, S., Wofsy, S.C. and Goulden, M. (2003) Effect of stand age on whole ecosystem CO_2 exchange in the Canadian boreal forest. *Journal of Geophysical Research* 108; doi:10.1029/2001JD000854.

Long, S.P., Garcia Moya, E., Imbamba, S.K., Kamnalrut, A., Piedade, M.T.F., Scurlock, J.M.O., Shen, Y.K. and Hall, D.O. (1989) Primary productivity of natural grass ecosystems of the tropics: a reappraisal. *Plant and Soil* 115, 155–166.

Ludwig, W.P., Amiotte-Suchett, P., Munhoven, G. and Probst, J.L. (1998) Atmospheric CO_2 consumption by continental erosion: present day controls and implications for the last glacial maximum. *Global and Planetary Change* 16, 107–120.

MacDonald, J.A., Dise, N.B., Matzner, E., Armbruster, M., Gundersen, P. and Forsius, M. (2002) Nitrogen input together with ecosystem nitrogen enrichment predict nitrate leaching from European forests. *Global Change Biology* 8, 1028–1033.

Malhi, Y., Nobre, A.D., Grace, J., Kruijt, B., Pereira, M.G.P., Culf, A. and Scott, S. (1998) Carbon dioxide transfer over a Central Amazonian rain forest. *Journal of Geophysical Research – Atmospheres* 103, 31593–31612.

Malhi, Y., Baker, T.R., Phillips, O.L., Almeida, S., Alvarez, E., Arroyo, L., Chave, J., Czimczik, C.I., Di Fiore, A., Higuchi, N., Killeen, T.J., Laurance, S.G., Laurance, W.F., Lewis, S.L., Mercado, L.M., Monteagudo, A., Neill, D.A., Núñez Vargas, P., Patiño, S., Pitman, N.C.A., Alberto Quesada, C., Silva, N., Torres Lezama, A., Vásquez Martínez, R., Terborgh, J., Vinceti, B. and Lloyd, J. (2004) The above-ground wood productivity and net primary productivity of 100 neotropical forests. *Global Change Biology* 10, 563–591.

Mouillot, F. and Field, C.B. (2005) Fire history and the global carbon budget: a 1° × 1° fire history reconstruction for the 20th century. *Global Change Biology* 11, 398–420.

Myneni, R.B., Keeling, C.D., Tucker, C.J., Asrar, G. and Nemani, R.R. (1997) Increased plant growth in the northern high latitudes 1981–1991. *Nature* 386, 698–702.

Nabuurs, G.J. (2004) Current consequences of past actions: how to separate direct from indirect. In: Field, C.B. and Raupach, M.R. (eds) *The Global Carbon Cycle*. Island Press, Washington, DC, pp. 317–326.

Nabuurs, G.J., Dolman, A.J., Verkaik, E., Kuikman, P.J. van Diepen, C.A., Whitmore, A.P., Daamen, W.P., Oenema, O., Kabat, P. and Mohren, G.M.J. (2000) Article 3.3 and 3.4 of the Kyoto Protocol: consequences for industrialised countries' commitment, the monitoring needs, and possible side effects. *Environmental Science and Policy* 3, 123–134.

Nabuurs, G.J., Schelhaas, M.J., Mohren, G.M.J. and Field, C.B. (2003) Temporal evolution of the European forest sector carbon sink 1950–1999. *Global Change Biology* 9, 152–160.

Nemani, R.R., Keeling, C.D., Hashimoto, H., Jolly, W.M., Piper, S.C., Tucker, C.J., Myneni, R.B. and Running, S.W. (2003) Climate-driven increases in global net primary productivity from 1982 to 1999. *Science* 300, 1560–1563.

Nepstad, D.C., de Carvalho, C.R., Davidson, E.A., Jipp, P.H., Lefebvre, P.A., de Negreiros, G.H., da Silva, E.D., Stone, T.A., Trumbore, S.E. and Vieira, S. (1994) The role of deep roots in the hydrological and carbon cycles of Amazonian forests and pastures. *Nature* 372, 666–669.

Norby, R.J. and Luo, Y. (2004) Evaluating ecosystem responses to rising atmospheric CO_2 and global warming in a multi-factor world. *New Phytologist* 162, 281–294.

Norby, R.J., Wullschleger, S.D., Gunderson, C.A., Johnson, D.W. and Ceulemans, R. (1999) Tree responses to rising CO_2 in field experiments: implications for the future forest. *Plant, Cell and Environment* 22, 683–714.

Odum, E.P. (1969) The strategy of ecosystem development. *Science* 164, 262–270.

Osborne, C.P., Mitchell, P.L., Sheehy, J.E. and Woodward, F.I. (2000) Modelling the recent historical impacts of atmospheric CO_2 and climate change in the Mediterranean vegetation. *Global Change Biology* 6, 445–458.

Osmond, B., Ananyev, G., Berry, J., Langdon, C., Kolber, Z., Lin, G., Monson, R., Nichol, C., Rascher, U., Schurr, U., Smith, S. and Yakir, D. (2004) Changing the way we think about global change research: scaling up in experimental ecosystem science. *Global Change Biology* 10, 393–407.

Pacala, S.W., Hurtt, G.C., Houghton, R.A., Birdsey, R.A., Heath, L., Sundquist, E.T., Stallard, R.F., Baker, D., Peylin, P., Moorcroft, P., Caspersen, J., Shevliakova, E., Harmon, M.E., Fan, S.-M., Sarmiento, J.L., Goodale, C., Field, C.B., Gloor, M. and Schimel, D. (2001) Consistent land- and atmosphere-based U.S. carbon sink estimates. *Science* 292, 2316–2320.

Papale, D. and Valentini, R. (2003) A new assessment of European forests carbon exchanges by eddy fluxes and artificial neural network spatialisation. *Global Change Biology* 9, 525–535.

Pregitzer, K.S. and Euskirchen, E. (2004) Carbon cycling and storage in world forests: biome patterns related to forest age. *Global Change Biology* 10, 2052–2077.

Prentice, C., Farqhuar, G.D., Fasham, M.J.R., Goulden, M.L., Heimann, M., Jaramillo, V.J., Kheshgi, H.S., Le Quéré, C., Scholes, R.J. and Wallace, W.R. (2001) The carbon cycle and atmospheric carbon dioxide. In: Houghton, J., Ding, Y., Griggs, D.J., Noguer, M., van der Linden, P.J., Dai, X., Maskell, K. and Johnson, C.A. (eds) *Climate Change 2001: The Scientific Basis (Contribution of Working Group I to the Third Assessment Report of the Intergovernmental Panel on Climate Change)*. Cambridge University Press, Cambridge.

Ramankutty, N. and Foley, J. (1999) Estimating historical changes in global land cover: croplands from 1700 to 1992. *Global Biogeochemical Cycles* 13, 997–1028.

Randerson, J.T. (2005) Terrestrial ecosystems and interannual variability in the global atmospheric budgets of $^{13}CO_2$ and $^{12}CO_2$. In: Flanagan, L.B., Ehleringer, R. and Pataki, D.E. (eds) *Stable Isotopes and Biosphere–Atmosphere Interactions: Processes and Biological Controls*. Academic Press, London, pp. 217–234.

Raupach, M.R., Canadell, J.G., Bakker, D.C.E., Ciais, P., Sanz, M.J., Fang, J.Y., Melillo, J.M., Romero Lankao, P., Sathaye, J.A., Schulze, E.D., Smith, P. and Tschirley, J. (2004) Interactions between CO_2 stabilization pathways and requirements for a sustainable earth system. In: Field, C.B. and Raupach, M.R. (eds) *The Global Carbon Cycle*. Island Press, Washington, DC, pp. 131–162.

Richards, J.F. (1990) Land transformation. In: Turner, B.L., Clark, W.C., Kates, R.W., Richards, J.F., Mathews, J.T. and Meyer, W.B. (eds) *The Earth as Transformed by Human Action*. Cambridge University Press, Cambridge, pp. 163–178.

Sabine, C.L., Heimann, M., Artaxo, P., Bakker, D.C.E., Chen, C.T.A., Field, C.B., Gruber, N., Le Quéré, C., Prinn, R.G., Richey, J.E., Lankao, P.R., Sathaye, J.A. and Valentini, R. (2004) Current status and past trends of the global carbon cycle. In: Field, C.B. and Raupach, M.R. (eds) *The Global Carbon Cycle*. Island Press, Washington, DC, pp. 17–44.

Saleska, S.R., Miller, S.D., Matross, D.M., Goulden, M.L., Wofsy, S.C., da Rocha, H., de Camargo, P.B., Crill, P.M., Daube, B.C., Freitas, C., Hutyra, L., Keller, M., Kirchhoff, V., Menton, M., Munger, J.W., Pyle, E.H., Rice, A.H. and Silva, H. (2003) Carbon in Amazon forests: unexpected seasonal fluxes and disturbance-induced losses. *Science* 302, 1554–1557.

Sanz, M.J., Schulze, E.D. and Valentini, R. (2004) International policy framework on climate change: sinks in recent international agreements. In: Field, C.B. and Raupach, M.R. (eds) *The Global Carbon Cycle*. Island Press, Washington, DC, pp. 431–438.

Sathaye, J. and Ravindranath, N.H. (1998) *Climate Change Mitigation in the Energy and Forestry Sectors of Developing Countries*. Lawrence Berkley National Laboratory Publication 42626.

Saugier, B., Roy, J. and Mooney, H.A. (2001) *Estimations of Global Terrestrial Productivity: Converging Towards a Single Number?* Academic Press, San Diego, California.

Schlesinger, W.H. (1997) *Biogeochemistry: An Analysis of Global Change*. Academic Press, San Diego, California.

Scholes, R.J. and Archer, S.R. (1997) Tree-grass interactions in savannas. *Annual Review of Ecology and Systematics* 28, 517–544.

Schulze, E.D., Wirth, C. and Heimann, M. (2000) Climate change: managing forests after Kyoto. *Science* 289, 2058–2059.

Schulze, E.D., Valentini, R. and Sanz, M.J. (2002) The long way from Kyoto to Marrakesh: implications of the Kyoto Protocol negotiations for global ecology. *Global Change Biology* 8, 505–518.

Schulze, E.D., Mollicone, D., Achard, F., Matteucci, G., Federici, S., Eva, H.D. and Valentini, R. (2003) Making deforestation pay under the Kyoto Protocol. *Science* 299, 1669.

Sturm, M., Racine, C. and Tape, K. (2001) Increasing shrub abundance in the Arctic. *Nature* 411, 546–547.

Thornton, P.E., Law, B.E., Gholz, H.L., Clark, K.L., Falge, E., Ellsworth, D.S., Goldstein, A.H., Monson, R.K., Hollinger, D., Falk, M., Chen, J. and Sparks, J.P. (2002) Modeling and measuring the effects of disturbance history and climate on carbon and water budgets in evergreen needleleaf forests. *Agricultural and Forest Meteorology* 113, 185–222.

Tilman, D., Reich, P., Phillips, H., Menton, M., Patel, A., Vos, E., Peterson, D. and Knops, J. (2000) Fire suppression and ecosystem carbon storage. *Ecology* 81, 2680–2685.

Trumbore, S.E. (1997) Potential responses of soil organic carbon to global environmental change. *Proceedings of the National Academy of Sciences USA* 94, 8284–8291.

Turner, I.I.B.L., Moss, R.H. and Skole, D.L. (1993) Relating land use and global land cover change: a proposal for an IGBP-HDP Core Project. IGBP Report no. 24, HDP Report no. 5. The International Geosphere–Biosphere Programme: A Study of Global Change and the Human Dimensions of Global Environmental Change Programme, Stockholm.

Valentini, R., Matteucci, G., Dolman, A.J., Schulze, E.D., Rebmann, C., Moors, E.J., Granier, A., Gross, P., Jensen, N.O., Pilegaard, K., Lindroth, A., Grelle, A., Bernhofer, C., Grunwald, T., Aubinet, M., Ceulemans, R., Kowalski, A.S., Vesala, T., Rannik, U., Berbigier, P., Loustau, D., Gudmundsson, J., Thorgeirsson, H., Ibrom, A., Morgenstern, K., Clement, R., Moncrieff, J., Montagnani, L., Minerbi, S. and Jarvis, P.G. (2000) Respiration as the main determinant of carbon balance in European forests. *Nature* 404, 861–864.

Vitousek, P.M., Aber, J.D., Howarth, R.W., Likens, G.E., Matson, P.A., Schindler, D.W., Schlesinger, W.H. and Tilman, D.G. (1997a) Human alteration of the global nitrogen cycle. *Ecological Applications* 7, 737–750.

Vitousek, P.M., Mooney, H.A., Lubchenco, J. and Melillo, J.M. (1997b) Human domination of earth's ecosystems. *Science* 277, 494–499.

Walter, B.P., Heimann, M. and Matthews, E. (2001) Modeling modern methane emissions from natural wetlands. 1. Model description and results. *Journal of Geophysical Research* 106, 34189–34206.

Waring, R. and Running, S.W. (1998) *Forest Ecosystems: Analysis at Multiple Scales*. Academic Press, San Diego, California.

Watson, R.T., Noble, I.R., Bolin, B., Ravindranath, N.H., Verado, D.J. and Dokken, D.J. (2000) *IPCC Special Report on Land Use, Land-use Change and Forestry*. Cambridge University Press, Cambridge.

3 The Oceanic Sink for Carbon Dioxide

Christopher L. Sabine and Richard A. Feely
NOAA Pacific Marine Environmental Laboratory, Seattle, Washington, USA

3.1 Introduction

Agricultural practices, fossil fuel burning and cement manufacturing have resulted in a more than 35% increase in atmospheric carbon dioxide (CO_2) concentrations over the last 200 years. However, this increase would have been much more dramatic if the ocean had not been absorbing a significant amount of the CO_2 resulting from human activity (anthropogenic CO_2). There are three major reservoirs with exchange rates fast enough to vary significantly on the timescale of decades to centuries: the atmosphere, terrestrial biosphere and the ocean. Of this three-component system, ~93% of the carbon is located in the ocean (Fig. 6.2). Although this means that the ocean has a tremendous capacity to influence future atmospheric CO_2 concentrations, it also means that it is very difficult to accurately constrain the oceanic sink for anthropogenic CO_2.

The anthropogenic signal that has accumulated in the ocean over the last 200 years is generally less than 3% of the natural carbon in surface sea water, making it difficult to distinguish the anthropogenic CO_2 from the observed natural variability. Unfortunately, ocean carbon measurements cannot directly distinguish between anthropogenic CO_2 and natural inorganic carbon; moreover, the historical measurements of ocean carbon are inadequate to document directly the increase in ocean inventories. Until recently, almost all estimates of oceanic CO_2 uptake have been based on modelling studies, which must be evaluated and improved based on geochemical tracers other than anthropogenic CO_2 (e.g. bomb radiocarbon, chlorofluorocarbons).

Concerns over the greenhouse effect and global climate change have inspired scientists to focus their attention on improving our understanding of the ocean's role in the global carbon cycle. Recently, oceanographers have made great advances in both ocean carbon observations and modelling, leading to an improved understanding of the cycling of carbon in the ocean and exchanges with other carbon reservoirs (e.g. Sabine *et al.*, 2004a). This chapter will focus primarily on the oceanic sink for CO_2 and the seasonal, interannual and decadal-scale variability in sea–air fluxes. In particular, it will discuss how the sea–air exchange of CO_2 relates to the oceanic uptake of anthropogenic CO_2, and potential feedbacks within the carbon cycle as well as the carbon–climate system that may change the oceanic uptake of CO_2 in the future.

3.2 Pre-industrial Carbon Fluxes

As there were no ocean carbon measurements during the pre-industrial period (prior to *c.* 1800), we have to use indirect geochemical evidence and our understanding of current carbon cycle dynamics to infer how the ocean carbon cycle operated prior to human intervention. One aspect of the global carbon cycle that helps us to better understand the pre-industrial period is the fact that atmospheric CO_2 concentration was remarkably stable, with variations in atmospheric CO_2 of <20 ppm, during at least the last 11,000 years prior to the anthropogenic perturbation (Joos and Prentice, 2004). Since global ocean circulation operates on timescales of ~1000 years and sea–air exchange has timescales of about 1 year (Broecker and Peng, 1982), the ocean must have been in a steady state with respect to the atmosphere and terrestrial biosphere (the only other sizable carbon reservoirs with comparable exchange timescales) during this period.

Figure 3.1 shows a recently compiled budget of the pre-industrial ocean carbon cycle (Sabine *et al.*, 2004a). It shows some of the complexity of the carbon transformations within the ocean. The gross exchanges across the sea–air interface are estimated to be ~70 Pg C/year. These fluxes are about ten times larger than the current emissions of CO_2 into the atmosphere from burning fossil fuels, but since the CO_2 release is very nearly balanced by an oceanic uptake there is little net effect on atmospheric CO_2 concentrations. The pre-industrial ocean is estimated to have a small net flux (~0.6 Pg C/year) out of the ocean to balance the carbon that enters from rivers and groundwater (Ridgwell and Edwards, Chapter 6, this volume). The river input comes from the terrestrial biosphere and the weathering of continental rocks, which in turn absorb the

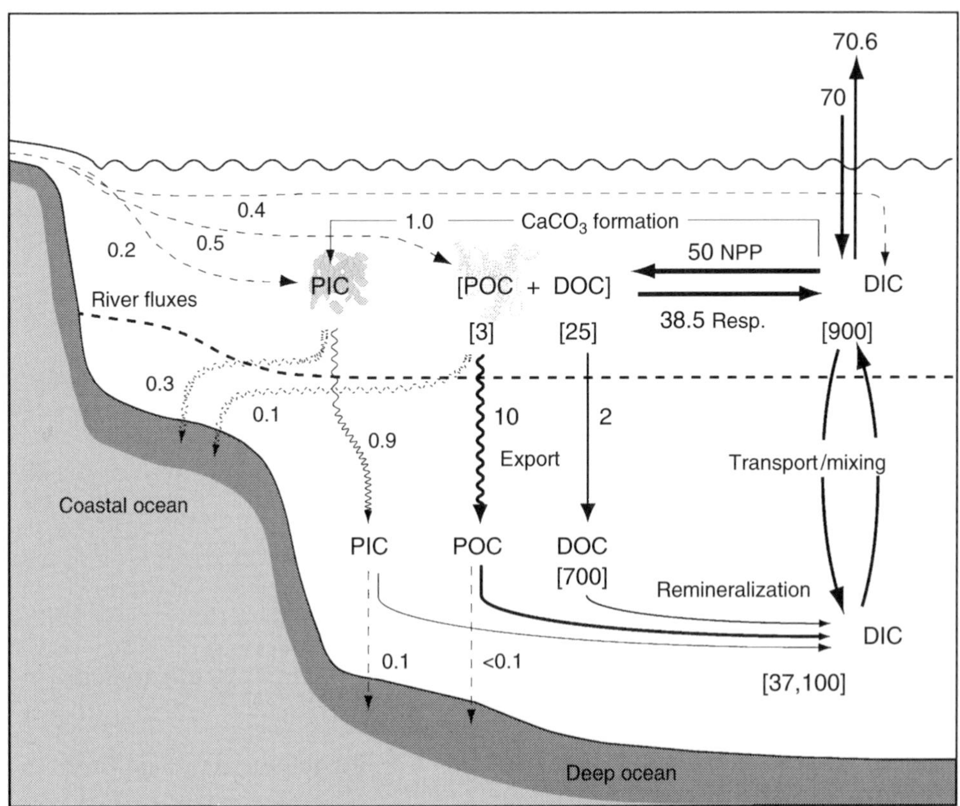

Fig. 3.1. Schematic representation of the pre-industrial ocean carbon cycle. Fluxes (arrows) are in Pg C/year and reservoir sizes (numbers in square brackets) are in Pg C. (Modified from Sabine *et al.*, 2004a.) Abbreviations: PIC, particulate inorganic carbon; DOC, dissolved organic carbon; DIC, dissolved inorganic carbon.

CO_2 from the atmosphere (Watson and Orr, 2003).

The partial pressure of CO_2 (pCO_2) in pre-industrial surface sea water is estimated to have roughly the same geographical and seasonal variability as the modern ocean, with values ranging from 150 to 750 ppm. Model studies suggest that prior to 1800 the northern hemisphere oceans were taking up atmospheric CO_2 while the Southern Ocean was a net source of CO_2 to the atmosphere. Recent transport calculations as part of the second Ocean Carbon Model Intercomparison Project (OCMIP-2), which included 12 ocean carbon general circulation models (GCMs), indicated that in pre-industrial times carbon was transported southward in the ocean interior across the equator at an average rate of just under 0.35 Pg C/year (Orr, 2004). This net transport was most likely lost back to the atmosphere in the Southern Ocean as a net atmospheric transport of carbon northward across the equator is required to balance the northern oceanic uptake. This is in contrast to the current scenario in which rising atmospheric CO_2 concentrations have made the modern Southern Ocean a net sink for CO_2 and the predominantly northern hemisphere sources of fossil fuel burning have resulted in a net atmospheric carbon transport from north to south, the same direction as the net ocean transport.

3.3 Modern Carbon Fluxes

Since the pre-industrial period, atmospheric CO_2 concentrations have increased from 280 ppm to nearly 380 ppm. This increase in CO_2 drives the sea water to absorb CO_2 from the atmosphere so that surface sea water is pushed to achieve thermodynamic equilibrium with the atmospheric partial pressure. Figure 3.2 shows a summary of the additional fluxes in the modern ocean resulting from human activity and rising atmospheric CO_2. The role of the ocean in the global carbon cycle has changed from being a net source of CO_2 to the atmosphere to a net sink for CO_2 of ~2 Pg C/year (Sabine *et al.*, 2004a).

Today, the average pCO_2 of the atmosphere is ~7 ppm higher than the global ocean pCO_2. This small air–sea difference, when spread across the entire surface of the ocean, is sufficient to account for the oceanic uptake of anthropogenic CO_2. The pCO_2 values in mixed-layer waters, which exchange CO_2 directly with the atmosphere, are affected primarily by changes in temperature, dissolved inorganic carbon (DIC) and total alkalinity (TAlk). While the water temperature is regulated by physical processes, including solar energy input, sea–air heat exchanges and mixed-layer thickness, DIC is primarily controlled by the physical processes of sea–air exchange and upwelling of subsurface waters as well as the biological processes of photosynthesis and respiration. Biological production removes carbon from surface waters to form organic material. As organisms die and sink to the ocean interior, they decompose, releasing the carbon once again to the water. This process contributes to higher pCO_2 and DIC concentrations in deep ocean waters relative to the surface waters. As pCO_2 increases when the water is warmed and decreases as a result of biological uptake, the oceanic uptake and release of CO_2 is governed by a balance between the changes in sea water temperature, net biological utilization of CO_2 and circulation processes in the upper ocean (Zeebe and Wolf-Gladrow, 2001).

Taro Takahashi of Lamont–Doherty Earth Observatory and his collaborators have amassed a database of more than 1.7 million surface ocean pCO_2 measurements, spanning more than 30 years, and derived a pCO_2 climatology for the global ocean (Takahashi *et al.*, 2002). These data have been used to determine global and regional sea–air CO_2 fluxes with an average annual global open-oceanic uptake of 1.5 ± 0.4 Pg C/year for a nominal year of 1995 (Takahashi *et al.*, 2002; revised by T. Takahashi, New York, 2005, personal communication). This flux estimate represents the total net flux in 1995. The total anthropogenic flux would be the difference between the 1995 net sea–air flux and the pre-industrial net sea–air flux (i.e. −1.5 − 0.6 = −2.1 Pg C/year) – a flux consistent with earlier estimates based on models.

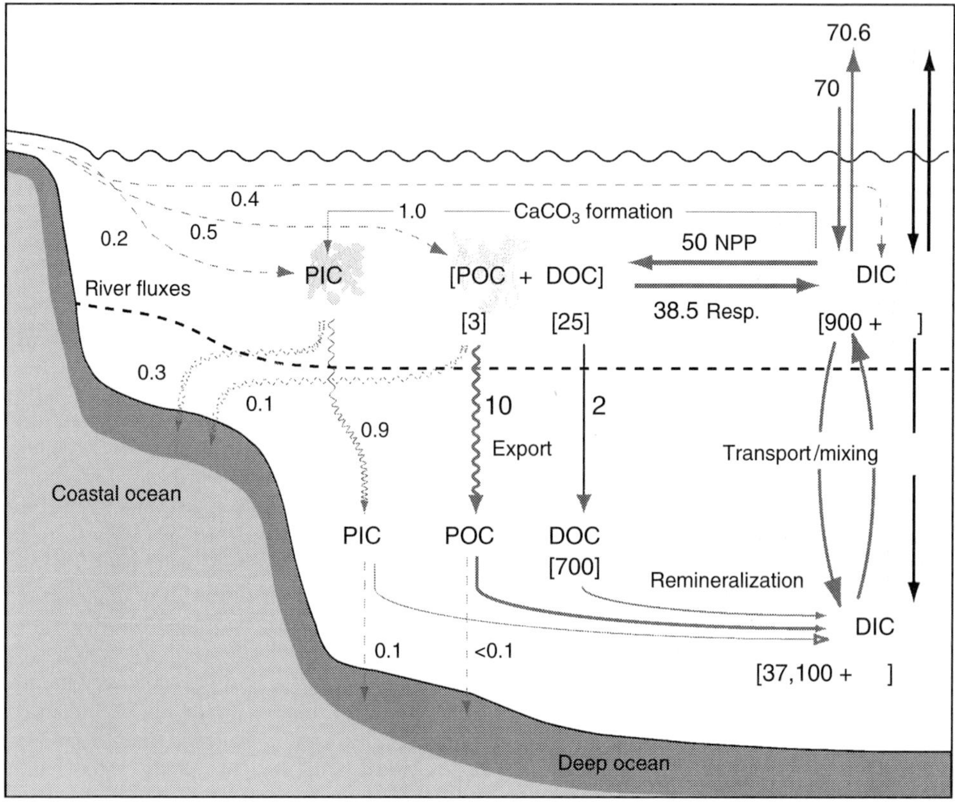

Fig. 3.2. Schematic representation of the ocean carbon cycle with pre-industrial fluxes and reservoir sizes (upright) and average values for the 1980s and 1990s (italic). Fluxes (arrows) are in Pg C/year and reservoir sizes (numbers in square brackets) are in Pg C. (Modified from Sabine *et al.*, 2004a.)

3.3.1 Effect of surface sea water pCO_2 variations on CO_2 fluxes

Figure 3.3 shows the distribution of total net sea–air CO_2 fluxes. The darker shades indicate oceanic areas where there is a net source of CO_2 to the atmosphere and the lighter shades indicate regions where there is a net sink of CO_2. The equatorial Pacific is a strong source of CO_2 to the atmosphere throughout the year as a result of upwelling that brings deep, high CO_2 waters to the surface in the central and eastern regions. This upwelling, and thus the CO_2 flux to the atmosphere, is heavily modulated by the El Niño–southern oscillation (ENSO) cycle. During strong El Niño years the equatorial Pacific CO_2 source can drop to zero. During La Niña the CO_2 source to the atmosphere is enhanced. High CO_2 outgassing fluxes are also observed in the tropical Atlantic and Indian oceans throughout the year. The Arabian Sea becomes a significant source of CO_2 to the atmosphere in the late summer and early fall months as the south-east monsoon generates intense upwelling off the Arabian peninsula.

Strong convective mixing also brings up high CO_2 values in the north-western sub-Arctic Pacific and Bering Sea during the northern winter. However, just outside this region there is a seasonal oscillation in CO_2 flux. The geochemical response of the ocean to changing temperatures is to decrease the pCO_2 by 4.23%/°C of sea water cooling (Takahashi *et al.*, 1993). In some regions, decreasing temperatures in the winter can lower the ocean's pCO_2 values

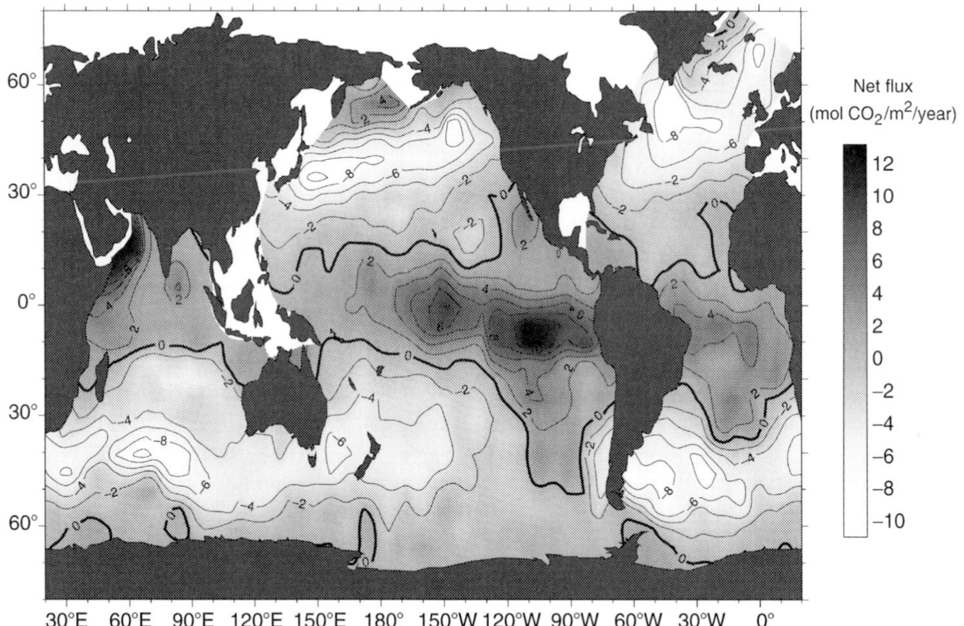

Fig. 3.3. Global climatology of the annual net sea–air CO_2 flux (mol CO_2/m^2/year) based on interpolation of sea–air pCO_2 differences as in 1995. (From Takahashi *et al.*, 2002.)

sufficiently to counteract the elevated CO_2 brought to the surface from stronger winter time mixing (e.g. temperate North Pacific and North Atlantic oceans). The fluxes out of the ocean from elevated temperatures during summer are limited by stratification, resulting in a small net annual flux into the ocean. Similar seasonal changes are observed in the southern temperate oceans, but are out of phase by half a year.

Intense regions of CO_2 uptake are seen in the high-latitude northern ocean in summer and in the high-latitude South Atlantic and southern Indian oceans in austral summer. This uptake is associated with high biological utilization of CO_2 in thin mixed layers. As the seasons progress, vertical mixing of deep waters eliminates this uptake of CO_2. These observations indicate that the CO_2 flux in high-latitude oceans is governed primarily by deep convection in winter and biological uptake during the spring and summer months, whereas in the temperate and subtropical oceans, the flux is governed primarily by water temperature. Outside the equatorial belt, the ΔpCO_2 (sea water pCO_2 – atmospheric pCO_2) is highest during winter in subpolar and polar waters, whereas it is highest during summer in the temperate regions. Thus, the seasonal variation of ΔpCO_2 and, consequently, the shift between net uptake and release of CO_2 in subpolar and polar regions are about 6 months out of phase with those in the temperate regions.

3.3.2 Effect of gas transfer velocity on CO_2 fluxes

The ΔpCO_2 maps are combined with solubility (s) in sea water and the kinetic forcing function, the gas transfer velocity (k), to produce the flux equation:

$$F = ks\Delta pCO_2 \tag{3.1}$$

where the gas transfer velocity, k, is controlled by near-surface turbulence in the liquid boundary layer. Laboratory studies in wind–wave tanks have shown that k is a strong but non-unique function of wind speed (Wanninkhof *et al.*, 2002). Results from various wind–wave tank investigations and field studies indicate that factors such as fetch, wave direction, atmospheric boundary layer stability and bubble entrainment influence the rate of gas transfer. Moreover, surfactants can inhibit gas exchange through their damping effect on waves. The commonly used gas transfer parameterizations have been based solely on wind speed, in large part because *k* is strongly dependent on wind, global and regional wind speed data are readily available and effects other than wind speed have not been well quantified (Wanninkhof *et al.*, 2002). Table 3.1 shows the regional variations of the climatological sea–air exchange fluxes.

Using an alternative gas exchange formulization, however, can suggest a different distribution of fluxes. For example, Wanninkhof and McGillis (1999) have suggested a cubic relationship to wind speed instead of the quadratic relationship of Wanninkhof (1992). The cubic relationship gives an uptake that is 45% larger than the quadratic relationship (Table 3.2). This primarily results from a larger CO_2 uptake in the high-latitude sink regions because of the stronger impact of the higher winds on the gas exchange (Feely *et al.*, 2001). More studies of gas exchange processes at high wind speed regimes are required before determining whether the quadratic, cubic or some other newly developed relationship is appropriate for high wind speeds.

3.3.3 CO_2 flux variability

Over the last 40 years, the growth rate of CO_2 in the atmosphere has experienced interannual variations as large as ±2 Pg C/year (Francey *et al.*, 1995; Keeling *et al.*, 1996). There is an ongoing controversy on the relative contributions of this variability from atmosphere–land and atmosphere–ocean exchanges (Fig. 3.4). Time series measurements of atmospheric CO_2, ^{13}C and O_2/N_2 sources have suggested that

Table 3.1. Regional distribution of net sea–air fluxes (Pg C/year) for 1995 based on NCEP 41-year U-10 reanalysis of wind data and the Wanninkhof (1992) gas exchange relationship and areas represented in each region. (Modified from Takahashi *et al.*, 2002; see http://www.ldeo.columbia.edu/res/pi/CO2/carbondioxide/pages/air_sea_flux_rev1.html.)

Latitude band	Pacific	Atlantic	Indian	Southern	Global
Flux (Pg C/year)					
North of 50°N	0.01	−0.31	–	–	−0.30
14°N–50°N	−0.49	−0.25	0.05	–	−0.69
14°S–14°N	0.65	0.13	0.13	–	0.91
14°S–50°S	−0.39	−0.21	−0.52	–	−1.12
South of 50°S	–	–	–	−0.30	−0.30
Total	−0.22	−0.64	−0.34	−0.30	−1.50
% of uptake	15	42	23	20	100
Area (10^6 km^2)					
North of 50°N	4.93	9.23	–	–	14.16
14°N–50°N	42.59	24.41	2.12	–	69.12
14°S–14°N	50.93	15.91	19.83	–	86.67
14°S–50°S	53.58	25.01	31.03	–	109.62
South of 50°S	–	–	–	41.10	41.10
Total	152.0	74.6	53.0	41.1	320.7
% of area	47	23	17	13	100

Table 3.2. Regional distribution of net sea–air flux (Pg C/year) for 1995 based on NCEP 41-year U-10 reanalysis of wind data and the Wanninkhof and McGillis (1999) gas exchange relationship. (Modified from Takahashi *et al.*, 2002; see http://www.ldeo.columbia.edu/res/pi/CO2/carbondioxide/pages/air_sea_flux_rev1.html.)

Latitude band	Pacific	Atlantic	Indian	Southern	Global
North of 50°N	0.03	−0.37	–	–	−0.34
14°N–50°N	−0.59	−0.31	0.06	–	−0.84
14°S–14°N	0.53	0.11	0.13	–	0.77
14°S–50°S	−0.45	−0.28	−0.64	–	−1.37
South of 50°S	–	–	–	−0.39	−0.39
Total	−0.48	−0.85	−0.45	−0.39	−2.17

ocean flux variations must be in the order of 1–2 Pg C/year (Francey *et al.*, 1995; Keeling *et al.*, 1996; Rayner *et al.*, 1999; Battle *et al.*, 2000; Bousquet *et al.*, 2000). However, ocean modelling and revised inverse models (Winguth *et al.*, 1994; Le Quéré *et al.*, 2000, 2003; Obata and Kitamura, 2003; McKinley *et al.*, 2004; Peylin *et al.*, 2005; Wetzel *et al.*, 2005) as well as empirical approaches (Lee *et al.*, 1998; Park *et al.*, 2006) have suggested a much smaller ocean variability of ~0.3–0.5 Pg C/year.

Of the few direct time series measurements made over large ocean regions so far, only the equatorial Pacific Ocean (Feely *et al.*, 1997, 1999) and the Greenland Sea (Skjelvan *et al.*, 1999) have shown large year-to-year variations in sea–air CO_2 flux. However, there are not many data-sets with which to evaluate such flux variability directly. The variability observed in the equatorial Pacific and North Atlantic oceans is not sufficient to account for all of the variability estimates, but other regions including the Southern Ocean and subtropical regions have not been studied sufficiently to determine their contributions to oceanic variability. Recent ocean model results have suggested that after the equatorial Pacific, the Southern Ocean and the northern extra-tropical regions are also important regions showing significant interannual variability in sea–air CO_2 flux (Peylin *et al.*, 2005; Wetzel *et al.*, 2005). Resolving this controversy and imposing stricter constraints on carbon cycle models will require more detailed observations of the magnitude and causes of variability in the sea–air CO_2 flux and other carbon-related species in the ocean, as well as continued atmospheric measurements of temporal and spatial distributions of CO_2, ^{13}C and O_2/N_2.

3.4 Uptake of Anthropogenic CO_2

The oceanic uptake of anthropogenic CO_2 is primarily a physical response to rising atmospheric CO_2 concentrations. Whenever the partial pressure of a gas is higher in the atmosphere over a body of water, the gas will diffuse into that water until the partial pressures across the air–water interface are equilibrated. There is no evidence that the rising atmospheric CO_2 concentrations have had a measurable impact on biological processes in the ocean. The growth rate of the primary producers in the ocean (phyto plankton) is generally limited by either light or nutrient availability, not carbon. It is possible that climate changes (e.g. ocean temperature or circulation changes) may be affecting the ocean carbon system, but these effects are thought to be small for the 19th and 20th centuries.

3.4.1 Modern CO_2 uptake rates

Several independent approaches have been used to estimate the modern oceanic uptake rate of anthropogenic CO_2. Table 3.3 shows a summary of the ocean observations

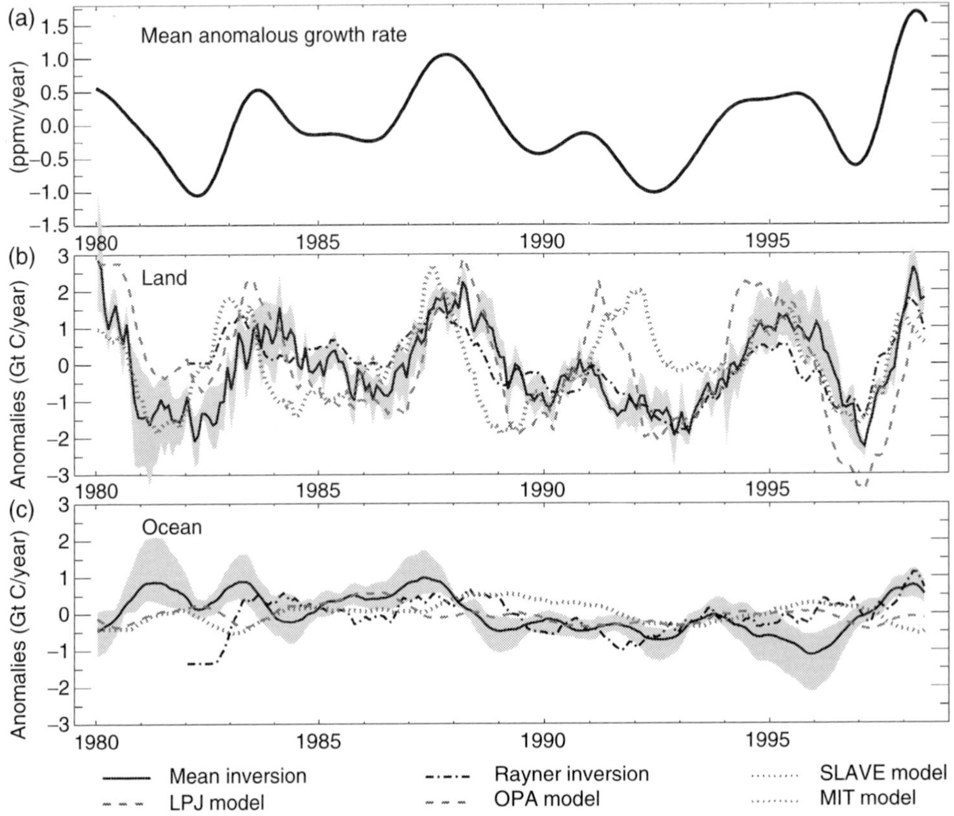

Fig. 3.4. Comparison of (a) atmospheric mean annual growth rate, (b) land CO_2 flux anomalies and (c) ocean CO_2 flux anomalies between 1980 and 1995 (in Gt C/year). The grey zone denotes the range of the inversion models, and the dark line denotes the mean. The coloured lines show the ocean models. (After Peylin *et al.*, 2005.)

and model estimates of the anthropogenic CO_2 uptake in the 1990s (in Pg C/year). Most of the models are in reasonably good agreement with the flux estimates from the observational data when corrected for the pre-industrial carbon flux.

Models suggest that anthropogenic CO_2 uptake occurs everywhere in the surface ocean, even in areas that have a total net CO_2 flux out of the ocean. For example, the pCO_2 of sea water in the eastern equatorial Pacific is much higher than atmospheric pCO_2. This sea–air difference in pCO_2 causes the ocean to release CO_2 to the atmosphere. As atmospheric CO_2 concentrations rise, the difference between the sea water pCO_2 and atmospheric pCO_2 is decreased and the rate of CO_2 loss drops. The additional CO_2 that is stored in the sea water, which would have been lost to an atmosphere with lower pCO_2 values, is referred to as the anthropogenic uptake.

It is important to note that there is a difference between CO_2 uptake and CO_2 storage. While there may be a large anthropogenic CO_2 uptake in the equatorial Pacific, the water is quickly transported off the equator and the anthropogenic CO_2 is actually stored in the subtropical gyres (Gloor *et al.*, 2003). Ocean carbon models suggest that the high-latitude Southern Ocean is also a region with large anthropogenic CO_2 uptake, but the anthropogenic CO_2 is stored further to the north where mode and intermediate water masses are formed. Water mass formation regions are areas in which

Table 3.3. Estimates of oceanic anthropogenic CO_2 uptake in Pg C/year.

Method	Carbon uptake (Pg C/year)	Reference
Measurements of sea–air pCO_2 difference	2.1 ± 0.5	Takahashi *et al.* (2002)
Inversion of atmospheric CO_2 observations	1.8 ± 1.0	Gurney *et al.* (2002)
Inversions based on ocean transport models and observed DIC	2.0 ± 0.4	Gloor *et al.* (2003)
Model simulations evaluated with CFCs and pre-bomb radiocarbon	2.2 ± 0.4	Matsumoto *et al.* (2004)
OCMIP-2 model simulations	2.4 ± 0.3	Orr (2004)
Based on measured atmospheric O_2 and CO_2 inventories corrected for ocean warming and stratification	2.3 ± 0.7	Bopp *et al.* (2002)
GCM model of ocean carbon	1.93	Wetzel *et al.* (2005)
CFC ages	2.0 ± 0.4	McNeil *et al.* (2003)

Notes: Fluxes are normalized to 1990–1999 and corrected for pre-industrial degassing flux of ~0.6 Pg C/year.

water is moved from the surface into the ocean interior. Once the waters leave the surface, the anthropogenic CO_2 is isolated from the atmosphere and stored until these waters return to the surface.

3.4.2 Long-term accumulation of anthropogenic CO_2

Recognizing the need to constrain the oceanic uptake, transport and storage of anthropogenic CO_2 during the anthropocene as well as to provide a baseline for future estimates of oceanic CO_2 uptake, two international ocean research programmes, the World Ocean Circulation Experiment (WOCE) and the Joint Global Ocean Flux Study (JGOFS), jointly conducted a comprehensive survey of inorganic carbon distributions in the global ocean in the 1990s (Wallace, 2001). After completion of the US field programme in 1998, a 5-year effort was started to compile and rigorously quality control the US and international data-sets including a few pre-WOCE data-sets in regions that had limited data (Key *et al.*, 2004). The final data-set, with 9618 hydrographic stations collected on 95 cruises, provides the most accurate and comprehensive view of the global ocean inorganic carbon distribution available (see http://cdiac.esd.ornl.gov/oceans/glodap/Glodap_home.htm). By combining these data with a back calculation technique (Gruber *et al.*, 1996) for isolating the anthropogenic component of the measured DIC, Sabine *et al.* (2004b) estimated that 118±19Pg C has accumulated in the ocean between 1800 and 1994. This inventory accounts for 48% of the fossil fuel and cement manufacturing CO_2 emissions to the atmosphere over this time frame.

A map of the anthropogenic CO_2 ocean column inventory (Fig. 3.5) shows that CO_2 is not evenly distributed in space. More than 23% of the inventory can be found in the North Atlantic, a region covering ~15% of the global ocean. By contrast, the region south of 50°S represents approximately the same ocean area but has only ~9% of

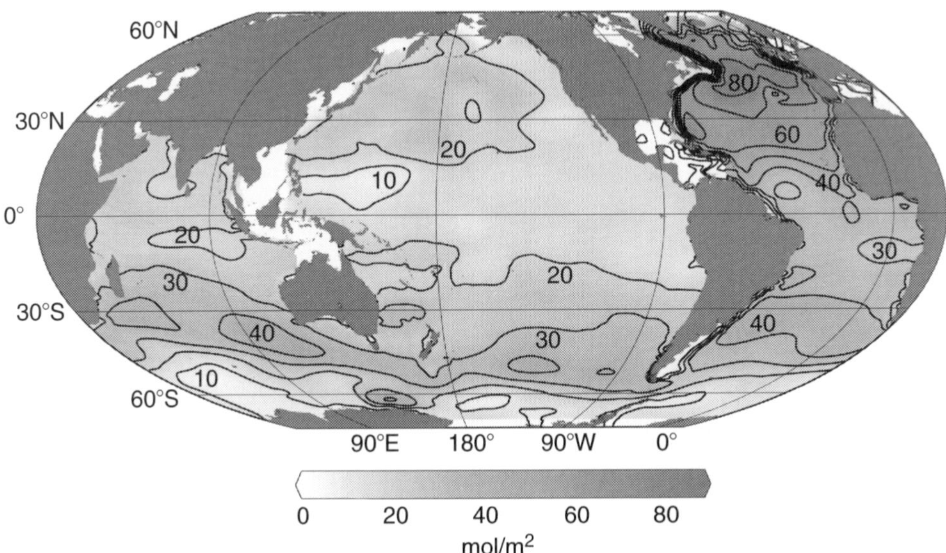

Fig. 3.5. Global map of anthropogenic CO_2 column inventory in mol/m^2. (From Sabine *et al.*, 2004b.)

the global inventory (Sabine *et al.*, 2004b). Despite the relatively slow equilibration rate for CO_2 in sea water (~1 year versus weeks for oxygen), uptake at the surface does not fully explain the spatial differences in storage. The primary reason for these differences is due to the slow mixing time in the ocean interior and the fact that waters move into the deep ocean only in a few locations. The highest inventories are found in locations where mode and intermediate waters move anthropogenic CO_2 into the ocean interior (e.g. the northern North Atlantic or in the southern hemisphere associated with the subtropical convergence zone at 40°S–50°S; Fig. 3.5).

One exception to the observation of higher inventories associated with water mass formation regions is the fact that no large inventories are associated with the formation of bottom waters around Antarctica. There are several possible reasons:

1. The anthropogenic signal has not been properly identified because of poor data coverage in these regions.
2. Low vertical stratification results in substantial mixing, invalidating a basic assumption of the technique used to estimate the anthropogenic CO_2 concentrations.
3. Newly formed bottom waters mix with old anthropogenic CO_2 free waters, diluting the signal below the limit of detection (~5 μmol/kg).
4. Short residence times of the waters at the surface and ice cover do not allow the CO_2 to equilibrate, resulting in incomplete uptake.
5. Carbon chemistry (i.e. high Revelle factor) makes the Southern Ocean very inefficient at taking up CO_2.

In reality, it is likely to be a combination of all these factors that has limited our ability to detect substantial anthropogenic CO_2 concentrations in the bottom waters around Antarctica.

Figure 3.6 shows sections of anthropogenic CO_2 in the Atlantic, Pacific and Indian oceans. One feature that clearly stands out in these examples is that most of the deep ocean has still not been exposed to elevated CO_2 levels. Nearly 50% of all the anthropogenic CO_2 is stored in the upper 10% of the global ocean (depths less than 400 m) and detectable concentrations of anthropogenic CO_2 average only as deep as 1000 m.

The global ocean is far from being saturated with CO_2. This further illustrates that the primary rate-limiting step for oceanic carbon uptake is not the exchange across the sea–air interface, but the rate at which that carbon is transported into the ocean interior. Model studies suggest that the ocean ultimately will absorb 70–85% of the CO_2 released from human activity, but given the slow mixing time of the ocean, this will take millennia to accomplish (Le Quéré and Metzl, 2004).

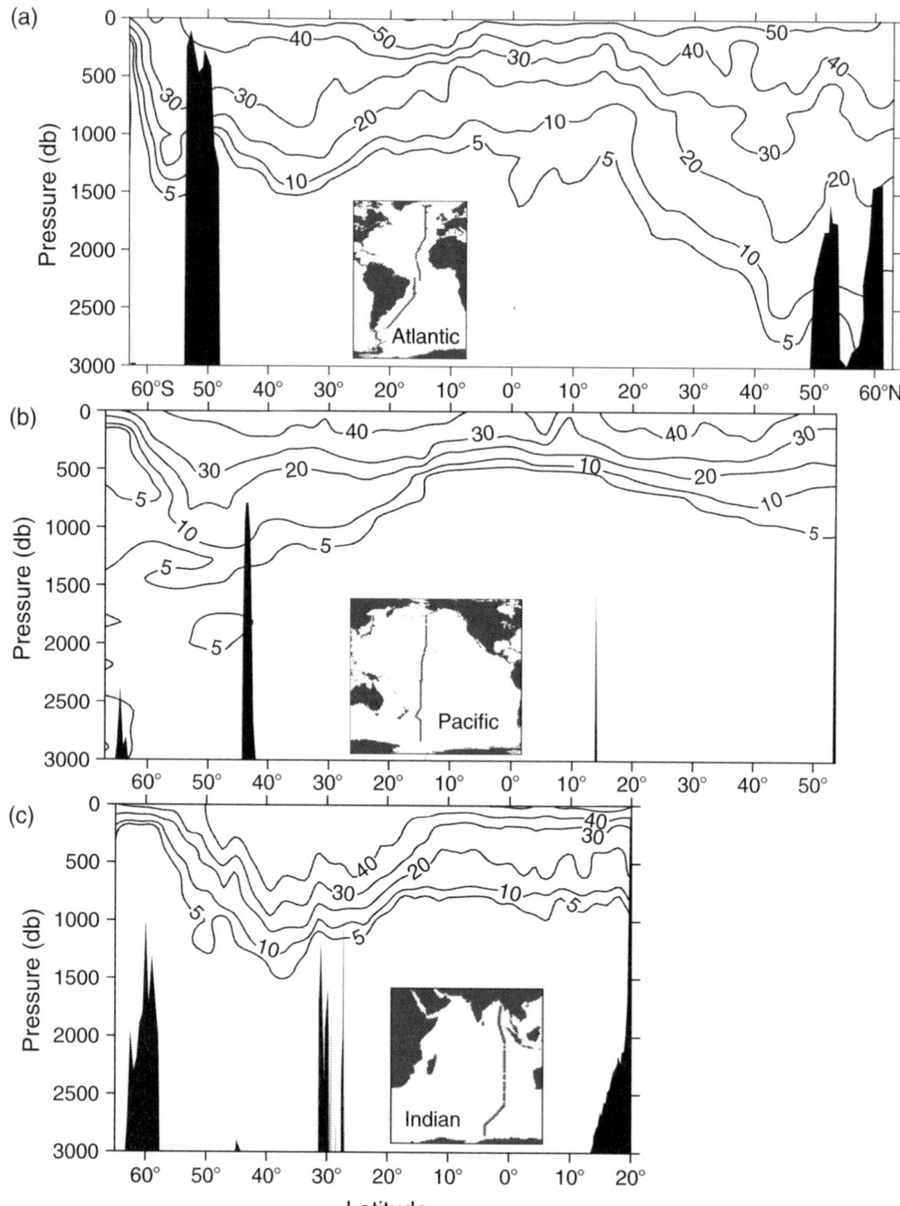

Fig. 3.6. Representative sections of anthropogenic CO_2 (μmol/kg) from the (a) Atlantic, (b) Pacific and (c) Indian oceans. Insets show maps of the station locations used to generate the sections.

3.4.3 Changes in ocean uptake rate

By combining the Sabine *et al.* (2004b) estimate of the anthropogenic CO_2 that has accumulated in the ocean between 1800 and 1994 with a synthesis of the average uptake estimate for the last 20 years (Sabine *et al.*, 2004a), we can evaluate potential changes in the decadal-scale uptake rate of anthropogenic CO_2 by the ocean. Table 3.4 shows the change in carbon inventories during the first 180 years of the anthropocene versus inventory changes over the last 20 years. These estimates suggest that the oceanic uptake of net CO_2 emissions decreased from ~44% during the first period to ~36% over the last two decades. Although this difference is not statistically significant, there is a suggestion that the oceanic uptake efficiency is decreasing with time.

Several countries have initiated programmes to evaluate decadal-scale changes in oceanic CO_2 uptake. For example, Ocean Station Papa in the north-eastern Pacific Ocean has been sampled for oceanic carbon on a semi-regular basis for the last 40 years (Signorini *et al.*, 2001). Within the USA, the Hawaii Ocean Time-series (HOT) programme and Bermuda Atlantic Time-series Study (BATS) have been measuring carbon concentrations in the water column for more than 15 years. These projects have focused most of their attention on seasonal to interannual variability, but are beginning to have records long enough to see longer-term variability in CO_2 uptake (e.g. Bates, 2001; Gruber *et al.*, 2002; Dore *et al.*, 2003; Keeling *et al.*, 2004). Additional sites are also being examined by European and Asian countries.

Table 3.4. The global carbon budget (Pg C).

	1800–1979	1980–1999
Atmospheric increase	+116 ± 4	+65 ± 1
Emissions (fossil fuel and cement)	+156 ± 20	+117 ± 5
Ocean inventory	−90 ± 19	−37 ± 8
Net terrestrial	+50 ± 28	−15 ± 9

Notes: Positive values represent atmospheric increase (or ocean/land sources); negative numbers represent atmospheric decrease (sinks).

Changes in the carbon concentrations along hydrographic sections sampled several years apart can also provide useful information on decadal-scale CO_2 uptake. At least seven countries have agreed to coordinate hydrographic survey cruises to monitor the decadal-scale changes in ocean carbon inventory. For example, the US CLIVAR/CO_2 Repeat Hydrography Program has outlined 19 cruises that will reoccupy sections that were last sampled in the 1990s. To date, six lines have been run. Preliminary results have suggested interesting basin-to-basin differences in the inferred uptake rates on these lines (e.g. 0.7 mol/m^2/year in the North Atlantic versus 1.1 mol/m^2/year in the North Pacific; see Feely *et al.*, 2005; Wanninkhof *et al.*, 2005). These changes in oceanic uptake may reflect changes in ocean circulation and/or the enactment of feedback mechanisms in the ocean that can serve to either enhance or reduce the uptake of anthropogenic CO_2 in the ocean.

3.5 Future Oceanic Uptake of Anthropogenic CO_2

In classical carbon cycle model studies, emissions from fossil fuel burning are prescribed and the model computes the time evolution of atmospheric CO_2 as the residual between emissions and uptake by land and ocean. Because the global carbon cycle is intimately embedded in the physical climate system, several feedback loops exist between the two systems (Friedlingstein *et al.*, 2003). For example, increasing CO_2 modifies the climate, which in turn impacts on ocean circulation and therefore on oceanic CO_2 uptake. Similar effects are expected to occur on land with rising temperatures (e.g. higher soil carbon respiration rates). When a climate or carbon cycle feedback results in an increase in the atmospheric CO_2 accumulation rate and thus in enhanced climate change, it is referred to as a positive feedback. A change that reduces atmospheric CO_2 is a negative feedback.

The quantitative assessment of these feedbacks necessitates the use of coupled carbon cycle climate models. Three coupled models, Hadley Centre, Institute Pierre-Simon Laplace (IPSL) and Climate System Modelling Initiative (CSMI 4), have recently examined the feedbacks based on Intergovernmental Panel on Climate Change (IPCC) scenarios between 1850 and 2100 (Cox *et al.*, 2000; Dufresne *et al.*, 2002; Fung *et al.*, 2005). By 2100 the results show dramatically different climate–carbon cycle sensitivities. These models simulate an enhanced increase of atmospheric CO_2 as a result of the climate change impacts on the carbon cycle. However, the magnitudes of the feedbacks vary by a factor of four between the simulations. Without the feedbacks the models reach an atmospheric concentration of ~700 ppm by 2100. When the feedbacks are operating, the Hadley Centre model (Cox *et al.*, 2000) reaches 980 ppm, leading to an average near-surface warming of +5°C, the IPSL model (Dufresne *et al.*, 2002) attains only 780 ppm and a warming of +3°C and the CSMI 4 model reaches an atmospheric CO_2 concentration of 792 ppm with a warming of +1.4°C. This different behaviour can be traced back to the land carbon cycle climate sensitivity of the Hadley Centre model being much larger than either the IPSL or CSMI 4 models as well as to the geochemical oceanic uptake being much larger in the IPSL model than in the Hadley Centre model.

These pioneering model simulations are subject to important limitations. In these models key biological processes on land and in the ocean are highly parameterized and poorly constrained (see Fung *et al.*, 2005). Proper modelling of the coupled carbon and climate system, however, requires an improved understanding of the two primary classes of feedbacks: that of the carbon cycle and that of the carbon–climate system.

3.5.1 Carbon cycle feedbacks

Carbon cycle feedbacks are processes that respond directly to increasing atmospheric CO_2, resulting in a change of the net land–air or sea–air exchange of CO_2. For example, the efficiency with which the ocean can absorb CO_2 at the surface is related to how much CO_2 can be converted to DIC. The measure of this is called the Revelle factor (RF) as given by Eq. 3.2 (Revelle and Suess, 1957):

$$(\Delta pCO_2 / \Delta DIC) \,/\, (pCO_2 / DIC) \qquad (3.2)$$

The RF of surface ocean waters varies from 8–9 in the subtropical gyres to 13–15 in the higher latitudes. Figure 3.7 shows the change in DIC concentration of the modern surface ocean in response to a uniform increase in pCO_2 of 10 ppm, plotted as a function of RF. It also shows that waters with low RF (~9) are four times more efficient at taking up CO_2 (ΔDIC) than waters with very high RF (~15). The RF of ocean waters is controlled by the distribution of the DIC species, including the pH of the ocean.

As the ocean takes up anthropogenic CO_2, the pH of the water decreases and the RF increases. With the anthropogenic CO_2 estimates of Sabine *et al.* (2004b), the global average RF of surface waters today appears about one unit higher than the pre-industrial values. Thus, the surface ocean today is less efficient at taking up CO_2 than the pre-industrial ocean providing a positive feedback. According to Fig. 3.7, the significance of this effect will vary depending on locations. Changing the RF by one in the high latitudes will have less effect than changes in the subtropics with relatively low RF. A further insight of these processes and their proper representation in ocean carbon models is important for understanding the ultimate long-term storage of anthropogenic CO_2 in the ocean.

Inorganic carbon thermodynamics are reasonably well understood, but some carbon cycle feedbacks, particularly those involving biological processes, are not well understood. One example of this is the effect of anthropogenic CO_2 on organisms that produce calcium carbonate ($CaCO_3$) shells. Shallow water environments, primarily coral reefs and carbonate shelves, produce ~0.3 Pg C/year, largely as metastable aragonite and

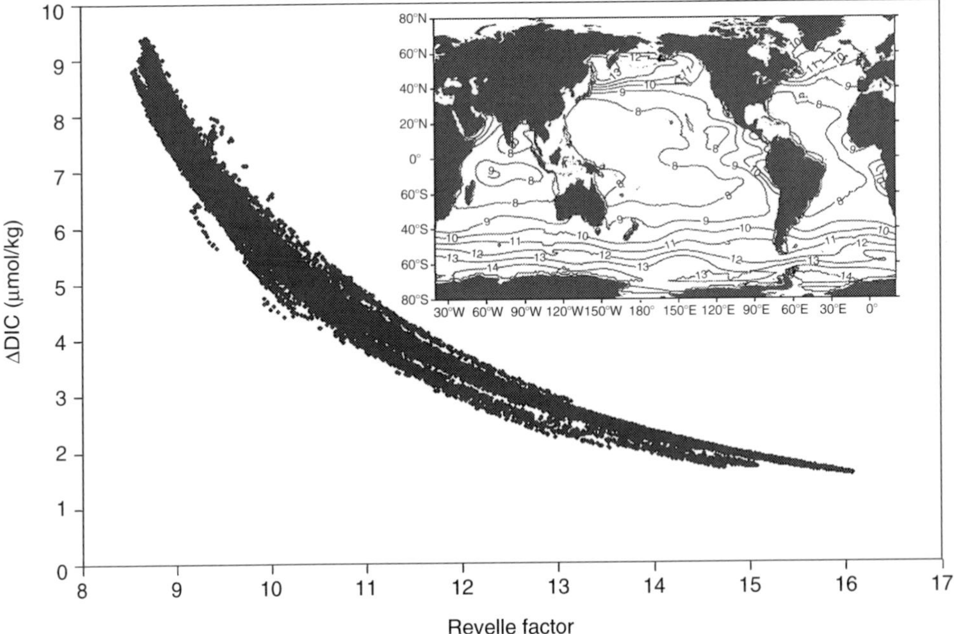

Fig. 3.7. Plot of the change in dissolved inorganic carbon for a 10 ppm change in pCO_2 as a function of Revelle factor for surface waters (<60 m) from the GLODAP bottle data-set. Inset shows a map of surface Revelle factor from the same data-set.

high-magnesian calcite. Open-ocean plankton produces an estimated 0.7–1.4 Pg C/year (Milliman, 1993; Lee, 2001), mostly as calcite but also some aragonite. These open-ocean calcifiers include phototrophic coccolithophorids and heterotrophic foraminifera as well as pteropods. Using Eq. 3.3, 1 mol of $CaCO_3$ produced releases 1 mol of CO_2:

$$Ca_2^+ + 2HCO_3^- \leftrightarrow CaCO_3 + CO_2 + H_2O \qquad (3.3)$$

Numerous studies have suggested that the rate of calcification in a wide variety of organisms is reduced when they are exposed to elevated CO_2 levels (see summary in Feely *et al.*, 2004). As atmospheric CO_2 levels increase, one might expect calcification to decrease, which would lead to a lower natural release of CO_2 from the ocean, providing a negative feedback.

The situation, however, is not that straightforward. A decrease in carbonate precipitation in the upper ocean would also lower the RF, increasing the capacity of the ocean to thermodynamically take up CO_2 from the atmosphere. A complete shutdown of surface ocean calcification would decrease surface ocean pCO_2 by ~20 ppm (Wolf-Gladrow *et al.*, 1999). On the other hand, if these organisms are primary producers, the decrease in organic matter production could result in a positive feedback. Furthermore, a decrease in $CaCO_3$ production would affect the ratio of organic/inorganic carbon delivery to the deep sea. If processes regulating this 'rain' of organic and inorganic carbon to deep-sea sediments are uncoupled, a decrease in $CaCO_3$ production would lead to increased dissolution of $CaCO_3$ in deep-sea sediments, which would raise the ocean pH and its capacity to store CO_2 (Archer and Maier-Reimer, 1994). However, if these two processes are coupled and the denser carbonate particles are necessary for transporting the organic matter into the deep ocean quickly (Armstrong *et al.*, 2002), reducing the carbonate production could result in shallower remineralization of organic carbon, producing a positive feedback (Klaas and Archer, 2002; Ridgwell, 2003), and a diminished role of sediments in the buffering of atmospheric CO_2 increases. It is also not clear how elevated CO_2 selection against a certain species (e.g. calcifying organisms) will affect the overall ecosystem structure and net CO_2 uptake by ocean biology in the future. Clearly, there is a need for more research on these mechanistic controls of the long-term changes in the carbonate system.

3.5.2 Carbon–climate feedbacks

In addition to the direct impacts of elevated CO_2 on the ocean carbon system, there are many possible indirect effects related to the climate changes associated with the atmospheric CO_2 increase. These feedback mechanisms include: (i) reduced CO_2 solubility due to the increase in sea water temperature; (ii) enhanced stabilization of the upper water masses of the water column that will lead to decreased exchange of DIC and nutrients from the ocean interior; and (iii) enhanced productivity in high-latitude regions (Table 3.5). The potential magnitude of these carbon–climate feedbacks has been examined in several modelling studies (Sarmiento and Le Quéré, 1996; Sarmiento *et al.*, 1998; Joos *et al.*, 1999; Matear and Hirst, 1999; Greenblatt and Sarmiento, 2004).

CO_2 solubility has a strong inverse relationship with temperature. Greenblatt and Sarmiento (2004) estimate that, as the surface ocean warms over this century, ~9–14% of the CO_2 that would have been stored in the ocean will be retained in the atmosphere by 2100 (a positive climate feedback). The thermodynamics of this process are well known and, consequently, the uncertainties are reasonably low.

However, there are other processes that are not fully understood. For example, increased stratification of the water column due to warming and changes in the hydrological cycle is expected to cause a decrease in the exchange of carbon and nutrients between water masses, particularly in high latitudes. The decreased carbon exchange makes it more difficult to move the anthropogenic CO_2 into

Table 3.5. Cumulative oceanic uptake of CO_2 (Pg C) due to different climate-induced feedback effects.

Scenario and years	Climate baseline	Solubility effect	Stratification effect	Biological effect	Net effect
1% CO_2/year[a]	554	−52	−117	+111	−58
for 100 years	–	(−9.4%)	(−21.1%)	(+20.0%)	(−10.5%)
IS92a-like,[b]	401	−56	−68	+108	−16
1765–2065	–	(−14.0%)	(−17.0%)	(+26.9%)	(−4.0%)
IS92a,[c]	376	−48	−41	+33	−56
1850–2100	–	(−12.8%)	(−10.9%)	(+8.8%)	(−14.9%)
WRE550,[d]	530	−68	−15	+33	−50
1765–2100	–	(−12.8%)	(−2.8%)	(+6.2%)	(−9.4%)
WRE1000,[e]	612	−58	−27	+36	−48
1765–2100	–	(−9.5%)	(−4.4%)	(+5.9%)	(−7.8%)

[a]From Sarmiento and Le Quéré (1996).
[b]From Sarmiento *et al.* (1998).
[c]From Matear and Hirst (1999).
[d]From Joos *et al.* (1999).
[e]From Plattner *et al.* (2001).
Notes: 'Climate baseline' refers to a simulation with anthropogenic CO_2 emissions but pre-industrial ocean temperatures and circulation. 'Effects' refer to uptake changes for various climate feedbacks and are expressed relative to the climate baseline. 'Net effect' is the uptake change when all climate feedbacks are present (i.e. full climate change simulation, after Greenblatt and Sarmiento, 2004).

the ocean interior, thus decreasing the oceanic uptake efficiency and providing a positive feedback (Table 3.5). This increased stratification, however, also increases CO_2 drawdown by biological activity in the Southern Ocean (negative feedback) where there is an excess of surface nutrients and the organisms are generally light-limited. The different model studies disagree on the magnitude of these two competing effects and, in some cases, do not even agree on whether the combination of these two effects will provide a positive or negative feedback (Table 3.5).

3.5.3 Variability in feedback timescales

Modelling studies have suggested that the ocean will ultimately absorb up to 70–85% of the CO_2 released by human activity (Le Quéré and Metzl, 2004). Including a carbon system feedback where carbonate sediments in the ocean are dissolved by the lowered pH of the waters suggests that the ocean may be the ultimate storage place for as much as 90% of the anthropogenic CO_2 (Archer *et al.*, 1997). The dissolution of carbonate sediments reverses Eq. 3.3 and increases the carbon storage capacity of the ocean. However, because of the slow mixing time to get the anthropogenic CO_2 into the deep ocean, this capacity may not be realized for hundreds or thousands of years. When considering the role of carbon system and carbon–climate feedbacks, it is important to understand the timescale of these processes. With typical lifetimes ranging from weeks to months, biological processes have the potential to respond very quickly to carbon system or climate changes. Large-scale circulation changes are likely to be relevant on annual to decadal timescales, and sediment dissolution processes are presumed to be relevant on centennial to millennial timescales.

As long as atmospheric CO_2 continues to rise, the ocean will continue to take up CO_2. Long-term feedbacks like dissolution of carbonate sediments may enhance the oceanic uptake but most indications are that the shorter-term feedbacks may reduce the rate of CO_2 uptake on the decadal to centennial timescales. Although there are considerable differences in the relative magnitudes of each of the individual feedback processes described in Table 3.5, all the models showed a net decrease in the overall uptake of CO_2 by the ocean over time (Greenblatt and Sarmiento, 2004). In the OCMIP-2 models, which utilized a constant biological activity and circulation, the estimated oceanic uptake of anthropogenic CO_2 by 2100 ranged from 4 to 8 Pg C/year, depending on the CO_2 emission scenario used in the model (Watson and Orr, 2003). This estimate has a factor of 2–4 times higher than the current value of 2 ± 0.5 Pg C/year but still lags behind the projected rate of CO_2 emissions. This means that a larger fraction of the CO_2 emissions will be retained by the atmosphere in the future, thus enhancing the overall climate change impact.

3.6 Conclusions

Of the ocean–atmosphere–land three-component system, ocean contains by far the most natural carbon. There is no realizable physical limit to the uptake capacity of the ocean and it is estimated that on millennial timescales the ocean will ultimately store up to 90% of the CO_2 released by human activity. However, on timescales more relevant to human society, the uptake rate of CO_2 is controlled by a complicated matrix of physical, chemical and biological processes. Studies suggest that the ocean has been the primary sink for excess CO_2 released to the atmosphere over the last 200 years, but the ocean's role may be changing over the next few decades to centuries.

Because the anthropogenic signal in the ocean is relatively small compared to the natural background concentrations and relative to the observed seasonal to interannual variations, it has been difficult to directly quantify the uptake and storage of anthropogenic CO_2 in the ocean. This has been further hampered by a paucity of data. The current estimates have been based primarily on indirect approaches or on a number of simplified assumptions, ignoring a number of potential carbon cycle and carbon–climate feedbacks. The potential role of these feedback processes in the ocean carbon cycle is just beginning to be understood and fully appreciated. As we obtain more data on processes and improve their representation in models, we will be better equipped to estimate the long-term role of the ocean in the global carbon cycle and its impact on future climate change.

References

Archer, D. and Maier-Reimer, E. (1994) Effect of deep-sea sedimentary calcite preservation on atmospheric CO_2 concentration. *Nature* 367(6460), 260–263.

Archer, D., Kheshgi, H. and Maier-Reimer, E. (1997) Multiple timescales for neutralization of fossil fuel CO_2. *Geophysical Research Letters* 24(4), 405–408.

Armstrong, R.A., Lee, C., Hedges, J.I., Honjo, S. and Wakeham, S.G. (2002) A new, mechanistic model for organic carbon fluxes in the ocean based on the quantitative association of POC with ballast minerals. *Deep-Sea Research Part II* 49(1–3), 219–236.

Bates, N.R. (2001) Interannual variability of oceanic CO_2 and biogeochemical properties in the Western North Atlantic subtropical gyre. *Deep-Sea Research Part II* 48(8–9), 1507–1528.

Battle, M., Bender, M.L., Tans, P.P., White, J.W.C., Ellis, J.T., Conway, T. and Francey, R.J. (2000) Global carbon sinks and their variability inferred from atmospheric O_2 and delta C-13. *Science* 287(5462), 2467–2470.

Bopp, L., Le Quéré, C., Heimann, M., Manning, A.C. and Monfray, P. (2002) Climate-induced oceanic oxygen fluxes: Implications for the contemporary carbon budget. *Global Biogeochemical Cycles* 16(2), doi: 10.1029/2001GB001445.

Bousquet, P., Peylin, P., Ciais, P., Le Quéré, C., Friedlingstein, P. and Tans, P.P. (2000) Regional changes in carbon dioxide fluxes of land and oceans since 1980. *Science* 290(5495), 1342–1346.

Broecker, W.S. and Peng, T.-H. (1982) *Tracers in the Sea*. Eldigio Press, Palisades, New York.

Cox, P.M., Betts, R.A., Jones, C.D., Spall, S.A. and Totterdell, I.J. (2000) Acceleration of global warming due to carbon-cycle feedbacks in a coupled climate model. *Nature* 408, 184–187.

Dore, J.E., Lukas, R., Sadler, D.W. and Karl, D.M. (2003) Climate-driven changes to the atmospheric CO_2 sink in the subtropical North Pacific Ocean. *Nature* 424(6950), 754–757.

Dufresne, J.-L., Friedlingstein, P., Berthelot, M., Bopp, L., Ciais, P., Fairhead, L., LeTreut, H. and Monfray, P. (2002) Effects of climate change due to CO_2 increase on land and ocean carbon uptake. *Geophysical Research Letters* 29(10); doi:10.1029/2001GL013777.

Feely, R.A., Wanninkhof, R., Goyet, C., Archer, D.E. and Takahashi, T. (1997) Variability of CO_2 distributions and sea–air fluxes in the central and eastern equatorial Pacific during the 1991–1994 El Niño. *Deep-Sea Research Part II* 44(9–10), 1851–1867.

Feely, R.A., Wanninkhof, R., Takahashi, T. and Tans, P. (1999) Influence of El Niño on the equatorial Pacific contribution of atmospheric CO_2 accumulation. *Nature* 398, 597–601.

Feely, R.A., Sabine, C.L., Takahashi, T. and Wanninkhof, R. (2001) Uptake and storage of carbon dioxide in the oceans: The global CO_2 survey. *Oceanography* 14(4), 18–32.

Feely, R.A., Sabine, C.L., Lee, K., Berelson, W., Kleypas, J., Fabry, V.J. and Millero, F.J. (2004) Impact of anthropogenic CO_2 on the $CaCO_3$ system in the oceans. *Science* 305(5682), 362–366.

Feely, R.A., Talley, L.D., Johnson, G.C., Sabine, C.L. and Wanninkhof, R. (2005) Repeat hydrography cruises reveal chemical changes in the North Atlantic. *EOS, Transactions, American Geophysical Union* 86(42), 399, 404–405.

Francey, R.J., Tans, P.P., Allison, C.E., Enting, I.G., White, J.W. and Trolier, M. (1995) Changes in oceanic and terrestrial carbon uptake since 1982. *Nature* 373(6512), 326–330.

Friedlingstein, P., Dufresne, J.L., Cox, P.M. and Rayner, P. (2003) How positive is the feedback between climate change and the carbon cycle? *Tellus Series B* 55(2), 692–700.

Fung, I.Y., Doney, S.C., Lindsay, K. and John, J. (2005) Evolution of carbon sinks in a changing climate. *Proceedings of the National Academy of Science* 102(32), 11201–11206.

Gloor, M., Gruber, N., Sarmiento, J.L., Sabine, C.L., Feely, R.A. and Rödenbeck, C. (2003) A first estimate of present and preindustrial air–sea CO_2 flux patterns based on ocean interior carbon measurements and models. *Geophysical Research Letters* 30(1), 1010; doi: 10.1029/2002GL015594.

Greenblatt, J.B. and Sarmiento, J.L. (2004) Variability and climate feedback mechanisms in ocean uptake of CO_2. In: Field, C.B. and Raupach, M.R. (eds) *The Global Carbon Cycle: Integrating Humans, Climate, and the Natural World*. Scope 62, Island Press, Washington, DC, pp. 257–275.

Gruber, N., Sarmiento, J.L. and Stocker, T.F. (1996) An improved method for detecting anthropogenic CO_2 in the oceans. *Global Biogeochemical Cycles* 10, 809–837.

Gruber, N., Keeling, C.D. and Bates, N.R. (2002) Interannual variability in the North Atlantic Ocean carbon sink. *Science* 298(5602), 2374–2378.

Gurney, K.R., Law, R.M., Denning, A.S., Rayner, P.J., Baker, D., Bousquet, P., Bruhwiler, L., Chen, Y.H., Ciais, P., Fan, S., Fung, I.Y., Gloor, M., Heimann, M., Higuchi, K., John, J., Maki, T., Maksyutov, S., Masarie, K., Peylin, P., Prather, M., Pak, B.C., Randerson, J., Sarmiento, J., Taguchi, S., Takahashi, T. and Yuen, C.W. (2002) Towards robust regional estimates of CO_2 sources and sinks using atmospheric transport models. *Nature* 415(6872), 626–630.

Joos, F. and Prentice, I.C. (2004) A paleo perspective on the future of atmospheric CO_2 and climate. In: Field, C.B. and Raupach, M.R. (eds) *The Global Carbon Cycle: Integrating Humans, Climate, and the Natural World*. Scope 62, Island Press, Washington, DC, pp. 165–186.

Joos, F., Plattner, G.K., Stocker, T.F., Marchal, O. and Schmittner, A. (1999) Global warming and marine carbon cycle feedbacks on future atmospheric CO_2. *Science* 284(5413), 464–467.

Keeling, C.D., Chin, J.F.S. and Whorf, T.P. (1996) Increased activity of northern vegetation inferred from atmospheric CO_2 measurements. *Nature* 382(6587), 146–149.

Keeling, C.D., Brix, H. and Gruber, N. (2004) Seasonal and long-term dynamics of the upper ocean carbon cycle at Station ALOHA near Hawaii. *Global Biogeochemical Cycles* 18(4), Art. No. GB4006.

Key, R.M., Kozyr, A., Sabine, C.L., Lee, K., Wanninkhof, R., Bullister, J.L., Feely, R.A., Millero, F.J., Mordy, C. and Peng, T.-H. (2004) A global ocean carbon climatology: Results from GLODAP. *Global Biogeochemical Cycles* 18, GB4031; doi:10.1029/2004GB002247.

Klaas, C. and Archer, D.E. (2002) Association of sinking organic matter with various types of mineral ballast in the deep sea: Implications for the rain ratio. *Global Biogeochemical Cycles* 16(4), Art. No. 1116.

Lee, K. (2001) Global net community production estimated from the annual cycle of surface water total dissolved inorganic carbon. *Limnology and Oceanography* 46(6), 1287–1297.

Lee, K., Wanninkhof, R., Takahashi, T., Doney, S.C. and Feely, R.A. (1998) Low interannual variability in recent oceanic uptake of atmospheric carbon dioxide. *Nature* 396(6707), 155–159.

Le Quéré, C. and Metzl, N. (2004) Natural processes regulating the oceanic uptake of CO_2. In: Field, C.B. and Raupach, M.R. (eds) *The Global Carbon Cycle: Integrating Humans, Climate, and the Natural World*. Scope 62, Island Press, Washington, DC, pp. 243–255.

Le Quéré, C., Orr, J.C., Monfray, P., Aumont, O. and Madec, G. (2000) Interannual variability of the oceanic sink of CO_2 from 1979 through 1997. *Global Biogeochemical Cycles* 14, 1247–1265.

Le Quéré, C., Aumont, O., Bopp, L., Bousquet, P., Ciais, P., Francey, R., Heimann, M., Keeling, C.D., Keeling, R.F., Kheshgi, H., Peylin, P., Piper, S.C., Prentice, I.C. and Rayner, P.J. (2003) Two decades of ocean CO_2 sink and variability. *Tellus* 55B, 649–656.

Matear, R.J. and Hirst, A.C. (1999) Climate change feedback on the future oceanic CO_2 uptake. *Tellus* 51B(3), 722–733.

Matsumoto, K., Sarmiento, J.L., Key, R.M., Aumont, O., Bullister, J.L., Caldeira, K., Campin, J.M., Doney, S.C., Drange, H., Dutay, J.C., Follows, M., Gao, Y., Gnanadesikan, A., Gruber, N., Ishida, A., Joos, F., Lindsay, K., Maier-Reimer, E., Marshall, J.C., Matear, R.J., Monfray, P., Mouchet, A., Najjar, R., Plattner, G.K., Schlitzer, R., Slater, R., Swathi, P.S., Totterdell, I.J., Weirig, M.F., Yamanaka, Y., Yool, A. and Orr, J.C. (2004) Evaluation of ocean carbon cycle models with data-based metrics. *Geophysical Research Letters* 31(7), Art. No. L07303.

McKinley, G.A., Rodenbeck, C., Gloor, M., Houweling, S. and Heimann, M. (2004) Mechanisms of air–sea CO_2 flux variability in the equatorial Pacific and the North Atlantic. *Global Biogeochemical Cycles* 18(2), Art. No. GB2011.

McNeil, B.I., Matear, R.J., Key, R.M., Bullister, J.L. and Sarmiento, J.L. (2003) Anthropogenic CO_2 uptake by the ocean based on the global chlorofluorocarbon data set. *Science* 299(5604), 235–239. Available at: http://corona.pmel.noaa.gov/j_pdf/2003/mcne2489.pdf

Milliman, J.D. (1993) Production and accumulation of calcium-carbonate in the ocean: Budget of a nonsteady state. *Global Biogeochemical Cycles* 7(4), 927–957.

Obata, A. and Kitamura, Y. (2003) Interannual variability of the sea–air exchange of CO_2 from 1961 to 1998 simulated with a global ocean circulation-biogeochemistry model. *Journal of Geophysical Research-Oceans* 108(C11), Art. No. 3337.

Orr, J.C. (2004) *Modelling of Ocean Storage of CO_2: The GOSAC Study*. Report PH4/37, IEA Greenhouse Gas R&D Programme.

Park, G.-H., Lee, K., Wanninkhof, R. and Feely, R.A. (2006) Empirical temperature-based estimates of variability in the oceanic uptake of CO_2 over the past two decades. *Journal of Geophysical Research* III(C7), C07S07, doi: 10.1029/2005JC003090.

Peylin, P., Bousquet, P., Le Quéré, C., Sitch, S., Friedlingstein, P., McKinley, G., Gruber, N., Rayner, P. and Ciais, P. (2005) Multiple, constraints on regional CO_2 flux variations over land and oceans. *Global Biogeochemical Cycles* 19, GB1011; doi:10.1029/2003GB002214.

Plattner, G.-K.G., Joos, F., Stocker, F. and Marchal, O. (2001) Feedback mechanisms and sensitivities of ocean carbon uptake under global warming. *Tellus* 53B, 564–592.

Rayner, P.J., Enting, I.G., Francey, R.J. and Langenfelds, R. (1999) Reconstructing the recent carbon cycle from atmospheric CO_2, delta C-13 and O_2/N_2 observations. *Tellus* 51B(2), 213–232.

Revelle, R. and Suess, H.E. (1957) Carbon dioxide exchange between the atmosphere and ocean and the question of an increase of atmospheric CO_2 during the past decade. *Tellus* 9, 18–27.

Ridgwell, A.J. (2003) An end to the 'rain ratio' reign? *Geochemistry Geophysics Geosystems* 4, Art. No. 1051.

Sabine, C.L., Heimann, M., Artaxo, P., Bakker, D., Chen, C.-T.A., Field, C.B., Gruber, N., Le Quéré, C., Prinn, R.G., Richey, J.E., Lankao, P.R., Sathaye, J. and Valentini, R. (2004a) Current status and past trends of the global carbon cycle. In: Field, C.B. and Raupach, M.R. (eds) *The Global Carbon Cycle: Integrating Humans, Climate, and the Natural World.* Scope 62, Island Press, Washington, DC, pp. 17–44.

Sabine, C.L., Feely, R.A., Gruber, N., Key, R.M., Lee, K., Bullister, J.L., Wanninkhof, R., Wong, C.L., Wallace, D.W.R., Tilbrook, B., Millero, F.J., Peng, T.-H., Kozyr, A., Ono, T. and Rios, A.F. (2004b) The oceanic sink for anthropogenic CO_2. *Science* 305(5682), 367–371.

Sarmiento, J.L. and Le Quéré, C. (1996) Oceanic carbon dioxide uptake in a model of century-scale global warming. *Science* 274(5291), 1346–1350.

Sarmiento, J.L., Hughes, T.M.C., Stouffer, R.J. and Manabe, S. (1998) Simulated response of the ocean carbon cycle to anthropogenic climate warming. *Nature* 393(6682), 245–249.

Signorini, S.R., McClain, C.R., Christian, J.R. and Wong, C.S. (2001) Seasonal and interannual variability of phytoplankton, nutrients, TCO_2, pCO_2, and O_2 in the eastern subarctic Pacific (ocean weather station Papa). *Journal of Geophysical Research-Oceans* 106(C12), 31197–31215.

Skjelvan, I., Johannessen, T. and Miller, L.A. (1999) Interannual variability of CO_2 in the Greenland and Norwegian Seas. *Tellus* 51B(2), 477–489.

Takahashi, T., Olafsson, J., Goddard, J.G., Chipman, D.W. and Sutherland, S.C. (1993) Seasonal variation of CO_2 and nutrients in high-latitude surface oceans: A comparative study. *Global Biogeochemical Cycles* 7, 843–878.

Takahashi, T., Sutherland, S.C., Sweeney, C., Poisson, A., Metzl, N., Tilbrook, B., Bates, N., Wanninkhof, R., Feely, R.A., Sabine, C. and Olafsson, J. (2002) Biological and temperature effects on seasonal changes of pCO_2 in global ocean surface waters. *Deep Sea Research* 49(9–10), 1601–1622.

Wallace, D.W.R. (2001) Storage and transport of excess CO_2 in the oceans: The JGOFS/WOCE Global CO_2 Survey. In: Siedler, G., Church, J. and Gould, J. (eds) *Ocean Circulation and Climate: Observing and Modelling the Global Ocean*. Academic Press, San Diego, California, pp. 489–521.

Wanninkhof, R. (1992) Relationship between gas exchange and wind speed over the ocean. *Journal of Geophysical Research* 97, 7373–7381.

Wanninkhof, R. and McGillis, W.R. (1999) A cubic relationship between air–sea CO_2 exchange and wind speed. *Geophysical Research Letters* 26, 1889–1892.

Wanninkhof, R.H., Doney, S.C., Takahashi, T. and McGillis, W.R. (2002) The effect of using time-averaged winds on regional air–sea CO_2 fluxes. In: Donelan, M.A., Drennan, W.M., Saltzman, E.S. and Wanninkhof, R.H. (eds) *Gas Transfer at Water Surfaces*. Geophysical Monograph Series, Vol.127, American Geophysical Union, Washington, DC, pp. 351–356.

Wanninkhof, R., Doney, S., Langdon, C., Bullister, J.L., Johnson, G.C., Warner, M. and Gruber, N. (2005) Decadal changes in inorganic carbon in the Atlantic Ocean. In: *Seventh International Carbon Dioxide Conference, Extended Abstracts*. Committee of the Seventh International Carbon Dioxide Conference, National Oceanic and Atmospheric Administration, Boulder, Colorado, pp. 417–418.

Watson, A.J. and Orr, J.C. (2003) Carbon dioxide fluxes in the global ocean. In: Fasham, M., Field, J., Platt, T. and Zeitzschel, B. (eds) *Ocean Biogeochemistry: The Role of the Ocean Carbon Cycle in Global Change (a JGOFS Synthesis)*. Springer, Berlin, pp. 123–141.

Wetzel, P., Winguth, A. and Maier-Reimer, E. (2005) Sea-to-air CO_2 flux from 1948 to 2003: A model study. *Global Biogeochemical Cycles* 19(2), Art. No. GB2005.

Winguth, A.M.E., Heimann, M., Kurz, K.D., Maier-Reimer, E., Mikolajewicz, U. and Segschneider, J. (1994) El Niño–southern oscillation related fluctuations of the marine carbon-cycle. *Global Biogeochemical Cycles* 8(1), 39–63.

Wolf-Gladrow, D.A., Riebesell, U., Burkhardt, S. and Bijma, J. (1999) Direct effects of CO_2 concentration on growth and isotopic composition of marine plankton. *Tellus* 51B(2), 461–476.

Zeebe, R.E. and Wolf-Gladrow, D. (2001) *CO_2 in Seawater: Equilibrium, Kinetics, Isotopes*. Elsevier Oceanographic Series 65, Elsevier, New York.

4 The Soil Carbon Dioxide Sink

Pete Smith[1] and Phil Ineson[2]
[1]School of Biological Sciences, University of Aberdeen, Aberdeen, UK; [2]Department of Biology (SEIY), University of York, UK

4.1 Introduction

Globally, soils contain ~2000 billion tonnes of organic carbon with ~1500 billion tonnes of carbon in the top metre of mineral soil (Batjes, 1996). This constitutes about three times the amount of carbon in vegetation and twice the amount in the atmosphere (IPCC, 2000a). About 300 billion tonnes can be found as detritus in the top soil, with this carbon rich material decomposing at varying rates depending on temperature and soil conditions. During this decomposition some of the carbon in soil detritus is respired by the decomposing organisms (often fungi and bacteria), and the carbon returned to the atmosphere as carbon dioxide (CO_2). The rest can be converted to soil organic matter, which decomposes more slowly and hence keeps the carbon away from the atmosphere for a longer period. A small amount of this carbon is further decomposed to forms that persist in the soil, sometimes for decades or centuries.

The conversion of soils from natural to agricultural use by humans has led to substantial loss in soil carbon stocks. Greater soil disturbance, such as that caused by ploughing, can result in rapid respiration and loss of large amount of soil carbon, which would otherwise decompose slowly. Sensitive land use practice is key for balancing the soil carbon sink, and perhaps reversing recent trends of loss of carbon from soils. Farming practices such as 'no-till', whereby agricultural land is used without the soil disturbance and carbon loss that comes with ploughing, are becoming more widespread, and land use remains a key area of research in studies of human-made greenhouse gas (GHG) emissions and strategies to reduce them.

4.2 Soil Carbon Stocks and Land Management

The annual fluxes of CO_2 from atmosphere to land – global Net Primary Productivity (NPP) – and land to atmosphere (respiration and fire) are each of the order of 60 billion tonnes C/year (IPCC, 2000a). During the 1990s, fossil fuel combustion and cement production emitted 6.3 ± 1.3 Pg C/year to the atmosphere, whilst land-use change emitted 1.6 ± 0.8 Pg C/year (IPCC, 2001; Schimel *et al.*, 2001). Atmospheric carbon increased at a rate of 3.2 ± 0.1 Pg C/year, the oceans absorbed 2.3 ± 0.8 Pg C/year with an estimated terrestrial sink of 2.3 ± 1.3 Pg C/year (IPCC, 2001; Schimel *et al.*, 2001).

The size of the pool of carbon in the soil is therefore large compared to gross and net annual fluxes of carbon to and from the terrestrial biosphere. Figure 4.1 (IPCC, 2001) shows a schematic diagram of the carbon cycle.

Soil carbon pools are smaller now than they were before human intervention. Historically, soils have lost between 40 and 90 Pg C globally through cultivation and disturbance (Schimel, 1995; Houghton, 1999; Houghton *et al.*, 1999; Lal, 1999). The goal of engineering carbon sequestration is to increase the carbon sink relative to the current stock.

Soil carbon sequestration can be achieved by increasing the net flux of carbon from the atmosphere to the terrestrial biosphere by increasing global NPP (thus increasing carbon inputs to the soil), by storing a larger proportion of the carbon from NPP in the longer-term carbon pools in the soil or by slowing decomposition. For soil carbon sinks, the best options are to increase carbon stocks in soils that have been depleted (i.e. agricultural soils and

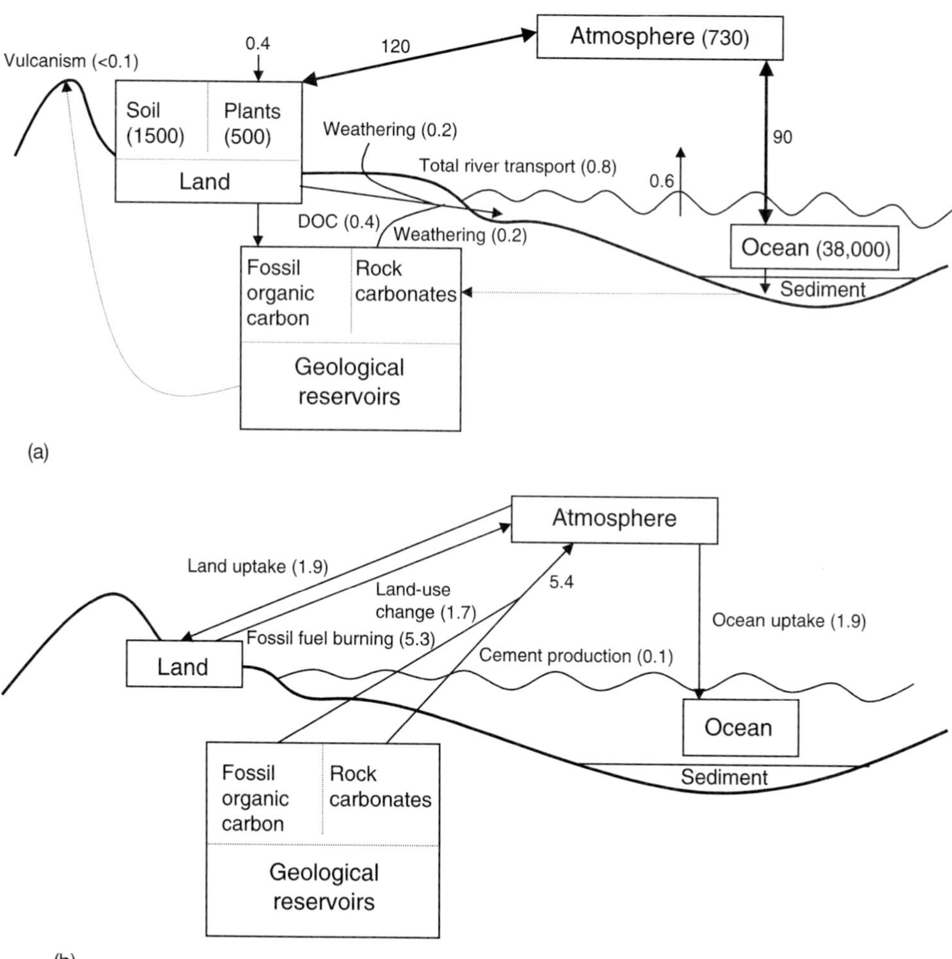

Fig. 4.1. The global carbon cycle for the 1990s (billion tonnes C). (a) The natural carbon cycle (DOC = dissolved organic carbon). (b) The human perturbation. (Redrawn from IPCC, 2001.)

degraded soils). In many regions, agricultural soils are net emitters of CO_2 as described in the following section.

4.3 Cropland and CO_2 Fluxes: An Example from Europe

Croplands (i.e. lands used for the production of arable crops) cover about one-third of Europe's land surface, and most cropland soils are out of equilibrium, as they have been affected by past land use and management practices. In Europe, cropland soils are estimated to be the largest biospheric source of carbon loss to the atmosphere each year and the cropland flux estimate is also the most uncertain among all land use types (Janssens *et al.*, 2003). It is estimated that croplands (in Europe as far east as the Urals) lose 0.12–0.3 Pg C/year (Janssens *et al.*, 2003, 2005), with the mean of 0.078 (SD: 0.037) Pg C/year for the European Union (EU-15) (Vleeshouwers and Verhagen, 2002). National estimates of cropland CO_2 fluxes for some EU countries are of similar magnitude on a per-area basis (Sleutel *et al.*, 2003) but other estimates are lower (Dersch and Boehm, 1997). The estimated carbon flux per hectare from cropland in Janssens *et al.* (2003) is similar to that measured when converting grassland to tilled cropland (as calculated from values in Johnston, 1973). Since this is an extreme land-use change, cropland fluxes may have been overestimated. Indeed, Janssens *et al.* (2005) reduced a previous estimate for cropland carbon efflux (Janssens *et al.*, 2003) for geographical Europe from 0.3 to 0.12 Pg C/year. The EU-15 estimates for the CO_2 cropland emissions (~0.078 Pg C/year) are of the same order of magnitude as the reported emissions of nitrous oxide (N_2O) from agricultural soil (~0.06 Pg C-eq. in 2000) and methane (CH_4) from agriculture (~0.05 Pg C-eq. in 2000; Smith *et al.*, 2004).

The values for CO_2 flux suggest that cropland soil carbon stocks in general are continuing to decline, perhaps as a result of recent (decadal) land-use change. However, values for net changes in land use during the last 20–30 years do not suggest a large-scale conversion to cropland from other land uses, but may not show all areas that have undergone a change as they report only net changes. An alternative reason for the high carbon loss from agricultural soils in some regions may be the changes of agricultural management (e.g. manure use) over recent decades (Sleutel *et al.*, 2003). The values for cropland soil carbon loss are highly uncertain (Janssens *et al.*, 2003). Clearly, there is potential to reduce the flux of carbon from the soil and even increase soil carbon stocks.

4.4 The Potential to Reduce Soil CO_2 Fluxes and to Store Carbon in Soils

Estimates of the potential for additional soil carbon sequestration vary widely. According to studies on European cropland (Smith *et al.*, 2000a), US cropland (Lal *et al.*, 1998), global degraded lands (Lal, 2003a) and global estimates (Cole *et al.*, 1996; IPCC, 2000a), global soil carbon sequestration potential is estimated at 0.9 ± 0.3 Pg C/year (Lal, 2003b), between one-third and one-fourth of the annual increase in atmospheric carbon levels. Over the next 50 years, the level of carbon sequestration suggested by Lal (2003b) would restore a large part of the carbon lost from soils historically. Estimates based only on Europe and North America by Smith *et al.* (2000b) suggest a lower potential (one-third to one-half of historical soil carbon loss over 100 years). Soil carbon sequestration rates have a limited duration and cannot be maintained indefinitely.

The estimates for carbon sequestration potential in soils are of the same order as for forest trees, which could sequester between 1 Pg C/year (the lower figure of IPCC, 1996) and 2 Pg C/year (Trexler, 1988, cited in Metting *et al.*, 1999), between one-third and two-thirds of the annual increase in atmospheric carbon levels.

4.5 Soil Carbon Sequestration Options

Many reviews have been published recently discussing options available for soil carbon sequestration and mitigation potentials (e.g. Lal *et al.*, 1998; Metting *et al.*, 1999; Nabuurs *et al.*, 1999; Follett *et al.*, 2000; IPCC, 2000a; Smith *et al.*, 2000a; Cannell, 2003; Lal, 2003b; Freibauer *et al.*, 2004). These options include the management practices that are described below.

For forests, carbon sequestration options include the increase of soil carbon stocks through afforestation, reforestation, improved forest management or revegetation. For croplands, options include zero or reduced tillage, set-aside or Conservation Reserve Program, conversion to permanent or deep-rooting crops, improved efficiency of animal manure use, improved efficiency of crop residue use, agricultural use of sewage sludge, application of compost to land, rotational changes, fertilizer use, irrigation, bioenergy crops, extensification or de-intensification of farming, organic farming (a combination of many different individual practices), conversion of cropland to grassland and management to reduce wind and water erosion. For grazing lands, soil carbon sequestration measures include improved efficiency of animal manure use, improved efficiency of crop residue use, improved livestock management to reduce soil disturbance, improved livestock management to maximize manure carbon returns, agricultural use of sewage sludge, conversion to deeper-rooting species, application of compost to land, fertilizer use, irrigation, extensification or de-intensification of farming and improved management to reduce wind and water erosion. On land undergoing revegetation, options include increasing soil carbon stocks by planting vegetation other than trees with higher carbon returns to soil or with litter more resistant to decomposition. Other potential soil sinks are found in wetlands (protection and creation), urban forest or grassland (protection and creation), deserts and degraded lands (improved management), sediments and aquatic systems (protection) and tundra and taiga (protection) (Smith, 2004).

Most of the estimates for the sequestration potential of the activities listed above range from ~0.1 to 0.8t C/ha/year, but some estimates are outside this range (Nabuurs *et al.*, 1999; Follett *et al.*, 2000; IPCC, 2000a; Smith *et al.*, 2000a; Lal, 2003b). When considering soil carbon sequestration options, it is also important to consider other side effects, including the emission of other GHGs. Smith *et al.* (2000c, 2001) have shown that as much as one-half of the climate mitigation potential of some carbon sequestration options could be lost when increased emissions of other GHGs (e.g. N_2O and methane) were included. Other authors have found a more extreme effect for zero-tillage agriculture (Six *et al.*, 2004). Robertson *et al.* (2000) also considered this issue, which is further discussed by Lemke and Janzen (Chapter 5, this volume).

4.6 The Political Context of Soil Carbon Sequestration

Soil carbon sequestration is perhaps more hotly debated in the political arena than in the scientific arena. Terrestrial sinks have received close political scrutiny since their inclusion in the Kyoto Protocol at the fourth-Conference of Parties (COP4) to the United Nations Framework Convention on Climate Change (UNFCCC). Under Articles 3.3 and 3.4 of the Kyoto Protocol, biospheric sinks and sources of carbon can be included by parties in meeting targets for the reduction of GHG emissions, by comparing emissions in the commitment period (CP; first CP = 2008–2012) with baseline (1990) emissions. The exact nature of the sinks and sources that would be allowed as well as the modalities and methods that could be used were addressed in subsequent meetings of the COP culminating in agreement at COP7 in 2001, leading to the publication of the Marrakesh Accords.

Under the Marrakesh Accords, Kyoto Article 3.3, activities limited to afforestation, reforestation and deforestation are included, but are subject to a complex series of rules to attempt to separate any impacts that are not directly human-induced (see UNFCCC web site at www.unfccc.de). Under the Marrakesh Accords, Kyoto Article 3.4, activities under forest management, cropland management, grazing land management and revegetation can be included. The details of the rules governing how sinks can be used are complex and contain some safeguards to avoid double-counting of sinks, to avoid carbon credit claims and the land subsequently reverting to low carbon stock levels (e.g. by deforestation), and to ensure that accounting is carried out in a transparent and verifiable way. Soil carbon changes occurring during afforestation, reforestation or deforestation are included under Article 3.3 of the Kyoto Protocol, whilst soil carbon sequestration in croplands, grazing lands, managed forests and land subject to revegetation is included under Kyoto Article 3.4.

An interesting outcome of the Marrakesh Accords is that cropland management, grazing land management and revegetation will be reported on a net–net basis, i.e. net emissions during the CP will be compared to net emissions during the baseline year. Under this form of accounting, a carbon credit will arise if a party reduces a source, even though a sink (i.e. a net removal of carbon from the atmosphere) has not been created. In this respect, decreasing a source is equivalent to creating a sink.

Scientific issues relating to the use of biospheric sinks in the Kyoto Protocol were addressed in the IPCC (2000a). Although many of these issues no longer apply (since the political negotiations have moved on), this text is still a useful source of information on the potential of biospheric sinks, including soils, for climate mitigation.

Soil carbon sequestration will remain in the political spotlight at least until the end of the first CP in 2012, but with further CPs set for negotiation in the near future, sinks are likely to remain politically important for the foreseeable future (see Reilly *et al.*, Chapter 8, this volume).

4.7 Duration of Soil Carbon Sequestration and Permanence of Soil Carbon Sinks

Soil carbon sinks resulting from sequestration activities are not permanent and will continue only for as long as appropriate management practices are maintained. If a land management or land-use change is reversed, the carbon accumulated will be lost, usually more rapidly than it was accumulated (Smith *et al.*, 1996). For the greatest potential of soil carbon sequestration to be realized, new carbon sinks, once established, need to be preserved in perpetuity. Within the Kyoto Protocol, mechanisms have been suggested to provide disincentives for sink reversal, i.e. when land is entered into the Kyoto process it has to continue to be accounted for and any sink reversal will result in a loss of carbon credits.

Soil carbon sinks increase most rapidly soon after a carbon-enhancing land management change has been implemented, but soil carbon levels may decrease initially if there is significant disturbance (e.g. when land is afforested). Sink strength (i.e. the rate at which carbon is removed from the atmosphere) in soil becomes lesser with time as the soil carbon stock approaches a new equilibrium. At equilibrium, the sink is saturated: the carbon stock may have increased, but the sink strength has decreased to zero.

The time taken for sink saturation (i.e. new equilibrium) to occur is highly variable. The period for soils in a temperate location to reach a new equilibrium after a land-use change is ~100 years (Jenkinson, 1988; Smith *et al.*, 1996) but tropical soils may reach equilibrium faster. Soils in boreal regions may take centuries to approach a new equilibrium. As a compromise, good practice guidelines give 20 years for soil carbon to approach a new equilibrium (IPCC, 1997; Paustian *et al.*, 1997).

4.8 The Role of Soil Carbon Sequestration in Climate Mitigation over the Next Century

The future trajectory of carbon emissions over the next century depends upon many factors. The IPCC recently developed a range of standard reference emission scenarios (SRES) to provide estimates of possible emissions under a range of different possible futures (IPCC, 2000b). These possible futures depend upon the degree to which society or policy becomes global and whether environmental or economic concerns take precedence in the next century.

Among the A1 family of scenarios (global – free market), a number of possible emission trajectories exist depending upon whether the energy sector remains fossil fuel-intensive (A1FI), the rapid introduction of new energy technologies allows a move away from carbon-intensive energy sources (A1T) or a balanced mix of fossil fuel and alternative energy sources (A1B) is achieved.

In all of these scenarios, the global population will grow, become wealthier and per capita energy demand will increase over the next century (IPCC, 2000b). The extent to which these changes will occur varies between different scenarios, with some showing larger increases than others, but in all of them, these trends are observed. For each of the scenarios carbon emission trajectories have been determined (IPCC, 2000b). Annual carbon emissions (billion tonnes C/year) by 2100 would be A1FI = ~30, A1B = ~17, A1T = ~7, A2 = ~28, B1 = ~6, and B2 = ~18.

Emission trajectories can also be calculated for a range of atmospheric CO_2 stabilization targets (e.g. 450, 550, 650 and 750 ppm). For each stabilization target, the allowed carbon emission trajectories, which cannot be exceeded if the target is to be reached, can be calculated. The difference between the allowed emission trajectory for stabilization at a given target concentration and the emissions associated with the estimated global energy demand is the carbon emission gap. For each of the IPCC scenarios, the carbon emission gaps by 2100 (Pg C/year) would be A1FI = 25, A1B = 12, A1T = 2, A2 = 22, B1 = 1 and B2 = 13 (IPCC, 2001).

The current annual emissions of CO_2 to the atmosphere are 6.3 ± 1.3 Pg C/year. Carbon emission gaps by 2100 could be as high as 25 Pg C/year, which means that the carbon emissions problem could be up to four times greater than at present. The maximum annual global carbon sequestration potential is 0.9 ± 0.3 Pg C/year, which means that even if these rates could be maintained until 2100, soil carbon sequestration would contribute a maximum of 2–5% towards reducing the carbon emissions gap under the highest emission scenarios. When we consider the limited duration of carbon sequestration options in removing carbon from the atmosphere, we see that carbon sequestration can play only a minor role in closing the emissions gap by 2100. It is clear from these figures that if we wish to stabilize atmospheric CO_2 concentrations by 2100, the increased global population and its increased energy demand can be supported only if there is a large-scale switch to non-carbon-emitting technologies for producing energy.

Given that soil carbon sequestration can play only a minor role in closing the carbon emissions gap by 2100, is there any role for carbon sequestration in climate mitigation in the future? The answer is 'yes'. If atmospheric CO_2 levels are to be stabilized at reasonable concentrations by 2100 (e.g. 450–650 ppm), drastic reductions in emissions are required over the next 20–30 years (IPCC, 2000b). During this critical period, all measures to reduce net carbon emissions to the atmosphere would play an important role – there will be no single solution (IPCC, 2000b). Given that carbon sequestration is likely to be most effective in its first 20 years of implementation, it should form a central role in any portfolio of measures to reduce atmospheric CO_2 concentrations over the next 20–30 years whilst new energy technologies are developed and implemented.

4.9 How Best to Implement Soil Carbon Sequestration Options

Soil carbon sequestration is a process under the control of human management and, as such, the social dimension needs to be considered when implementing these practices. Since there will be increasing competition for limited land resources in the next century, soil carbon sequestration cannot be isolated from other environmental and social needs. The IPCC (2001) has noted that global, regional and local environmental issues such as climate change, loss of biodiversity, desertification, stratospheric ozone depletion, regional acid deposition and local air quality are inextricably linked. Soil carbon sequestration measures clearly belong to this list. The importance of integrated approaches to sustainable environmental management is becoming clearer.

In any scenario, there will be winners and losers. The key to increasing soil carbon sequestration, as part of wider programmes to enhance sustainability, is to maximize the number of winners and minimize the number of losers. One possibility for improving the social and/or cultural acceptability of soil carbon sequestration measures would be to include compensation costs for losers when costing implementation strategies. However, the best option is to identify win–win measures that increase carbon stocks whilst improving other aspects of the environment (e.g. improved soil fertility, decreased erosion) or enhancing profitability (e.g. improved yield of agricultural or forestry products). A number of management practices are available that can be implemented to protect and enhance existing carbon sinks now and in the future (i.e. a no-regrets policy). Smith and Powlson (2003) developed these arguments for soil sustainability, but the no-regrets policy option is equally applicable to soil carbon sequestration. Since such practices are consistent with, and may even be encouraged by, many current international agreements and conventions, their rapid adoption should be encouraged.

References

Batjes, N.H. (1996) Total carbon and nitrogen in the soils of the world. *European Journal of Soil Science* 47, 151–163.

Cannell, M.G.R. (2003) Carbon sequestration and biomass energy offset: theoretical, potential and achievable capacities globally, in Europe and the UK. *Biomass and Bioenergy* 24, 97–116.

Cole, V., Cerri, C., Minami, K., Mosier, A., *et al.* (1996) Agricultural options for mitigation of greenhouse gas emissions. In: Watson, R.T., Zinyowera, M.C., Moss, R.H. and Dokken D.J. (eds) *Climate Change 1995. Impacts, Adaptations and Mitigation of Climate Change: Scientific–Technical Analyses*. Cambridge University Press, New York, pp. 745–771.

Dersch, G. and Boehm, K. (1997) Bundesamt und Forschungszentrum fuer Landwirtschaft, Österreich. In: Blum, W.E.H., Klaghofer, E., Loechl, A. and Ruckenbauer P. (eds) *Bodenschutz in Österreich*. Federal Environment Agency, Vienna, Austria, pp. 411–432.

Follett, R.F., Kimble, J.M. and Lal, R. (2000) The potential of U.S. grazing lands to sequester soil carbon. In: Follett, R.F., Kimble, J.M. and Lal, R. (eds) *The Potential of U.S. Grazing Lands to Sequester Carbon and Mitigate the Greenhouse Effect*. Lewis Publishers, Boca Raton, Florida, pp. 401–430.

Freibauer, A., Rounsevell, M., Smith, P. and Verhagen, A. (2004) Carbon sequestration in the agricultural soils of Europe. *Geoderma* 122, 1–23.

Houghton, R.A. (1999) The annual net flux of carbon to the atmosphere from changes in land use 1850 to 1990. *Tellus* 50B, 298–313.

Houghton, R.A., Hackler, J.L. and Lawrence, K.T. (1999) The US carbon budget: contributions form land-use change. *Science* 285, 574–578.

IPCC (1996) *Climate Change 1995: The Second Assessment Report*. Cambridge University Press, Cambridge.

IPCC (1997) *IPCC (Revised 1996) Guidelines for National Greenhouse Gas Inventories. Workbook*. Intergovernmental Panel on Climate Change, Paris.

IPCC (2000a) *Special Report on Land Use, Land use change, and Forestry*. Cambridge University Press, Cambridge.

IPCC (2000b) *Special Report on Emissions Scenarios*. Cambridge University Press, Cambridge.

IPCC (2001) *Climate Change: The Scientific Basis*. Cambridge University Press, Cambridge.

Janssens, I.A., Freibauer, A., Ciais, P., Smith, P., Nabuurs, G.J., Folberth, G., Schlamadinger, B., Hutjes, R.W.A., Ceulemans, R., Schulze, E-D., Valentini, R. and Dolman, H. (2003) Europe's biosphere absorbs 7–12% of anthrogogenic carbon emissions. *Science* 300, 1538–1542.

Janssens, I.A., Freibauer, A., Schlamadinger, B., Ceulemans, R., Ciais, P., Dolman, A.J., Heimann, M., Nabuurs, G.J., Smith, P., Valentini, R. and Schulze E.D. (2005) The carbon budget of terrestrial ecosystems at country-scale. A European case study. *Biogeosciences* 2, 15–27.

Jenkinson, D.S. (1988) Soil organic matter and its dynamics. In: Wild, A. (ed.) *Russell's Soil Conditions and Plant Growth*, 11th edn. Longman, London, pp. 564–607.
Lal, R. (1999) Soil management and restoration for C sequestration to mitigate the accelerated greenhouse effect. *Progress in Environmental Science* 1, 307–326.
Lal, R. (2003a) Soil erosion and the global carbon budget. *Environment International* 29, 437–450.
Lal, R. (2003b) Soil carbon sequestration to mitigate climate change. *Geoderma* 123, 1–22.
Lal, R., Kimble, J.M., Follet, R.F. and Cole, C.V. (1998) *The Potential of U.S. Cropland to Sequester Carbon and Mitigate the Greenhouse Effect.* Ann Arbor Press, Chelsea, Michigan.
Metting, F.B., Smith, J.L. and Amthor, J.S. (1999) Science needs and new technology for soil carbon sequestration. In: Rosenberg, N.J., Izaurralde, R.C. and Malone, E.L. (eds) *Carbon Sequestration in Soils: Science, Monitoring and Beyond.* Battelle Press, Columbus, Ohio, pp. 1–34.
Nabuurs, G.J., Daamen, W.P., Dolman, A.J., Oenema, O., Verkaik, E., Kabat, P., Whitmore, A.P. and Mohren, G.M.J. (1999) *Resolving Issues on Terrestrial Biospheric Sinks in the Kyoto Protocol.* Dutch National Programme on Global Air Pollution and Climate Change, Report 410 200 030.
Paustian, K., Andrén, O., Janzen, H.H., Lal, R., Smith, P., Tian, G., Tiessen, H., van Noordwijk, M. and Woomer, P.L. (1997) Agricultural soils as a sink to mitigate CO_2 emissions. *Soil Use and Management* 13, 229–244.
Robertson, G.P., Paul, E.A. and Harwood, R.R. (2000) Greenhouse gases in intensive agriculture: contributions of individual gases to the radiative forcing of the atmosphere. *Science* 289, 1922–1925.
Schimel, D.S. (1995) Terrestrial ecosystems and the carbon-cycle. *Global Change Biology* 1, 77–91.
Schimel, D.S., House, J.I., Hibbard, K.A., Bousquet, P., Ciais, P., Peylin, P., Braswell, B.H., Apps, M.J., Baker, D., Bondeau, A., Canadell, J., Churkina, G., Cramer, W., Denning, A.S., Field, C.B., Friedlingstein, P., Goodale, C., Heimann, M., Houghton, R.A., Melillo, J.M., Moore, B., Murdiyarso, D., Noble, I., Pacala, S.W., Prentice, I.C., Raupach, M.R., Rayner, P.J., Scholes, R.J., Steffen, W.L. and Wirth, C. (2001) Recent patterns and mechanisms of carbon exchange by terrestrial ecosystems. *Nature* 414, 169–172.
Six, J., Ogle, S.M., Breidt, F.J., Conant, R.T., Mosier, A.R. and Paustian, K. (2004) The potential to mitigate global warming with no-tillage management is only realized when practised in the long term. *Global Change Biology* 10, 155–160.
Sleutel, S., De Neve, S. and Hofman, G. (2003) Estimates of carbon stock changes in Belgian cropland. *Soil Use and Management* 19, 166–171.
Smith, P. (2004) Soils as carbon sinks – the global context. *Soil Use and Management* 20, 212–218.
Smith, P. and Powlson, D.S. (2003) Sustainability of soil management practices a global perspective. In: Abbott, L.K. and Murphy, D.V. (eds) *Soil Biological Fertility: A Key to Sustainable Land Use in Agriculture.* Kluwer Academic Publishers, Dordrecht, The Netherlands, pp. 241–254.
Smith, P., Powlson, D.S. and Glendining, M.J. (1996) Establishing a European soil organic matter network (SOMNET). In: Powlson, D.S., Smith, P. and Smith, J.U. (eds) *Evaluation of Soil Organic Matter Models using Existing, Long-Term Datasets.* NATO ASI Series I, Vol. 38. Springer, Berlin, pp. 81–98.
Smith, P., Powlson, D.S., Smith, J.U., Falloon, P.D. and Coleman, K. (2000a) Meeting Europe's climate change commitments: quantitative estimates of the potential for carbon mitigation by agriculture. *Global Change Biology* 6, 525–539.
Smith, P., Powlson, D.S., Smith, J.U., Falloon, P.D., Coleman, K. and Goulding, K.W. (2000b) Agricultural carbon mitigation options in Europe: improved estimates and the global perspective. *Acta Agronomica Hungarica* 48, 209–216.
Smith, P., Goulding, K.W., Smith, K.A., Powlson, D.S., Smith, J.U., Falloon, P. and Coleman, K. (2000c) Including trace gas fluxes in estimates of the carbon mitigation potential of UK agricultural land. *Soil Use and Management* 16, 251–259.
Smith, P., Goulding, K.W., Smith, K.A., Powlson, D.S., Smith, J.U., Falloon, P. and Coleman, K. (2001) Enhancing the carbon sink in European agricultural soils: including trace gas fluxes in estimates of carbon mitigation potential. *Nutrient Cycling in Agroecosystems* 60, 237–252.
Smith, P. *et al.* (2004) *CarboEurope GHG: Greenhouse Gas Emissions from European Croplands.* CarboEurope GHG, Specific Study Number 2, University of Tuscia, Viterbo, Italy.
Vleeshouwers, L.M. and Verhagen, A. (2002) Carbon emission and sequestration by agricultural land use: a model study for Europe. *Global Change Biology* 8, 519–530.

5 Implications for Increasing the Soil Carbon Store: Calculating the Net Greenhouse Gas Balance of No-till Farming

Reynald L. Lemke[1] and H. Henry Janzen[2]
[1]*Agriculture and AgriFood Canada, Swift Current, Saskatchewan, Canada;* [2]*Agriculture and AgriFood Canada, Lethbridge, Alberta, Canada*

5.1 Introduction

No-till (NT) farming has been widely advocated as a way to enhance soil carbon stores (e.g. Lal, 2004a,b; Pacala and Socolow, 2004). Because the carbon in soil organic matter ultimately comes from photosynthesis, building soil carbon withdraws CO_2 from the air, helping to slow the accumulation of atmospheric CO_2 from fossil fuel burning and land-use change.

Until recently, scientists studying NT systems for greenhouse gas (GHG) mitigation have focused largely on soil carbon. But increasingly we are aware that tillage treatment may also influence emissions of other GHGs, notably N_2O. The flows of carbon through soil are inextricably linked to those of nitrogen; thus, a practice that alters carbon storage will almost certainly alter the dynamics of nitrogen, thereby affecting N_2O emissions. Tillage can also affect rates of CH_4 removal and CO_2 emission from energy use. To understand the potential of NT to reduce radiative forcing we need to consider all three gases, ideally over a period of decades.

Our objective is to review how adoption of NT farming affects the emission or removal of individual gases, and then consider the overall net effect, in both the short and long terms, using one case study as an illustration. The estimates of net emissions from this case study may not apply elsewhere directly, but our intent is to demonstrate the importance of considering all of the emissions together.

5.2 No-till Farming as a CO_2 Sink

The world's croplands hold ~160 Pg C in soil organic matter to a depth of 1 m (Paustian *et al.*, 1998; Jobbágy and Jackson, 2000). The amount of carbon stored, however, is subject to management influence through effects on carbon inputs and losses via decomposition: any practice that reduces input or hastens decomposition will reduce carbon reserves; and a practice that adds more carbon or slows decomposition will build soil organic carbon (SOC).

Historically, agricultural practices have favoured SOC loss, especially in the decades after initial cultivation. Typically, arable soils have lost ~20–40% of initial reserves in the soil solum (Mann, 1986; Davidson and Ackerman, 1993). Globally, cumulative losses of carbon from cultivated soils amount to ~50 Pg C or more (Paustian *et al.*, 1998; Lal, 2003, 2004a,c; Janzen, 2005). Why do these losses occur? First, agriculture deliberately removes large amounts of photosynthetically fixed carbon, leaving less litter carbon for replenishment of SOC reserves. Second, arable farming often accelerates decomposition by physically mixing residues into the soil, disrupting protective aggregates, or creating more favourable moisture and temperature regimes. Arable agriculture can also hasten SOC loss via erosion, but much of the eroded SOC may be deposited locally, and the net effect on overall carbon storage is still debated.

Recently, there has been growing optimism that at least some of the lost SOC can be recovered with improved farming practices, notably through the adoption of NT techniques (Paustian *et al.*, 1998, 2000; Machado and de Silva, 2001; Díaz-Zorita *et al.* 2002; Lal, 2004a,b). Although intensive tillage was once necessary to control weeds and prepare the land for seeding, effective herbicides and new seeding implements now allow many crops to be grown without any tillage beyond the placement of the seed itself. Such NT practices can increase SOC because they minimize soil disturbance, avoiding disruption of protective soil aggregates and the mixing of litter into soil. In some cases, notably where yields are limited by drought, NT can further increase SOC by enhancing carbon inputs through better moisture use efficiency.

Typically, soils under NT will gain carbon at rates of ~0.2–0.5 t C/ha/year for about two decades after starting the practice (Paustian *et al.*, 1997; Smith *et al.*, 1998; Six *et al.*, 2002, 2004; West and Marland, 2002; West and Post, 2002; Smith, 2004a,b; Alvarez 2005; Puget and Lal, 2005). But rates of carbon accrual can vary widely. For example, rates of SOC gain as high as 1 t C/ha/year or more have been reported for cropping systems managed under NT in Brazil (Sá *et al.*, 2001; Diekow *et al.*, 2005). Climate seems to exert a strong influence on SOC response to tillage. In a review of Canadian studies, VandenBygaart *et al.* (2003) found average rates of 0.32 t C/ha/year in cool, dry regions (western Canada) but no consistent SOC gain under NT in cool, humid regions (eastern Canada). Franzluebbers and Steiner (2002) similarly observed a strong climatic influence on SOC gain under NT across North America; rates were highest in subhumid regions, and lowest in cold or dry climates. In a comprehensive review, Ogle *et al.* (2005) estimated mean SOC accrual under NT (relative to conventional tillage (CT)) over 20 years of 1.23 ± 0.05 in tropical moist climates, 1.17 ± 0.05 in tropical dry climates, 1.16 ± 0.02 in temperate moist climates and 1.10 ± 0.03 in temperate dry climates. These and other studies demonstrate that, while NT often elicits significant soil carbon accrual, the SOC gains are not consistent or even assured.

Much of the carbon accumulation under NT occurs near the soil surface. In some cases, this surface accumulation is offset by reduced carbon storage lower in the soil profile so that when the entire solum is considered, net accrual of SOC may be minimal (VandenBygaart and Kay, 2004; Gregorich *et al.*, 2005). Considering only the surface layer could lead to an overestimate of SOC gains under NT (Royal Society, 2001; Puget and Lal, 2005).

How much SOC could be stored globally by adopting NT agriculture? According to Lal (2004c), 'conservation tillage' might store 0.1–1.0 Pg C/year. The wide range indicates that such estimates are tentative because of uncertainty over achievable rates of SOC gains and the extent to which such practices can be adopted. Because SOC gain can only continue for a few decades, soil carbon sequestration under NT (and other cropland practices) can only make a minor contribution to reaching long-term CO_2 mitigation targets (Smith, 2004a,b). However, in view of its potential for immediate implementation, and many other benefits to soil conservation and improved farming, NT farming remains a practice widely encouraged and increasingly adopted.

5.3 No-till Farming and Soil-emitted N_2O

While the effects of tillage on soil carbon have been documented extensively, the effects on N_2O emissions are obscure. To understand how tillage might affect N_2O

emissions, we first review the processes affecting N_2O emission, and then examine how tillage might affect these processes.

5.3.1 Sources of N_2O

N_2O is generated in soils through various chemical and biochemical pathways. Chemical processes, collectively referred to as chemodenitrification, involve the reaction of nitrite (NO_2^-) with organic matter (Bremner and Nelson, 1968; Nelson, 1982) and hydroxylamine (NH_2OH) with organic matter or exchangeable cations (Bremner *et al.*, 1980). However, production of N_2O via chemical pathways only proceeds under select circumstances, and is not likely to be a significant source of N_2O from soil (Bremner, 1997).

Biochemical processes, in contrast, proceed under most soil environmental conditions because soil microbial communities are remarkably diverse and adaptable. N_2O emitted from soils is therefore mostly of biological origin, predominantly from two processes: nitrification and denitrification (Bremner, 1997). Broadly stated, nitrifiers oxidize ammonium (NH_4^+) to nitrate (NO_3^-), while denitrifiers reduce NO_3^- to N_2O or dinitrogen (N_2) during anaerobic respiration. Even these very broad categories tend to be blurred as nitrifiers can also denitrify (nitrifier-denitrification) (Wrage *et al.*, 2001). The various microbial groups occupy different ecological niches in soil (Fig. 5.1). Some N_2O might also be generated by dissimilatory reduction of NO_3^- to NH_4^+ (DNRA) (DeCatanzaro *et al.*, 1987; Stevens *et al.*, 1998) and other biochemical pathways (Robertson and Tiedje, 1987), but the contribution from these pathways is likely negligible in upland agricultural soils.

5.3.2 Controls on N_2O production

N_2O can be both produced and consumed during denitrification; the magnitude and direction of N_2O exchange between the soil and atmosphere therefore reflect the net amount of these two opposing processes. The rate of production depends not only on the amount of nitrogen that is nitrified and/or

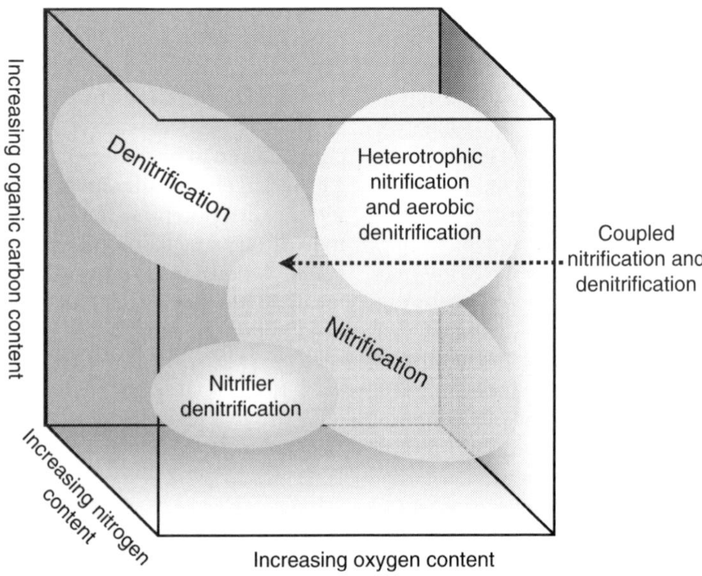

Fig. 5.1. Possible ecological niches for nitrogen transformation pathways in fertilized soils. (Redrawn from Wrage *et al.*, 2001.)

denitrified, but also on the ratio of N_2O generated per unit of nitrogen processed. The amount of nitrogen that is nitrified and/or denitrified is related to the size and activity of the nitrifier and denitrifier populations, the amount of nitrogen available and the competitive pressure from other biological activity (e.g. nitrogen uptake by plants). Nitrogen availability is governed by external nitrogen additions (e.g. organic and inorganic fertilizers) and, because microbial decomposition of organic matter provides NH_4^+ for nitrification and nitrification provides NO_3^- for denitrification (Fig. 5.2), by the rate and magnitude of antecedent microbial processes.

Conditions that govern the rate of nitrification tend also to influence the ratio of N_2O produced per unit of NH_4^+ (N_2O/NH_4^+ ratio) consumed, but often in opposing directions. For example, high NH_4^+ and oxygen availability favours nitrification but reduces the N_2O/NH_4^+ ratio. Although the temperature range of nitrification is typically reported as 5–40°C (Bremner, 1997), populations may adapt to prevailing climatic conditions. In the cool subhumid region of Canada, for example, Malhi and McGill (1982) observed measurable nitrification at –4°C and peak rates at 20°C. The N_2O/NH_4^+ ratio appears to increase as soil temperature and soil pH increase (Bremner, 1997).

Denitrification is favoured by the absence of oxygen and by adequate levels of soluble carbon and NO_3^-. The optimum pH range for denitrifiers is ~6.0–8.0, although denitrifier populations appear to adapt to long-term pH conditions (Parkin *et al.*, 1985) and activity can occur at pH as low as 3.5 (Aulakh *et al.*, 1992). Denitrifiers favour temperatures between 5°C and 40°C but, as with nitrifiers, populations may adapt to prevailing climatic conditions (Malhi *et al.*, 1990). The N_2O/NO_3^- ratio is very dynamic and sensitive

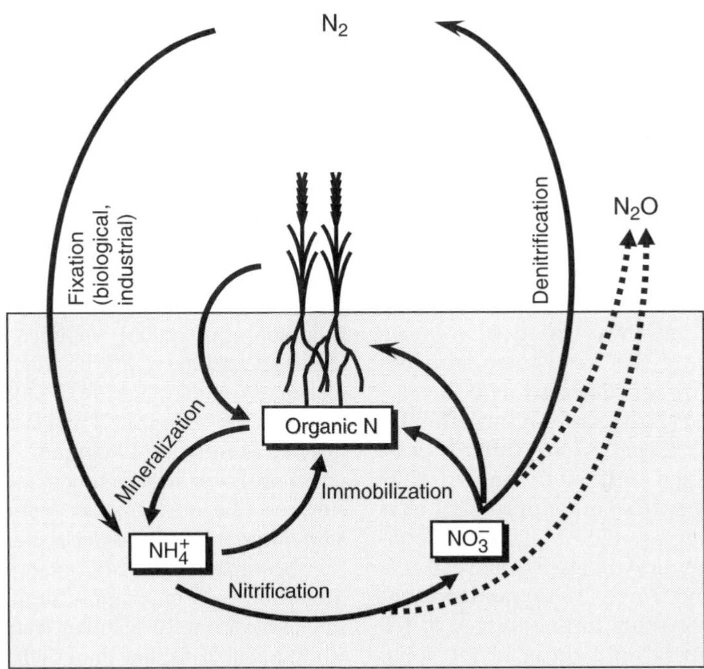

Fig. 5.2. A simplified view of the nitrogen cycle in a cropping system. Adoption of no-till (NT) can influence numerous processes in this cycle: (i) industrial fixation (via amount of nitrogen fertilizer needed); (ii) mineralization (via reduced disturbance); (iii) immobilization (via residue placement); (iv) nitrification (via effects on soil moisture and processes generating NH_4^+); and (v) denitrification (via effects on aeration).

to the factors controlling denitrification rate. The ratio tends to be related positively to the NO_3^-/soluble carbon ratio and oxygen level, and negatively to temperature and pH.

N_2O production at the cellular level is determined by the interaction of many regulating factors. Oxygen status, for example, is strongly governed by soil water content; since oxygen travels more slowly through water than through air, oxygen diffusion into the soil decreases as soil water-filled pore space increases. Emissions of N_2O from nitrification peak between 50% and 60% water-filled pore space (WFPS), while emissions from denitrification peak between 70% and 90% WFPS (Aulakh *et al.*, 1984; Davidson *et al.*, 1986; Stevens *et al.*, 1997). Emission rates drop rapidly when WFPS exceeds 90% because denitrification goes to completion (N_2) in the absence of oxygen, and the transmission of N_2O to the soil surface is impeded, resulting in greater opportunity for further reduction.

5.3.3 Effect on N_2O emissions

Assessing the influence of tillage systems on N_2O emissions is not straightforward. Farming systems include a complex mix of tillage tools, timings and frequencies, combined with variations in fertilizer and residue management and crop type, all interacting with local climate, topography and soil type. Soil conditions in NT systems differ from those in tilled systems in several ways: SOC and microbial biomass tend to be concentrated near the surface because residues are not buried; bulk density and aggregation are often higher, affecting oxygen diffusion; and surface moisture may be higher, suppressing oxygen because higher microbial activity consumes oxygen and air-filled porosity is lower. Linn and Doran (1984) reported that counts of denitrifiers in the surface of NT soils were several-fold those in CT soils. These factors may favour gaseous nitrogen loss via denitrification.

The influence on N_2O emissions of the higher soil water under NT likely depends on the range of WFPS typical for the location. If other factors are not limiting, N_2O emissions from denitrification only begin to increase sharply at WFPS of 60% or higher. If, as is likely in the subhumid and semiarid regions, NT does not increase soil water content to this level, the impact on N_2O emissions from denitrification may be negligible. Indeed nitrification may be the dominant source of N_2O in these drier regions.

Tillage tends to favour decomposition and nitrification by disrupting soil aggregates and exposing physically protected SOC to decomposers. In addition, residues left on the surface in NT systems may be desiccated for prolonged periods in dry climates, slowing decomposition. In cool semiarid and subhumid regions, therefore, soil CO_2 emissions are often lower in NT than in tilled systems (Cochran *et al.*, 1997; Lupwayi *et al.*, 1999, 2004; Curtin *et al.*, 2000), indicating slower decomposition and, presumably, nitrification. Lower nitrification rates may partly explain the similar or lower N_2O emissions reported on NT compared to tilled soils in cool subhumid and semiarid regions (Cochran *et al.*, 1997; Lemke *et al.*, 1999; Malhi *et al.*, 2006). Conversely, where rainfall is higher, N_2O emissions in NT often exceed those in tilled systems (Linn and Doran, 1984; MacKenzie *et al.*, 1998; Ball *et al.*, 1999; Grageda-Cabrera *et al.*, 2004), most likely due to increased denitrification. However, this relationship is not always consistent; in high rainfall areas N_2O emissions in NT can also be similar to, or lower than, those in tilled soils (Kaharabata *et al.*, 2003; Helgason *et al.*, 2005). Venterea *et al.* (2005) observed an interaction between fertilizer N placement and tillage. N_2O emissions were lower on NT than on CT when nitrogen was injected as anhydrous ammonia, higher on NT when urea was surface-broadcast and similar between tillage treatments when liquid urea ammonium nitrate was surface-applied.

According to selected examples from the literature, N_2O emissions from NT systems can range from 20% lower (ratio of NT/CT = 0.8) to 600% higher than their tilled counterparts (Table 5.1). The influence on N_2O can significantly offset the benefits of NT on soil carbon storage, or appreciably augment these benefits, depending on location.

Table 5.1. The influence of tillage on N_2O emissions in various studies, selected to illustrate diverse environmental conditions. Also shown is the annual change in soil carbon storage required to offset the increased N_2O emissions from adoption of no-till (NT). In some cases, where NT reduces N_2O emissions, the effects on carbon sequestration and on N_2O emissions are additive, rather than offsetting.

Location	Nitrogen applied (kg N/ha)	MAT (°C)	MAP (mm)	Ratio NT/Till	Annual/ seasonal N_2O loss (kg N/ha)	Δ SOC required to offset Δ N_2O (kg C/ha/year)
Delta Junction, Alaska, USA[a]	90	−1.9	303	~1.0	na	na
Ellerslie, Alberta, Canada[b]	56	3.5	450	0.7	1.4–2.1	(50–80)[j]
Sidney, Nebraska, USA[c]	0	8.2	411	0.8	na	na
Piketon, Ohio, USA[d]	180	4.4	400	0.8	0.9–3.7	(20–90)
Ormstown, Quebec, Canada[e]	0 and 180	6.4	949	1.2	3.4–4.2	80–90
Celaya, Guanajuato, Mexico[f]	180	18	650	2.5–4.0	0.7–19.8	50–1880
Turitea, New Zealand[g]	68	13	1305	1.3[i]	9.2–12.0	270–360
Oxford, England[h]	70 and 140	10.4	642	2.1–6.0	0.5–8.6	330–910

[a]Cochran *et al.*, 1997;
[b]Lemke *et al.*, 1999; [c]Kessavalou *et al.*, 1998;
[d]Jacinthe and Dick, 1997; [e]MacKenzie *et al.*, 1998;
[f]Grageda-Cabrera *et al.*, 2004;
[g]Choudhary *et al.*, 2002;
[h]Burford *et al.*, 1981.
[i]Difference between tillage systems not significant.
[j]Numbers in brackets indicate negative values.

The radiative forcing from a kilogram of N_2O, over a 100-year period, is about 296 times that of a kilogram of CO_2 (IPCC, 2001). Adjusting for this greater 'global warming potential' (GWP) and the molar proportions of nitrogen and carbon in N_2O and CO_2, the change in N_2O emissions from adopting NT can be expressed as equivalent kg C/ha/year (Table 5.1). This provides an indication of how much soil carbon would need to be sequestered to offset an increase in N_2O emissions. The values calculated in Table 5.1 frequently match, or even greatly surpass, the rate of soil carbon gain typically reported after adoption of NT practices. For example, at Turitea, New Zealand (Table 5.1), a carbon sequestration rate of ~0.3 t C/ha would be required to offset the increased N_2O emissions from adoption of NT. In some locations, however, the effect on N_2O amplifies the benefits on carbon sequestration (e.g. Ellerslie, Canada).

The influence of tillage on N_2O may also depend on other concurrent practices. On the semiarid North American Great Plains, for example, summer fallow is often used to avert risk of drought. During the fallow year, weeds are controlled but no crop is grown, allowing a 21-month period for soil water recharge before the next crop is planted. Fallow may be included in rotations once every 2–4 years. Adopting NT, which improves moisture conservation, has allowed farmers to reduce fallow frequency, or to eliminate fallow entirely. How this affects N_2O emissions is unclear. Fertilizer N requirements are usually higher under reduced fallow frequency, resulting in higher N_2O emissions, but eliminating plant growth during the fallow phase increases soil moisture, favouring denitrification, both from the higher water content itself and by increasing soil NO_3^- from enhanced

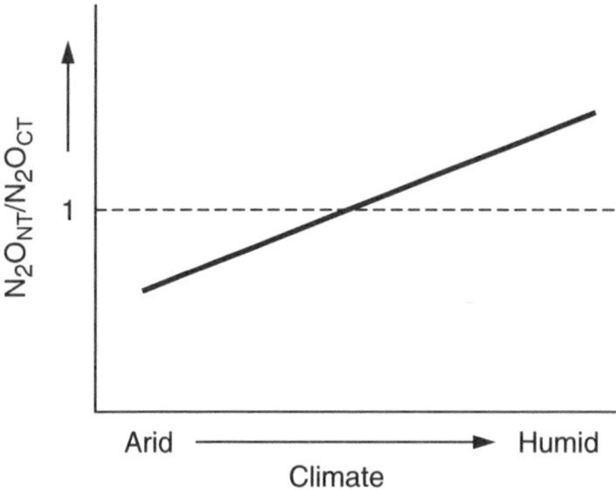

Fig. 5.3. A conceptual graph, illustrating a possible relationship between climate and the ratio of N_2O emissions under no-till (NT) to that under conventional tillage (CT) (N_2O_{NT}/N_2O_{CT}). This relationship is likely oversimplified, but may point to a useful working hypothesis.

mineralization. In western Canada, for example, annual N_2O emissions from fallow plots were as high or higher than those from fertilized cereal plots (Lemke *et al.*, 1999).

How will N_2O emissions from soil change if NT practices are adopted? It may depend on region. In cool subhumid to semi-arid regions, similar to those found in western Canada, direct soil-emitted N_2O emissions in NT are likely to be similar to, or less than, those in CT systems (Fig. 5.3). In more humid regions, conversely, emissions under NT may be similar to, or higher than, those under CT. Reversals have, however, been reported (e.g. Helgason *et al.*, 2005), and these conclusions remain tentative. Results to date emphasize that N_2O merits concerted attention in assessing the potential of NT practices for reducing GHG emissions. While the benefits of reduced tillage for soil carbon sequestration are important, they may not always override effects of tillage on N_2O emissions.

5.4 No-till Farming and Other Greenhouse Gases

5.4.1 Methane consumption by soils

Methane consumption by soils has been measured in a variety of ecosystems (e.g. Steudler *et al.*, 1989; Whalen and Reeburgh, 1990; Yavitt *et al.*, 1990; Mosier *et al.*, 1991; Striegl *et al.*, 1992). In well-aerated soils, methanotrophs can use CH_4 as a source of carbon and, by oxidation to CO_2, as a source of energy. Globally, this oxidation represents a significant sink for CH_4 (Smith and Conen, 2004). Converting native ecosystems to agricultural use tends to reduce the soil sink strength (Ojima *et al.*, 1993; Willison *et al.*, 1995; Dobbie *et al.*, 1996; Prieme *et al.*, 1997; Smith *et al.*, 2000). Resolving the reason for this reduction is not easy because many changes occur simultaneously when native ecosystems are converted to agricultural use, but application of NH_4^+-containing fertilizers (Mosier *et al.*, 1991; Hütsch *et al.*, 1994; King and Schnell, 1994) and physical disturbance have been implicated as likely causes (Hütsch, 2001).

5.4.2 Effect on soil methane consumption

The influence of tillage on CH_4 consumption is not always consistent; nor is it easy to distinguish from that of concurrent management changes. Tillage may suppress CH_4 oxidation by several mechanisms: it enhances immediate mineralization, generating NH_4^+, which may inhibit CH_4 oxidation; it tends to dry the soil, suppressing microbial activity; it reduces gas diffusion into the soil via macropores; and it disturbs microbial habitat by affecting soil structure (Hütsch, 1998). Adopting NT, therefore, may sometimes enhance CH_4 oxidation.

The influence of NT on CH_4 consumption, however, may depend also on other, interactive factors. For example, Cochran *et al.* (1997) noted an interaction between tillage and soil water contents. Early in the season when soil water contents were high, CH_4 consumption was lower in NT than in tilled plots. These high water contents, particularly in the NT plots, may have restricted gas diffusion into the soil, reducing CH_4 supply to methanotrophs or favouring CH_4 production by restricting oxygen availability. During the remaining season, soils were quite dry, apparently suppressing methanotroph activity more in tilled than in NT soils. Kessavalou *et al.* (1998) attributed short-term reductions in CH_4 oxidation immediately after tillage partly to soil drying. Venterea *et al.* (2005) observed that tillage increased or decreased CH_4 uptake, depending on the fertilizer N treatment. After conducting a limited number of studies, Six *et al.* (2002) estimated that CH_4 consumption would be 20% greater in NT than in tilled soils.

5.4.3 CO_2 emissions from fossil fuel use

In mechanized farming, fossil fuel is used both to power farm implements and to manufacture and transport fertilizers, pesticides and machinery. Adopting NT usually conserves energy by eliminating energy-intensive tillage and reducing the wear on tillage equipment. The amount of energy saved depends on the previous tillage intensity. For a subhumid site in western Canada, Zentner *et al.* (2004) found that NT, compared to CT, reduced on-farm fuel and lubricants by 25–31%. In wetter regions where the intensity of CT systems is higher, NT may reduce tillage-related on-farm fossil energy use by up to 60% (West and Marland, 2002).

Tillage accounts for only 30% or less of total energy use, so energy saved by reducing tillage can easily be offset by increased herbicide and particularly by fertilizer inputs. At sites in western Canada, for example, energy savings from reduced tillage intensity were completely offset by increased amounts of fertilizer N and herbicides (Zentner *et al.*, 1998). In the USA, West and Marland (2002) reported that energy gains due to reduced tillage were negated by increased amounts of fertilizer N and herbicides for maize, but not for winter wheat or soybean.

A critical question, clearly, is whether increased use of fertilizer or herbicides is needed to maintain crop yields under NT. According to Elliot and Coleman (1988), herbicide requirements were actually lower in NT than in CT, after the initial transition. At two sites in western Canada, yields in cereal monocultures receiving equivalent rates of nitrogen were lower in NT than in CT (Nyborg *et al.*, 1995), but at other sites in this region, yields under NT were similar to, or higher than, those under CT, particularly when rotations included pulse crops (Carefoot *et al.*, 1990; Izaurralde *et al.*, 1995). Kupusta *et al.* (1996) reported that maize yields under NT were similar to those under CT when equivalent rates of nitrogen were broadcast, but were slightly lower under NT when starter nitrogen was used.

If soil carbon increases upon adopting NT, higher nitrogen inputs might be needed, at least in the short-term, assuming that the carbon/nitrogen ratio of organic matter is constant. The increased requirement for nitrogen inputs should diminish as soil carbon levels reach steady state. Many factors interact to determine how tillage affects yield, and an assumption that nitrogen requirements will be higher for NT than for CT may not always be justified, especially in the long term. If crop yields can be maintained without increased use of fertilizer N, NT can often significantly reduce CO_2 from fossil fuel use.

5.5 No-till Farming: The Net Result

To calculate the net effect of NT farming on radiative forcing, we need to consider jointly the emissions and removals of all the GHGs (Robertson and Grace, 2004; Mosier *et al.*, 2005). To allow for differences in radiative forcing, all gases are expressed as CO_2 equivalents (CO_2e), assuming that N_2O has a 100-year GWP of 296 (i.e. 1 kg of N_2O has the same radiative forcing as 296 kg of CO_2) and CH_4 has a GWP of 23 (IPCC, 2001). Rather than describe generalized trends, we consider a specific example to illustrate the processes to be considered. Before proceeding to that example, however, there is another complicating factor that deserves mention – the issue of time.

5.5.1 The issue of time

In any given year, for any given site, the net benefit of NT on annual GHG emissions will probably depend on time elapsed since adoption of NT. This may be especially true for CO_2. Rates of SOC accrual under NT are usually highest soon after adoption, and then gradually diminish as the soil organic matter approaches a new steady state, perhaps after several decades (West and Post, 2002; Alvarez, 2005). Thus, NT systems can be initially a significant sink for CO_2, but the magnitude of that sink wanes with time.

N_2O emissions, too, will probably change with time elapsed since the adoption of NT, though the nature of this response is not yet well defined. If supplemental nitrogen is required to offset nitrogen immobilization in the first years after adoption, emissions might initially increase. If NT results in persistent changes to soil conditions such as bulk density or moisture, the effect on N_2O emissions, if any, would likely also persist. Finally, N_2O fluxes might change in response to evolving nitrogen mineralization patterns – enhanced immobilization in the early stages, and increasing mineralization later as the system matures with higher organic matter.

CH_4 uptake by soil may also change with time from inception of NT, increasing gradually over years or decades. Since CH_4 oxidation plays only a small role in net GHG emissions, this effect will likely be small. Finally, CO_2 emissions from fossil energy use may change with time, particularly if dependence on fertilizer and herbicide evolves as the NT system matures.

This brief overview demonstrates that any effect of NT on net GHG emissions is not fixed; it will almost certainly vary with time, perhaps in a complex, somewhat unpredictable pattern. This complexity may be further enhanced by the lingering influence of history – at any point in time, the GHG emissions from a given site depend not only on the practices currently imposed on the ecosystem, but also on how the land was managed in previous years (or decades).

5.5.2 The net result: a simple example

How then do we assess the net GHG balance after a change in tillage intensity? We explore this question by presenting an example from semiarid agriculture like that in the northern Great Plains of North America. The estimates may not apply directly to cropping systems elsewhere, but the approach and the inherent insights derived from such an exercise may be more broadly applicable.

As shown in Table 5.2, the 30 years after adoption of NT are divided into three 10-year phases: an initial decade where soils are gaining carbon and N_2O emissions are influenced by higher fertilizer N requirements; a second decade where carbon accumulation continues but at a lower rate, and N_2O emissions subside as fertilizer N inputs are reduced; and a third decade where soil carbon is assumed to have reached a new steady state, and the NT system is mature.

Mean non-renewable energy inputs for a continuous spring wheat rotation under CT and NT management were calculated from values presented by Zentner *et al.* (1998). For this example, fertilizer N rates on NT were increased by 10% during the first decade to account for the additional nitrogen immobilized, and were equivalent to CT thereafter. Fertilizer-induced emissions (FIE) of N_2O were calculated assuming that 1.25% of applied nitrogen is emitted as N_2O (IPCC, 1997). Recent research suggests that N_2O emissions tend to be lower from NT compared to tilled soils for the semi-arid region of western Canada (Lemke *et al.*, 1999; Helgason *et al.*, 2005; Malhi *et al.*, 2006); with these and other data (Lemke, 2006), we assumed that emissions decrease by 20%. Mean CH_4 consumption rates were assumed to be 1.6 kg CH_4/year for CT and 2.0 kg CH_4/year for NT (Six *et al.*, 2002). Because it takes time for rates to adjust, we assumed an intermediate value (1.8 kg CH_4/year) for NT during the first decade. Soil carbon was assumed to increase by an overall average of 320 kg/ha/year (VandenBygaart *et al.*, 2003) for the first 20 years, with no further increases thereafter, but we assumed that rates were higher in the first decade (420 kg/ha/year) than in the second (220 kg/ha/year).

In our example (Table 5.2), emissions from fuel and machinery are slightly lower

Table 5.2. Sample budget of CO_2 equivalents for a hypothetical site in the cool semiarid region of western Canada before and after adopting no-till (NT) farming practices.

	Emissions (kg CO_2e/ha/year)				
	Conventional till	No-till (phase 1)[a]	No-till (phase 2)[b]	No-till (phase 3)[c]	No-till (30-year mean)
CH_4	−37	−41	−46	−46	−44
CO_2 (fuel and machinery)	133	125	125	125	125
CO_2 (pesticides and fertilizer)	174	201	185	185	190
Sub-total	270	285	264	264	271
FIE N_2O[d]	209	184	167	167	173
Indirect N_2O[e]	81	90	81	81	84
CO_2 (soil)	0	−1541	−807	0	−783
Total	560	−982	−295	512	−255

[a]Fertilizer N rates 10% higher than CT (40kg N/ha) to offset immobilization due to SOC increases of 420 kg C/ha/year.
[b]SOC increases by 220 kg/ha/year and fertilizer N rates reduce to match CT.
[c]Soil carbon is assumed to have reached a new steady state.
[d]Direct fertilizer-induced emissions.
[e]Off-site emissions resulting from volatilized nitrogen (10% of fertilizer N applied) and nitrogen leached (15% of fertilizer N remaining after volatilization). N_2O–N = volatilized N × 1.0% and leached N × 2.5% (IPCC, 1997).

from NT compared to CT systems, but this is more than offset by increased emissions associated with herbicides and fertilizers. Excluding changes in direct and indirect N_2O emissions and soil carbon status, emissions actually increase slightly under NT during the first phase, but decrease slightly during the last two phases, resulting in a 30-year mean that is essentially equal to CT.

While the increased amount of fertilizer N used on NT during the first phase results in more FIE of N_2O, we assumed for this example that the proportion of nitrogen lost as N_2O from NT is less than from CT, resulting in a net reduction of emissions. In addition to changes to SOC, over a 30-year period NT would contribute ~1.0 t CO_2e/ha less to the atmosphere than CT. Increases in SOC would be an additional net benefit. After 30 years, in this hypothetical illustration, NT would have removed more than 7 t CO_2e/ha while its CT counterpart would have produced more than 16 t CO_2e/ha.

The influence of NT adoption on overall CO_2e emissions will be site-specific. Converting from CT to NT farming in the cool subhumid and semiarid regions of western Canada, particularly if pulse crops are included in rotation, will probably result in lower CO_2e emissions from fossil fuel use, similar or lower N_2O emissions, similar or higher CH_4 consumption rates and modest gains in soil carbon status. These changes are 'synergistic', reducing overall CO_2e emissions. In wetter regions, adopting NT may reduce fossil fuel use, but N_2O emissions may be markedly increased. If an increase in soil carbon status occurs, overall CO_2e emissions may still be lower on NT compared to CT, but this benefit would vanish over time as soils reach a new steady state. In some cool humid regions such as eastern Canada, where N_2O emissions are likely to increase but soils do not appear to sequester carbon after adoption of NT (VandenBygaart *et al.*, 2003), NT practices would likely increase overall CO_2e emissions compared to CT. Clearly, the influence of NT on N_2O emissions at a given site is a prominent factor in determining how this practice affects overall, long-term emissions of GHGs.

5.6 Implications for Other Practices

This brief look at NT farming has implications also for broader questions of GHG mitigation from agroecosystems. Perhaps the most obvious is the urgency of accounting for all GHGs in assessing how well a proposed practice might reduce emissions (Robertson and Grace, 2004; Mosier *et al.*, 2005). Although much of the early focus, justifiably, was on soil carbon sequestration, agriculture is a major contributor of N_2O and CH_4, both potent GHGs. Because of the high GWPs of these gases, small shifts in their emissions can substantially augment or offset the benefits from any soil carbon gain. So we cannot consider only the soil carbon accrual from reduced tillage; we have to estimate the effects on N_2O emission. We cannot limit our attention only to the SOC gains from planting grasses; we have to think about the CH_4 emitted when those grasses are fed to livestock. We cannot examine only the SOC benefits of practices that favour higher yields; we have to quantify the N_2O emitted from higher fertilizer rates needed to support those yields and the energy consumed in making that fertilizer. Thus, GHG mitigation in agriculture depends increasingly on looking at farms as ecosystems, considering the entire web of intertwined flows of carbon, nitrogen and energy in that system. Perhaps the best way of doing this, quantitatively, is by building models of varying sophistication. Some progress has already been made in that direction (e.g. Flessa *et al.*, 2002; Soussana *et al.*, 2004; Janzen *et al.*, 2006; Schils *et al.*, 2005), but the complexity of such analyses suggests that immediate success in such ventures is unlikely. A particular challenge, and an aim worth pursuing, is to link the nutrient and energy flows between livestock and cropping facets of farming systems, facets that heretofore have often been examined separately.

A second finding is that mitigation practices such as NT farming cannot be advocated blindly without taking into account local conditions. NT practices offer a powerful opportunity to reduce GHG emissions in some settings; elsewhere they have minimal net benefits, and may even *enhance* GHG emissions. While NT and other practices have been widely recommended as mitigation practices – with good reason – the next step may be to identify those conditions and complementary farming practices where they are most beneficial (or at least not counterproductive).

Further, the preceding analysis implies that the choice of mitigation practices may involve trade-offs. Although NT farming is often promoted as a win–win opportunity, it (like other proposed mitigation practices) may not be without possible costs. For example, while NT has many benefits for soil conservation, what does the farmer do if there is animal manure to apply? Does the farmer maintain strict NT or compromise aversion to tillage by incorporating the manure, thereby minimizing environmental effects of that manure? What if NT farming reduces net GHG emissions but poses higher risk to water quality via leaching through macropores – who decides the relative costs and merits of each? The value judgements needed to resolve such trade-offs may need to consider not only scientific data but also social factors (Lubchenco, 1998; Ludwig *et al.*, 2001).

Finally – and perhaps most daunting – is the challenge to include time into the analysis of GHG mitigation. As observed for NT, the net benefit of a practice depends on how much time has elapsed since the practice was adopted. Eventually the SOC gains must cease as soil carbon approaches a new equilibrium, and in the long run – over many decades – the net benefit of NT may depend less on the SOC gained than on the continued emissions of N_2O and energy-derived CO_2 needed to continue the practice (Fig. 5.4). Further, the effectiveness of a proposed practice will depend on history – the net accrual of carbon under NT, for example, depends not only on the practice now in place on a farm, but also on the practices imposed on that land years (even generations) before. Lastly, time needs to be included to acknowledge future global changes. Conditions a half-century from now may be different than those observed in our current careful studies – the climate may be different, atmospheric CO_2 certainly will be higher, energy availability may be altered. What will happen, in that changed world, to the carbon that we have so carefully extracted from the atmosphere and stored away in our soils? Do we risk losing the soil carbon

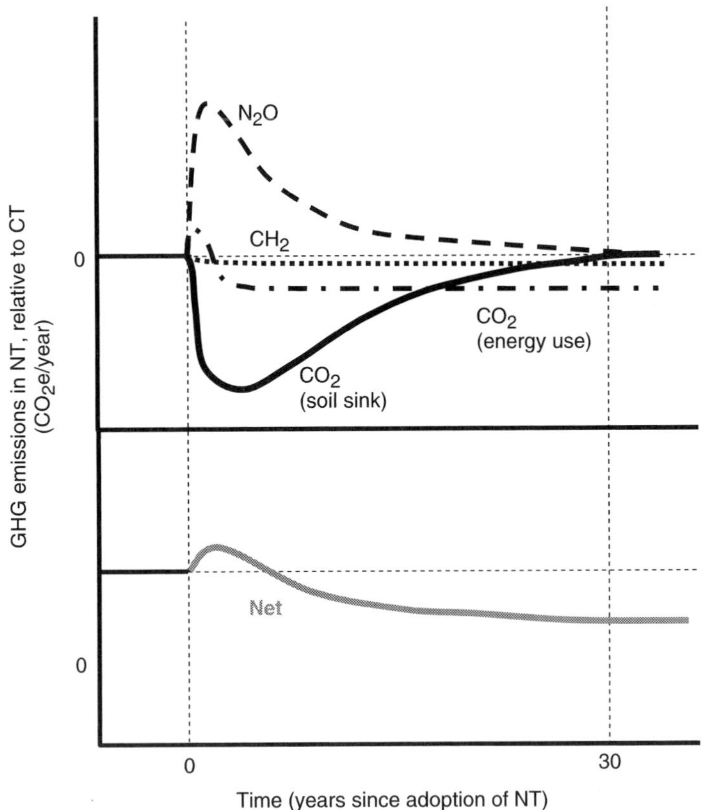

Fig. 5.4. Conceptual illustration of how the rates of greenhouse gas (GHG) emissions, individually and together, might change with time after the adoption of no-till (NT) farming. Arbitrary rates are expressed as the increase (or decrease) over those in the preceding tilled system. A value less than zero denotes reduced emission (or enhanced removal) compared to the tilled system. The actual rates, as well as the pattern over time, will vary among sites.

so carefully sequestered, perhaps accentuating CO_2 emissions in future decades (e.g. Knorr *et al.*, 2005)?

These examples of challenges, and doubtless many others, demonstrate the need to develop new ways of studying and understanding our ecosystems. We will need to study them as systems, with complex interactions within their borders, and many ties to those outside. Such efforts, likely aided by building better simulation models, will help us not only reduce GHG emissions from farms, but also preserve and augment other ecosystem functions that are no less important.

References

Alvarez, R. (2005) A review of nitrogen fertilizer and conservation tillage effects on soil organic carbon storage. *Soil Use and Management* 21, 38–52.

Aulakh, M.S., Rennie, D.A. and Paul, E. (1984) Acetylene and N-serve effects upon N_2O emissions from NH_4^+ and NO_3^- treated soils under aerobic and anaerobic conditions. *Soil Biology and Biochemistry* 16, 351–356.

Aulakh, M.S., Doran, J.W. and Mosier, A.R. (1992) Soil denitrification – significance, measurement, and effects of management. *Advances in Soil Science* 18, 1–57.

Ball, B.C., Scott, A. and Parker, J.P. (1999) Field N_2O, CO_2 and CH_4 fluxes in relation to tillage, compaction and soil quality in Scotland. *Soil and Tillage Research* 53, 29–39.

Bremner, J.M. (1997) Sources of nitrous oxide in soils. *Nutrient Cycling in Agroecosystems* 49, 7–16.

Bremner, J.M. and Nelson, D.W. (1968) Chemical decomposition of nitrite in soil. *Transactions of the 9th International Congress of Soil Science* 2, 495–503.

Bremner, J.M., Blackmer, A.M. and Waring, S.A. (1980) Formation of nitrous oxide and dinitrogen by chemical decomposition of hydroxylamine in soils. *Soil Biology and Biochemistry* 12, 263–269.

Burford, J.R., Dowdell, R.J. and Crees, R. (1981) Emissions of nitrous oxide to the atmosphere from direct-drilled and ploughed clay soils. *Journal of the Science of Food and Agriculture* 32, 219–223.

Carefoot, J.M., Nyborg, M. and Lindwall, C.W. (1990) Tillage-induced soil changes and related grain yield in a semi-arid region. *Canadian Journal of Soil Science* 70, 203–214.

Choudhary, M.A., Akramkhanov, A. and Saggar, S. (2002) Nitrous oxide emissions from a New Zealand cropped soil: tillage effects, spatial and seasonal variability. *Agriculture Ecosystems and Environment* 93, 33–43.

Cochran, V.L., Sparrow, E.B., Schlentner, S.F. and Knight, C.W. (1997) Long-term tillage and crop residue management in the subarctic: fluxes of methane and nitrous oxide. *Canadian Journal of Soil Science* 77, 565–570.

Curtin, D., Wang, H., Selles, F., McConkey, B.G. and Campbell, C.A. (2000) Tillage effects on carbon fluxes in continuous wheat and fallow-wheat rotations. *Soil Science Society of America Journal* 64, 2080–2086.

Davidson, E.A. and Ackerman, I.L. (1993) Changes in soil carbon inventories following cultivation of previously untilled soil. *Biogeochemistry* 20, 161–193.

Davidson, E.A., Swank, W.T. and Perry, T.O. (1986) Distinguishing between nitrification and denitrification as sources of gaseous nitrogen production in soil. *Applied and Environmental Microbiology* 52, 1280–1286.

DeCatanzaro, J.B., Beauchamp, E.G. and Drury, C.F. (1987) Denitrification vs. dissimilatory nitrate reduction in soil with alfalfa, straw, glucose and sulfide treatments. *Soil Biology and Biochemistry* 19, 583–587.

Díaz-Zorita, M., Duarte, G.A. and Grove, J.H. (2002) A review of no-till systems and soil management for sustainable crop production in the subhumid and semiarid Pampas of Argentina. *Soil and Tillage Research* 65, 1–18.

Diekow, J., Mielniczuk, J., Knicker, H., Bayer, C., Dick, D.P. and Kögel-Knabner, I. (2005) Soil C and N stocks as affected by cropping systems and nitrogen fertilisation in a southern Brazil Acrisol managed under no-tillage for 17 years. *Soil and Tillage Research* 81, 87–95.

Dobbie, K.E., Smith, K.A., Prieme, A., Christensen, S., Degorska, A. and Orlanski, P. (1996) Effect of land use on the rate of methane uptake by surface soils in northern Europe. *Atmospheric Environment* 30, 1005–1011.

Elliot, E.T. and Coleman, D.C. (1988) Let the soil do the work for us. *Ecological Bulletins* 339, 23–32.

Flessa, H., Ruser, R., Dorsch, P., Kamp, T., Jimenez, M.A., Munch, J.C. and Beese, F. (2002) Integrated evaluation of greenhouse gas emissions (CO_2, CH_4, N_2O) from two farming systems in southern Germany. *Agriculture, Ecosystems and the Environment* 91, 175–189.

Franzluebbers, A.J. and Steiner, J.L. (2002) Climatic influences on soil organic carbon storage with no tillage. In: Kimble, J.M. *et al.* (eds) *Agricultural Practices and Policies for Carbon Sequestration in Soil.* Lewis Publishers, Boca Raton, Florida, pp. 71–85.

Grageda-Cabrera, O.A., Medina-Cázares, T., Aguilar-Acuña1, J.L., Hernández-Martínez, M., Solís-Moya, E., Aguado-Santacruz, G.A. and Peña-Cabriales, J.J. (2004) Gaseous nitrogen loss by N_2 and N_2O emissions from different tillage systems and three nitrogen sources. *Agrociencia* 38, 625–633.

Gregorich, E.G., Rochette, P., VandenBygaart, A.J. and Angers, D.A. (2005) Greenhouse gas contributions of agricultural soils and potential mitigation practices in Eastern Canada. *Soil and Tillage Research* 83, 53–72.

Helgason, B.L., Janzen, H.H., Chantigny, M.H., Drury, C.F., Ellert, B.H., Gregorich, E.G., Lemke, R.L., Pattey, E., Rochette, P. and Wagner-Riddle, C. (2005) Toward improved coefficients for predicting direct N_2O emissions from soil in Canadian agroecosystems. *Nutrient Cycling in Agroecosystems* 72, 87–99.

Hütsch, B.W. (1998) Tillage and land use effects on methane oxidation rates and their vertical profiles in soil. *Biology and Fertility of Soils* 27, 284–292.

Hütsch, B.W. (2001) Methane oxidation in non-flooded soils as affected by crop production – invited paper. *European Journal of Agronomy* 14, 237–260.

Hütsch, B.W., Webster, C.P. and Powlson D.S. (1994) Methane oxidation in soil as affected by land use, soil pH and N fertilization. *Soil Biology and Biochemistry* 26, 1613–1622.

IPCC (1997) Revised 1996 IPCC guidelines for national greenhouse gas inventories. In: Houghton, J.T. *et al.* (eds) Reference Vol. 3. Available at: http://www.ipcc-nggip.iges.or.jp/public/gl/inversus6.htm (verified 9 December 2005).

IPCC (2001) Climate change 2001: the scientific basis. In: Houghton, J.T. *et al.* (eds) *Intergovernmental Panel on Climate Change*. Available at: http://www.grida.no/climate/ipcc_tar/wg1/ (verified 9 December 2005).

Izaurralde, R.C., Choudhary, M., Juma, N.G., McGill, W.B. and Haderlein, L. (1995) Crop and nitrogen yield in legume-based rotations practiced with zero tillage and low-input methods. *Agronomy Journal* 87, 958–964.

Jacinthe, P.A. and Dick, W.A. (1997) Soil management and nitrous oxide emissions from cultivated fields in southern Ohio. *Soil and Tillage Research* 41, 221–235.

Janzen, H.H. (2005) Soil carbon: a measure of ecosystem response in a changing world? *Canadian Journal of Soil Science* 85, 467–480.

Janzen, H.H., Angers, D.A., Boehm, M., Bolinder, M., Desjardins, R.L., Dyer, J., Ellert, B.H., Gibb, D.J., Gregorich, E.G., Helgason, B.L., Lemke, R.L., Massé, D., McGinn, S.M., McAllister, T.A., Newlands, N., Pattey, E., Rochette, P., Smith, W., VandenBygaart, A.J. and Wang, H. (2006) A proposed approach to estimate and reduce net greenhouse gas emissions from whole farms. *Canadian Journal of Soil Science* 86, 401–418.

Jobbágy, E.G. and Jackson, R.B. (2000) The vertical distribution of soil organic carbon and its relation to climate and vegetation. *Ecological Applications* 10, 423–436.

Kaharabata, S.K., Drury, C.F., Priesack, E., Desjardins, R.L., McKenney, D.J., Tan, C.S. and Reynolds, D. (2003) Comparing measured and expert-N predicted N_2O emissions from conventional till and no till corn treatments. *Nutrient Cycling in Agroecosystems* 66, 107–118.

Kessavalou, A., Mosier, A.R., Doran, J.W., Drijber, R.A., Lyon, D.J. and Heinemeyer, O. (1998) Fluxes of carbon dioxide, nitrous oxide, and methane in grass sod and winter wheat-fallow tillage management. *Journal of Environmental Quality* 27, 1094–1104.

King, G.M. and Schnell, S. (1994) Effect of increasing atmospheric methane concentration on ammonium inhibition of soil methane consumption. *Nature* 370, 282–284.

Knorr, W., Prentice, I.C., House, J.I. and Holland, E.A. (2005) Long-term sensitivity of soil carbon turnover to warming. *Nature* 433, 298–301.

Kupusta, G., Krausz, R.F. and Matthews, J.L. (1996) Corn yield is equal in conventional, reduced and no tillage after 20 years. *Agronomy Journal* 88, 812–817.

Lal, R. (2003) Global potential of soil carbon sequestration to mitigate the greenhouse effect. *Critical Reviews in Plant Science* 22, 151–184.

Lal, R. (2004a) Soil carbon sequestration to mitigate climate change. *Geoderma* 123, 1–22.

Lal, R. (2004b) Agricultural activities and the global carbon cycle. *Nutrient Cycling in Agroecosystems* 70, 103–116.

Lal, R. (2004c) Soil carbon sequestration impacts on global climate change and food security. *Science* 304, 1623–1627.

Lemke, R.L., Izaurralde, R.C., Nyborg, M. and Solberg, E.D. (1999) Tillage and N-source influence soil-emitted nitrous oxide in the Alberta Parkland region. *Canadian Journal of Soil Science* 79, 15–24.

Linn, D.M. and Doran, J.W. (1984) Effect of water-filled pore space on carbon dioxide and nitrous oxide production in tilled and non tilled soils. *Soil Science Society of America Journal* 48, 1267–1272.

Lubchenco, J. (1998) Entering the century of the environment: a new social contract for science. *Science* 279, 491–497.

Ludwig, D., Mangel, M. and Haddad, B. (2001) Ecology, conservation, and public policy. *Annual Review of Ecology and Systematics* 32, 481–517.

Lupwayi, N.Z., Rice, W.A. and Clayton, G.W. (1999) Soil microbial biomass and carbon dioxide flux under wheat as influenced by tillage and crop rotation. *Canadian Journal of Soil Science* 79, 273–280.

Lupwayi, N.Z., Clayton, G.W., O'Donovan, J.T., Harker, K.N., Turkington, T.K. and Rice, W.A. (2004) Decomposition of crop residues under conventional and zero tillage. *Canadian Journal of Soil Science* 84, 403–410.

Machado, P.L.O. and de Silva, C.A. (2001) Soil management under no-tillage systems in the tropics with special reference to Brazil. *Nutrient Cycling in Agroecosystems* 61, 119–130.

MacKenzie, A.F., Fan, M.X. and Cadrin, F. (1998) Nitrous oxide emission in three years as affected by tillage, corn-soybean-alfalfa rotations, and nitrogen fertilization. *Journal of Environmental Quality* 27, 698–703.

Malhi, S.S. and McGill, W.B. (1982) Nitrification in three Alberta soils: effect of temperature, moisture, and substrate concentration. *Soil Biology and Biochemistry* 14, 393–399.

Malhi, S.S., McGill, W.B. and Nyborg, M. (1990) Nitrate losses in soils: effect of temperature, moisture and substrate concentration. *Soil Biology and Biochemistry* 22, 733–737.

Malhi, S.S., Lemke, R.L., Wang, Z.H. and Chhabra, B.S. (2006) Tillage, nitrogen and crop residue effects on crop yield and nutrient uptake, soil quality and greenhouse gas emissions. *Soil and Tillage Research* 90, 171–183.

Mann, L.K. (1986) Changes in soil carbon storage after cultivation. *Soil Science* 142, 279–288.

Mosier, A., Schimel, D., Valentine, D., Bronson, K. and Parton, W. (1991) Methane and nitrous oxide fluxes in native, fertilized and cultivated grasslands. *Nature* 350, 330–332.

Mosier, A.R., Halvorson, A.D., Peterson, G.A., Robertson, G.P. and Sherrod, L. (2005) Measurement of net global warming potential in three agroecosystems. *Nutrient Cycling in Agroecosystems* 72, 67–76.

Nelson, D.W. (1982) Gaseous losses of nitrogen other than through denitrification. In: Stevenson, F.J. *et al.* (eds) *Nitrogen in Agricultural Soils*. Soil Science Society of America, Madison, Wisconsin, pp. 327–363.

Nyborg, M., Solberg, E.D., Izaurralde, R.C., Malhi, S.S. and Molina-Ayala, M. (1995) Influence of long-term tillage, straw and N fertilizer on barley yield, plant-N uptake and soil-N balance. *Soil and Tillage Research* 36, 165–174.

Ogle, S.M., Breidt, F.J. and Paustian, K. (2005) Agricultural management impacts on soil organic carbon storage under moist and dry climatic conditions of temperate and tropical regions. *Biogeochemistry* 72, 87–121.

Ojima, D.S., Valentine, D.W., Mosier, A.R., Parton, W. J. and Schimel, D.S. (1993) Effect of land-use change on methane oxidation in temperate forest and grassland soils. *Chemosphere* 26, 675–685.

Pacala, S. and Socolow, R. (2004) Stabilization wedges: solving the climate problem for the next 50 years with current technologies. *Science* 305, 968–972.

Parkin, T.B., Sexstone, A.L. and Tiedje, J.M. (1985) Adaptation of denitrifying populations to low soil pH. *Applied and Environmental Microbiology* 49, 1053–1056.

Paustian, K., Andren, O., Janzen, H.H., Lal, R., Smith, P., Tian, G., Tiessen, H., Van Noordwijk, M. and Woomer, P.L. (1997) Agricultural soils as a sink to mitigate CO_2 emissions. *Soil Use Management* 13, 230–244.

Paustian, K., Cole, C.V., Sauerbeck, D. and Sampson, N. (1998) CO_2 mitigation by agriculture: an overview. *Climatic Change* 40, 135–162.

Paustian, K., Six, J., Elliot, E.T. and Hunt, H.W. (2000) Management options for reducing CO_2 emissions from agricultural soils. *Biogeochemistry* 48, 147–163.

Prieme, A., Christensen, S., Dobbie, K.E. and Smith, K.A. (1997) Slow increase in rate of methane oxidation in soils with time following land-use change from arable agriculture to woodland. *Soil Biology and Biochemistry* 29, 1269–1273.

Puget, P. and Lal, R. (2005) Soil organic carbon and nitrogen in a Mollisol in central Ohio as affected by tillage and land use. *Soil and Tillage Research* 80, 201–213.

Robertson, G.P. and Grace, P.R. (2004) Greenhouse gas fluxes in tropical and temperate agriculture: the need for a full-cost accounting of global warming potentials. *Environment, Development, and Sustainability* 6, 51–63.

Robertson, G.P. and Tiedje, J.M. (1987) Nitrous oxide sources in aerobic soils: nitrification, denitrification, and other biological processes. *Soil Biology and Biochemistry* 19, 187–193.

Royal Society (2001) The role of land carbon sinks in mitigating global climate change. *The Royal Society (Science Advice Section)*, London.

Sá, J.C. de M., Cerri, C.C., Dick. W.A., Lal, R., Filho, S.P.V., Piccolo, M.C. and Feigl, B.E. (2001) Organic matter dynamics and carbon sequestration rates for a tillage chronosequence in a Brazilian Oxisol. *Soil Science Society of America Journal* 65, 1486–1499.

Schils, R.L.M., Verhagen, A., Aarts, H.F.M. and Šebek, L.B.J. (2005) A farm level approach to define successful mitigation strategies for GHG emissions from ruminant livestock systems. *Nutrient Cycling in Agroecosystems* 71, 163–175.

Six, J., Feller, C., Denef, K., Ogle, S.M., Sá, J.C. de M. and Albrecht, A. (2002) Soil organic matter, biota and aggregation in temperate and tropical soils: effects of no-tillage. *Agronomie* 22, 755–775.

Six, J., Ogle, S.M., Breidt, F.J., Conant, R.T., Mosier, A.R. and Paustian, K. (2004) The potential to mitigate global warming with no-tillage management is only realized when practised in the long term. *Global Change Biology* 10, 155–160.

Smith, K.A. and Conen, F. (2004) Impacts of land management on fluxes of trace greenhouse gases. *Soil Use and Management* 20, 255–263.

Smith, K.A., Dobbie, K.E., Ball, B.C., Bakken, L.R., Sitaula, B.K., Hansen, S., Brumme, R., Borken, W., Christensen, S., Priemé, Â., Fowler, D., MacDonald, J.A., Skiba, U., Klemedtsson, L., Kasimir-Klemedtsson, A., Degórska, A. and Orlanski, P. (2000) Oxidation of atmospheric methane in Northern European soils, comparison with other ecosystems, and uncertainties in the global terrestrial sink. *Global Change Biology* 6, 791–803.

Smith, P. (2004a) Carbon sequestration in croplands: the potential in Europe and the global context. *European Journal of Agronomy* 20, 229–236.

Smith, P. (2004b) Soils as carbon sinks: the global context. *Soil Use and Management* 20, 212–218.

Smith, P., Powlson, D.S., Glendining, M.J. and Smith, J.U. (1998) Preliminary estimates of the potential for carbon mitigation in European soils through no-till farming. *Global Change Biology* 4, 679–685.

Soussana, J.-F., Loiseau, P., Vuichard, N., Ceschia, E., Balesdent, J., Chevallier, T. and Arrouays, D. (2004) Carbon cycling and sequestration opportunities in temperate grasslands. *Soil Use and Management* 20, 219–230.

Steudler, P.A., Bowden, R.D., Melillo, J.M. and Aber, J.D. (1989) Influence of nitrogen fertilization on methane uptake in temperate forest soils. *Nature* 341, 314–316.

Stevens, R.J., Laughlin, R.J., Burns, L.C., Arah, J.R.M. and Hood, R.C. (1997) Measuring the contributions of nitrification and denitrification to the flux of nitrous oxide from soil. *Soil Biology and Biochemistry* 29, 139–151.

Stevens, R.J., Laughlin, R.J. and Malone, J.P. (1998) Soil pH affects the processes reducing nitrate to nitrous oxide and di-nitrogen. *Soil Biology and Biochemistry* 30, 1119–1126.

Striegl, R.G., McConnaughey, T.A., Thorstenson, D.C., Weeks, E.P. and Woodward, J.C. (1992) Consumption of atmospheric methane by desert soils. *Nature* 357, 145–147.

VandenBygaart, A.J. and Kay, B.D. (2004) Persistence of soil organic carbon after plowing a long-term no-till field in southern Ontario, Canada. *Soil Science Society of America Journal* 68, 1394–1402.

VandenBygaart, A.J., Gregorich, E.G. and Angers, D.A. (2003) Influence of agricultural management on soil organic carbon: a compendium and assessment of Canadian studies. *Canadian Journal of Soil Science* 83, 363–380.

Venterea, R.T., Burger, M. and Spokas, K.A. (2005) Nitrogen oxide and methane emissions under varying tillage and fertilizer management. *Journal of Environmental Quality* 34, 1467–1477.

West, T.O. and Marland, G. (2002) A synthesis of carbon sequestration, carbon emissions, and net carbon flux in agriculture: comparing tillage practices in the United States. *Agriculture Ecosystems and Environment* 91, 217–232.

West, T.O. and Post, W.M. (2002) Soil organic carbon sequestration rates by tillage and crop rotation: a global data analysis. *Soil Science Society of America Journal* 66, 1930–1946.

Whalen, S.C. and Reeburgh, W.S. (1990) Consumption of atmospheric methane by tundra soils. *Nature* 346, 160–162.

Willison, T.W., Webster, C.P., Goulding, K.W.T. and Powlson, D.S. (1995) Methane oxidation in temperate soils: effects of land use and the chemical form of nitrogen fertilizer. *Chemosphere* 30, 539–546.

Wrage, N., Velthof, G.L., van Beusichem, M.L. and Oenema, O. (2001) Role of nitrifier denitrification in the production of nitrous oxide. *Soil Biology and Biochemistry* 33, 1723–1732.

Yavitt, J.B., Downey, D.M., Lancaster, E. and Lang, G.E. (1990) Methane consumption in decomposing Sphagnum-derived peat. *Soil Biology and Biochemistry* 22, 441–447.

Zentner, R.P., McConkey, B.G., Stumborg, M.A., Campbell, C.A. and Selles, F. (1998) Energy performance of conservation tillage management for spring wheat production in the brown soil zone. *Canadian Journal of Plant Science* 78, 553–563.

Zentner, R.P., Lafond, G.P., Derksen, D.A., Nagy, C.N., Wall, D.D. and May, W.E. (2004) Effects of tillage method and crop rotations on non-renewable energy use efficiency for a thin black chernozem in the Canadian prairies. *Soil and Tillage Research* 77,125–136.

6 Geological Carbon Sinks

Andy Ridgwell[1] and Ursula Edwards[2]
[1]*School of Geographical Sciences, University of Bristol, Bristol, UK;* [2]*Occidental Oil and Gas Corporation, Houston, Texas, USA*

6.1 Introduction

The sequestering (locking up) of carbon in geological formations and removal of carbon dioxide (CO_2) from the atmosphere is not a unique, human-driven invention thought up for ameliorating (reducing) the degree of greenhouse gas-driven climate change in the future. CO_2 has been spewing from volcanoes on land and the spreading ridges of the ocean throughout geological time. Yet, CO_2 levels are not thought to have risen inexorably since the Earth was formed some 4.5 billion years ago. Quite the opposite – geological evidence suggests that we live in a period in which atmospheric CO_2 concentrations are probably amongst the lowest to have occurred on Earth, at least for the past 600 million years (Royer *et al.*, 2004) (Fig. 6.1a). Indeed, compared to the 'pre-Industrial' atmosphere (i.e. immediately prior to the industrial revolution *c.* 1765 and the onset of industrialization and increasingly rapid fossil fuel consumption), which was characterized by a CO_2 concentration of 278 ppm (Enting *et al.*, 1994), geological periods such as the Jurassic (200–145 million years ago) and Devonian (416–359 million years ago) saw about ten times as much carbon residing in the atmosphere. The ocean carbon reservoir would also have been much larger at times in the past (Ridgwell, 2005) (Fig. 6.1b). But where has this carbon gone? There must exist greenhouse gas sinks that are able not only to sequester large amounts of carbon but also to keep a tight hold of it for extended periods of time. The amount of carbon stored in vegetation and soils has varied significantly through time, with as much as ~1000 Pg C less at the height of the last glacial period associated with major shifts in vegetation type and coverage, compared to the ~2200 Pg C in the modern terrestrial biosphere (Fig. 6.2). However, the fall in atmospheric CO_2 between the Devonian and Carboniferous (369–299 million years ago) equates to ~5000 Pg C reduction in carbon stored in the atmosphere (and a considerably greater reduction in the ocean inventory) (Fig. 6.1a). It is not easy to envisage how the terrestrial biosphere could possibly have accommodated this increase in carbon storage. The terrestrial biosphere is also not a 'safe' long-term store for carbon. For instance, the devastation wrought by the Indonesian wildfires of 1997 has been estimated to have resulted in the release of 0.8–2.6 Pg C (Page *et al.*, 2002). This is equivalent to 13–40% of the annual emissions from anthropogenic fossil fuel combustion, and could help explain why the growth rate of CO_2 in the atmosphere approximately doubled during the 1997–1998 period (Schimel and Baker, 2002).

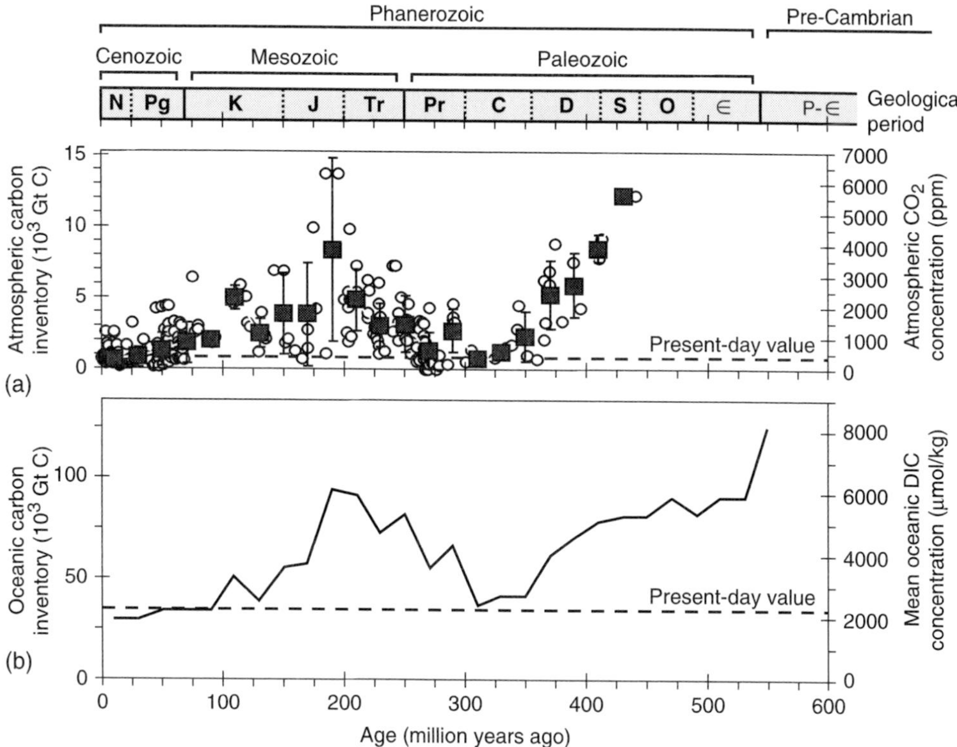

Fig. 6.1. Evolution of atmospheric and oceanic carbon reservoirs through time. (a) Phanerozoic evolution of atmospheric CO_2 reconstructed from proxy records (Royer *et al*., 2004). The filled squares show the data binned into intervals of 20 million years, with one standard deviation of the error shown as a vertical black line for each point. The raw proxy data are plotted as open circles. There are two vertical scales; atmospheric concentration (right) and the corresponding total carbon inventory (left). (b) Model-estimated evolution of the ocean carbon reservoir (Ridgwell, 2005). The vertical scales are mean dissolved inorganic carbon (DIC) concentration (right) and total ocean carbon inventory (left). The horizontal lines in both panels indicate the size of the present-day carbon inventories. The geological timescale abbreviations for the periods are: D, Devonian; C, Carboniferous; Pr, Permian; J, Jurassic; N, Neogene; P, Paleogene; K, Cretaceous; S, Silurian; O, Ordovician; ∈, Cambrian; P-∈, Pre-Cambrian.

Clues as to the ultimate fate of CO_2 released to the atmosphere lie in the rocks around us. Fossil fuel deposits such as coal measures, oil and gas reservoirs, as well as oil shales and other organic matter rich sedimentary rocks, all hold substantial quantities of carbon (Fig. 6.2). The relationship of these reservoirs to atmospheric CO_2 is conceptually fairly straightforward – sequestration of organic matter in geological formations must result in less carbon in the atmosphere and ocean. Past increases in organic carbon burial driven by evolutionary and tectonic factors have been linked to decreases in atmospheric CO_2, particularly the CO_2 'trough' during the Carboniferous and Permian periods (Berner, 1990) (Fig. 6.1a). We are all too familiar with the converse link: the burning of deposits of ancient carbon and increasing CO_2 concentrations in the atmosphere. However, if the rate of burial of organic matter were to increase in the future, perhaps in response to climate change, there would presumably be an additional removal of fossil fuel CO_2 from the atmosphere. Is this likely to occur, and how important might this be? To answer this question we are going to have to look at how organic matter is deposited and preserved in accumulating sediments.

The formation of carbonate rocks such as limestones (the remains of ancient reefs) and

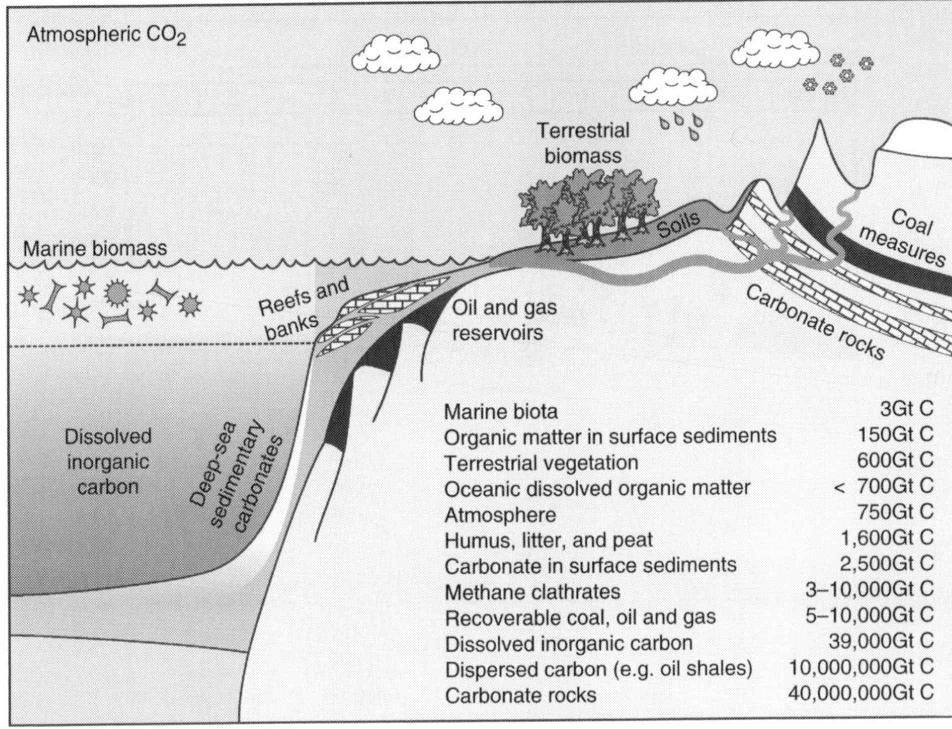

Fig. 6.2. Estimated inventories of the various carbon reservoirs on Earth. (Adapted from Kump *et al.*, 2004.) Units are in Gt C; 1 Gt C = 1 Pg (10^{15} g) of carbon or about 8.3° × 10^{13} mol carbon. To put this into some perspective, the 750 Gt C in the atmosphere is equivalent to an average concentration of CO_2 of 351 ppm, and the average emission of anthropogenic CO_2 (from fossil fuels and cement production) during the 1990s was 6.3 Gt C/year. (From Houghton *et al.*, 2001.)

chalks (the microscopic shells of dead calcifying marine plankton) also represents a sink of carbon from the Earth's surface. This turns out to be an even more important reservoir of carbon than fossil fuels and other forms of ancient organic matter (Fig. 6.2). The relationship between carbonate deposition and atmospheric CO_2 is also less straightforward – when marine carbonates are precipitated from solution, the concentration of CO_2 in the atmosphere actually goes up. How the burial of carbonate can at the same time be a geologic sink for carbon needs some explanation. We also need to quantify how this sink might change in the future, and whether it will be important on the (century and shorter) timescales that matter to us most.

In this chapter we examine the role played by 'geologic' sinks for fossil fuel CO_2 emitted to the atmosphere. We tell the inorganic (carbonate) carbon and organic carbon sides of the story separately (in Sections 6.2 and 6.3, respectively), and look in detail at the underlying mechanisms involved. We also consider how human (anthropogenic) activities and climate change may perturb these processes in ways that will affect the rate at which fossil fuel CO_2 is removed from the atmosphere. Finally, we provide a summary and perspective in Section 6.4. Box 6.1 contains a basic '101' tutorial on carbonate chemistry and the key geochemical reactions involved in the geologic carbon sink (see Zeebe and Wolf-Gladrow, 2001 for a more detailed primer).

Box 6.1. Carbonate chemistry '101'.

The mineral calcium carbonate ($CaCO_3$) has a crystal lattice consisting of calcium ions (Ca^{2+}) ionically bound to carbonate ions (CO_3^{2-}). The lattice can take one of several different 'polymorphic' forms (i.e. the same chemical composition but different crystalline structure) such as *calcite*, or a higher symmetry *aragonite* phase. Calcite is the more abundant of the two polymorphs that are biologically precipitated in the open ocean. Because it is also more thermodynamically stable than aragonite it is the phase responsible for almost all carbonate burial in the deep sea. In contrast, aragonite is abundant amongst shallow water carbonates (e.g. corals). Biogenic carbonates are not always pure $CaCO_3$ and a range of substitutions of magnesium (Mg^{2+}) for Ca^{2+} are possible in the crystal structure to give natural carbonates a generic formula: $Mg_x \times Ca_{(1-x)} \times CO_3$. We will, however, focus on the more abundant $CaCO_3$ end member here.

Precipitation of calcium carbonate may be described by the following reaction:

$$Ca^2 + 2HCO_3^- \rightarrow CaCO_3 + CO_{2(aq)} + H_2O$$

Because Ca^{2+} has a 'residence' time in the ocean counted in millions of years, we can assume that the concentration of Ca^{2+} does not change on the (100–100,000 year) timescales we are interested in, and that ocean mixing homogenizes its concentration throughout the ocean. The *bicarbonate ion* (HCO_3^-) required in the precipitation reaction is formed through the hydration and dissolution of CO_2 gas ($CO_{2(g)}$) to form a proton (H^+) and a bicarbonate ion (HCO_3^-):

$$H_2O + CO_{2(aq)} \rightarrow H^+ + HCO_3^-$$

(The first, hydration step is the formation of *carbonic acid* (H_2CO_3), but this is only present in very small concentrations and is commonly ignored.) A fraction of the HCO_3^- dissociates to form a *carbonate ion* (CO_3^{2-}) and another proton:

$$HCO_3^- \rightarrow H^+ + CO_3^{2-}$$

The sum total of carbon in all its dissolved (inorganic) forms (i.e. $CO_{2(aq)} + HCO_3^- + CO_3^{2-}$) is collectively termed *dissolved inorganic carbon* (DIC). The relationship between these different forms of DIC, and to hydrogen ion (H^+) concentrations and to pH is shown in Fig. 6.3.

When $CaCO_3$ is precipitated from solution, although the total sum of DIC is reduced, the remaining carbon is re-partitioned in favour of the $CO_{2(aq)}$ species. One way of thinking about this is in terms of removing CO_3^{2-} and shifting the aqueous carbonate equilibrium reaction:

$$CO_{2(aq)} + CO_3^{2-} + H_2O \rightarrow 2HCO_3^-$$

to the left to compensate. The counterintuitive consequence of this is that the precipitation of carbonate drives an *increase* in the *partial pressure* of CO_2 (pCO_2) in the surface ocean, despite there being a reduction in total carbon (DIC). (pCO_2 is the variable that determines the exchange of CO_2 between ocean and atmosphere – if atmospheric pCO_2, which is equal to the CO_2 molar ratio at a surface pressure of 1 atmosphere is greater than the pCO_2 of the surface ocean, there will be a net transfer of CO_2 from the atmosphere to the ocean, and vice versa.)

Whether $CaCO_3$ precipitates or dissolves is dictated by the stability of its crystal structure relative to the ambient environmental conditions. This can be directly related to the concentrations of Ca^{2+} and CO_3^{2-} and written in terms of the 'saturation state' (also known as the solubility ratio) Ω of the solution, defined as:

$$\Omega = (Ca^{2+}) \times (CO_3^{2-}) / K_{sp}$$

where K_{sp} is a solubility constant. The precipitation of calcium carbonate from sea water is thermodynamically favourable when Ω is greater than unity. Conversely, $CaCO_3$ will tend to dissolve at $\Omega < 1.0$. In addition to the concentrations of Ca^{2+} and CO_3^{2-}, depth is also important because K_{sp} scales with increasing pressure (as well as with decreasing temperature). Thus, the greater the depth in the ocean, the more the ambient environment will tend to be undersaturated (i.e. $\Omega < 1.0$) and the less likely that carbonate will be present in the sediments.

continued

Box 6.1. *Continued*

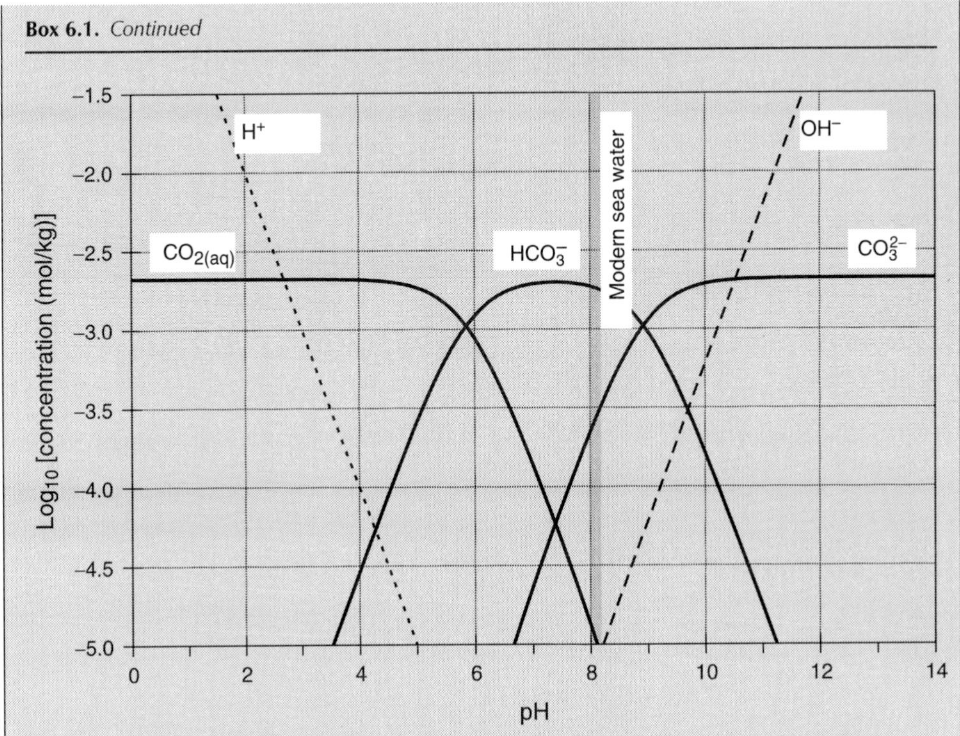

Fig. 6.3. The concentrations of the dissolved carbonate species as a function of pH (referred to as the Bjerrum plot; cf. Zeebe and Wolf-Gladrow, 1999): dissolved carbon dioxide ($CO_{2(aq)}$), bicarbonate (HCO_3^-), carbonate ion (CO_3^{2-}), hydrogen ion (H^+) and hydroxyl ion (OH^-). At modern sea water pH, most of the dissolved inorganic carbon is in the form of bicarbonate.

6.2 The Inorganic (Carbonate) Carbon Sedimentary Sink for Fossil Fuel CO_2

To see where carbonate rocks come into the greenhouse sink picture, we recap on the sequence of different fates that befall CO_2 released to the atmosphere through anthropogenic activities such as the burning of fossil fuels and cement production (Fig. 6.4). Some of the added CO_2 may be relatively quickly removed from the atmosphere and taken up by the terrestrial biosphere as a result of 'CO_2 fertilization' of plant productivity (although nutrient limitation may limit the importance of this effect; see Hymus and Valentini, Chapter 2, this volume) as well as forest regrowth and changes in land use practice. Current estimates suggest that 100–180 Pg C may already have been removed in this way, equivalent to 28–50% of total emissions from fossil fuels and cement production (Sabine *et al.*, 2004). The timescale for this CO_2 sink to operate is years to decades (for the aboveground vegetation response) to centuries (for the soil carbon inventory to adjust). At the same time, CO_2 dissolves in the surface ocean. If the ocean did not circulate or mix and if dissolved CO_2 remained as $CO_{2(aq)}$ (see Box 6.1), the ocean surface would quickly come into equilibrium with the atmosphere without having absorbed much anthropogenic CO_2. Fortunately, neither ocean dynamics nor CO_2 chemistry is simple, making the ocean a powerful carbon sink (Sabine and Feely, Chapter 3, this volume). However, once fossil fuel emissions to the atmosphere have ceased, sequestration by this means, termed 'ocean invasion', cannot continue indefinitely. First, the ocean becomes less efficient at storing additional dissolved carbon at higher atmospheric CO_2 concen-

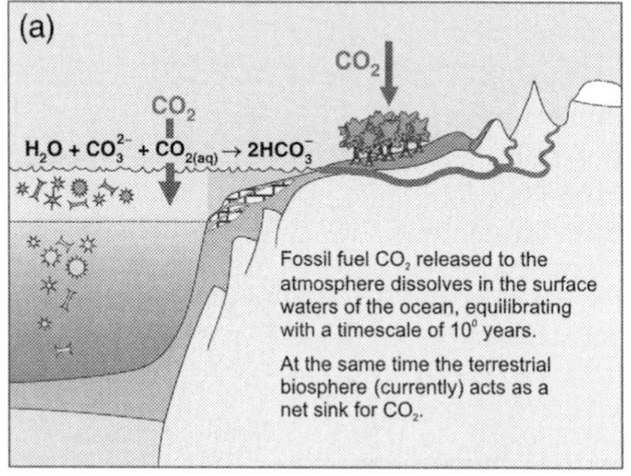

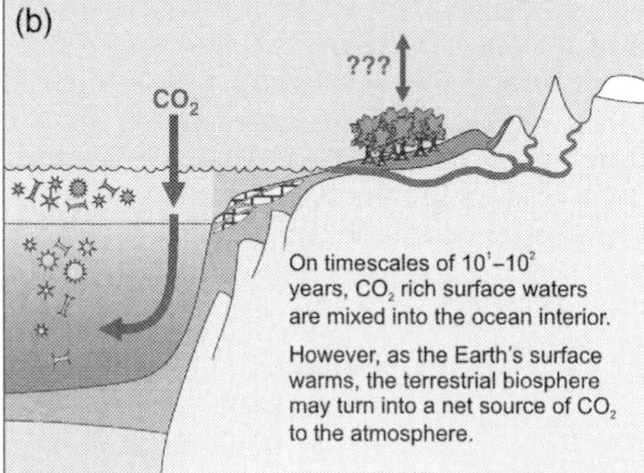

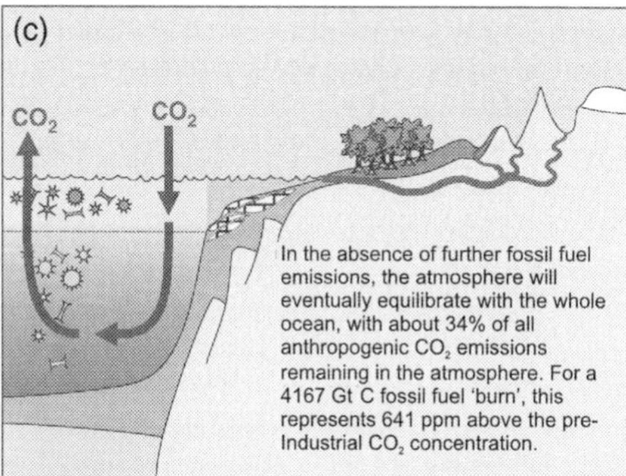

Fig. 6.4. Mechanisms of carbon sequestration (I). Panels (a) through (c) illustrate the pathways of carbon uptake operating on timescales of years (10^1) to centuries (10^2) – ocean invasion and 'CO_2 fertilization' of the terrestrial biosphere.

trations. Secondly, once CO_2-enriched surface waters have been mixed down to depth and throughout the ocean, outgassing of previously absorbed CO_2 will then tend to balance the rate of uptake from the atmosphere. Once this happens, the removal of anthropogenic CO_2 by the ocean will cease and the ocean and atmosphere can then be said to be in 'equilibrium'.

We use a computer model of ocean–atmosphere carbon cycling (see Box 6.2 for details) to illustrate the importance of ocean invasion and the processes discussed in Chapter 3. This will give us something of a benchmark with which to compare the relative importance of the geologic sinks. The predicted evolution of atmospheric CO_2 in response to a future CO_2 emissions trajectory (Fig. 6.5a) and the operation of the ocean invasion sink is shown in Fig. 6.5b. For a total release (burn) of 4167 Pg C (Fig. 6.5a) and with global climate held in the modern state (i.e. global temperatures are not allowed to respond to rising CO_2), the final atmospheric CO_2 concentration reached is 919 ppm. This is equivalent to 2007 Pg C; ~1400 Pg C more than the amount prior to the Industrial Revolution (in *c.* 1765). Thus, the ocean has taken up a little over 66% of the total release, storing it mainly in the form of bicarbonate ions (HCO_3^-) (in Fig. 6.5c), with the remainder in the atmosphere. We have not taken into account any net uptake (or release) by the terrestrial biosphere in this calculation.

One caveat to this assessment is that the fraction of fossil fuel CO_2 that is sequestered by ocean invasion actually declines with increasing total burn. In other words, if we had chosen a smaller quantity of carbon than 4167 Pg C, the proportion taken up by ocean invasion would be greater than 66%. Ocean invasion was found to account for 80.7% when the fossil fuel release was just 874 Pg C, but 69.7% when the fossil fuel release was 4550 Pg C in an ocean carbon cycle general circulation model (GCM) (Archer *et al.*, 1997, 1998). The results of the high-end CO_2 release of Archer *et al.* (1998) are thus comparable to what we obtain here, with the ~3% difference reflecting variations in the representation of ocean circulation and marine carbon cycling between the two models. A second caveat is that no change in ocean circulation, marine biological productivity or surface temperatures has yet been taken into account.

Because CO_2 solubility decreases at higher temperatures (Zeebe and Wolf-Gladrow, 2001), warming of the surface ocean due to a

Box 6.2. The 'genie' carbon cycle model.

The details of the computer model used to illustrate the different pathways and mechanisms for CO_2 sequestration are not central to the discussions in this chapter: 'To retain respect for either models of sausages, one must not watch too carefully what goes into either of them' (Ridgwell, 2001, adapted from a remark attributed to Otto Von Bismarck, 1st Chancellor of the German Empire, 1871–1890). However, we include a brief overview of the model for completeness.

To quantify the uptake of atmospheric CO_2 by the ocean we use the GENIE-1 (Ridgwell *et al.*, 2006a) coupled carbon–climate model, developed as part of the 'genie' Earth system modelling initiative (www.genie.ac.uk). The climate model component is fully described in Edwards and Marsh (2005, and references therein). In a nutshell, it is a coarse (i.e. low) resolution of a 'frictional geostrophic' general circulation model (Edwards and Shepherd, 2002) coupled to a 2D energy/moisture balance model atmospheric component (Weaver *et al.*, 2001) including a simple thermodynamic and dynamic representation of sea ice. An 'ensemble Kalman filter' has been used to calibrate this model and thereby achieve a reasonable simulation of the modern climate (Hargreaves *et al.*, 2004).

The (ocean) biogeochemical component of the GENIE-1 model calculates the (mainly vertical) redistribution of tracer concentrations occurring rapidly relative to transport by the large-scale circulation of the ocean. This happens through the removal from solution of nutrients (PO_4) together with dissolved inorganic carbon (DIC) and alkalinity (ALK) in the sunlit surface ocean layer (euphotic zone) by biological activity. The resulting export of particulate matter to the ocean interior is subject to remineralization processes, releasing dissolved constituent species back to the ocean (but at greater depth). Further redistribution of tracers occurs through gas exchange with the atmosphere as well as due to the creation and destruction of dissolved organic matter. An ensemble Kalman filter is also used to calibrate the biogeochemical model (and reproduce the observed 3D distributions of phosphate and alkalinity in the ocean) (Ridgwell *et al.*, 2006a).

To quantify the importance of carbonate burial and the role of the 'geologic' carbon sink, the GENIE-1 model is further extended by including a representation of the geochemical interaction between the ocean and deep-sea sediments (Ridgwell, 2001). This extension calculates the fraction (if any) of $CaCO_3$ reaching the ocean floor that is preserved and buried in the sediments, and described in full in Ridgwell and Hargreaves (in press). It also calculates the amount (and rate) of carbonate previously buried in the sediments that can be dissolved to neutralize fossil fuel CO_2.

The GENIE-1 model is uniquely suited for the analysis of the long-term fate of fossil fuel CO_2 because it can simulate over 1000 years in less than 1 h of CPU time, and achieves this speed on a 'normal' Linux-based PC. (Some much higher-resolution and more detailed climate models would literally take a year of supercomputer time to do this.) Another important feature is that climate can interact with the carbon cycle (i.e. climate is responsive to changes in atmospheric CO_2), allowing the importance of 'feedbacks' to be quantified (e.g. Ridgwell *et al.*, 2006b). For instance, if the positive feedback between increasing atmospheric CO_2 and sea surface temperature warming is not taken into account, the amount of CO_2 taken up from the atmosphere by the ocean could be overpredicted by ~10% (see Fig. 6.5).

CO_2 is added to the atmosphere in the model to simulate anthropogenic CO_2 emissions. We chose a hypothetical time-history of fossil fuel consumption (and combustion) of 4167 Pg C (Fig. 6.5a), similar to the '4kfast' scenario of Lenton (2000). This follows IS92a 'business as usual' to 2100, followed by a linear decline in emissions to use up all 'conventional' fossil fuel reserves (coal, oil, gas) of ~4000 Pg C. The total amount of carbon released to the atmosphere is 4167 Pg C, with 3784 Pg C released from year 2000 onwards. This scenario falls midway between the 'A22' (3028 Pg C) and 'A23' (4550 Pg C) scenarios analysed by Archer *et al.* (1998), and is slightly less than the 5270 Pg C scenario employed by Caldeira and Wickett (2003). If 'exotic' fossil fuels, including methane clathrates, are exploited, total fossil fuel release could be as much as 15,000 Pg C (Hasselmann *et al.*, 2003). Fossil fuel CO_2 emissions to the atmosphere up to year 2000 are estimated from the increase in ocean + atmosphere carbon inventory in the experiment when atmospheric CO_2 was forced to conform to the observed CO_2 concentration trajectory – Enting *et al.* (1990) up until 1994, and Keeling and Whorf (2005) thereafter.

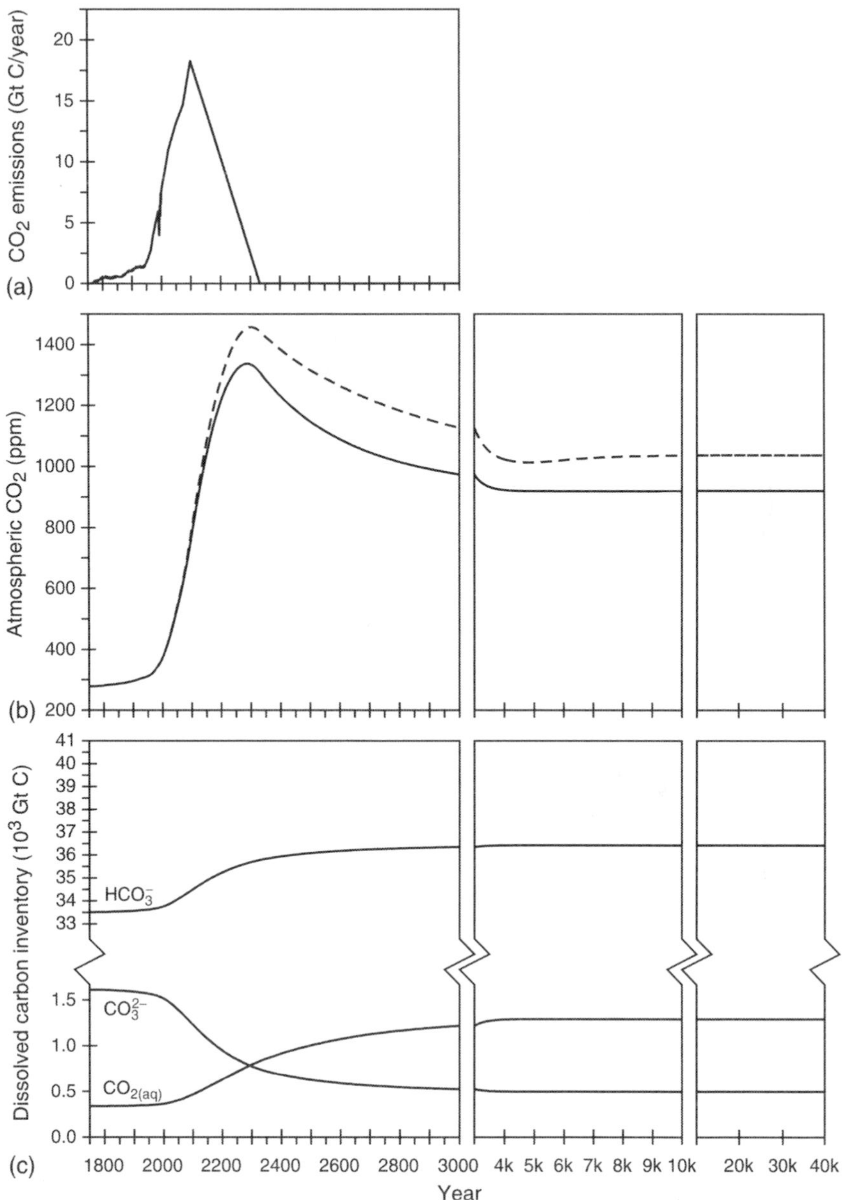

Fig. 6.5. Model analysis of the uptake of fossil fuel CO_2 by the ocean. (a) Time-history of the rate of CO_2 emissions to the atmosphere prescribed in the model (see Box 6.2). From 1765 to 2000, the emissions trajectory is deduced to be consistent with the observed trajectory of atmospheric CO_2 (Enting *et al.*, 1994; Keeling and Whorf, 2005). Note that these calculated anthropogenic emissions are *net* of any uptake by the terrestrial biosphere (i.e. if the terrestrial 40 biosphere acts as a carbon sink over this interval, the CO_2 emissions will be underestimated). (b) Model-predicted trajectory of atmospheric CO_2 (assuming no change in the terrrestrial biosphere). The solid line shows predicted atmosphere CO_2 with climate (and ocean surface temperatures) held constant. The dotted line shows the impact of allowing the carbon cycle and climate to interact and produce 'feedback' between CO_2 and surface temperatures. (c) Predicted evolution of the different components of the ocean DIC reservoir: $CO_{2(aq)}$, HCO_3^- and CO_3^{2-}. Note the different *y*-axis scales used for $CO_{2(aq)}$ and CO_3^{2-} (bottom of panel c) and HCO_3^- (top of panel c).

stronger greenhouse effect will also render ocean invasion a less effective sink (Plattner *et al.*, 2001; see Chapter 3). This effect creates a 'feedback loop' (see Berner, 1999; Ridgwell, 2003); higher temperatures result in more CO_2 left in the atmosphere, which drives a stronger greenhouse effect causing yet higher temperatures, and so on. In this case the feedback has a 'positive' sign, and acts to amplify the impact of an initial perturbation (in our example, the release of fossil fuel CO_2). If we now take into account feedback between CO_2 and climate in the model, the final (steady-state) fraction of total emissions taken up by the ocean declines to 60%, with a residual 1037 ppm remaining in the atmosphere (Fig. 6.5b). The 2050 weakening of the ocean invasion sink due to feedback between CO_2 and climate in the GENIE-1 model used here (Box 6.2) is ~13%, compared with ~24% found by Plattner *et al.* (2001), most likely reflecting differences in the ocean circulation response to surface warming (and freshening).

With no other carbon sinks operating, our long-term future is thus looking decidedly on the warm side; the mean global ocean surface temperature is 23.1°C, ~4.5°C warmer than the pre-Industrial state of the model (18.6°C). To put this into some perspective, the year 2005 value is 19.2°C in the model; just 0.6°C above the pre-Industrial estimate. This situation would persist indefinitely. There are important implications of the residual atmospheric fossil fuel fraction and degree of long-term greenhouse warming for stability of Greenland and Antarctic ice caps and of methane hydrates present in continental margin sediments (Archer and Buffett, 2004), as well as for the timing of the onset of the next ice age (Archer and Ganopolski, 2005).

6.2.1 Geologic carbon sinks: reaction with sedimentary carbonates

Where does the 'geologic' part come into the picture? In shallow water environments, carbonates are precipitated by corals and benthic shelly animals, with a smaller 'abiotic' contribution occurring as fill-in cements and coatings on mineral grains and biogenic matter. Approximately 0.3 Pg C equivalent of calcium carbonate ($CaCO_3$) is produced annually in these environments (Milliman and Droxler, 1996). Carbonate is also precipitated biologically in the open ocean (i.e. away from the continental shelf) by plankton such as coccolithophores and foraminifera, as well as by pteropods. About 0.8 Pg C/year of $CaCO_3$ is thought to be produced here (Milliman and Droxler, 1996; Feely *et al.*, 2004). What happens to this carbonate?

The shallow waters of the ocean margins are everywhere oversaturated with respect to the solid $CaCO_3$ phase (i.e. the saturation state, $\Omega > 1.0$; see Box 6.1). The dissolution loss of carbonate is therefore relatively small; of the ~0.3 Pg C/year produced, about 0.17 Pg C/year is thought to be buried virtually *in situ* while another 0.04 Pg C/year is exported to the adjoining continental slopes (Milliman, 1993; Milliman and Droxler, 1996). The total neritic accumulation of $CaCO_3$ today is therefore ~0.2 Pg C/year, and thousands of years of build-up of this material has given rise to large-scale topographical features such as barrier reefs and carbonate banks and platforms. The shallow-water accumulation rate encapsulated in these estimates is probably significantly higher than the long-term (glacial–interglacial, or >100,000 years) average because reef growth rates are still adjusting to the rise in sea level that accompanied the termination of the last glacial period (e.g. Ryan *et al.*, 2001; Vecsei and Berger, 2004).

The situation is quite different in the open ocean because oceanic waters become increasingly less saturated at greater depth (and increased pressure). When the ambient environment becomes undersaturated ($\Omega < 1.0$) carbonates will start to dissolve. This occurs at ~4500 m in the Atlantic Ocean and ~3000 m in the Pacific Ocean. At more than 1000 m deeper than this, sediments are typically completely devoid of any carbonate particles. Topographic 'highs' on the ocean floor such as the mid-Atlantic ridge (where the ocean floor is 'only' ~3000 m deep and $\Omega > 1.0$) can thus be picked out by sediments rich in $CaCO_3$ while the adjacent deep basins ($\Omega < 1.0$) are low in $CaCO_3$ content (Fig. 6.6); an effect likened to 'snow-capped

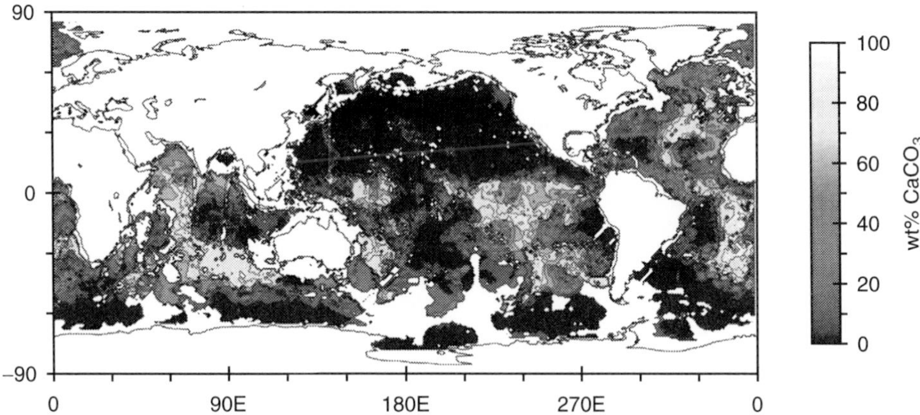

Fig. 6.6. Observed distribution of the calcium carbonate content (as percentage dry weight (wt%) in each sample) of the surface sediments of the deep sea. (From Archer, 1996a.) Areas with no data coverage (parts of the Southern Ocean, and many of the continental margins) are left blank.

mountains'. The situation is actually much more complicated than this and other factors are required to completely explain the whole observed pattern of sediment composition, such as the breakdown of organic matter in surface sediments and the release of metabolic CO_2 into the sediment porewaters (see Archer, 1996b; and Ridgwell and Zeebe, 2005). At a global scale, only ~10–15% (equal to 0.1 Pg C/year) of carbonate produced at the surface ever escapes dissolution to be buried in accumulating deep-sea sediments (Milliman and Droxler, 1996; Archer, 1996b; Feely *et al.*, 2004).

Although $CaCO_3$ burial in shallow waters currently appears to be greater than in deep-water sediments, we will focus on the latter carbon sink. We justify this simplification because carbonate preservation and burial in the deep sea is much more important in regulating atmospheric CO_2 (Archer, 2003). Although it is beyond the scope of this chapter, note that sea-level and evolutionary changes occurring during the Phanerozoic period have altered the balance between shallow- and deep-water $CaCO_3$ burial. This must have had a profound impact on the regulation of ocean chemistry and climate (Ridgwell *et al.*, 2003; Ridgwell, 2005).

Carbonate burial (in the deep ocean) represents a 'geologic' sink of carbon. How does this relate to the long-term fate of fossil fuel CO_2? The first mechanism we will discuss is illustrated in Fig. 6.7a. Dissolution of CO_2 in surface waters results in a decrease in ambient carbonate ion (CO_3^{2-}) concentration (see Figs 6.4c, 6.7c). Think about shifting the aqueous carbonate equilibrium reaction $CO_{2(aq)} + CO_3^{2-} + H_2O \leftrightarrow 2HCO_3^-$ to the right to (partly) compensate for the addition of CO_2. Because a reduction in CO_3^{2-} reduces the stability of carbonates (see Box 6.1), the invasion of fossil fuel CO_2-enriched waters into the deep ocean will drive an increase in the rate of dissolution of $CaCO_3$ in the sediments (Sundquist, 1990; Archer *et al.*, 1997, 1998). If this rate of dissolution overtakes the rate of supply of new biogenic carbonate from above, previously deposited carbonate will start to dissolve (erode) (Fig. 6.8a).

Each mole of $CaCO_3$ that dissolves removes one mole of $CO_{2(aq)}$ to form two moles of bicarbonate:

$$CaCO_3 + CO_{2(aq)} + H_2O \rightarrow Ca^{2+} + 2HCO_3^-$$

(the difference between dotted and dashed HCO_3^- inventory trajectories in Fig. 6.8c). Thus, water masses that have passed over carbonate-rich sediments become, in a sense, 'recharged', and are able to absorb more CO_2 from the atmosphere (Fig. 6.7a). One can think of this as anthropogenic CO_2 being 'neutralized' by the reaction with sedimentary $CaCO_3$. We will refer to this carbon

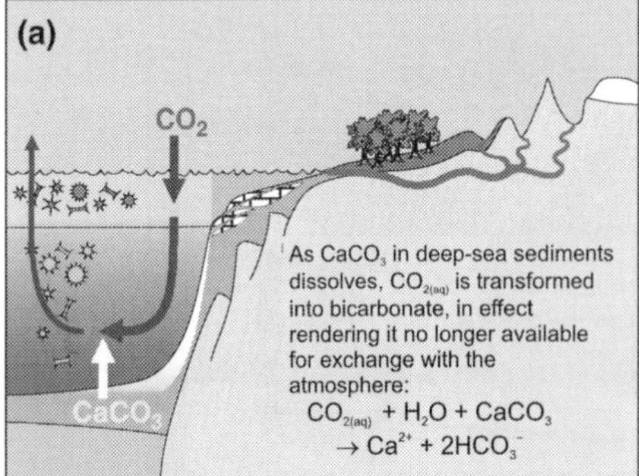

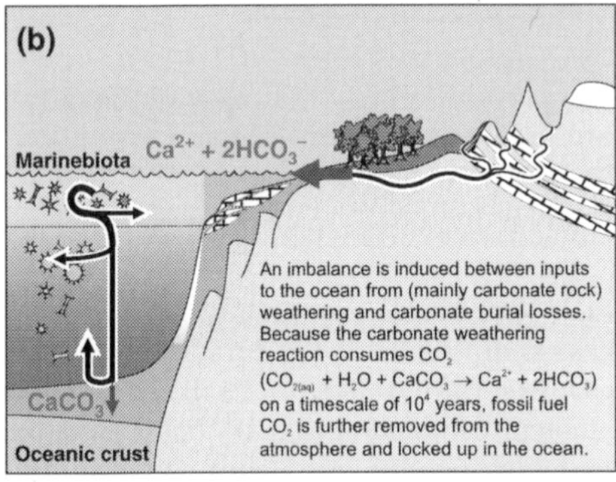

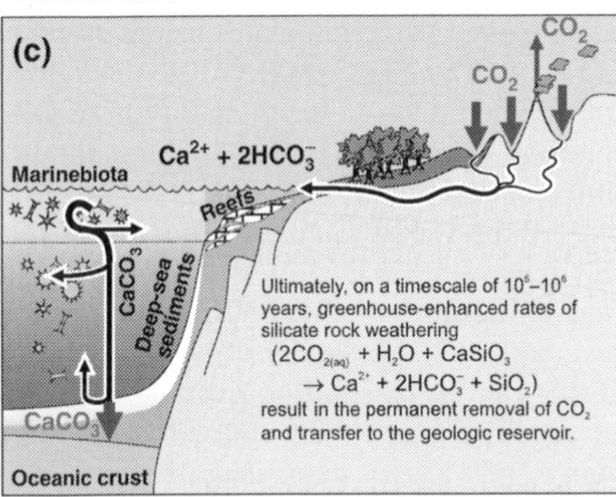

Fig. 6.7. Mechanisms of carbon sequestration (II). Panels (a) through (c) illustrate the pathways of carbon uptake occurring on timescales of millennia (103 years) and beyond – the 'geologic' carbon sinks. (a) Operation of the sea-floor $CaCO_3$ neutralization; (b) operation of terrestrial $CaCO_3$ neutralization; and (c) operation of the silicate weathering carbon sink.

sequestration process as the 'sea-floor $CaCO_3$ neutralization' sink. It should be carefully considered that although the dissolution of sedimentary $CaCO_3$ results in an *increase* in the *total* amount of carbon dissolved in the ocean, the proportion of dissolved inorganic carbon (DIC) that is in the form of $CO_{2(aq)}$ actually decreases (compare dotted and dashed lines in Fig. 6.8c) as a result of reaction with $CaCO_3$. It is the associated reduction in ambient pCO_2 that allows further transfer of CO_2 from the atmosphere to the ocean.

Clearly we need to quantify how much $CaCO_3$ will dissolve from the sediments and what effect it will have on the removal of CO_2 from the atmosphere. We therefore use an extended carbon cycle model that includes the relevant interaction with carbonates in deep-sea sediments (Box 6.2). The evolution of atmospheric CO_2 is shown in Fig. 6.8b. Now, CO_2 is declining slightly faster at year 3000 compared with when the ocean invasion sink is operating alone. It is important to recognize, however, that the peak atmospheric CO_2 value reached (and thus the maximum extent of 'global warming') is virtually unaffected by the inclusion of the buffering of ocean chemistry by carbonate-rich sediments. The effect of sediment dissolution is rather more pronounced over the following few thousand years, and atmospheric CO_2 reaches a new, lower steady state of 715 ppm not long after year 10,000. Thus, neutralization with sea-floor carbonates eventually results in the additional removal of 444 Pg C from the atmosphere, or about 11% of the initial fossil fuel burn that we assumed. Again, the relative importance and fraction of CO_2 sequestered by this mechanism depends on the magnitude of the fossil fuel burn – sea-floor carbonate neutralization has been estimated to account for 9.0% of an 874 Pg C fossil fuel release, rising to 14.8% for a 4550 Pg C release (Archer *et al.*, 1998). Thus, the buffering response in the GENIE-1 model is slightly less than in a previous model study.

6.2.2 Geologic carbon sinks: the weathering of carbonate rocks on land

CO_2 in the atmosphere dissolves in rainwater to form a weak carbonic acid solution, which dissolves carbonate minerals in rocks exposed at the land surface and mineral grains in soils:

$$CO_2 + H_2O + CaCO_3 \rightarrow Ca^{2+} + 2HCO_3^-$$

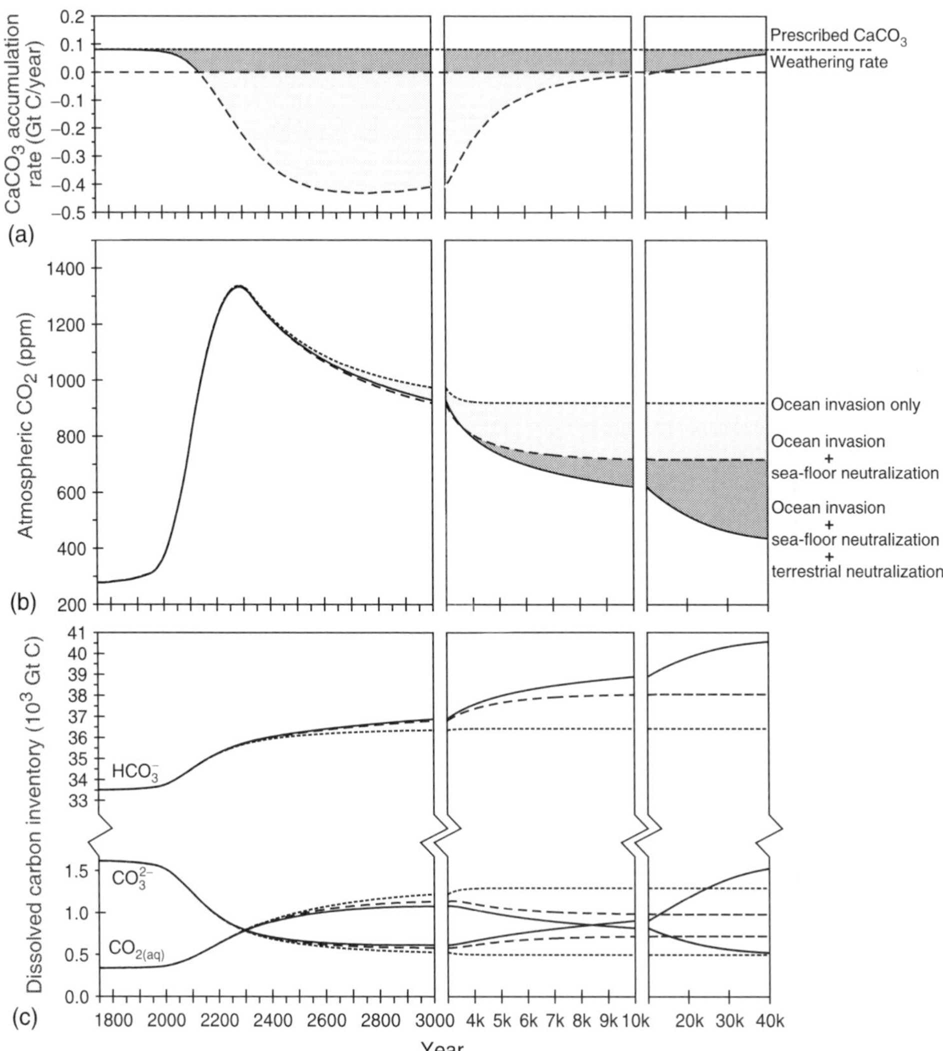

Fig. 6.8. Model analysis of the role of 'geologic' carbon sinks in the sequestration of fossil fuel CO_2. (a) Model-predicted global accumulation rate of $CaCO_3$ in deep-sea sediments. The lighter shaded area under the curve (negative $CaCO_3$ accumulation rate) represents the net erosion of carbonates previously deposited in deep-sea sediments – sea-floor $CaCO_3$ neutralization. The darker shaded region represents periods characterized by a positive accumulation rate of $CaCO_3$, but at a rate lower than the supply by carbonate weathering on land – terrestrial neutralization. (b) Trajectory of atmospheric CO_2. The dotted line represents anthropogenic uptake by the ocean only (same as the curve shown in Fig. 6.5 solid line). The dashed line shows the effect of seafloor neutralization (Fig. 6.7a) in addition to the ocean invasion carbon sink. The lighter shaded region thus indicates the reduction of atmospheric CO_2 due to sea-floor neutralization alone. The solid line shows the effect of terrestrial neutralization (Fig. 6.7b) in addition to ocean invasion and sea-floor neutralization. The darker shaded region thus indicates the reduction of atmospheric CO_2 due to terrestrial neutralization alone. Note that weathering rates are held constant for this experiment, meaning that the ultimate CO_2 sequestration mechanism of silicate weathering (Fig. 6.7c) is not 'switched on'. (c) Evolution of the different components of the ocean dissolved inorganic component (DIC) reservoir: $CO_{2(aq)}$, HCO_3^- and CO_3^{2-}.

The solutes that result from this reaction are carried by rivers to the ocean. All the while that anthropogenic acidification of the ocean is causing carbonate accumulation in the deep ocean to be reduced and then reversed (Fig. 6.8a), the input of solutes derived from carbonate weathering on land continues. The accumulation rate of new marine carbonates is thus slower than the terrestrial weathering rate. The consequence of this is a net removal of CO_2 from the atmosphere and transformation into HCO_3^- (Fig. 6.8b). At the same time, the carbonate ion concentration in the ocean increases (the solid $CO_3^=$ inventory line in Fig. 6.8c), raising the saturation state of the ocean (higher Ω) and increasing carbonate preservation in deep-sea sediments. Eventually, the preservation and burial of $CaCO_3$ in deep-sea sediments will once again balance the weathering input (Fig. 6.8a) and this second 'geologic' sequestration process comes to an end. We will refer to this process as the 'terrestrial $CaCO_3$ neutralization' sink.

How much CO_2 can be sequestered by reaction with terrestrial carbonates? The evolution of atmospheric CO_2 with this additional process enabled is shown in Fig. 6.8b. Now, even at year 40,000, steady state has not quite been attained and CO_2 is still continuing to fall slightly. (A model run of 120,000 years' duration (not shown) reveals that atmospheric CO_2 would fall by only another 18 ppm.) At year 40,000, the concentration of CO_2 in the atmosphere is 435 ppm, which equates to an atmospheric inventory of 951 Pg C. Thus, terrestrial $CaCO_3$ neutralization has removed 15% of the original 4167 Pg C burn (in addition to the initial ocean invasion and erosion of $CaCO_3$ in deep-sea sediments). Just 344 Pg C (8%) of anthropogenic CO_2 is then left in the atmosphere – a finding that is consistent with previous estimates (with a range of 7.4–7.9%, depending on the magnitude of the assumed fossil fuel burn (Archer *et al.*, 1998)).

6.2.3 Geologic carbon sinks: the weathering of silicate rocks on land

Some tens of thousands of years after the burning of fossil fuels has ceased, ~8% of fossil fuel CO_2 emissions (assumed to be 4167 Pg C here) will remain in the atmosphere. The atmospheric CO_2 concentration is 435 ppm, compared with 376 ppm in 2003 (Keeling and Whorf, 2005) and a pre-Industrial value of 278 ppm (Enting *et al.*, 1994). This would probably give half as much climate change as has already occurred to date. Is this the 'end of the road', or does the geologic carbon sink have any further cards to play?

Estimates of the evolution of the amount of carbon in the ocean and atmosphere through Earth history have both reservoirs generally paralleling each other over very long periods of time (>1 million years); i.e. CO_2 and DIC tend to increase and decrease together (Fig. 6.1). In contrast, our model has so far predicted that when atmospheric CO_2 declines, the ocean inventory increases (Fig. 6.8b and c); an antiphased relationship. It would not be unreasonable to conclude from this that we are missing (at least) one important mechanism. We now come to the final geologic (carbonate) carbon sink and one of the most fundamental regulatory mechanisms of the Earth system – the weathering of silicate rocks.

The reaction involved in the weathering of calcium silicate minerals (particularly the feldspar family, which are the most abundant group of minerals in continental rocks) can be written as:

$$2CO_2 + 3H_2O + CaAl_2Si_2O_8 \rightarrow Ca^{2+} + 2HCO_3^- + A1_2Si_2O_5(OH)_4$$

This differs from the weathering of carbonate rocks (in contrast to the weathering reaction listed in Section 6.2.2) in one fundamental regard; it takes two moles of CO_2 to weather each mole of $CaAl_2Si_2O_8$ and release a single mole of calcium ions (plus 2 of bicarbonate ions). The calcium ion is subsequently removed from solution in the same precipitation reaction as before, meaning that only one mole of CO_2 is released back to the ocean (and atmosphere). The weathering of silicate rocks is thus a net sink for atmospheric CO_2 (Berner, 1992) (Fig. 6.7c) – i.e. one mole of CO_2 is being sequestered for each mole of calcium silicate mineral weathered. In the long term, the rate of silicate weathering should balance the rate of volcanic release of CO_2 to the atmosphere (Berner and Caldeira, 1997). If this mechanism is then already busy removing volcanic CO_2 emissions, how can it help in removing the final fraction of anthropogenic CO_2 from the atmosphere?

The rate at which the weathering reaction proceeds depends on a variety of variables. The ones that interest us here are ambient temperature and CO_2 concentration (which is enhanced in soils through the metabolic activity of plants, animals and microbes) (Berner, 1990, 1992). Now we can see how the ultimate fate of fossil fuel CO_2 and the final 'geologic' (carbonate) carbon sink arises – a faster rate of weathering of silicate rocks under a fossil fuel-elevated CO_2 atmosphere (and a warmer, wetter climate), which acts to remove the excess carbon from the atmosphere and sequesters it in marine carbonates. In fact, silicate weathering could remove not only the remaining ~8% fraction of fossil fuel CO_2 left in the atmosphere, but also the fossil fuel CO_2 stored (as bicarbonate ions – see Fig. 6.8c). No trace of our meddling with the environment would remain, except for a slightly more weathered continental surface than before and a fresh thick layer of carbonates covering the ocean floor. Unfortunately, the planetary cleaners will not finish their work any time soon – the timescale for this process is counted in hundreds of thousands of years (Berner and Caldeira, 1997).

6.2.4 Other considerations in the geologic (carbonate) carbon sink

The reduction in pH and carbonate ion (CO_3^{2-}) concentration in the ocean caused by anthropogenic CO_2 emissions has another important consequence, in addition to the dissolution of carbonates deposited on the sea floor (Section 6.2.2). This arises because even if surface waters do not quite become undersaturated ($\Omega < 1.0$), the marine organisms that produce carbonate shells and skeletons will be affected (Royal Society, 2005). If $CaCO_3$ precipitation becomes less thermodynamically favourable, the metabolic (energy) cost of making shells and skeletons will rise. The result is that organisms will precipitate less carbonate and/or will be disadvantaged in the ecosystem. The implications of this for coral reef ecosystems and associated biodiversity and economic impacts are already being widely recognized (Kleypas *et al.*, 2001; Hughes *et al.*, 2003). There is also increasing evidence that calcifying plankton could also be affected by higher atmospheric CO_2 (Bijma *et al.*, 1999; Riebesell *et al.*, 2000; Zondervan *et al.*, 2001; Delille *et al.*, 2005) as well as pteropods, which make aragonite shells (Orr *et al.*, 2005). To understand the implications of this effect for geologic carbon sequestration one must first recognize that the precipitation of $CaCO_3$ by calcifying plankton in the surface ocean and its subsequent removal through gravitational settling raises the partial pressure of CO_2 (pCO_2) at the surface (see Box 6.1). This acts to reduce the rate of fossil fuel CO_2 uptake from the atmosphere. Thus, if carbonate production were to decrease, surface ocean pCO_2 would fall and the rate of CO_2 invasion into the ocean would increase (Zondervan *et al.*, 2001; Barker *et al.*, 2003; Zeebe and Westbroek, 2003). Secondly, a reduction in the flux of $CaCO_3$ to deep-sea sediments brings forward the year in which the net accumulation of carbonate first becomes negative and 'erosion' starts to occur. This means that neutralization by sea-floor carbonates would have an earlier and potentially more extensive impact compared to the case where $CaCO_3$ production does not change.

These effects are illustrated with the help of the model. Carbonate production in the open ocean is now allowed to decrease in response to anthropogenic acidification and reduced surface carbonate ion concentrations. The result is that atmospheric CO_2 is 82 ppm lower in year 3000 compared to the control run (solid line = no calcification change in Fig. 6.9), and 36 ppm lower in year 10,000. Interestingly, the final (steady-state) CO_2 concentration is virtually identical (Ridgwell and Hargreaves, in press). The important point is that the maximum CO_2 value attained, and thus the maximum

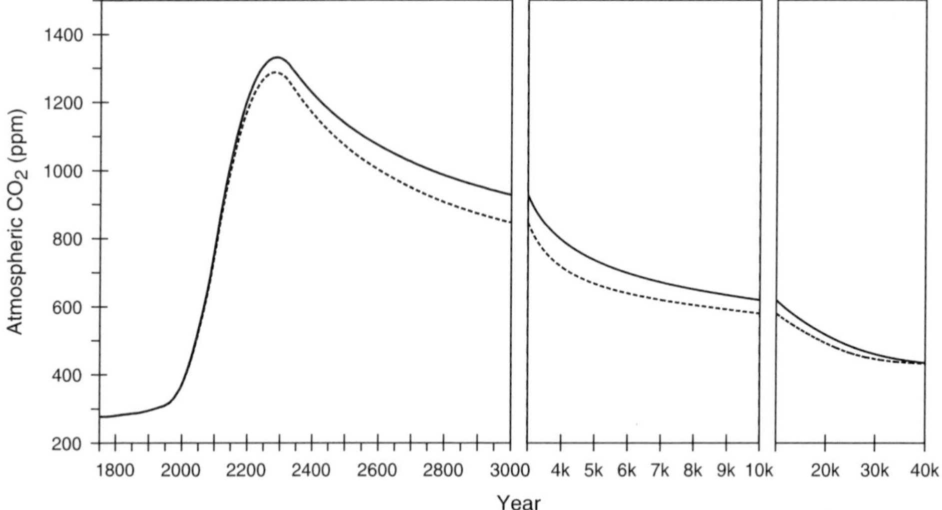

Fig. 6.9. Model analysis of the impact of a reduction in marine calcification (Ridgwell and Hargreaves, in press) on the 'geologic' carbon sink and sequestration of fossil fuel CO_2. The CO_2 trajectory resulting from a combination of ocean invasion, sea-floor neutralization and terrestrial neutralization is shown as a solid line (i.e. the same as the solid line in Fig. 6.8b). The dashed line shows the impact of a reduction in calcification rates in the open ocean.

degree of greenhouse warming, is lower in the presence of CO_2-calcification feedback (Ridgwell *et al.*, 2006b). Changes in the production and burial of carbonates in shallow waters (not represented in the model) are likely to drive an additional reduction in the CO_2 maximum. However, we have not taken into account in this analysis any impact of reduced carbonate production on the transport of organic matter into the deep ocean – the 'ballast hypothesis' (Armstrong *et al.*, 2002; Klaas and Archer, 2002). If this hypothesis is correct, the amount of CO_2 sequestered due to the CO_2-calcification feedback will be less than we have predicted (Barker *et al.*, 2003; Ridgwell, 2003; Heinze, 2004). There are also significant uncertainties as to just how sensitive biogenic calcification is to a reduction in CO_3^{2-} (and saturation state), particularly at the ecosystem (and global) level (Ridgwell *et al.*, 2006b).

6.2.5 Even the geologic sink is not 'for ever'

On the longest timescales we actually come full circle – carbonates deposited in marine settings are destined to be weathered (i.e. dissolved) or undergo 'decarbonation', a reaction that proceeds at high temperatures (and pressures) according to:

$$CaCO_3 + SiO_2 \rightarrow CO_{2(g)} + CaSiO_3$$

The carbon that was formerly locked up in the form of $CaCO_3$ is thus released back to the atmosphere and ocean. In the case of carbonates laid down in shallow seas such as limestones or chalks, these can be exposed to weathering during mountain-building episodes and as a result of sea-level fall. In contrast, carbonates deposited in deep-sea sediments are only infrequently exposed at the Earth's surface. Instead, the primary recycling of deep-sea $CaCO_3$ occurs through the decarbonation reaction when sea-floor sediments are subducted under (i.e. dragged beneath) continental margins (Caldeira, 1991; Berner, 1999; Edmond and Huh, 2003).

While virtually all fossil fuel CO_2 will be sequestered in ocean floor carbonates over the next million years or so, the ultimate fate of this material is subduction followed by decarbonation and the release of stored carbon back to the atmosphere. Fortunately, the sequestered CO_2 will not be released all at once, or anytime soon, because the opening and closing of ocean basins and recycling of oceanic crust takes place on 100 million year timescales.

Interestingly, we are currently short-circuiting the 'natural' geologic process of decarbonation through cement manufacture, although limestones are quarried as the reactant rather than deep-sea carbonate oozes. This process is something of a 'double whammy' (two problems at the same time) to the environment because fossil fuels are used to create the high temperatures needed for the CO_2-releasing decarbonation reaction to proceed. Annually, ~0.2 Pg C is released as a result of this activity (Houghton *et al.*, 2001), which is about twice the rate of burial of carbonates in the entire deep ocean (~0.1 Pg C/year).

6.3 The Geologic Organic Carbon Sink

The burial of organic carbon produced as a result of biological photosynthetic activity represents a net sink for CO_2, and can be encapsulated in the reaction $CO_2 + H_2O$ (+ sunlight) $\rightarrow CH_2O + O_2$. This reaction explains how burial of organic matter could have driven the changes in atmospheric oxygen concentrations during the Phanerozoic period (see Berner *et al.*, 2000, 2003 and references therein). Considerably more organic carbon is 'fixed' from CO_2 by photosynthesis than is ever eventually buried. For instance, global primary productivity by phytoplankton in the ocean is estimated to be ~45 Pg C/year (Houghton *et al.*, 2001). Of this, 11 Pg C/year escapes consumption and respiration (called 'remineralization') by zooplankton grazers and bacteria at the surface, and sinks in particulate form to the ocean interior. In turn, only a tiny fraction of this flux, –0.05 Pg C/year, is ever buried in marine sediments. The remainder (>99%) is remineralized either in the water column or in the surface sediments. This gives us some clues as to ways in which the strength of the geologic organic carbon sink could be enhanced and thus help sequester fossil fuel CO_2 from the atmosphere. We will outline the individual mechanisms that could give rise to a higher burial flux in the following sections, and discuss their possible response (if any) to future global change. The terrestrial geologic organic carbon sink (coal) is discussed separately at the end of the chapter.

6.3.1 Mechanisms of organic carbon burial: marine productivity and sedimentation

We can expect that the sedimentary burial flux of carbon will scale in some way with the strength of productivity in the overlying ocean. All we have to do in order to obtain a stronger geologic sink is to increase primary productivity. One way would be to increase the rate of upwelling of deep waters, thus supplying more nutrients such as phosphate (PO_4) to the ocean surface where photosynthesis takes place. However, most ocean circulation models predict that the ocean is likely to become more stratified in the future as the surface warms, reducing rather than enhancing ocean productivity (Sarmiento *et al.*, 1998; Plattner *et al.*, 2001; Schmittner, 2005).

As an alternative to increasing the rate of upwelling, we could increase the total amount of PO_4 dissolved in the ocean (i.e. raising the concentration everywhere). Phosphate is supplied to the ocean by the dissolution of phosphate-bearing minerals (e.g. apatites) in exposed rocks and soils on land. An increase in weathering rate, which we already expect might occur as a result of higher CO_2 and surface temperatures (Section 6.2.3), will thus act in the 'right' direction. However, PO_4 has a relatively long residence time in the ocean – estimated to be 10,000–80,000 years (Benitez-Nelson, 2000). Thus, even if the global weathering rate were to be instantaneously doubled, we would have to wait at least until the year 10,000 for PO_4 concentrations (and productivity) to have increased by 10–55% (depending on the residence time chosen). Assuming that burial scaled linearly with productivity, this equates to an increase in the strength of the geologic carbon sink of 0.005–0.027 Pg C/year, which is insignificant compared to a total fossil fuel CO_2 release of 4176 Pg C. An increase in weathering rate would also increase the rate of erosion and oxidation of ancient organic matter (kerogens) sequestered in sedimentary rocks and, rather unhelpfully, release additional CO_2 to the atmosphere.

Finally, not everywhere in the ocean is all the PO_4 that is supplied to the ocean surface fully utilized by the phytoplankton. In places such as the Southern Ocean, and Eastern Equatorial and North Pacific, insufficient iron availability (a micronutrient essential for parts of the photosynthetic machinery of cells) limits productivity (Jickells *et al.*, 2005). Supply of this iron ultimately comes from dust deposited to the ocean surface. Thus, if dust supply were to increase, presumably so would productivity (e.g. Watson *et al.*, 2000), and with it, an increase in the rate of burial of organic carbon in sediments and sequestration of CO_2. Although the factors affecting dust production, transport and deposition are complex (Jickells *et al.*, 2005; Ridgwell and Kohfeld, in press), it seems that on balance, dust supply is likely to decrease rather than increase in response to future climate change, further restricting biological productivity. Overall, therefore, future productivity changes in the open ocean would seem to be of little help in the sequestering of fossil fuel CO_2.

Bulk sedimentation rate also appears to be an important control on carbon burial (see Arthur and Sageman, 1994; Hedges and Keil, 1995). This is because a faster accumulation rate reduces the residence time of organic matter in the surface sediments, giving bacteria and benthic animals less chance to consume and metabolize it. However, if productivity were to decline in the future, there is little a priori reason to expect an increase in bulk sedimentation rate, particularly if carbonate production were also to be suppressed (which could give rise to a decrease in the efficiency of transport of organic matter to the sediments if the 'ballast hypothesis' is correct – see Section 6.2.4). In contrast, soil erosion and poor land use management might be expected to contribute to increased sedimentation rates on the continental margins, which could result in increased rates of organic matter burial. For instance, increased Himalayan erosion during the Neogene has been hypothesized to have driven lower atmospheric CO_2 by just such a mechanism (France-Lanord and Derry, 1997). However, the damming of rivers for irrigation and power generation could prevent much of the clay mineral supply from reaching the continental margins, thus reducing the potential importance of this effect.

6.3.2 Mechanisms of organic carbon burial: marine anoxia

Most of the organic carbon that is consumed in the water column and surface sediments is preferentially utilized by bacteria and small animals that metabolize aerobically. Oxygen is used by these organisms because it allows the maximum energy to be extracted out of each molecule of organic matter (represented by 'CH_2O'): $CH_2O + O_2 \rightarrow CO_2 + H_2O$ (+ metabolic energy). If oxygen runs out and conditions become 'anoxic', other bacteria that can utilize nitrate (NO_3^-) or even sulphate (SO_4^{2-}) as the electron receptor for the reaction take over. It is less efficient doing it this way and toxic by-products such as H_2S can build up. The result is that the fraction of organic matter that escapes degradation should be greater under anoxic conditions (Hedges and Keil, 1995). Another effect of anoxia is that burrowing animals, efficient scavengers and consumers of particulate organic matter from the surface sediments are excluded. This will also tend to increase the preservation and burial of organic carbon. In the geological record, the occurrence of 'organic-rich' (defined as sediments containing >1% carbon by dry weight) sedimentary formations called 'black shales' (of which 'sapropels' are a specific case) have often been interpreted as being caused by local anoxia (Arthur and Sageman, 1994). A contemporary example is the Black Sea, where anoxic conditions prevail deeper than about 150 m in the water column and the organic carbon content of the underlying sediments is typically 1–3% (Arthur and Dean, 1998).

Any increase in the occurrence of anoxia in the ocean should therefore enhance the geologic organic carbon sink. The solubility of oxygen decreases with increasing water temperatures, so a warmer Earth should have a less oxygenated ocean (all other things being equal). Increased stratification will also reduce the downward transport of oxygenated surface waters to depth. An increase in carbon preservation and burial in marine sediments in response to global warming is therefore starting to look like a distinct possibility.

The story gets more complicated because NO_3^- is consumed in anoxic regions, and lower ocean oxygen concentrations means less NO_3^-. Because nitrate, like PO_4 and Fe, is an essential nutrient for plant growth, lower NO_3^- concentrations will restrict productivity, and thus carbon burial. Decreased productivity will result in less oxygen consumed during remineralization, and ocean waters will be driven back away from anoxia. This is a 'negative' feedback loop, helping to stabilize oceanic oxygen concentrations. The net outcome of all these competing processes is not obvious, and this is still not including all the relevant factors (see Berner, 1999). This is where the importance of computer models is felt that can calculate the net impact on oceanic oxygen concentrations due to the interacting effects of surface warming, circulation changes and reduced productivity. The results of long-term climate experiments with such models suggest that no deep anoxia is likely to develop in the future, although in regions such as the eastern equatorial Pacific Ocean, an increase in the volume of hypoxic (i.e. not completely anoxic, but with <10 μmol/kg oxygen) thermocline waters is possible (Matear and Hirst, 2003). The oceanic distributions of dissolved oxygen predicted in the GENIE-1 model under our fossil fuel carbon dioxide release experiment also remained oxygenated, although hypoxia became more extensive within the thermocline of the northern Indian Ocean in this particular model (not shown).

To put the potential importance of this mechanism into some sort of perspective, even if ocean anoxia doubled the global rate of carbon sequestration (adding another 0.05 Gt C/year) in the future, it would still take almost 84,000 years for all of the 4176 Pg C of emitted fossil fuel CO_2 to be removed by this mechanism operating alone.

6.3.3 Mechanisms of organic carbon burial: trapping in clay mineral interlayers

There is a third proposed controlling mechanism for the preservation and burial of organic matter in sedimentary rocks – one involving a certain type of clay mineral. Dissolved organic molecules are ubiquitous in the ocean and in the pore spaces of accumulating sediments, and are attracted to the charged surfaces of clay minerals. Therefore, by burying clay minerals in sedimentary formations one would a priori expect a sink of carbon (Hedges and Keil, 1995). Not all clay minerals are created equally, and a mineral called smectite turns out to be 1–2 orders of magnitude more effective than other mineral grains. This is because unlike other clay minerals such as illite, kaolinite or chlorite, which have external surface areas of approximately 20–30 m^2/g, smectite also has an 'internal' (interlayer) surface, giving it a combined surface area of ~800 m^2/g (Kennedy *et al.*, 2002). These internal surfaces arise because of the particular way in which the individual crystalline sheets of smectite are stacked with 'large' interlayer gaps (Fig. 6.10). Polar (and even non-polar) organic molecules such as humic acids and proteins become sorbed into the interlayer sites from the surrounding pore water environment, thereby effectively protecting them from degradation by bacteria (Keil *et al.*, 1994).

Clay mineral formation is the result of the weathering of (mainly feldspar) minerals, such as the hydrolysis of orthoclase feldspar to form kaolinite:

$$4KAlSi_3O_8 + 4H^+ + 2H_2O \rightarrow 4K + Al_4Si_4O_{10}(OH)_8 + 8SiO_2$$

Chemical weathering of feldspars to form clays is complex, and is dependent on weathering intensity, rate and starting composition

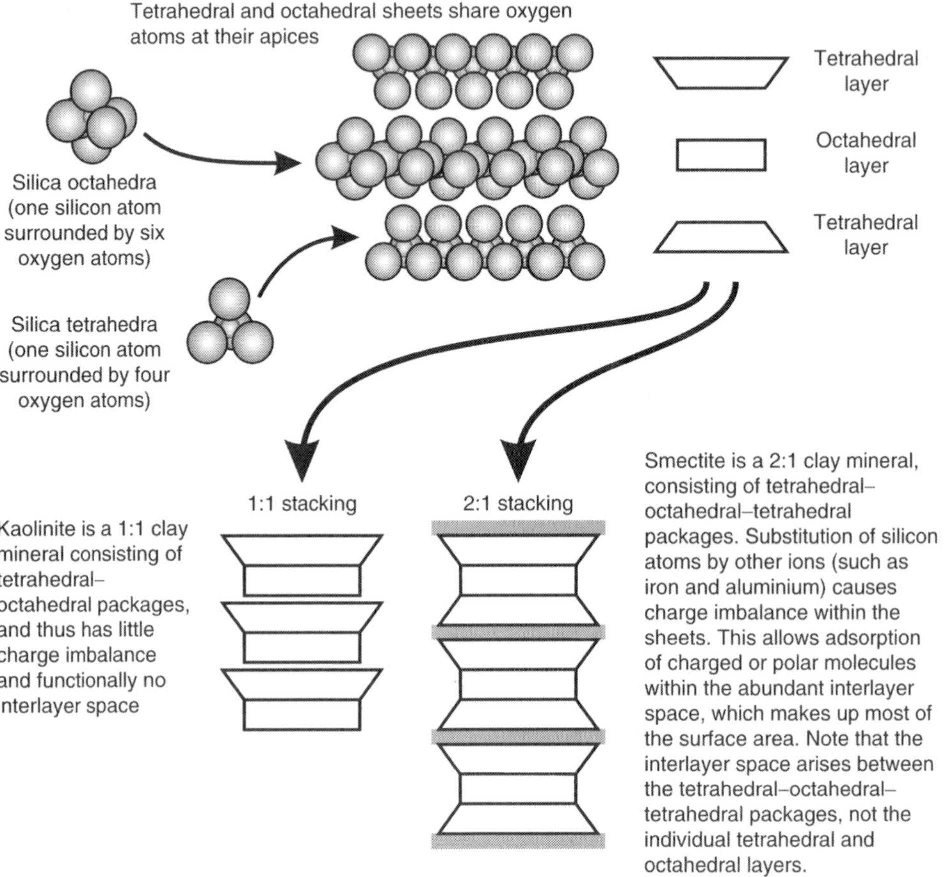

Fig. 6.10. Illustration of the structural units and stacking of two different common clay minerals – kaolinite and smectite. (Unlike the schema, real clays have vastly greater lateral dimension than shown: typically 3–8 layers thick (3–8 nm) and 1 μm in lateral extent. Therefore, for smectite, most surface is internal.) Because clays are formed at surface temperatures and pressures, there is abundant substitution of ions such as aluminium and iron for silica. This results in a net charge imbalance, which is why clay minerals adsorb charged and polar molecules (organic molecules and water most commonly), and exchangeable cations such as calcium and potassium. However, simple substitution is not the only control on the degree to which exchangeable cations and polar molecules are attracted to, and sorbed onto, interlayer surfaces: the location of the substitution (tetrahedral versus octahedral sheet) also has a large influence on the effectiveness of interlayer adsorption. Thus, 2:1 clays such as smectite have more substitution than 1:1 clays such as kaolinite, in addition to having greater interlayer space.

of the silicate minerals undergoing reaction. However, it is interesting to note that if the feldspar in question is calcium-rich plagioclase ($CaAl_2Si_2O_8$), the weathering reaction produces calcium ions in solution, which could then react with bicarbonate ions, forming inorganic carbonate.

An increase in the rate of formation and supply of smectite clays would strengthen the geologic organic carbon sink and help to sequester fossil fuel CO_2. However, smectites are preferentially formed at high weathering *intensity*, which does not necessarily mean a high overall weathering *rate*. For instance, while a change in weathering regime in the Himalayas during the late Miocene resulted in significantly more smectite deposition, it is thought to have resulted in reduced physical erosion at the same time (Derry and France-Lanord, 1996). Thus, decreased

CO_2 consumption because of reduced silicate weathering might offset some (or all) of the 'gains' made by greater carbon burial associated with smectites.

6.3.4 Mechanisms of organic carbon burial: peatlands and coal formation

It is briefly worth mentioning one more geologic carbon sink. Carbon is buried on land in peatlands and swamps, eventually to form coal measures. Peat accumulation depends sensitively on seasonal temperatures and the height of the water table, as well as on biological productivity. With a warmer, wetter future climate we might therefore expect increased rates of peat accumulation. This effect would be enhanced by higher plant productivity induced (fertilized) by rising CO_2 concentrations in the atmosphere. How important this mechanism might be is difficult to quantify, and to date there has not been a systematic attempt to estimate the large-scale future peatland response. However, we can attempt to place it into context by considering past changes.

During the late Holocene (the past few thousand years) there appears to have been a significant increase in carbon storage in peatlands, particularly in the boreal latitudes of the northern hemisphere (Gorham, 1991; Laine *et al.*, 1996; Gajewski *et al.*, 2001). The estimates that have been made suggest an average accumulation rate over the late Holocene of 0.096 Pg C/year (Gorham, 1991). If true, this would make peatlands a potentially important additional long-term carbon sink. However, it is far from clear how much of this expansion of peatlands is in response to the retreat of the northern hemisphere ice sheets following the end of the last glacial period and the creation of new areas suitable for wetlands to develop, and what might have been a response to Holocene warming (and an increase in CO_2). Although 'mining' and drainage in the recent past have probably reversed the global rate of peatland and carbon accumulation, an important contribution from this mechanism to fossil fuel CO_2 sequestration cannot be ruled out, particularly since peatlands are increasingly coming under protection, and in some cases formerly drained peatland forests are being reflooded to promote long-term carbon sequestration and other beneficial reasons.

6.4 Summary and Perspectives

We have looked at the long-term fate of CO_2 released to the atmosphere and used an Earth system model to illustrate (and quantify) the role of some of the major geological carbon sinks. After the ocean has absorbed ~66% of the total release, reaction with sea-floor carbonates results in the sequestration of another 11% on a timescale of a few thousand years. Then, reaction with $CaCO_3$ on land consumes a further 15%, but at a slightly slower pace (tens of thousands of years). Finally, the fate of the remaining 8% in the atmosphere (as well as the 92% now in the ocean as bicarbonate) is removal through silicate rock weathering and burial as marine carbonates.

The situation is not quite this straightforward. Rising surface temperatures and changes in ocean circulation could reduce the effectiveness of the initial ocean invasion sink by 10–20%. Additional and as yet poorly quantified complications arise due to a probable reduction in biogenic calcification in the surface ocean that could lead to an acceleration of ocean invasion and the sea-floor $CaCO_3$ neutralization of CO_2.

We have also considered different mechanisms of organic carbon deposition and burial in accumulating sediments. There is much greater uncertainty here in what the response will be to future global change. A greater rate of clay deposition and of continental margin sedimentary burial seems possible, and would favour a stronger carbon sink. Acting against this will be a reduction in open-ocean productivity caused by increased ocean stratification. However, regardless of the net impact of these factors, it seems unlikely that the organic carbon sink will be as important in the removal of fossil fuel CO_2 from the atmosphere and ocean as changes in carbonate deposition. In contrast, the potential role for peatlands (and incipient coal formation) in the future may be greater.

A clue as to how the geologic carbon sinks might react in the future can be found in the distant past – there is good evidence for a 'catastrophic' release of carbon associated with an event called the 'Paleocene/Eocene thermal maximum' (PETM), some 55.5 million years ago (Dickens *et al.*, 1995). Associated with this was a CO_2 (and maybe CH_4)-driven greenhouse warming of the Earth's surface (Bains *et al.*, 1999; Tripati and Elderfield, 2005). Of particular interest to us is that sediment cores document how carbonate accumulation in the deep sea at first declined, and then later recovered in response to this event (Zachos *et al.*, 2005). It has also been suggested that the observed timescale for this recovery (~60,000 years) was accelerated by increased productivity in the ocean (Bains *et al.*, 2000). This would imply that the action of organic carbon sinks in the marine environment might be more important in the future than we speculated earlier (Section 6.3). Paleo-analogues for future global change such as the PETM have a critical role in helping to elucidate the role that the geologic carbon sinks might play in the future.

In the geological carbon sinks, particularly involving silicate rock weathering, the Earth possesses powerful feedback mechanisms that are able to regulate the surface environmental conditions of the planet. There is therefore no doubt about the long-term survival of the biosphere, despite the currently accelerating rate of greenhouse gas emissions. Indeed, by transforming fossil fuel CO_2 into carbonates that are buried on the ocean floor, the geological carbon sink will eventually clean up our mess and return atmospheric composition to the pre-Industrial state. Can we rely on these 'geologic' sinks to stabilize atmospheric CO_2 and climate without any societal intervention, leaving us free to continue to burn fossil fuels in a 'business as usual' fashion? Unfortunately, 'Mother Earth' is not a quick healer and the geologic carbon sinks will be of little help in damping the maximum future 'greenhouse' warming of the planet. This warming, along with other global environmental impacts will occur much quicker than the geological sinks can cope with.

Acknowledgement

Andy Ridgwell acknowledges support from Canada Research Chairs, the Canadian Foundation for Climate and Atmospheric Sciences, and the Trusthouse Charitable Foundation. Development of the model was supported by the NERC e-Science programme (NER/T/S/2002/00217) through the Grid ENabled Integrated Earth system modelling (GENIE) project (www.genie.ac.uk) and by the Tyndall Centre for Climate Change Research (Project TC IT 1.31).

References

Archer, D. (1996a) An atlas of the distribution of calcium carbonate in sediments of the deep sea. *Global Biogeochemical Cycles* 10, 159–174.

Archer, D. (1996b) A data-driven model of the global calcite lysocline. *Global Biogeochemical Cycles* 10, 511–526.

Archer, D. (2003) Who threw that snowball? *Science* 302, 791–780.

Archer, D. and Buffett, B. (2004) Time-dependent response of the global ocean clathrate reservoir to climatic and anthropogenic forcing. *Geochemical, Geophysics, Geosystems* 6; doi:10.1029/2004GC000854.

Archer, D. and Ganopolski, A. (2005) A movable trigger: fossil fuel CO_2 and the onset of the next glaciation. *Geochemical, Geophysics, Geosystems* 6; doi:10.1029/2005GL022449.

Archer, D., Kheshgi, H. and Maier-Reimer, E. (1998) Dynamics of fossil fuel CO_2 neutralization by marine $CaCO_3$. *Global Biogeochemical Cycles* 12, 259.

Armstrong, R.A., Lee, C., Hedges, J.I., Honjo, S. and Wakeham, S.G. (2002) A new, mechanistic model for organic carbon fluxes in the ocean: based on the quantitative association of POC with ballast minerals. *Deep-Sea Research II* 49, 219–236.

Arthur, M.A. and Dean, W.E. (1998) Organic-matter production and preservation and evolution of anoxia in the Holocene Black Sea. *Paleoceanography* 13, 395–411.

Arthur, M.A. and Sageman, B.B. (1994) Marine black shales: depositional mechanisms and environments of ancient-deposits. *Annual Review of Earth and Planetary Sciences* 22, 499–551.

Bains, S. *et al.* (1999) Mechanisms of climate warming at the end of the Paleocene. *Science* 285, 724–727.

Bains, S., Norris, R.D., Corfield, R.M. and Faul, K.L. (2000) Termination of global warmth at the Palaeocene/Eocene boundary through productivity feedback. *Nature* 407, 171–174.

Barker, S., Higgins, J.A. and Elderfield, H. (2003) The future of the carbon cycle: review, calcification response, ballast and feedback on atmospheric CO_2. *Philosophical Transactions of the Royal Society A* 361, 1977.

Benitez-Nelson, C.R. (2000) The biogeochemical cycling of phosphorus in marine systems. *Earth-Science Reviews* 51, 109–135.

Berner, R.A. (1990) Atmospheric carbon dioxide levels over Phanerozoic time. *Science* 249, 1382–1386.

Berner, R.A. (1992) Weathering, plants, and the long-term carbon cycle. *Geochimica et Cosmochimica Acta* 56, 3225–3231.

Berner, R.A. (1999) A new look at the long-term carbon cycle. *GSA Today* 9, 1–6.

Berner, R.A. and Caldeira, K. (1997) The need for mass balance and feedback in the geochemical carbon cycle. *Geology* 25, 955–956.

Berner, R.A. *et al.* (2000) Isotope fractionation and atmospheric oxygen: implications for Phanerozoic O_2 evolution. *Science* 287, 1630–1633.

Berner, R.A. *et al.* (2003) Phanerozoic atmospheric oxygen. *Annual Review of Earth and Planetary Sciences* 31, 105–134.

Caldeira, K. (1991) Continental-pelagic carbonate partitioning and the global carbonate-silicate cycle. *Geology* 19, 204–206.

Caldeira, K. and Wickett, M.E. (2003) Anthropogenic carbon and ocean pH. *Nature* 425, 365.

Derry, L.A. and France-Lanord, C. (1996) Neogene Himalayan weathering history and river 87Sr/86Sr: impact on the marine Sr record. *Earth and Planetary Science Letters* 142, 59–74.

Dickens, G.R. *et al.* (1995) Dissociation of oceanic methane hydrate as a cause of the carbon-isotope excursion at the end of the Paleocene. *Paleoceanography* 10, 965–971.

Edmond, J.M. and Huh, Y. (2003) Non-steady state carbonate recycling and implications for the evolution of atmospheric pCO_2. *Earth and Planetary Science Letters* 216, 125–139.

Edwards, N.R. and Marsh, R. (2005) Uncertainties due to transport-parameter sensitivity in an efficient 3-D ocean-climate model. *Climate Dynamics* 24, 415–433.

Edwards, N.R. and Shepherd, J.G. (2002) Bifurcations of the thermohaline circulation in a simplified three-dimensional model of the world ocean and the effects of interbasin connectivity. *Climate Dynamics* 19, 31–42.

Enting, I.G., Wigley, T.M.L. and Heimann, M. (1994) Future Emissions and Concentrations of Carbon Dioxide: Key Ocean/Atmosphere/Land Analyses. CSIRO Division of Atmospheric Research Technical Paper No. 31, Commonwealth Scientific and Industrial Research Organisation, Aspendale, Australia.

Feely, R.A. *et al.* (2004) Impact of anthropogenic CO_2 on the $CaCO_3$ system in the oceans. *Science* 305, 362–366.

France-Lanord, C. and Derry, L.A. (1997) Organic carbon burial forcing of the carbon cycle from Himalayan erosion. *Nature* 390, 65–67.

Gajewski, K., Viau, A., Sawada, M., Atkinson, D. and Wilson, S. (2001) Sphagnum peatland distribution in North America and Eurasia during the past 21,000 years. *Global Biogeochemical Cycles* 15, 297–310.

Gorham, E. (1991) Northern peatlands: role in the carbon cycle and probable responses to climatic warming. *Ecological Applications* 1, 182–195.

Hargreaves, J.C., Annan, J.D., Edwards, N.R. and Marsh, R. (2004) An efficient climate forecasting method using an intermediate complexity Earth system model and the ensemble Kalman filter. *Climate Dynamics* 23, 745–760.

Hasselmann, K. *et al.* (2003) The challenge of long-term climate change. *Science* 302, 1923–1925.

Hedges, J.I. and Keil, R.G. (1995) Sedimentary organic-matter preservation: an assessment and speculative. *Marine Chemistry* 49, 81–115.

Houghton, J.T. *et al.* (eds) (2001) *Climate Change 2001: The Scientific Basis*. Contribution of Working Group I. Third Assessment Report of the Intergovernmental Panel on Climate Change, Cambridge University Press, Cambridge.

Hughes, T.P. *et al.* (2003) Climate change, human impacts, and the resilience of coral reefs. *Science* 301, 929–933.

Jickells, T.D. *et al.* (2005) Global iron connections between desert dust, ocean biogeochemistry and climate. *Science* 308, 67.

Keeling, C.D. and Whorf, T.P. (2005) Atmospheric CO_2 records from sites in the SIO air sampling network. In: *Trends: A Compendium of Data on Global Change*. Carbon Dioxide Information Analysis Center, Oak Ridge National Laboratory, US Department of Energy, Oak Ridge, Tennessee. Available at: http://cdiac.esd.ornl.gov/trends/co2/sio-mlo.htm

Keil, R.G., Montlucon, D.B., Prahl, F.G. and Hedges, J.I. (1994) Sorptive preservation of labile organic-matter in marine-sediments. *Nature* 370, 549–552.

Kennedy, M.J., Peaver, D.R. and Hill, R.J. (2002) Mineral surface control of organic carbon in black shale. *Science* 295, 657–660.

Klaas, C. and Archer, D.E. (2002) Association of sinking organic matter with various types of mineral ballast in the deep sea: implications for the rain ratio. *Global Biogeochemical Cycles* 16; doi: 10.1029/2001GB001765.

Kleypas, J.A., Buddemeier, R.W. and Gattuso, J.-P. (2001) The future of coral reefs in an age of global change. *International Journal of Earth Sciences* 90, 426–437.

Kump, L.R., Kasting, J.F. and Crane, R.G. (2004) *The Earth System*. Prentice-Hall, Upper Saddle River, New Jersey.

Laine, J. *et al.* (1996) Effect of water-level drawdown on global climatic warming: northern peatlands. *Ambio* 25, 179–184.

Lenton, T.M. (2000) Land and ocean carbon cycle feedback effects on global warming in a simple Earth system model. *Tellus* B 52, 1159–1188.

Matear, R.J. and Hirst, A.C. (2003) Long-term changes in dissolved oxygen concentrations in the ocean caused by protracted global warming. *Global Biogeochemical Cycles* 17; doi:10.1029/2002GB001997.

Milliman, J.D. (1993) Production and accumulation of calcium carbonate in the ocean: budget of a nonsteady state. *Global Biogeochemical Cycles* 7, 927–957.

Milliman, J.D. and Droxler, A.W. (1996) Neritic and pelagic carbonate sedimentation in the marine environment: ignorance is not bliss. *Geologische Rundschau* 85, 496–504.

Orr, J.C. *et al.* (2005) Anthropogenic ocean acidification over the twenty-first century and its impact on calcifying organisms. *Nature* 437, 681–686.

Page, S.E. *et al.* (2002) The amount of carbon released from peat and forest fires in Indonesia during 1997. *Nature* 420, 61–65.

Plattner, G.K., Joos, F., Stocker, T.F. and Marchal, O. (2001) Feedback mechanisms and sensitivities of ocean carbon uptake under global warming. *Tellus B* 53, 564–592.

Ridgwell, A. (2005) Changes in the mode of carbonate deposition: implications for Phanerozoic ocean chemistry. *Marine Geology* 217, 193.

Ridgwell, A. and Zeebe, R.E. (2005) The role of the global carbonate cycle in the regulation and evolution of the Earth system. *EPSL* 234, 299.

Ridgwell, A., Hargreaves, N., Edwards, N., Annan, T., Lenton, R., Marsh, R., Yool, A. and Watson, A. (2006a) Marine geochemical data assimilaton in an efficient earth system model of global biogeochemical cycling. *Biogeosciences Discussions* 3, 1313–1354.

Ridgwell, A., Zondervan, I., Hargreaves, J., Bijma, J. and Lenton, T. (2006b) Significant long-term increase of fossil fuel CO_2 uptake from reduced marine calcification. *Biogeosciences Discussions* 3, 1763–1780.

Ridgwell, A.J. (2001) Glacial–interglacial perturbations in the global carbon cycle. PhD thesis, University of East Anglia at Norwich, UK. Available at: http://andy.seao2.org/publications/ridgwell_2001.pdf

Ridgwell, A.J. (2003) Implications of the glacial CO_2 'iron hypothesis' for quaternary climate change. *Geochemical, Geophysics, Geosystems* 4, 1076; doi:10.1029/2003GC000563.

Ridgwell, A.J., Kennedy, M.J. and Caldeira, K. (2003) Carbonate deposition, climate stability, and Neoproterozoic ice ages. *Science* 302, 859–862.

Riebesell, U., Zondervan, I., Rost, B., Tortell, P.D., Zeebe, R.E. and Morel, F.M.M. (2000) Reduced calcification of marine plankton in response to increased atmospheric CO_2. *Nature* 407, 364–367.

Royal Society (2005) *Ocean acidification due to increasing atmospheric carbon dioxide*. Royal Society Document 12/05, Royal Society, London.

Royer, D.L. *et al.* (2004) CO_2 as a primary driver of Phanerozoic climate. *GSA Today* 14, 4–10.

Ryan, D.A., Opdyke, B.N. and Jell, J.S. (2001) Holocene sediments of Wistari Reef: towards a global quantification of coral reef related neritic sedimentation in the Holocene. *Palaeogeography, Palaeoclimatology, Palaeoecology* 175, 173–184.

Sabine, C.L. *et al.* (2004) The oceanic sink for anthropogenic CO_2. *Science* 305, 367–371.
Sarmiento, J.L., Hughes, T.M.C., Stouffer, R.J. and Manabe, S. (1998) Simulated response of the ocean carbon cycle to anthropogenic climate warming. *Nature* 393, 245–249.
Schimel, D. and Baker, D. (2002) Carbon cycle: the wildfire factor. *Nature* 420, 29–30, 2002.
Schmittner, A. (2005) Decline of the marine ecosystem caused by a reduction in the Atlantic overturning circulation. *Nature* 434, 628–633.
Tripati, A. and Elderfield, H. (2005) Deep-sea temperature and circulation changes at the Paleocene-Eocene thermal maximum. *Science* 308, 1894–1898.
Vecsei, A. and Berger, W.H. (2004) Increase of atmospheric CO_2 during deglaciation: constraints on the coral reef hypothesis from patterns of deposition. *Global Biogeochemical Cycles* 18; doi:10.1029/2003GB002147.
Watson, A.J., Bakker, D.C.E., Ridgwell, A.J., Boyd, P.W. and Law, C.S. (2000) Effect of iron supply on Southern Ocean CO_2 uptake and implications for glacial atmospheric CO_2. *Nature* 407, 730–733.
Weaver, A.J. *et al.* (2001) The UVic Earth system climate model: model description, climatology and application to past, present and future climates. *Atmosphere-Ocean* 39, 361–428.
Zachos, J.C. *et al.* (2005) Rapid acidification of the ocean during the Paleocene-Eocene thermal maximum. *Science* 308, 1611–1615.
Zeebe, R.E. and Westbroek, P. (2003) A simple model for the $CaCO_3$ saturation state of the ocean: the 'Strangelove,' the 'Neritan,' and the 'Cretan' ocean. *Geochemical, Geophysics, Geosystems* 4; doi:10.1029/2003GC000538.
Zeebe, R.E. and Wolf-Gladrow, D. (2001) *CO_2 in Seawater: Equilibrium, Kinetics, Isotopes*. Elsevier Oceanographic Series 65, Elsevier, New York.
Zondervan, I., Zeebe, R.E., Rost, B. and Riebesell, U. (2001) Decreasing marine biogenic calcification: a negative feedback on rising atmospheric pCO_2. *Global Biogeochemical Cycles* 15, 507–516.

7 Artificial Carbon Sinks: Utilization of Carbon Dioxide for the Synthesis of Chemicals and Technological Applications

Michele Aresta and Angela Dibenedetto
Department of Chemistry and CIRCC, University of Bari, Bari, Italy

7.1 Introduction

7.1.1 Reduction of the atmospheric loading of CO_2

The reduction of the accumulation of atmospheric CO_2 is the objective of international and national programmes. A number of technologies are available, each characterized by a different level of: (i) knowledge of the effects generated upon application; (ii) scale; (iii) cost; and (iv) experimentation and exploitation (Table 7.1). Such technologies may either reduce the production of CO_2 or be used for disposing the produced CO_2; in the latter case, the permanence of disposed CO_2 becomes a key factor.

Currently, it is impossible to designate a single technology that alone may be able to reduce the CO_2 atmospheric loading to such a level that may guarantee that 'non-return' points are avoided. More likely, a combination of technologies may help to reach the objective. It is foreseeable that each of the technologies listed in Table 7.1 may vary its amplitude of application with time. By making the right choices now, the amount of CO_2 that will be removed from the atmosphere will reasonably increase with time with respect to the actual potential of each technology. However, implementing ten different technologies, each having a reduction potential of 100 million tonnes per year, it will be possible to cut 1 Pg CO_2/year, which is a significant amount as a starting point. In the short, medium and long term, each technology will increase its potential for application at a different rate and this will make it meaningful to exclude some possibilities.

This chapter deals with the use of CO_2 in the synthesis of chemicals or for technological applications. A specific attention will be paid in the discussion to the assessment of the potential of such options for the control of the CO_2 atmospheric loading.

7.1.2 Actual uses of CO_2

CO_2 is already industrially utilized as:

1. Feedstock for the synthesis of chemicals;
2. Technological fluid;
3. Source of carbon for enhanced production of biomass (mostly marine).

The use of CO_2 in the synthesis of chemicals is today applied in only a few cases, with a total amount of CO_2 converted equal to ~110 Tg per year, most of which goes into the production of urea (~70 Tg per year) (Ricci, 2003). The use as an additive to CO for the synthesis of methanol is another important

Table 7.1. Technologies that may be used in the control of the atmospheric level of CO_2.

Technologies based on the utilization of fossil fuels	Main issues that generate problems
Efficiency in the production of other forms of energy (thermal, mechanical and electric energy) from fossil fuels	Level of the actual implementation
Disposal of CO_2 in aquifers, geological cavities and ocean	Site specificity (request of specific conditions)
Disposal of CO_2 in oil and gas spent fields	Energy penalty (energy spent in application of the technology with CO_2 generation)
Utilization of CO_2 in gaseous or liquid hydrocarbons extraction, e.g. enhanced oil recovery (EOR)	Cost of exploitation additional to energy cost
Enhanced biological utilization of CO_2 (production of aquatic biomass)	Scale (from kt to Gt)
Chemical utilization of CO_2	Environmental impact (of the exploitation of the technology)
Technological utilization of CO_2	Permanence of CO_2 (for the disposal technologies)
Fixation of CO_2 into inorganic carbonates	
Alternative energies	
Geothermal energy	Site specificity, cost, total power
Solar energy	Site specificity, cost, low density and intensity
Wind energy	Site specificity, continuity, power per unit, impact
Hydropower	Site specificity
Nuclear energy	Existing concerns

use (5–6 Tg per year). Minor applications are the synthesis of molecular organic carbonates (0.1 Tg per year) and the synthesis of specialty chemicals such as 2-hydroxybenzoic acid (40,000 t/year). Specialty inorganic carbonates also represent a large mass application (30 Tg per year). In addition to the utilization in the production of chemicals, CO_2 is used as technological fluid or as an additive to beverages (Vansant, 2003) at a level of ~18 Tg per year, with a steady increase in the past years and a 5–10% market expansion rate estimated for the near future. Such CO_2 is either collected from natural wells or recovered from chemical processes (mainly the synthesis of NH_3) or (in lower amounts) from fermentation processes. Sources such as power plants are not yet exploited. The latter source produces CO_2 that is accompanied by pollutants that have to be separated for most applications, as they may cause negative effects on humans or animals, or more generally to biotic and abiotic systems. Large amounts of very pure CO_2 (several Tg per year) produced in the sugarcane and other similar industries are vented because of the seasonality of the production, which does not assure a continuous feed over the year to a potential CO_2 user.

The relevance of the use of CO_2 to the reduction of its atmospheric loading and the estimation of its potential contribution to this loading are key issues for determining the role of such an option in the panorama of CO_2 reduction technologies. This aspect will now be discussed in order to put the rest of this chapter in context. It is worth recalling that most of the synthetic processes currently in use produce large amounts of waste as by-products, inorganic and organic salts, as well as waste solvents. The ratio of the mass of waste to the mass of the marketable products is the 'waste factor'. It may range from 2 to 50, or more, according to the process complexity (number of steps and type of reaction). One of the processes with high 'waste factor' is the synthesis of carboxylates. This is not a straightforward procedure, but the ability to introduce the CO_2 moiety into an organic substrate using CO_2 would greatly reduce the waste production.

The industrial utilization of CO_2 includes the use of the entire molecule as a building block in the chemical industry or of its reduced forms (HCOOH, HCHO, CO, CH_3OH and, ultimately, CH_4). These two uses have different energy requirements and this is a key point in the utilization of CO_2. If the recovery of CO_2 is implemented on a large scale, CO_2 will be more readily available. Two options are then open: its disposal or its utilization. All technologies aimed at disposal are characterized by an energy penalty and this may equal 0.2–0.45 t of CO_2 emitted per tonne of disposed CO_2. According to some authors (e.g. Mann and Spath, 1997), the implementation of such disposal technologies will cause a net expansion of the extraction of fuels, thus shortening their availability with questionable benefits to the climate. Conversely, chemical, biological and technological uses may avoid CO_2 emissions and reduce the need for fossil fuel extraction. For the sake of correctness, one must be aware that this utilization option is not the solution to the CO_2 problem by itself, but can at least contribute to wider efforts to reduce global CO_2 emissions.

It must also be considered that most of the compounds into which CO_2 can be converted will release CO_2 on a timescale of months to years, according to their nature and use. Polymers (polycarbonates and polyurethanes) will have a good, but limited for the amount, potential for CO_2 storage (decades to centuries). Other chemicals will lead to a reduction in emissions through carbon recycling or, better yet, the implementation of innovative synthetic technologies that will reduce the use of energy and fossil carbon relative to existing ones. Therefore, the reduction of CO_2 emission with respect to the existing situation in the chemical industry, will be represented by the amount of avoided CO_2 produced by the implementation of new synthetic strategies that use CO_2. From this point of view, the utilization of CO_2 can be interpreted as a peculiar case of 'efficiency technology'. Interestingly, such technologies may also produce net economic benefits, as they convert a spent chemical into a valuable product. Therefore, the utilization of recovered CO_2 in the synthesis of specific chemicals may occur with the reduction of the energy and carbon use associated with traditional chemical synthesis.

The potential benefits of using CO_2 can be assessed by comparing innovative to existing technologies. In general, it can be said that the utilization of CO_2 will contribute to a reduction in emissions wherever the process meets the following conditions. It must be exoergonic (free energy should be considered more than enthalpy alone, as CO_2-based reactions are in general accompanied by a significant change of entropy) and must minimize: (i) the processing energy; (ii) the total carbon balance; (iii) the other materials input; (iv) the CO_2 emission; and (v) other emissions.

The patent literature shows that a very large number of reactions utilizing CO_2 have been developed in recent years as alternatives to existing processes (Fig. 7.1). These processes are at a range of developmental stages and degrees of exploitation. As Fig. 7.1 shows, two classes of reactions can be distinguished for the conversion of CO_2 into other chemicals (Aresta, 1987): the carboxylation and the reduction reactions. The former produce an increase in the C/H ratio, which has been considered by some authors as an effective pathway for CO_2 fixation and emission reduction (Pechtl, 1991; Audus and Oonk, 1997).

The quantification of the amount of CO_2 avoided is not as straightforward a procedure as it may be for simple CO_2 disposal in that it requires a complex calculation methodology, such as a life cycle assessment, to be applied to the processes under assessment. Similar considerations are valid when CO_2 is used as a technological fluid (i.e. used directly) rather than it being converted into other chemicals. Whenever it substitutes for other chemicals, the effect of this product substitution can be evaluated by comparing the global warming potential (GWP) of CO_2 with that of the chemicals it substitutes – the GWP of these chemicals can often be several thousand times that of CO_2 (CO_2 has a GWP of 1).

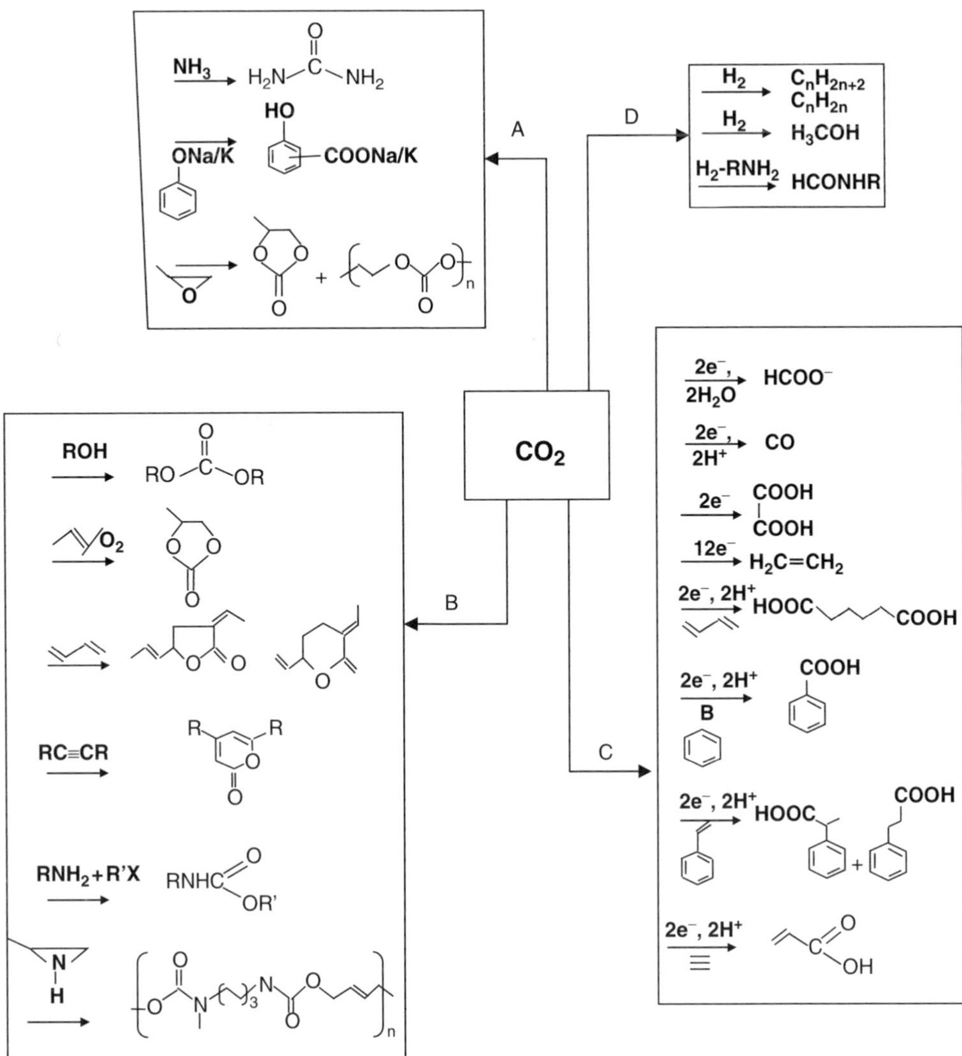

Fig. 7.1. Chemicals obtained from CO_2.

7.1.3 Life cycle methodology for the assessment of reduced climate forcing through CO_2 utilization

The life cycle assessment (LCA) may be applied for the evaluation of either the economic, energetic or environmental convenience of a product system. It may be advantageously used for comparing a CO_2-based process to one that does not use CO_2. Although LCA is usually applied 'from the cradle to the grave' for a product system, the analysis 'from cradle to distribution' can also be used when comparing two different synthetic methodologies for a given chemical. In this case, the fate of the product will be the same. Complete energy and mass balances (considering the yield, selectivity and waste production) for each step of a process starting from the extraction of fossil carbon (coal, oil, liquified natural gas) and consideration of all products implied in the process are necessary. The confidentiality of industrial process data and the uncertainty of some published data represent the most serious barriers to an extended use of LCA for the evaluation of product systems.

Estimate of CO_2 emission reduction associated with the utilization of CO_2: the case of the synthesis of carbonates

The following scheme gives three routes comparing CO_2 emissions for three different technologies that produce ethylene carbonate:

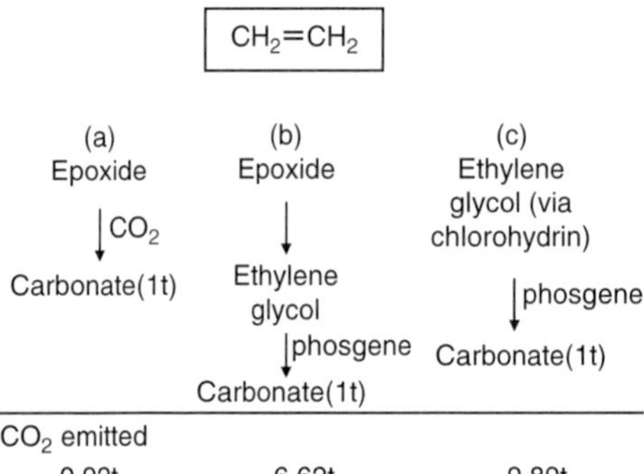

Scheme 7.1. Synthesis of ethylene carbonate using CO_2.

Organic carbonates (linear and cyclic, see below) are prepared according to various technologies, often using toxic reagents. However, there are ways of synthesizing them using CO_2. The synthesis of ethylene carbonate (EC, monomer of polyethylene carbonate) is a good example of the application of mass and energy balance for estimating the net CO_2 emissions of alternative processes. The direct carboxylation of ethylene oxide, shown by route (a) (currently in-use), is compared with the old process based on the use of ethylene glycol and phosgene, the traditional way to EC. Ethylene glycol can be made from the epoxide, shown by route (b) (currently in use), or from ethylene chlorohydrin, shown by route (c) (an older process, now rarely used).

Ethylene is the common starting chemical for routes (a)–(c). In this comparative exercise, the energy necessary for its production will not be considered as this can be assumed to be the same for all routes. The CO_2-based process (a) emits less CO_2 per unit product with respect to either of the other processes: a reduction in emissions of 5.7 or 8.97 $t_{CO_2}/t_{carbonate}$ is observed compared to (b) or (c), respectively, as both these routes are based on phosgene. The reduction of the toxicity factors (phosgene substitution) and environmental impacts (avoidance of waste chlorides) will further enhance the benefit of this substitution. This example illustrates in a simple way the concept of reduced climate forcing based on the use of CO_2 with respect to energy efficiency, compared to processes not making use of CO_2.

In the following sections the beneficial effects of adding CO_2 to CO in the synthesis of methanol will also be discussed.

7.2 Developing Artificial Processes Based on Natural Processes

Nature uses CO_2 in thousands of reactions for the synthesis of a large number of products. These reactions can be categorized into two major classes:

- The reduction of CO_2;
- The carboxylation of substrates, either inorganic or organic. Note that the 'weathering of silicates' is a special case of carboxylation that fixes CO_2 into long-living compounds such as inorganic carbonates (see Chapter 6).

The *reduction reactions*, in which CO_2 is reduced to other C_1 (CO, CH_3OH) or C_n molecules (block D in Fig. 7.1) that can be used as fuels, use dihydrogen as co-reagent. Again, such CO_2 utilization may contribute to an overall reduction in CO_2 emissions (Song, 2002).

The carboxylation reactions, which occur in mild conditions, are exoergonic (blocks B and C in Fig. 7.1). In such reactions the entire CO_2 moiety is incorporated into molecular or polymeric compounds containing moieties like COOR (carboxylates, esters, lactones), N-COOR (carbamates) or NCO (isocyanates or ureas) and ROCOOR (carbonates).

7.2.1 Reduction of CO_2: methane reforming with CO_2 and synthesis of methanol

Methane and CO_2 are constituents of liquid natural gas (LNG). According to conventional technologies, the two gases are separated and methane is used for the production of Syngas, an H_2–CO mixture (Eq. 7.1) that has been used for more than 50 years for the synthesis of gasoline (Fischer Tropsch process). Methanol is also produced from Syngas (Eq. 7.2). The gas-to-liquid (GTL) conversion is currently a process of great interest as its implementation at the site of LNG extraction site would considerably reduce transportation costs:

$$CH_4 + H_2O \rightarrow CO + 3H_2 \quad \Delta H_f^\circ = +206 \text{ kJ/mol} \quad (7.1)$$

$$CO + 2H_2 \rightarrow CH_3OH \quad (7.2)$$

The production of Syngas from methane and water (Eq. 7.1) is a strongly endothermic process (methane is burned to furnish the necessary energy) and produces an H_2–CO mixture with a higher H_2/CO ratio than required for the synthesis of methanol (Eq. 7.2) or gasoline. The excess hydrogen is not recovered and therefore as a whole the process is not energetically optimized (Lange, 1997).

CO_2 can, in principle, either partially or totally substitute CO in the synthesis of methanol. It can either be added to the steam reformer (Eq. 7.3) or to the methanol unit (Eq. 7.4) (an ICI process in use for several decades), resulting in a better utilization of the energy and hydrogen:

$$CH_4 + CO_2 \rightarrow 2CO + 2H_2 \quad \Delta H_f^\circ = +247 \text{ kJ/mol} \quad (7.3)$$

$$CO_2 + 3H_2 \rightarrow CH_3OH + H_2O \quad (7.4)$$

In the CO–CO_2–H_2 system the direct conversion of CO_2 to methanol has been shown by Rozovskii (1989) by tracer analysis. In a standard technology based on methane wet-reforming and subsequent conversion of Syngas into methanol (Eqs 7.1 and 7.2), the energy consumption is 31 kJ/t methanol with a thermal yield (lower heating value or (LHV)) of 64.3% (Lange, 1997); the thermal yield based on natural gas is just 42.3% (M. Ricci and F. Podestà, Novara, Italy, 2004, personal communication). Substitution of 5% methane with CO_2 increases the thermal yield to 66.5% (LHV) (Lange, 1997; Lemonidou *et al.*, 2003) as a consequence of the better use of hydrogen (Maroto-Valer *et al.*, 2002; Song, 2002). This feed substitution can be further increased up to a maximum of 30%. In order to partially compensate the two endothermic processes of wet- (Eq. 7.1) and dry-reforming (Eq. 7.3) of methane, its partial oxidation is used (Eq. 7.5):

$$CH_4 + {}^1/_2O_2 \rightarrow CO + 2H_2 \quad \Delta H_f^\circ = -36 \text{ kJ/mol} \quad (7.5)$$

The combination of Eqs 7.1, 7.3 and 7.5 is known as 'tri-reforming' (Song, 2002) and it is this that produces the correct H_2/CO ratio (1.7:2.0) for methanol or hydrocarbon synthesis (Fox, 1993; Rostrup-Nielsen, 1994; Ross *et al.*, 1996). The tri-reforming has been recently reviewed by Halmann and Steinfeld (2005) and shown to be of practical interest for the treatment of flue gases from coal and gas-fired power stations within the framework of better energy management. Progress in catalyst development (Mirodatos *et al.*, 2005) for dry-reforming may foster the further exploitation of tri-reforming and CO_2 utilization.

Addition of CO_2 to methane for the synthesis of methanol, despite being an endothermic process and the fact that CO_2 consumes 1 mol of H_2 in excess with respect to CO (compare Eqs 7.2 and 7.4), reduces CO_2 emissions by 0.2 $t_{CO_2}/t_{methanol}$ as a consequence of better energy and carbon management.

An interesting innovation is the coupling of cold plasmas with catalysts in methane dry-reforming (Eq. 7.3) for the synthesis of methanol (Eliasson and Kogelschatz, 1991) or oxy-fuels (Zhang *et al.*, 2003). Interestingly, in 2002 a pilot plant for the conversion of natural gas into liquid fuel under plasma conditions began operation in Alberta, Canada (Czernichowski *et al.*, 2002).

The catalytic hydrogenation of CO_2 to methanol has attracted the attention of several research groups and remarkable progress has been made in the last 10 years in terms of catalyst development (reaching a 100% selectivity and high turnover frequency (TOF)). Such excellent performances are due to a different reaction mechanism (Kieffer *et al.*, 1997) of CO_2 with respect to CO. The first methanol synthesis from H_2–CO_2 at the demonstration pilot plant scale (50kg/day) has already been achieved in Japan (Ushikoshi *et al.*, 1998), using a Cu–ZnO-based catalyst at 250°C and 5 MPa. Direct methanol synthesis from an H_2–CO_2 feed has been reviewed by several authors (Saito *et al.*, 1996; Arakawa, 1998; Halmann and Steinberg, 1999). Experimental results show a higher yield of methanol from H_2–CO_2 at 260°C, with respect to H_2–CO, with a further improvement when Pd-modified Cu–ZnO (Inui *et al.*, 2000) is used as catalyst. Nevertheless, if the H_2 is produced from fossil fuels, the direct production of methanol from H_2–CO_2 is uneconomic and makes a very limited contribution to reducing net CO_2 emissions. Conversely, integrated systems can also be used such as:

1. Generation of dihydrogen via electrolysis of water using electric energy from nuclear power plants coupled with CO_2 reduction (the synthesis of methanol is exoergonic);
2. Coupling water oxidation with CO_2 reduction using solar energy.

In this case, a new route can be developed for recycling carbon through the conversion of captured CO_2 into methanol, which may find a practical application during the transition period to the 'H_2-economy'. In fact, such methanol could be used either as a fuel or as a starting material in the chemical industry.

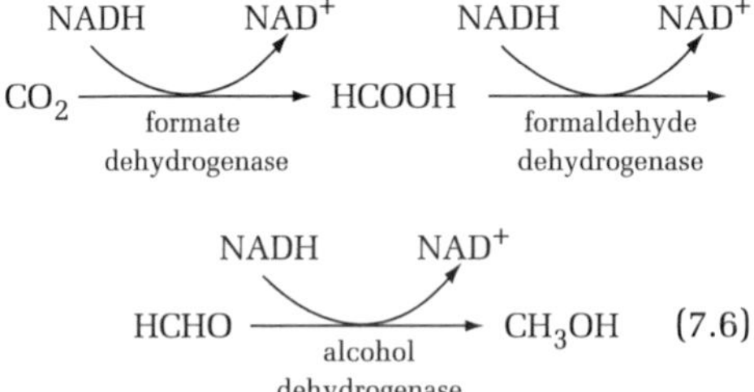

Interestingly, CO_2 can be converted into methanol at room temperature in an aqueous environment by using a cascade of reactions catalysed by enzymes under electron transfer conditions (see Eq. 7.6).

CO_2 is first reduced to HCOOH, which is converted into formaldehyde (CH_2O), and eventually into methanol (Obert and Dave, 1999; Jiang *et al.*, 2003). Such research is still in its infancy, but it has been demonstrated that the three enzymes can be encapsulated and used together for an easy conversion of CO_2 into methanol (Jiang *et al.*, 2003; Xu *et al.*, 2005). The current limitation is in the use of NADPH/NADP couple as an electron source. The hurdle could be overcome by coupling the reaction with the electrochemical re-activation of the reducing agent or with the use of other cheap reductants. An increased availability of enzymes, and the use of existing knowledge of chemical engineering of bioreactors, may allow further exploitation of this technology.

7.2.2 Other reduction reactions

There is industrial interest in more selective and efficient technologies with respect to the use of CO_2 for the direct synthesis of ethanol (C_2) or higher (C_n) alcohols. Efficient catalysts for ethanol formation from H_2–CO_2 have been developed by Takagawa *et al.* (1998), while higher alcohols have been produced by Kieffer *et al.* (1997), but with a yield too low for practical application.

The catalytic hydrogenation of CO_2 has also been applied to the selective synthesis of C_{5+} olefins. Iron carbide (Fe_5C_2) has been patented by the Exxon Corporation (Fiato *et al.*, 1992; Choi *et al.*, 2000). The C_2–C_4 olefin synthesis is made using a Fe–K/alumina catalyst with a selectivity of about 44%, at a CO_2 conversion of 68% under 2 MPa at 400°C. Alternatively, and more efficiently, methanol produced from H_2–CO_2 can be selectively converted using solid acid catalysts. Inui (1996) has shown that in a two-stage reactor, C_2–C_4 olefins are produced with >90% selectivity using the same Cu-ZnO catalyst as in the synthesis of methanol in the first stage and a solid acid catalyst in the second stage.

The photochemical conversion of H_2O–CO_2 into fuels is of interest if solar energy can be used effectively. Several authors (Inoue *et al.*, 1979; Mackor *et al.*, 1987; Arakawa, 2003) have reported that HCOOH, HCHO and CH_3OH are produced in the reduction of CO_2 with H_2O under solar irradiation of an aqueous suspension of a variety of semiconductors such as TiO_2 and $SrTiO_3$. The barrier to this application is the low quantum yield that is achieved (only 0.1%). In order to find a practical application new selective photocatalysts must be developed that are characterized by a light efficiency higher than 2% (Halmann and Steinberg, 1999). This efficiency may be improved by using sacrificial hole traps or electron donors, such as *n*-propanol, tertiary amines or ethylenediaminetetraacetic acid (EDTA), but this approach is not economic. An interesting photocatalytic system for the reduction of CO_2 with H_2O has been reported by Anpo *et al.* (1997), who have described a selectivity of 30% for ethanol production using a Ti-modified mesoporous silica catalyst, compared with just 1.4% over bulk TiO_2.

Overall though, the electrochemical reduction of CO_2 to hydrocarbons is not yet practicable given the low selectivity and the cost, and demands further research (Fujishima, 1998).

7.2.3 Artificial conversion of CO_2 into methane

Another interesting biotechnological approach is the methanation of CO_2 by methanogens, a class of bacteria known to operate under anaerobic conditions (Aresta *et al.*, 1998). Such a reaction can be carried out *in vitro* and can also occur electrochemically when cell membranes are supported over electrodes (Graetzel *et al.*, 1987). However, such artificial conversion of CO_2 into methane, though scientifically interesting, has no practical application currently.

7.2.4 CO_2 as a mild oxidant: hydrocarbon dehydrogenation

An innovative technology is the use of CO_2 as a mild oxidant under controlled conditions. This CO_2-based process operates at much lower temperature than those thermal processes currently in use, while also being more selective. Such an approach can be applied to great environmental benefit in terms of energy savings and associated greenhouse gas (GHG) emissions, as a result of both hydrocarbon oxidation and hydrocarbon valorization. CO_2 is conveniently used, for example, as a dehydrogenating agent in the conversion of $C_6H_5CH_2CH_3$ into styrene, $C_6H_5CH = CH_2$, using TiO_2, ZrO_2 or TiO_2–ZrO_2 (Eq. 7.7) (Park and Yoo, 2004):

$$C_6H_5CH_2CH_3 + CO_2 \rightarrow C_6H_5CH{=}CH_2 + CO + H_2O \quad (7.7)$$

$$CO + H_2O \rightarrow CO_2 + H_2 \quad (7.8)$$

$$C_6H_5CH_2CH_3 \rightarrow C_6H_5CH{=}CH_2 + H_2 \quad (7.9)$$

One of the most interesting aspects of this reaction is that CO could be converted back into CO_2 using an active water–gas shift reaction (WGSR) catalyst (Eq. 7.8) so that the resulting reaction would be an effective dehydrogenation of the starting hydrocarbon (Eq. 7.9) promoted by CO_2. The combination of oxidation catalysts with systems that may promote the WGSR would represent a real breakthrough in this area.

7.3 Carboxylation Reactions

The carboxylation of a substrate, either inorganic or organic, is an exoergonic process, as shown in Table 7.2. Although thermodynamically feasible, such reactions may be characterized by adverse kinetics, so that they result in too slow or too inconvenient an application to be practicable.

7.3.1 Inorganic carbonates

The fixation of CO_2 into natural minerals rich in basic oxides such as CaO or MgO is a methodology for perennial disposal of CO_2 (see Chapter 16). Basic minerals are very abundant as components of the Earth's crust, and can be extracted, milled and used for fixing CO_2 (Zevenhoven and Kavaliauskaite, 2003). In order for this process to have an appreciable rate, it must be carried out in an aqueous medium. The overall process implies the mining, milling, dissolution and reaction of the minerals with a total energy and chemical

Table 7.2. Energy of formation of some chemicals relevant to CO_2 chemistry.

Compound	ΔH_f° (kJ/mol)	ΔG_f° (kJ/mol)	S_f° (cal/K)	Compound	ΔH_f° (kJ/mol)	ΔG_f° (kJ/mol)	S_f° (cal/K)
CO (g)	−110.53	−137.2	197.7	CH_3CNO	−92		
				(l)methylisocyanate			
CO_2 (g)	−393.51	−394.4	213.8	CH_3OCH_3 (g)	−184.1	−112.6	266.4
CO_2 (l)		−386					
CO_2 (aq)	−413.26			C_2H_5OH (l)	−277.7	−174.8	160.7
				(g)	−235.1	−168.5	282.7
CO_3^{2-} (aq)	−675.23			$HOCH_2CH_2OH$ (l)	−455.3		163.2
				(g)	−387.5		303.8
CaO (s)	−634.92			$CH_3CH_2NH_2$ (l)	−74.1		
				(g)	−47.5	36.3	283.8
HCO_3^- (aq)	−689.93	−603.3	38.1	$CH_2CH–CH$ (g)cyclopropene	277.1		
H_2O (l)	−285.83			C_3H_6 (l)	1.7		
				propene (g)	20		
H_2O (g)	−241.83			$CH_2CH_2CH_2$ (g)cyclopropane	53.3		
$CaCO_3$ (s)	−1207.6	− 1129.1	91.7	CH_3COCH_3 (l)	−248.1		
	(calcite)	1128.2	88	(g)	−217.3		
	−1207.8						
	(aragonite)						
$COCl_2$ (g)	−219.1	−204.9	283	OCH_2CH_2CO (l)	−329.9		
CS_2 (l)	89	64.6	151.3	$OCH_2CH_2OC(O)$ (s)	–		
(g)	116.6	67.1	237.8	ethenecarbonate	586.30		
HCN (l)	108.9	125.0	112.8	CH_3CHCH_2O (l)	−122.6		
(g)	135.1	124.7	201.8	(g)	−94.7		
CH_2O (g)	−108.6	−102.5	218.8	C_3H_6O (l)	−215.3		
				propanal (g)	−185.6		
HCOOH (l)	−424.7	−361.4	129	$C_4H_4O_2$ (l) diketene	−233.1		
(g)	−378.6						

CH_4 (g)	−74.4	−50.3	186.3	C_4H_6 (1:2) (l)	139	
				(g)	162	
CH_3Cl (g)	−81.9			C_4H_6 (1:3) (l)	87.9	
				(g)	110	
H_2NCONH_2 (s)	−333.6			COOMe (s)	−756.3	
CH_3OH (l)	−239.1	−166.6	126.8	C_4H_8O (l)	−216.2	
(g)	−201.5	−162.6	239.8	tetrahydrofurane (g)	−184.2	
CaC_2O_4 (s)	−1360.6			$(C_2H_5O)_2CO$ (l)	−681.5	
				ethylcarbonate (g)	−637.9	
MgC_2O_4 (s)	−1269.0			C_5H_{10} (l) 1-pentene	−46.9	
$K_2C_2O_4$ (s)	−1346.0			$C_6H_{12}O_2$ (l) hexanoic acid	−583.8	
PbC_2O_4 (s)	−851.4	−750.1	146	C_6H_6 (l)	49	
				(g)	82.6	
C_2H_2 (g)	228.2	210.7	200.9	C_6H_5OH (s) phenol	−165.1	
C_2H_2O (l) Ketene	−67.9			$C_6H_5NH_2$ (l) aniline	31.3	
(g)	−47.5	−48.3	247.6			
$H_2C_2O_4$ (s)	−821.7		109.8	$C_6H_{12}O$ (l)	−348.2	
(g)	−723.7			1-hexene-oxide		
$Sr(HCOO)_2$ (s)	−1393.3			C_6H_5COOH (s) benzoic acid	−385.2	167.6
CH_3COONa (s)	−708.8	−607.2	123	HO-C_2H_4COOH (s)	−589.9	
C_2H_4 (g)	52.5	68.4	219.6	$C_6H_4(COOH)_2$ (s)	−782.0	
C_2H_4O (l)	−77.8	−11.8	153.9			
Ethene-oxide (g)	−52.6	−13.0	242.5			
CH_3COOH (l)	−484.5	−389.9	159.8	*o*-$H_3CC_6H_4COOH$ (s)	−416.5	
(g)	−432.8	−374.5	282.5	otoluic acid	−426.1	
				m-$H_3CC_6H_4COOH$ (s)	−429.2	
				p-$H_3CC_6H_4COOH$ (s)		
$HCOOCH_3$ (l)	−386.5			$H_3CC_6H_5$ (g) toluene	12.4	
(g)	−355.5		285.3			

cost that is too high for any practical application, unless some other benefits come in support of the technique such as when carbonation is used for the remediation of sites.

7.3.2 Synthesis of organic carbonates, carbamates and isocyanates

CO_2 can advantageously substitute phosgene in the synthesis of carbonates, carbamates, isocyanates, polycarbonates and polyurethanes. The following scheme (Scheme 7.2) shows the network of reactions based on CO_2 that correlates the chemicals mentioned previously. The carboxylation of epoxides and the aminolysis of carbonates are already well established. Other processes are at different states of development (Aresta *et al.*, 2003c). Carbonates are good target molecules due to their favourable thermodynamics and their widespread use as solvents or reagents:

$$COCl_2 + 2ROH + 2NaOH \rightarrow (RO)_2CO + 2NaCl + 2H_2O \quad (7.10)$$

$$CO + {}^{1}/{}_{2}O_2 + 2ROH \rightarrow (RO)_2CO + H_2O \quad (7.11)$$

$$2ROH + CO_2 \rightarrow (RO)_2CO + H_2O \quad (7.12)$$

Equations 7.10–7.12 represent the first, second and third generation processes, respectively, for the synthesis of dimethylcarbonate (DMC). The utilization of CO_2 (Eq. 7.12) is a winning strategy as it is a clean process that also complies with the atom economy principle (for R = CH_3, the atom efficiency is 54.5% for Eq. 7.10, and 80% for Eqs 7.11 and 7.12). Water is the only co-product, and the reaction is selective (100%). For thermodynamic isues, the reaction is slightly shifted to the left. New catalysts are needed and new reactors for shifting the equilibrium to the right (Aresta and Dibenedetto, 2002; Aresta and Dibenedetto, 2003; Aresta *et al.*, 2003a,b, 2005a; Ballivet-Tkatchenko and Sorokina, 2003; Ricci, 2003).

The production of polycarbonates by reacting epoxides and CO_2 is well understood and industrially exploited. Only a limited number of epoxides have been used so far (essentially propene oxide and cyclohexene oxide) whereas many others could be used to generate materials with new properties (Beckman *et al.*, 1997). Polycarbonates find a large utilization in

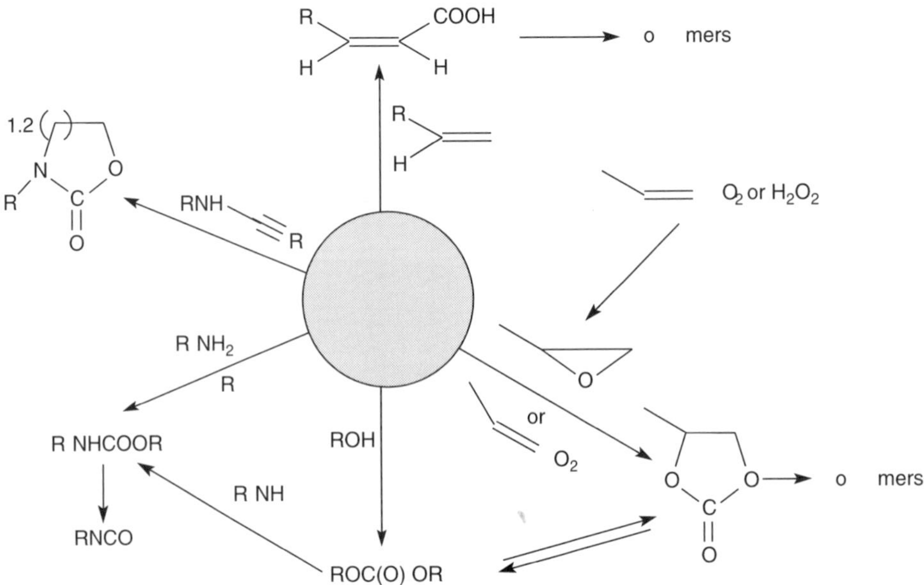

Scheme 7.2. Carboxylation reactions based on CO_2.

several industrial sectors, with an expanding market (Burridge, 2002a).

Urea could find a new use as a chemical building block for the synthesis of urethanes and carbonates, with a significant utilization of CO_2 as a result (Eqs 7.13 and 7.14). Ammonia can be recycled and converted into urea:

$$H_2NCONH_2 + ROH \rightarrow H_2NCOOR + NH_3 \quad (7.13)$$

$$H_2NCOOR + ROH \rightarrow (RO)_2CO + NH_3 \quad (7.14)$$

Similarly, carbonates can be used as starting materials for the synthesis of carbamates by aminolysis with primary or secondary amines (Aresta *et al.*, 2004).

Other applications with large potential include the production of polyurethanes from isocyanates (Burridge, 2002b,c) and polyesters (co-polymers CO_2-ethene or propene), the polymers with the largest market.

7.3.3 Synthesis of carboxylates

The direct carboxylation of an organic substrate to produce carboxylic acids and their derivatives is an interesting application of CO_2. Such reactions have been attempted using 'residual' or 'waste' ionizing radiations that promote CO_2 fixation at room temperature. Applications of this technology include the direct synthesis of carboxylates from hydrocarbons and the polymerization in or with supercritical CO_2 (sc-CO_2) (Fujita *et al.*, 1994, 1996). The carboxylation of active C–H bonds has been achieved with 100% selectivity at room temperature using carboxylated ionic liquids (Aresta *et al.*, 2003e).

Specialty chemicals, having a small market (several thousand tonnes per year) and high value, such as pharmaceuticals, amino acids and asymmetric compounds, tend to be synthesized through complex routes, and produce a large amount of waste, often 50 times their mass. As such, they may represent an attractive application of CO_2-based electrochemical syntheses (Dunach *et al.*, 2001; Augustinski *et al.*, 2003).

7.4 Use of Supercritical CO_2 as a Solvent and Reagent

CO_2 exists as a supercritical fluid above 31°C and 7.38 MPa. Its density and viscosity can be modulated over quite a wide range by changing the pressure and temperature. The 'dense phase fluid' has properties close to those of a non-polar organic solvent, such as pentane or dichloromethane. The benefits derived from its use are well known today, so that its utilization is spreading in various industrial sectors. Environmental concerns about the use of organic solvents have also propelled research efforts towards the replacement of traditional organic solvents with sc-CO_2. Such CO_2 has been in use for a long time in many chemical applications, which include:

- Decaffeination of coffee beans (Anastas *et al.*, 1996; Arakawa *et al.*, 2001);
- Extraction of fragrances and essences from plants, or proteins (Wahbeh, 1997), or fatty acids and hydrocarbons from algae (Aresta *et al.*, 2003d);
- Use as a solvent for reactions, crystallizations, prepartions of solid thermal-sensitive phamaceuticals having controlled size distribution (Van Ginneken and Weyten, 2003a);
- Catalysis, both homogeneous (Holmes *et al.*, 2003) and heterogeneous (Poliakoff *et al.*, 2003);
- Synthesis (Filardo *et al.*, 2003a) and modification of polymers including perfluoropolymers (Filardo *et al.*, 2003b);
- As a mobile phase for supercritical fluid chromatography (SFC) (Van Ginneken and Weyten, 2003b);
- For dyeing (Penninger, 2002), dry-cleaning (DeSimone, 2001) and even nuclear waste treatment (RUCADI Project Final Report).

The most important feature of sc-CO_2 use is that it can be easily recovered at the end of the process (by thermal decompression), recompressed and recycled. The market demand is growing and may reach several million tonnes per year in the short term. It is worth recalling here that for each tonne of organic solvent (global market of the

Table 7.3. Compounds extracted from algae.

Classes of products	Compounds	Uses
Proteins		Food industry
Amino acids		Food industry
Lipids		Biofuel, food industry, pharmaceutical industry
Essences, fragrances and oil	Geraniol, formates, acetates, citronellol, nonanol, eucalyptol	Various industries
Alkaloids		Pharmaceutical industry
Sterols	Cholesterol	
Pigment, chlorophille, carotenoids	Isoprenoids	
Amines	Mono-, di-, trimethylamines, ethylamine, propylamine, isobutylamine	Pharmaceutical industry
Inorganic compounds	Iodides, bromides, sulphates, nitrates	Pharmaceutical industry

order of 35 Tg per year) replaced with sc-CO_2, there is a reduction of the CO_2 emission of the order of 3 t due to the fact that waste solvents will not be burned.

7.5 Indirect Production of Chemicals from CO_2: Enhanced Aquatic Biomass Production and Their Conversion

The use of biomass, mainly aquatic, as a source of chemicals brings together the interests of scientists and industrialists. Programmes are ongoing all over the world aimed at establishing the potential of the use of terrestrial and aquatic biomass as a source of chemicals (biorefinery) and fuels (biodiesel). Aquatic biomass has the advantage of less land requirement: 1 Pg/year of CO_2 capture would require 6 million hectares, which is much less than terrestrial biomass. Also, the growth of micro-algae (Benemann and Oswald, 1996; Benemann *et al.*, 2005) or macro-algae (Dibenedetto and Tommasi, 2003; Aresta *et al.*, 2005b) can be associated with water reclamation (fisheries, sewage, some industrial waters), with added environmental and economic benefits. The widespread implementation of such technologies has been prevented so far by the low price of fossil fuels. By combining the actual rise of oil price with the convenience represented by water treatment, one can foresee a possible large-scale (>100 Tg per year) exploitation of technologies based on the utilization of biomass in the medium term. Several technologies are available for biomass treatment including extraction with sc-CO_2, which is particularly well suited to obtaining liquid fuels and other thermolabile compounds; gasification, for the production of Syngas; pyrolysis, which gives crude products; and anaerobic fermentation for the production of biogas. Several compounds with a significant market (>1 Tg per year) are also produced from macro-algae on an almost zero carbon emission base (Table 7.3).

7.6 Conclusions

The industrial utilization of CO_2 can contribute to reducing global CO_2 emissions, provided that the CO_2-based innovative technologies are more energy- and carbon-efficient than existing ones. This requires the development of more direct syntheses, with high selectivity and conversion yields. With the exception of polymers, the products derived from CO_2 will not have a 'storage' capacity, if used. The amount of 'avoided' CO_2 can be estimated using the LCA methodology. The implementation of all possible technologies based on CO_2 (synthetic and other uses) discussed here may allow a reduction in emissions of 250–300 Tg CO_2/year in the medium term.

A large potential for reducing CO_2 exists in photochemical processes if solar energy is used for CO_2 reduction and water oxidation. In such a case, the amount of recycled carbon would be well beyond the actual estimate (just 7%) of the potential for carbon recovery in this way (IEA, 1990).

The most promising applications appear to be: the production of organic carbonates, carbamates, isocyanates, polymers (polycarbonates and polyurethans), carboxylates and methanol (CO_2 as additive to CO); the use as mild oxidant; and the use as technological fluid (the use in air conditioners may be a new important application).

The development of such new technologies may, directly or indirectly, reduce the amount of CO_2 released to the atmosphere, while further benefits (security, health) may come from the substitution of toxic compounds currently used in the chemical industry and other applications (fluids in refrigeration apparatus).

The amount of avoided CO_2 may not represent a large share of the reduction required at the world level, but it is important to note that it will also allow a reduction in fossil fuel extraction. The exploitation of the utilization option will generate a profit that may also pay for other disposal options.

References

Anastas, P.T., Webster, L.C. and Williamson, T.C. (1996) Environmentally benign production of commodity chemicals through biotechnology: recent progress and future potential. In: *Green Chemistry*. ACS Symposium Series 626, pp. 198–211.

Anpo, M., Zhag, S.G., Fujii, Y., Yamashita, H., Koyano, K. and Tatsumi, K.T. (1997) Photocatalytic reduction of CO_2 with H_2O on Ti-MCM-41 and Ti-MCM-48 mesoporous zeolites at 328K. *Chemistry Letters* 7, 659–660.

Arakawa, H. (1998) Research and development on new synthetic routes for basic chemicals by catalytic hydrogenation of CO_2. In: *Advances in Chemical Conversions for Mitigating Carbon Dioxide*. Elsevier Science BV, Amsterdam, The Netherlands, pp. 19–30.

Arakawa, H., Zou, Z., Sayama, K., Abe, R. and Kusama, H. (2003) Solar hydrogen production for carbon dioxide fixation. *Book of Abstracts, 7th ICCDU*, 12–16 October, Seoul, Korea, 133–134.

Aresta, M. (1987) The carbon dioxide problems: mankind energy needs and environmental pollution. In: Aresta, M. and Forti, G. (eds) *Carbon Dioxide as a Source of Carbon: Biochemical and Chemical Uses*. D. Reidel Publishing Company, Dordrecht, The Netherlands, pp. 1–22.

Aresta, M. and Dibenedetto, A. (2002) Carbon dioxide as building block for the synthesis of organic carbonates: behavior of homogeneous and heterogeneous catalysts in the oxidative carboxylation of olefins. *Journal of Molecular Catalysis A: Chemical* 182–183, 399–409.

Aresta, M., Tommasi, I., Giannoccaro, P., Quaranta, E. and Fragale, C. (1998) Bioinorganic chemistry of nickel and carbon dioxide: a Ni-complex behaving as a model system for carbon monoxide dehydrogenase enzyme. *Inorganica Chimica Acta* 272, 38–42.

Aresta, M. and Dibenedetto, A. (2003) Carbon dioxide fixation into organic compounds. In: Aresta, M. (ed.) *Recovery and Utilization of Carbon Dioxide*. Kluwer Academic Publishers, The Netherlands, pp. 211–260.

Aresta, M., Dibenedetto, A., Gianfrate, L. and Pastore, C. (2003a) Enantioselective synthesis of organic carbonates promoted by Nb(IV) and Nb(V) catalysts. *Applied Catalysis A: General* 255(1), 5–11.

Aresta, M., Dibenedetto, A., Gianfrate, L. and Pastore, C. (2003b) Nb(V) Compounds as epoxides carboxylation catalysts: the role of the solvent. *Journal of Molecular Catalysis A: Chemical* 204–205, 245–252.

Aresta, M., Dibenedetto, A., Dileo, C., Tommasi, I. and Amodio, E. (2003c) The first synthesis of a cyclic carbonate from a ketal in sc-CO_2. *Journal of Supercritical Fluids* 25(2), 177–182.

Aresta, M., Dibenedetto, A., Tommasi, I., Cecere, E., Narracci, M., Petrocelli, A. and Perrone, C. (2003d) The use of marine biomass as renewable energy source for reducing CO_2 emissions. In: Gale, J. and Kaya, Y. (eds) *Greenhouse Gas Control Technologies*, Vol. 11. Elsevier Science Ltd, The Netherlands, pp. 1497–1502.

Aresta, M., Tkatchenko, I. and Tommasi, I. (2003e) Unprecedented synthesis of 1,3-dialkylimida zolium-2-carboxylate: applications in the synthesis of halogen-free ionic liquids and reactivity as carbon dioxide transfer agent to active C–H bonds. In: *Ionic Liquids as Green Solvents: Progress and Prospects*. ACS Symposium Series 856, 93–99.

Aresta, M., Dibenedetto, A., Devita, C., Bourova, O.A. and Chupakin, O.N. (2004) New catalysts for the conversion of urea into carbamates and carbonates with C1 and C2 alcohols. *Studies in Surface Science and Catalysis* 153, 213–220.

Aresta, M., Dibenedetto, A., Pastore, C., Giannoccaro, P., Papai, I. and Schubert, G. (2005a) Mechanism of formation of organic carbonates from aliphatic alcohols and carbon dioxide under mild conditions promoted by carbodiimides: DFT calculation and experimental study. *Journal of Organic Chemistry* 70, 6177–6186.

Aresta, M., Alabiso, G., Cecere, E., Carone, M., Ribenedetto, A. and Petrocelli, A. (2005b) Influence of the CO_2 concentration on the abundance of fatty acids in the macroalga *C. linum*. *Proceedings of the International Conference on CO_2 Utilization*, 206, Oslo, 20–23 June.

Audus, H. and Oonk, H. (1997) An assessment procedure for chemical utilization schemes intended to reduce CO_2 emission to atmosphere. *Energy Conversion and Management* 38 (Suppl. Proceedings of the Third International Conference on Carbon Dioxide Removal, 1996), S409–S414.

Augustinski, J., Sartoretti, C.J. and Kedzierzawski, P. (2003) Electrochemical conversion of carbon dioxide. In: Aresta, M. (ed.) *Recovery and Utilization of Carbon Dioxide*. Kluwer Academic Publishers, The Netherlands, pp. 280–292.

Ballivet-Tkatchenko, D. and Sorokina, S. (2003) Linear organic carbonates. In: Aresta, M. (ed.) *Recovery and Utilization of Carbon Dioxide*. Kluwer Academic Publishers, The Netherlands, pp. 261–277.

Beckman, E., Super, M., Berluche, E. and Ostello, C. (1997) Copolimerization of 1,2-epoxycyclohexane and carbon dioxide using carbon dioxide as both reactant and solvent. *Macromolecules* 30(3), 368–372.

Benemann, J.R. and Oswald, W.J. (1996) Systems and economic analysis of microalgae ponds for conversion of CO_2 to biomass. *Final Report*, Pittsburg Energy Technology Center, 260.

Benemann, J.R., Manancourt, A. and Pedroni, P.M. (2005) Biofixation of CO_2 and greenhouse gas abatement with microalgae. *Proceedings of the International Conference on Carbon Diozide Utilization*, 46, Oslo, 20–23 June.

Burridge, E. (2002a) Polycarbonate. *European Chemical News* June, 14, 24–30.

Burridge, E. (2002b) MDI. *European Chemical News* July, 15, 24–30.

Burridge, E. (2002c) TDI. *European Chemical News* July, 14, 24–30.

Choi, M.J., Kikim, K., Lee, H., Kim, S.B., Nam, J.S. and Lee, K.W. (2000) *Proceedings of the Fifth International Conference on Greenhouse Gas Control Technologies*, 607–612.

Czernichowski, A., Czernichowski, M., Czernichowski, P. and Cooley, T.E. (2002) Reforming of methane into syngas in a plasma-assisted reactor. *Fuel Chemistry Division Preprints* 47(1), 280–281.

DeSimone, J. (2001) The CO_2 technology platform: surfactants for increased CO_2 utility. *Proceedings of ICCDU VI*, 9–14 September, 33.

Dibenedetto, A. and Tommasi, I. (2003) Biological utilization of carbon dioxide: the marine biomass option. In: Aresta, M. (ed.) *Recovery and Utilization of Carbon Dioxide*. Kluwer Academic Publishers, The Netherlands, pp. 315–324.

Dunach, E., Tascedda, P., Weidmann, M. and Dinjus, E. (2001) Nickel-catalyzed electrochemical carboxylation of epoxides: mechanistic aspects. *Applied Organometallic Chemistry* 15, 141–144.

Eliasson, B. and Kogelschatz, U. (1991) Non equilibrium volume plasma chemical processing. *IEEE Transaction on Plasma Science* 19(6), 1063–1077.

Fiato, R.A., Soled, S.L., Rice, G.B. and Miseo, S. (1992) Preparation of olefins by hydrogenation of carbon dioxide. *US Patent* 5, 140,049.

Filardo, G., Galia, A. and Giaconia, A. (2003a) Polymer synthesis in supercritical carbon dioxide. In: Aresta, M. (ed.) *Recovery and Utilization of Carbon Dioxide*. Kluwer Academic Publishers, The Netherlands, pp. 181–195.

Filardo, G., Galia, A. and Giaconia, A. (2003b) Modification of polymer in supercritical carbon dioxide. In: Aresta, M. (ed.) *Recovery and Utilization of Carbon Dioxide*. Kluwer Academic Publishers, The Netherlands, pp. 197–207.

Fox, J.M. III (1993) The different catalytic routes for methane valorization: an assessment of process for liquid fuels. *Catalysis Review–Science and Engineering* 35(2), 169–212.

Fujishima, A. (1998) Research and development on new synthetic routes for basic chemicals by catalytic hydrogenation of CO_2. In: *Advances in Chemical Conversions for Mitigating Carbon Dioxide*. Elsevier Science BV, Amsterdam, The Netherlands, pp. 31–42.

Fujita, N., Morita, H., Matsuura, C. and Hiroishi, D. (1994) Radiation-induced CO_2 reduction in an aqueous-medium suspended with iron-powder. *Radiation Physical Chemistry* 44(4), 349–357.

Fujita, N., Fukuda, Y., Matsuura, C. and Saigo, K. (1996) Changes in pH and redox potential during radiation-induced CO_2 reduction in an aqueous solution containing iron powder. *Radiation Physical Chemistry* 47(4), 543–549.

Graetzel, M., Thampi, K.R. and Kiwi, J. (1987) Methanation and photomethanation of CO_2 at room temperature and atmospheric pressure. *Nature* (327), 506–508.

Halmann, M.M. and Steinberg, M. (eds) (1999) *Greenhouse Gas Carbon Dioxide Mitigation Science and Technology*. Lewis Publishers, Boca Raton, Florida.

Halmann, M.M. and Steinfeld, A. (2005) *Proceedings of the International Conference on Carbon Dioxide Utilization*, 24, Oslo, 20–23 June.

Holmes, A.B., Quaranta, E., Early, T.R., Lee, J.K. and Stamp, L.M. (2003) Homogeneous catalysis in supercritical carbon dioxide. In: Aresta, M. (ed.) *Recovery and Utilization of Carbon Dioxide*. Kluwer Academic Publishers, The Netherlands, pp. 149–168.

IEA (International Energy Agency) (1990) Executive Conference on 'Solar Photoconversion Processes for Recycling Carbon Dioxide from the Atmosphere: An Assessment of Energy, Research and Industrial Implication', Colorado Springs, Colorado, 13–16 March.

Inoue, T., Fujisima, A., Konishi, S. and Honda, K. (1979) Photoelectorocatalytic reduction of carbon dioxide in aqueous suspensions of semiconductor powders. *Nature* 277, 637–638.

Inui, T. (1996) Highly effective conversion of carbon dioxide to valuable compounds on composite catalysts. *Catalysis Today* 29(1–4), 329–337.

Inui, T., Takeguchi, T., Yanagisawa, K. and Inoue, M. (2000) Effect of the property of solid acid upon syngas-to-dimethyl ether conversion on he hybrid catalysts composed of Cu–Zn–Ga and solid acids. *Applied Catalysis A: General* 192, 201–209.

Jiang, Z., Xu, S. and Wu, H. (2003) Novel conversion of carbon dioxide to methanol catalyzed by sol-gel immobilized dehydrogenases. *Book of Abstract, 7th ICCDU*, 12–16 October, Seoul – Korea, 21–22.

Kieffer, R., Fujiwara, M., Udron, L. and Souma, Y. (1997) Hydrogenation of CO and CO_2 toward methanol, alcohols and hydrocarbons on promoted copper-rare earth oxide catalysts. *Catalysis Today* 36(1), 15–24.

Lange, J.P. (1997) Perspectives for manufacturing methanol at fuel value. *Industrial Engineering Chemical Research* 36, 4282–4290.

Lemonidou, A.A., Valla, J. and Vasalos, I.A. (2003) Methanol production from natural gas. Assessment of CO_2 utilization in natural gas reforming. In: Aresta, M. (ed.) *Recovery and Utilization of Carbon Dioxide*. Kluwer Academic Publishers, The Netherlands, pp. 379–394.

Mackor, A., Tinnemans, A.H.A. and Koster, T.P.M. (1987) Reduction of CO_2 at titanate powders in sunligth and at electrodes in the light or dark. In: Aresta, M. and Forti, G. (eds) *Carbon Dioxide as a Source of Carbon: Biochemical and Chemical Uses*. D. Reidel Publishing Company, Dordrecht, The Netherlands.

Mann, M.K. and Spath, P.L. (1997) *Life Cycle Assessment of a Biomass Gasification Combined-cycle Power System*. National Renewable Energy Laboratory, Golden, Colorado, December 1997.

Maroto-Valer, M.M., Song, C. and Soong, Y. (eds) (2002) *Environmental Challenges and Greenhouse Gas Control for Fossil Fuel Utilization in the 21st Century*. Kluwer Academic/Plenum Publishers, New York.

Mirodatos, C., Olafsen, A., Schuurman, C., Daniel, C., Raberg, L.B., Jensen, M.B. and Olsbye, U. (2005) Light alkanes CO_2 reforming to synthesis gas over Ni-based catalysts. *Proceedings of the International Conference on Carbon Dioxide Utilization*, 36, Olso, 20–23 June.

Obert, R. and Dave, B.C. (1999) Enzymatic conversion of carbon dioxide to methanol: enhanced methanol production in silica sol-gel matrices. *Journal of American Chemical Society* 121, 12192–12193.

Park, S.-E. and Yoo, J.S. (2004) New CO_2 chemistry: recent advances in utilizing CO_2 as an oxidant and current understanding of its role. *Studies in Surface Science and Catalysis* 153, 303–314.

Pechtl, P.A. (1991) Carbon dioxide emission control: evaluation of carbon dioxide control measures in fossil fuel-fired power plants. *Erdoel & Kohle, Erdgas, Petrochemie* 44(4), 159–165.

Penninger, J. (2002) EU RUCADI Project BRRT-CT98-5089. Final Report.

Poliakoff, M., Amandi, R. and Hyde, J. (2003) Heterogeneous reactions in supercritical carbon dioxide. In: Aresta, M. (ed.) *Recovery and Utilization of Carbon Dioxide*. Kluwer Academic Publishers, The Netherlands, pp. 169–179.

Ricci, M. (2003) Carbon dioxide as a building block for organic intermediates: an industrial perspective. In: Aresta, M. (ed.) *Recovery and Utilization of Carbon Dioxide*. Kluwer Academic Publishers, The Netherlands, pp. 395–402.

Ross, J.R.H., van Keulen, A.N.J., Hegarty, M.E.S. and Seshan, K. (1996) The catalytic conversion of natural gas to useful products. *Catalysis Today* 30(1–3), 193–199.

Rostrup-Nielsen, J.R. (1994) Aspects of CO_2-reforming of methane. In: Curry-Hyde, H.E. and Howe, R.F. (eds) *Natural Gas Conversion II. Proceedings of the 3rd Natural Gas Conversion Symposium*. Elsevier, Amsterdam, The Netherlands, pp. 25–41.

Rozovskii, A. (1989) Modern problems in the synthesis of methanol. *Russian Chemical Review* 58(1), 41–56.

Saito, M., Fujitani, T., Takeuchi, M. and Watanabe, T. (1996) Development of copper/zinc oxide-based multicomponent catalysts for methanol synthesis from carbon dioxide and hydrogen. *Applied Catalysis A: General* 138(2), 311–318.

Song, C. (2002) CO_2 Conversion and utilization: an overview. In: Song, C., Gaffney, A.M. and Fujimoto, K. (eds) *CO_2 Conversion and Utilization*. ACS Symp Series 809, USA, pp. 2–30.

Takagawa, M., Okamoto, A., Fujimura, H., Izawa, Y. and Arakawa, H. (1998) Ethanol synthesis from carbon dioxide and hydrogen. In: *Advances in Chemical Conversions for Mitigating Carbon Dioxide*. Elsevier Science BV, Amsterdam, The Netherlands, pp. 525–528.

Ushikoshi, K., Mori, K., Watanabe, T., Takeuchi, M. and Saito, M. (1998) A 50 kg/day class test plant for methanol synthesis from CO_2 and H_2. In: *Advances in Chemical Conversions for Mitigating Carbon Dioxide*. Elsevier Science BV, Amsterdam, The Netherlands, pp. 357–362.

Van Ginneken, L. and Weyten, H. (2003a) Particle formation using supercritical carbon dioxide. In: Aresta, M. (ed.) *Recovery and Utilization of Carbon Dioxide*. Kluwer Academic Publishers, The Netherlands, pp. 123–136.

Van Ginneken, L. and Weyten, H. (2003b) Supercritical fluid chromatography. In: Aresta, M. (ed.) *Recovery and Utilization of Carbon Dioxide*. Kluwer Academic Publishers, The Netherlands, pp. 137–148.

Vansant, J. (2003) Carbon dioxide emission and merchant market in the European Union. In: Aresta, M. (ed.) *Recovery and Utilization of Carbon Dioxide*. Kluwer Academic Publishers, The Netherlands, pp. 3–50.

Wahbeh, M.I. (1997) Amino acid and fatty acid profiles of four species of macroalgae from aqaba and their suitability for use in fish diets. *Aquaculture* 159, 101–109.

Xu, S.-W., Jiang, K.-Y., Lu, Y. and Wu, H. (2005) Three dehydrogenases coencapsulated in novel Alg-SiO_2 hybrid composite for efficient conversion of CO_2 to methanol. *Proceedings of the International Conference on Carbon Dioxide Utilization*, 50, Oslo, 20–23 June.

Zevenhoven, R. and Kavaliauskaite, I. (2003) Mineral carbonation for long-term CO_2 storage: an energy analysis. *Proceedings of ECOS 2003 – s 16th International Conference on Efficiency, Costs, Optimisation, Simulation and Environmental Impact of Energy Systems*, Copenhagen, Denmark, 30 June to 2 July (also published in 2004 in *International Journal of Thermodynamics* 1, 127–134).

Zhang, Y., Li, Y., Wang, Y., Liu, C. and Eliasson, B. (2003) Plasma methane conversion in the presence of carbon dioxide using dielectric-barrier discharges. *Fuel Processing Technology* 83(1), 101–109.

8 Prospects for Biological Carbon Sinks in Greenhouse Gas Emissions Trading Systems

John Reilly[1], Benjamin Felzer[2], David Kicklighter[2], Jerry Melillo[2], Hanqin Tian[3] and Malcolm Asadoorian[1]

[1]*Joint Program on the Science and Policy of Global Change, Massachusetts Institute of Technology, Cambridge, Massachusetts, USA;* [2]*The Ecosystems Center, Marine Biological Laboratory, Massachusetts, USA;* [3]*School of Forestry and Wildlife Sciences, Auburn University, Auburn, Alabama, USA*

8.1 Introduction

The role of sinks in climate policy has been controversial and confused. The major supporters for including sinks in an international climate policy under the Kyoto Protocol were the Umbrella Group of countries, led by the USA and including Australia, Canada, Japan and Russia. This group also pushed strongly for international emissions trading, imagining that countries would distribute emissions allowances to private sector emitters, who would then be required to have an allowance for each tonne of greenhouse gas (GHG) they emitted. With emissions trading, emitters who found they could cheaply reduce their emissions might have allowances to sell, or those who could not easily reduce these could purchase allowances to cover their emissions. With international trading, these permits could be exchanged among allowance holders anywhere among the parties subject to an emissions cap.

In principle, accounting and crediting sinks under a cap-and-trade system should be straightforward: (i) measure the stock of carbon at an initial year; (ii) measure the stock of carbon in subsequent years; (iii) if the carbon stock rises from one period to the next, the increased sequestration is added to the allowances or cap on emissions of the country or entity, and if the stock declines, the net release to the atmosphere is subtracted from the allowances or cap. This simplicity has eluded designers of carbon policy. For various reasons, a desire has developed to identify specific types of sink-enhancement actions that may or may not be included under agreed caps as well as an unwillingness to bring the entire terrestrial biosphere carbon stock within a policy target. The result has been thousands of pages of attempts to define a forest, the difference between afforestation and reforestation, what constitutes 'management', if a change in carbon stocks is due to human action, and spatial and temporal leakage. Most of this would be irrelevant if a simple accounting framework and broad coverage of land use emissions and uptake were adopted in the design of carbon policy. How and why did we get from a simple and straightforward idea to the complex design and controversial issues now discussed as part of sinks policy? Are there good reasons why the problem is not as simple as it at first seems? Is it possible (or desirable) to now try to work towards fairly simple mechanisms for sinks in a carbon policy? These are the questions we hope to address in this chapter.

We first review existing policies with attention to issues that arise with regard to terrestrial sinks, and how sinks are to be included. We then show some of the important aspects of managing sinks that arise because they depend on environmental conditions that are largely outside the control of the land owner. Next we work through a very simple example of two hypothetical countries, and show the effects of including sinks. Finally we address several issues that have arisen as countries have negotiated the inclusion of sinks in GHG mitigation policies. Some of these are important and real issues that must be addressed if climate policy design is to create incentives for efficiently managing carbon in the terrestrial biosphere. However, many of the issues arise from, or in response to, the tangled policy approaches we have designed for sinks enhancement, and attempts to straighten it out seem only to further tangle the issue.

8.2 Current Climate Policies, Emissions Trading and the Role of Sinks

After fighting hard for sinks and emissions trading in the Kyoto Protocol, the USA and Australia are among the few countries that, while having signed the Protocol initially, have now expressed their intention of not ratifying it. Thus, key Conference of the Party (COP) members who had pushed hardest for inclusion of sinks are now not part of the Protocol. Canada has ratified and is perhaps most active among ratifying parties in developing measuring and monitoring techniques that they hope would allow expanded inclusion of sinks. After much uncertainty, the Protocol entered into force because Russia ratified the agreement. With Russia, emissions of ratifying Annex B members – who took on binding caps under the Protocol – exceed 55% of the 1990 emissions of the original Annex B list. This was the key threshold for entry into force as set out in the Protocol, and Russia's ratification brought the parties across that threshold (UNFCCC, 1997, 2005).

Australia has indicated that, while not ratifying Kyoto, it would meet its obligations under the agreement. What this means is unclear, but it may affect the sinks issue. If not formally under Kyoto, Australia could pursue a strategy of meeting the numerical target while defining credits for land-use change beyond the limits of the Protocol. The possibility of crediting reduced rates of deforestation against Australia's target was identified in at least one study of the pros and cons of Australian ratification (Kyoto Ratification Advisory Group, 2003). At this point, the chance that the USA would meet the numerical target set out under Kyoto seems remote. The Bush administration announced instead an emissions-intensity target that would allow emissions to rise somewhat from 2000 levels, which contrasts with the Kyoto requirement that the USA return to 93% of 1990 levels (White House, 2002). For the time being the administration believes its intensity target will be met through voluntary measures. The administration pressured industries to identify emissions-reducing actions, and attempted to publicize these promises. Given the nature of the target and the promised actions, varying changes in emissions intensity in different industries, and changing industry composition over time, it is hard to determine with any rigour whether the actions proposed are sufficient to meet the intensity goal. The other continuing effort in the USA is a programme whereby entities can receive recognition for reducing emissions or enhancing sinks by being awarded 'registered reductions' (Federal Register, 2002). The 'incentive' to do so is either goodwill or, more likely, expectations that at some point, there will be a mandatory cap on at least some entities and that registered reductions could be applied to them.

Despite the absence from the Protocol of the USA and Australia, the key supporters of trading, the push for emissions trading appears to have taken hold to some extent under Kyoto. The EU has developed an emissions trading system, and introduced a test phase (2005–2007) that will run prior to the first commitment period of the Kyoto Protocol (EC, 2003, 2005; Betz *et al.*, 2004). So even though the EU was initially hesitant on emissions trading, it now appears to be a major force in designing a domestic system that could be a model for other parties, making the vision of an international market for permits a reality (Ellerman, 2001). There is, however, still a long way to go to extend such a trading system among all ratifying parties. The EU's test programme is limited to large emitting point sources (>10,000t CO_2/year) and thus covers less than half of the EU's total CO_2 emissions. Other key parties including Canada, Japan and Russia have not yet moved to establish emissions trading systems. Performance in the EU's trading system from 2005 to early 2006 resulted in prices on the order of ~20–28/t CO_2, surprisingly high to many analysts because the required reduction was estimated to be only ~1% (Pew Center, 2005; Point Carbon, 2005a,b). In May 2006, the price dropped to as low as ~9/t CO_2 and trading was generally in the range of ~10–20/t CO_2.

While emissions' trading remains alive under the Protocol, the Umbrella group's push for sinks was not nearly as successful. Part of the reason why agreement was not reached at the 6th meeting of the COP in The Hague in November 2001 was that the EU held out for limits on the total quantities of sinks credits that could be applied against each country's emissions cap, and this was unacceptable to the USA. Even before George W. Bush was elected President and announced the USA's rejection of the Protocol, the failure at The Hague was essentially the death knell of Kyoto in the USA (Reiner, 2001; Reilly, 2003). The difference in willingness to embrace sinks appears to derive from different views of the nature of the climate change issue as a societal problem. Many in Europe saw the response to the climate change problem as part of an even broader agenda of switching from fossil fuels to 'renewable' sources of energy. The use of sinks was, in the view of some at least, denial or avoidance of these necessary steps to turn the economy away from fossil fuels. In the language of economics, this view might be cast as one in which markets had failed to price fossil fuels to include all of the social costs associated with them (everything from security, air pollution, other health and safety issues, their nature as exhaustible resource, etc.). Rather than try to correct each of these problems, renewable energy proponents see the answer as simply switching away from fossil fuels. Sinks credits were thus seen as a loophole, allowing continued fossil fuel use. This view has appeared to influence the EU's climate change – negotiating positions and the formulation of its domestic policies.

In contrast, the perspective of the USA and Umbrella group concentrated directly on the climate GHG problem, and for carbon this meant focusing on actions that would limit atmospheric concentrations of CO_2. It mattered little whether fossil fuel emissions were reduced or carbon uptake by vegetation and soils was increased, or carbon was otherwise sequestered. One tonne removed from the atmosphere was just as good as reducing emissions by a tonne. This focus, along with a desire for cost-effectiveness, led to a desire for maximum flexibility in choosing the least costly way to reduce atmospheric CO_2 levels. Separate quantity limits on the use of sinks, if binding, would result in a two-tier permit market – a higher price for emissions reduction and lower price for sinks, reflecting the fact that there were more cheap sinks options available than allowed by the restriction on their use. Most analysts believe that the sinks quantities allowed in the Kyoto Protocol as finally negotiated at COP 7 in Marrakesh in 2001 are so limiting, and the definition of what can be counted so loose that the agreement has been widely modelled as simply a relaxation of the constraint on emissions (Babiker *et al.*, 2002). Underlying this view is the calculation that most countries are likely to have enough carbon uptake in forests without doing anything more than they would have done anyway to fill their sinks limit. In particular, the Protocol allows consideration of forest uptake anytime from 1990 through 2012 to be credited against emissions in the first commitment period of 2008–2012.

The language of the agreement requires some sort of active management to get credit for carbon uptake. For some countries, the implicit definition of 'management' was very narrow: identified tracts of land that were replanted, or planted, and managed with the express intent of storing carbon. Analysts began referring to 'Kyoto forests' to represent the idea that lands earning credits for forest activities would be specifically identified. However, the USA, leading up to the COP meeting at The Hague in 2000, proposed that 'forest management is an activity involving the regeneration, tending, protection, harvest, access and utilization of forest resources to meet goals defined by the forest landowner' (UNFCCC, 2000).

This broad definition would bring in essentially all forestland, at least in the USA and the most developed countries, if not in most of the world, if only because property laws that limit access would, under this definition, seem to qualify the land as 'managed'. Taking this interpretation would essentially mean that all carbon accumulated by forest regrowth during the 2008–2012 period would be creditable against a country's Kyoto target, up to the limits set at Marrakesh.

While the logical basis for including sinks in climate policy is strong, the weak link in the argument is the lack of proven methods for measuring and monitoring them. At the time (and still today) a complete inventory of carbon sinks for all the major parties is not available, and there are legitimate questions about the accuracy of even the best of these inventories. Negotiating the Kyoto Protocol caps with sinks broadly included would have meant that the negotiating countries did not know their own 1990 net emissions baselines, nor did they have much idea of what they would be in 2010. With a cap on fossil emissions there was some certainty, or so it seemed, that the parties to the agreement would lower emissions by about 5% below the 1990 level. Because the forest area of the capped parties together formed a substantial net sink, if all of that could have been credited against emissions, the end result would be that emissions would be higher in 2008–2012 than in 1990 rather than lower. Lacking resolution on how to interpret key terms in the Protocol, the quantified limits on sinks finally agreed upon gave up some of the 5% reduction that would have been achieved, but with the exact sink quantities specified it was not an open-ended amount.

The specific numerical limits on which Europe insisted at The Hague, and those finally reached at Marrakesh, ended a very confusing and complex discussion of just how to include sinks. With these numerical limits the other language that would limit sinks is far less important if not irrelevant. The 'success' of the negotiated limits is that countries can stretch the bounds of plausibility of sinks accounting, if they so desire, but clever interpretation and accounting can never get more credits than the numerically limited amount. The 'failure' is that if it is possible to easily fill up the sinks limit with sinks that would have occurred anyway, the strict limits remove any incentive to actually enhance biological sinks. That is, in a cap-and-trade system, no matter what the allowance price, the credit price for sinks credits would be limited and could approach zero because the use of credits was limited far below the amount that could be forthcoming. For example, the Energy Information Administration (EIA, 2003) analysis of a cap-and-trade system in the USA, with a limit on credits, projected a two-tier pricing result with a lower price for credits than for allowances.

The combination of several factors – (i) the withdrawal of the USA and Australia where emissions are growing rapidly; (ii) a target for Russia and the transition economies of Eastern Europe well above expected emissions (so-called hot air); and (iii) the sinks quantities that were ultimately allowed – has led many analysts to conclude that the cap on the remaining parties may be non-binding in the first commitment period anyway (e.g. Bohringer, 2001; Manne and Richels, 2001; Babiker *et al.*, 2002). So even without generous sinks accounting, it is far from certain that the emissions target in Kyoto will lead to real environmental gains. If it does, it will be the result of countries doing more than they pledged under the agreement (by implementing domestic policies and not fully availing themselves of the excess credits above reference emissions from Russia – so-called hot air). Sinks credits can also be brought into play under the clean development mechanism (CDM) and a number of proposed projects are now undergoing the review process.

As noted above, the current stated policy of the USA is to reduce GHG intensity by 18% over the decade. Most analysis shows that emissions intensity has historically improved at 14%, and so achieving 18% would be a modest reduction below the reference growth. Others dispute this, forecasting that an 18% improvement would occur if nothing were done (e.g. Reilly, 2002). Given the uncertainty, this is probably well within projection error, even if one accepts 14% improvement as the median estimate.

Other unilateral policies have been proposed in the USA, most notably the Climate Stewardship Act of 2003 (S. 139), widely known as the McCain–Lieberman Bill after the senators who co-sponsored the legislation. As introduced legislation, it produced some specific details of what a mitigation programme would look like if this Bill had become law. It was a cap-and-trade with year 2000 emissions as the benchmark, and fairly broadly covered emissions of GHGs. The cap did not cover land use sources or sinks or small sources (<10,000 t CO_2 equivalent), although it did cover transport fuel by bringing it under control at the refinery. Small sources and terrestrial sinks of any size were covered under a crediting system, but the total number was limited to a percentage of the total allowances. Paltsev *et al.* (2003) and EIA (2003) analysed at some length the economic implications of the Bill, and discussed its provisions. While numerically different from the Kyoto target for the USA, the mechanism for sinks – project credits produced outside the cap with a limit on how many could be applied under the cap – is essentially the mechanism of the Kyoto Protocol. Not straying too far from the existing international agreement is perhaps good news if one has hopes that the USA would at some point join it, but bad news if one is looking for innovative policy design that leads to effective and efficient management of carbon in the biosphere. The Bill failed to pass in the Senate, but once-drafted Bills are often reintroduced or the language in them borrowed for succeeding attempts to draft a Bill. Thus, even in failure it provides some guide as to how the US Congress might approach the problem of mitigating climate change.

Having described the complexity of sinks inclusion in the Kyoto Protocol, a final requirement here is to review the language of the Protocol that includes sinks. The complexity derives (apparently) from a compromise among those wanting to limit sinks and those wanting broad coverage. Thus, we end up with an attempt to limit sinks offsets by defining specific sinks projects on which all could agree. These are 'Article 3.3 sinks' as the language is laid out there. It allows 'removals by sinks resulting from direct human-induced land-use change and forestry activities, limited to afforestation, reforestation, and deforestation since 1990, measured as verifiable changes in carbon stocks in each commitment period' to be used to meet commitments under the Article.

Defining reforestation versus afforestation has required people to imagine how far back in history or prehistory one might go to determine whether a forest was there or not. Defining a forest has required consideration of the minimum density and height of the woody vegetation (Birdsey *et al.*, 2000). The debate has a tendency to become philosophical as analysts grapple with attributing some part of sink increase to 'direct human-induced' change apart from that due to natural causes or indirect actions by humans.

Those pushing for broader inclusion of sinks hold out hope for the so-called 'Article 3.4 sinks'. The language here opens up consideration at the first meeting of the parties (MOP), to occur upon entry into force of the Protocol, or as soon as practicable thereafter of the 'modalities, rules and guidelines as to how, and which, additional human-induced activities related to changes in greenhouse gas emissions by source and removals by sinks in the agricultural soils and the land-use change and forestry categories shall be added to, or subtracted from, the assigned amounts for Parties'.

Apparently, as it became clear that the language of Article 3.3 could be interpreted to render the limits not very binding, the absolute numerical limits on sinks were brought to the table, and then limits were ultimately agreed on. If, in fact, the original motivation for the narrow definition of Article 3.3 was concerned that excessive sinks would be credited, the eventual agreement to strict numerical limits would seem to make the entire distinction among these different categories irrelevant, yet the language persists. We turn now to biophysical aspects of the sinks issue that relate to the policy discussion above, and to our practical suggestions for how sinks might be included in a cap-and-trade system in later sections.

8.3 Important Biophysical Aspects of Sinks that Shape Their Inclusion in Trading

Forest and soil sinks depend jointly on natural processes and the actions of humans. A farmer or forester manages the land; that management affects the rate of vegetation growth (carbon uptake) and decomposition (return of carbon to the atmosphere). Carbon storage occurs if for some reason the uptake exceeds decomposition, and for the storage to be meaningful, average uptake must exceed decomposition for some number of years, otherwise one is simply tracking diurnal and seasonal cycles. The joint dependence on actions by humans and on the response of natural systems raises some issues in considering biological carbon management. We will argue later that these differences do not pose major problems for inclusion of sinks in a cap-and-trade system providing reasonable procedures can be established for measuring carbon storage and tracking changes in it. It should be noted that energy systems also have a joint dependence on nature and human management. Energy demand for space conditioning is weather- and climate-dependent, as severe weather can damage or interfere with energy infrastructure, and renewable energy such as hydro, solar, wind and biomass is at least as dependent on nature as is carbon storage in biological systems. A severe drought might disrupt vegetation growth or lead to a forest fire and thus to unplanned carbon emissions. That same drought might lead to low hydro capacity, increase demand for electricity for air conditioning and be associated with a lack of wind to power wind turbines. An electricity generator might then need to rely on existing fossil-generating capacity more than expected, leading to unplanned emissions of CO_2. The unique features of the interaction between nature and management are important considerations in the design of a carbon trading system, and how it will work, but they are not a barrier to establishing markets for carbon. If anything, they enhance the case for a market – it is in just these cases of unexpected changes that markets are able to allocate goods to their highest use, and find goods at their least cost.

Given our focus on vegetation and soil sinks, we review here some of the evidence on the interaction of management and nature as it affects carbon. The intent is to illustrate the magnitude and nature of these interactions rather than to assess them comprehensively. Understanding these issues leads to some practical guidance on how to include sinks in a carbon trading system. Among the important features we identify: (i) the effect of management can be extremely site-specific; (ii) carbon storage is highly variable from season to season as it depends on weather even if management is unchanged; and (iii) earth system feedbacks blur the line between nature and human action. Here we rely on previously published results or results from models that have been previously published.

8.3.1 Management effects

The site-specific nature of carbon storage is illustrated in Fig. 8.1 simulating the Terrestrial Ecosystem Model (TEM) (Melillo *et al.*, 1993; Xiao *et al.*, 1997, 1998; Tian *et al.*, 1999, 2003; Felzer *et al.*, 2004, 2005) for two sites and under three man-

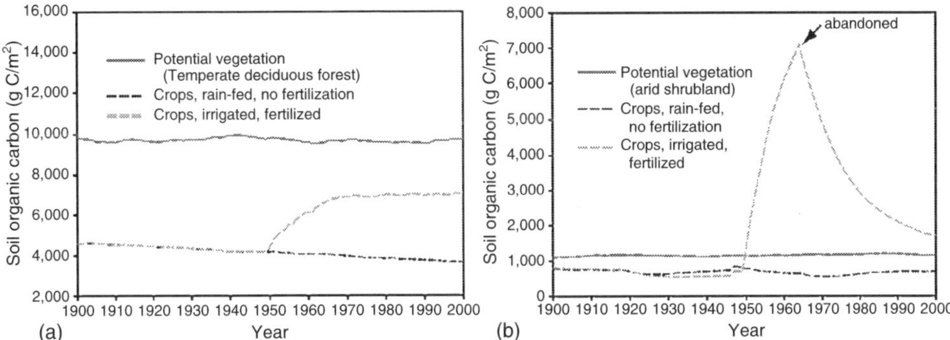

Fig. 8.1. Simulated historical changes in reactive soil organic carbon (RSOLC) at agricultural sites in (a) Buffalo, New York, and (b) Bakersfield, California, under three hypothetical management scenarios. Note that fertilizer application did not occur until 1950 in the fertilized scenario and that cropland at the Bakersfield site was abandoned in 1965.

agement regimes. Several observations are worth making. First, simulated reactive soil organic carbon (RSOLC) based on natural conditions (potential vegetation) varies by an order of magnitude between the sites. The arid Bakersfield site holds only about 1000 g RSOLC per square meter (g C/m^2), whereas the Buffalo site under natural conditions was estimated to hold about 10,000 g C/m^2.[1] Second, cropping was estimated to significantly reduce carbon storage at both sites, but the reduction was far greater at the Buffalo site in absolute as well as percentage terms of RSOLC. Third, while it is often assumed that the difference between the carbon in currently degraded soil and that prior to degradation represents the potential amount of carbon that could be stored, that difference is largely irrelevant to estimates of increased storage when a different management practice is applied. In particular, the Buffalo site with the addition of fertilizer and irrigation only gains back somewhat more than half of the RSOLC lost when converted to crops. In contrast, irrigation and fertilization leads to an increase in RSOLC at the arid Bakersfield site several times that under natural conditions. Fourth, as illustrated for the Bakersfield site, if management is removed, carbon storage can change substantially. In this case, much of the modelled increase in RSOLC due to irrigation and fertilization was lost in just a few years once the site was abandoned. Interestingly, it appears that some of the additional carbon stored may remain even after being abandoned for as many as 35 years. Even though it fell after abandonment, RSOLC remained on the order of 80% above the natural level. The management regime (and abandonment) was set to represent the actual historical management at these sites. Abandonment of cropping at the Buffalo site would likely lead to a further increase in carbon, perhaps back to near the predisturbed level. Other management practices alone or in combination may lead to other results, but our conclusions from just these two sites are that the impact of different management practices on carbon storage can differ by an order of magnitude, and that the 'predisturbed' soil carbon level is not always a clear guide to how much carbon could be stored.

8.3.2 Annual variability

Figure 8.2 shows total carbon in vegetation and its allocation among plant parts from a TEM simulation for maize for a site in China for two different years, a wet year (1995) and a dry year (1997). Here we see the substantial difference in carbon accumulation for two different years driven by the different weather conditions. Carbon in each plant part accumulated over the season in

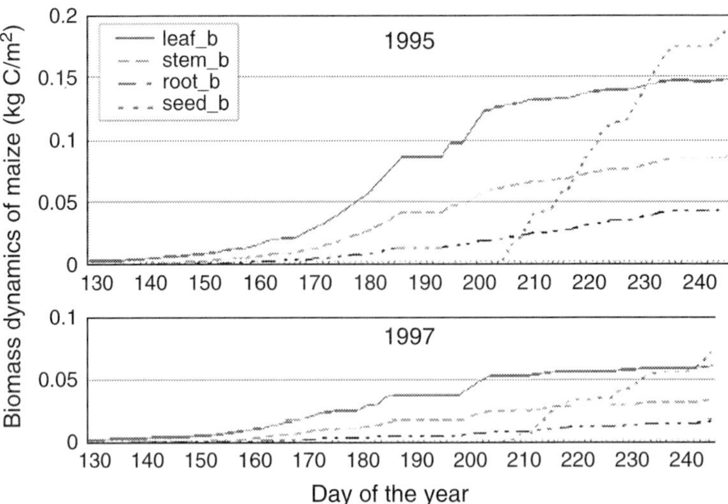

Fig. 8.2. Simulated carbon allocation among major plant parts using the Terrestrial Ecosystem Model (TEM) for maize grown in a north-eastern site in China (117° 12′ W 39° 06′ N). Actual daily climate data for a wet year (1995) and a dry year (1997) were used.

the dry year is only about half that accumulated in the wet year, with the difference in seed even greater.

8.3.3 Separating human and natural changes

Figure 8.3 illustrates a simulation of the different sources of change in carbon storage from 1950 through 2000 for the USA. Felzer *et al.* (2004) estimated that the USA was a net carbon sink but there were multiple factors, some offsetting and others interacting, that explain the net effect. The factors involve feedbacks from natural systems, natural variability and those that could be attributed to direct management. For example, land-use change and fertilization of crops (with nitrogen) are related directly to management decisions. Climate, shown to have a varying effect, is both naturally variable and may be changing because of human influence – sorting how much is natural variability and how much is due to human influence is a complex issue and not completely resolvable, particularly at smaller scales. Here the scale is near continental, but to create incentives for carbon management the scale needs to be at the level of parcels owned by specific individuals or companies. Increased tropospheric ozone damage is mainly due to increased precursor emissions from anthropogenic sources but these emissions are from energy use, over which the forest or farm manager has no direct control. Note that Felzer *et al.* (2004) also show an interaction effect between nitrogen fertilization and ozone damage: there is increased damage from ozone when there is nitrogen fertilization, that cannot be attributed to management alone or to the earth system feedback alone. In general, interactive effects are likely to be more important at smaller scales. The type of vegetation grown will interact with climate and CO_2 concentrations. When there are fundamental interactions of this type it is not possible to clearly attribute carbon changes to one or the other factor. Thus, attempts to base policy on whether the change in carbon is due to direct management, natural variability or some indirect anthropogenic factors are futile.

We clearly have a complex policy environment and a complex natural system with multiple feedbacks and interactions

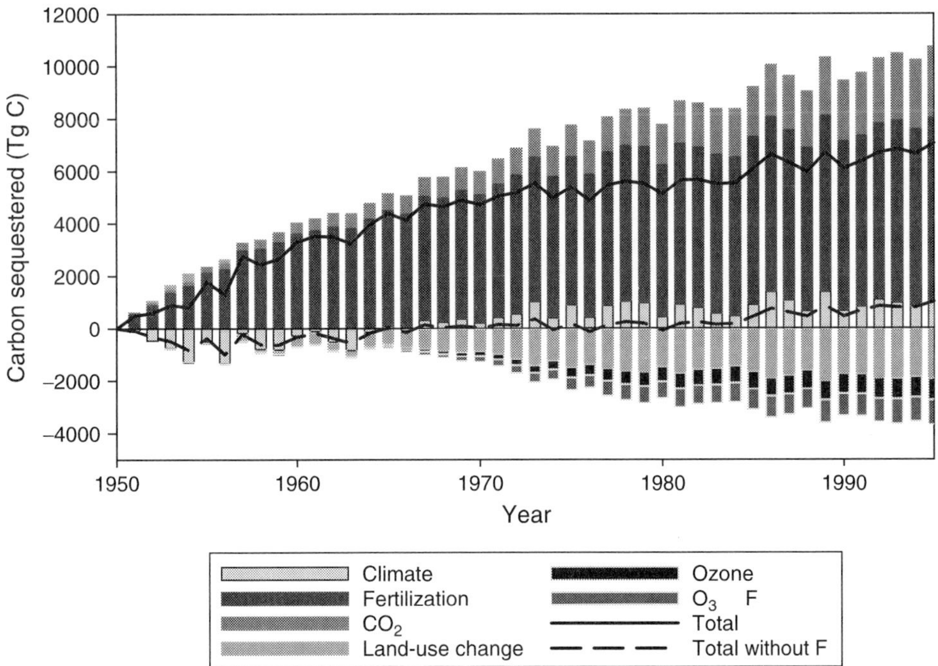

Fig. 8.3. Simulated effects on carbon storage in the USA attributable to different sources. (From Felzer *et al.*, 2004.)

between management and nature. Is there a way out of this complexity? We turn now to imagining how a very simple emissions trading system could work, if we could escape the inertia of recent negotiations and policy thinking and add some more information on the rough magnitude of sinks potential.

8.4 An Idealized CO_2 Cap-and-trade System with Land Use Sinks

Table 8.1 depicts a completely fictional situation in two imaginary countries. Both have the same fossil fuel CO_2 emissions in 1990 and the same projected level of fossil fuel emissions in 2010, a year chosen to be representative of the Kyoto commitment period. Country A is a large net land use sink, and country B is a moderate land use source. In the absence of any policy, country A's sink is projected to rise and country B's land use source is also projected to rise. If we imagine a hypothetical Kyoto-type target of returning to 90% of 1990 emission levels, we can see that in this example the target is different when we apply it only to fossil emissions or to total net fossil and land use emissions. Looking at the row labelled 1, country B appears to gain by including land use emissions because its allowance level based on total net emissions is 99, up from 90, when only fossil emissions are included. Country A gets only 72 allowances compared with 90, and therefore looks worse off with the total CO_2 accounting. But comparing the projected situation in 2010 as shown in the row labelled 2, it is actually country A that benefits from the total accounting because it needs to reduce only 23 below projected reference (compared to 30 if applied only to fossil fuels), and country B that is worse off, requiring a reduction of 41 with the total accounting compared with 30. This occurs because in the absence of any policy, sinks are projected to increase in country A, thus reducing the need to lower emissions or further enhance sinks.

In country B, however, the land use source is growing, thus putting more pressure over time to reduce emissions or increase uptake to offset the land use source.

What if, instead of the sink and source growing in these countries, it was projected to fall to a zero net sink and source? The reduction for country A from projected 2010 net total emissions would then be 48 in the total accounting example if they needed to return to 90% of 1990 emissions, whereas for country B the required reduction would be 21.

The first lesson is that moving from a fossil-only accounting to a total accounting including both land use and fossil emissions does not necessarily benefit the country that gets more allowances or the country with the large net sink in the base year. What is important is whether in the absence of policy the land use sink or source is projected to grow or to reduce.

The second lesson is that including sinks in a total accounting framework does not necessarily lower the real reduction. In the situation portrayed in Table 8.1, including sinks results in a real reduction below reference, totalling across both countries, of 64 compared with only 60 if sinks were excluded. Net emissions to the atmosphere in the fossil-only case in 2010 are country A (90 – 25) + country B (90 + 20) = 175, with the total accounting net emissions restricted to country A (72) + country B (99) = 171. In other situations, which readers can invent if they desire, total accounting could lead to an increase in net emissions.

The third lesson is that any effect on net emissions to the atmosphere due to the inclusion of land use sinks can be eliminated by adjusting the target reduction. If one has a projection of land use emissions for 2010, it is easy to calculate an adjusted percentage reduction from 1990 that will lead to exactly the same 'reduction of 30 from reference' in 2010 for each country with a total accounting. Adjusted percentages below 1990 (0.81 for country A; 1.00 for country B) are simply 2010 reference emissions less the 30 reduction from 2010 estimated in the fossil-only policy and then divided by the 1990 total net emissions. This results in a 'differentiated' reduction percentage for the two countries. Even though they are identical in terms of fossil emissions, differentiation occurs because they differ in terms of land use emissions. This assumes that the total reduction of 60 and the equal split of 30 in each country had some special merit.[2] The calculation depends on having a pro-

Table 8.1. Two hypothetical countries' fossil and land use carbon, 1990 and 2010. Relative units are given.

	Country A	Country B				
	Fossil emissions	Net land use emissions (+) or uptake (–)	Total net emissions	Fossil emissions	Net land use emissions (+) or uptake (–)	Total net emissions
1990	100	–20	80	100	10	110
2010 Reference	120	–25	95	120	20	140
Allowance allocations and real reductions: fossil-only compared with total net accounting						
1. Hypothetical target of 0.90 of 1990	Fossil-only		Total net	Fossil-only		Total net
	90		72	90		99
2. Reduction from 2010 projected emissions	30		23	30		41

jection of net land use emissions, and so one could argue that you cannot be sure you would get the same reduction in both cases because of this uncertainty. However, the same differentiation concerns arose for fossil emissions where there were recognized differences in projected growth among countries, and those projections were also highly uncertain at the time the targets were set.

To implement a trading system that operates among private parties the countrywide allocation must be distributed to private parties. This raises some additional issues about the implications for land use owners participating in a cap-and-trade system that often seem not well understood. We have thus invented, in Table 8.2, another hypothetical situation, focusing on the domestic situation in country A. Here we imagine two fossil fuel users with identical emissions in 1990, and two landowners: one with net sequestration and one that is a net source. One of the fossil fuel emitters has a projected decline in emissions in the reference for 2010 while the other's reference path will increase substantially. Landowner 1 remains a sink but the sink declines, while landowner 2, a source in 1990, becomes a small sink in 2010.

It is almost inevitable that the gross sink amount in the country (adding together the sink for only those landowners or parcels that are net sinks) is much greater than the country's net sink amount. Earlier, we used the example of the estimated 902 million tonnes as the annual net sink in the USA to illustrate how that could be used to offset US emissions. If one keeps to the simple strategy of measuring all terrestrial sinks and sources, the net sink is the offset. Much of the discussion of sinks credits, at least in the USA and Canada and as expressed in the Kyoto Protocol, focuses exclusively on credits for carbon uptake. This leaves out of the programme those landowners who

Table 8.2. Hypothetical situation for emissions sources and sinks in country A, with a target of 0.90 reduction from 1990 below total net emissions. Units are arbitrary.

	Fossil source 1	Fossil source 2	Landowner 1	Landowner 2
Emissions in 1990 and reference emissions projections for 2010				
1990	50	50	−40	20
2010 Reference	40	80	−20	−5
Possible allowance allocation, within parentheses required reduction (+) or excess allowances (−) that could be sold				
1. Grandfather to 1990 with proportional reductions	45 (−5)	45 (35)	−36 (16)	18 (−23)
2. Estimate 2010 reference with proportional reductions	30.3 (4)	60.7 (4.5)	−15.2 (−4.8)	−3.8 (−1.2)
3. Proportional responsibility from 2010 reference	33.6 (6.4)	67.4 (12.6)	−23.2 (3.2)	−5.8 (.8)
4. Credit for 1990 baseline sink, proportional reduction for all sources from 1990	30 (10)	30 (50)	0 (−20)	12 (−17)
5. Credit for any sink in 1990, allowance to match any land use source, proportional reductions for fossil sources from 1990	26 (14)	26 (54)	0 (−20)	20 (−25)

are net sources. If that is the background, the amount of sink credit is not limited by the net sink but by the gross sink amount, which is a much larger number. The gross sink amount is not even well defined unless the parcels of land are well defined and unchanging over time. In the example of Table 8.2, consider the possibility that landowner 1 has some land that is a net source, emitting 20 annually in 1990. To have a net sink of 40 in 1990, the remaining areas are thus a sink of 60. If there is an incentive to count only net sinks, landowner 1 might sell the parcel that was a net source, and thus get credit for 60 instead of 40. The ability to divide parcels into sinks and sources is nearly fractal in nature, making the potential gross sink huge.

This issue of deforestation is not ignored in the Kyoto Protocol – reducing deforestation is a potentially creditable action – but the failure to include the entire terrestrial biosphere in tracking compliance with the policy targets creates problems. There is an incentive for those who might reduce deforestation or who have sinks or might increase them to register credits, depending on how the baselines are established, but no accounting in compliance with the target cap of those that are likely to remain a source or become a bigger source. Lack of full coverage creates the problem of leakage – reductions among credited sinks being offset by increases in non-covered land areas. However, allowing landowners to voluntarily register credits when it is in their interest worsens the problem because it is almost certain to enlist mostly those who intended to increase sinks anyway, while producing no incentive to control for those who had intended to become a large source.

In the second part of Table 8.2, some hypothetical allowance allocation principles have been considered. Supposing that a cap would cover terrestrial biosphere sources and sinks as well as fossil emissions. The common implicit assumption in most discussions of sinks allowances is that one can only cap sources, not sinks. Of the five allowance allocation principles in Table 8.2, the first two do not distinguish between sources or sinks in setting allowances, i.e. they give no special consideration of the zero point on the number line.

Principle 1 is essentially the Kyoto allocation rule as applied to a country's emissions, used to allocate the allowances among entities within a country and including both fossil sources and land use owners. Like the Kyoto national allocations, it uses 1990 as a benchmark and allocates the reduction to individual entities proportionally – country A's target is 90% of 1990 net emissions, and thus each entity receives an allocation proportional to its 1990 emissions and/or sinks. For the land use source, this is a reduction of its emissions to 18 from the 20 in 1990. For the land use sink, the allocation is −36 compared with emissions of −40 (i.e. a sink of 40 in 1990). Rather than getting an allowance of zero, 40 more than it 'needs' as of 1990, landowner 1 starts in a debit position but at least as of 1990 there was enough sink to cover this debit (and indeed an excess). The very different rates of growth of emissions and changes in sinks for the 2010 reference conditions reveal the same issue that has plagued the Kyoto national allocations. If the targets are undifferentiated, these different expected growth rates lead to very different burdens. So even though the Kyoto allocations refer to 1990, differentiated reductions for individual countries take into account to some degree expected differential growth in emissions. Here we see that allocation principle 1 results in very different required reductions for individual entities. Fossil source 1 and landowner 2 have allocations well above the projection of reference emissions. They could sell these allowances, and probably reduce further at low cost and sell even more. The burden of buying permits would fall on fossil source 2 and landowner 1, even though landowner 1 is a large sink. How big this differential growth effect can be obviously depends on how differently emissions and sinks are expected to change for different entities.

Allocation principle 2 corrects this differential growth problem by making the allocations proportional to the projected reference for 2010. This reveals a different issue that arises with simply multiplying the base by 0.9. The mathematics works to produce the right national cap of 72 but the rule means that any entity that is a sink will necessarily have excess credits to sell, and this occurs for both landowners in this example. The algebra of this reduction is

$$\text{National target} = \%\text{RED} * \text{EMISS} + \%\text{RED} * \text{SINKS} \quad (8.1)$$

where %RED is the required national reduction from reference, EMISS is emissions from fossil sources and SINKS is net emissions (sink or source) from land use.

Allocation principle 3 seeks to make the burden of reduction proportional to the level of emission source or sink by altering Eq. 8.1 slightly:

$$\text{National target} = (1 - \%\Delta) * \text{EMISS} + (1 + \%\Delta) * \text{SINKS} \quad (8.2)$$

Knowing the national target, projected emissions and projected sinks, one can then solve this for the %Δ. This formulation simply generates an allowance allocation that, without trading, would require sources to proportionally reduce their emissions and sinks to proportionally increase their sink. For the example we have created %Δ = 0.159. Emissions sources get an allowance that is 15.9% below projected reference emissions, and sinks get an allowance debit of 15.9% more than their projected sink. This again leads to an allocation that meets the national cap, but now no entity has allowances that would allow them to sell credits without taking some additional action beyond what is projected to occur in the reference. Each entity bears a 'proportional' burden.

This formulation is far from perfect. Note that landowner 2 has a small net sink, and so the equiproportional change results in a small absolute change. Consider a landowner who coincidentally is at zero, neither a source nor a sink. This landowner would get away without any burden, even though he or she may be in a position to become a significant sink without much effort. At first look, this is not so different from the problem faced by fossil emitters – reductions may be costly and difficult for one and easy for another, and so equiproportional reductions need not imply the same cost burden. However, for landowners it is not unreasonable to imagine an owner of 100 acres and an owner of 1,000,000 acres. If the latter coincidentally has zero net emissions, no burden exists under this allowance principle. Yet, other things being equal, there would be much more scope for increasing sinks on the 1,000,000 acres than on the 100.

The principles for allocation rules 4 and 5 are closer to what appears to be the view of the land use community. The implicit equity principle is that coincidentally being a sink means that one should be able to sell all of the sink allowances. In both of these, landowner 1 gets zero allowances rather than a debit as in allocation rules 1–3. Even though this entity's sink is declining, he or she has allowances to sell even without reversing the decline. Allocation principle 4 treats the landowner source symmetrically with the fossil emission source, requiring a proportional reduction in emissions. Again, however, the zero problem is likely to occur. Big landowners with a source approaching zero would have a very small required reduction, with the potential to easily become a large sink. This could be considered an asymmetric treatment with that of fossil emitters but it is a symmetric treatment with landowner 1, the net sink. Being a net source is villainous, but crossing zero on the number line makes you virtuous with a generous allocation of allowances as your reward.

Allocation rule 5 further distinguishes between land use emissions and/or sinks and fossil emission sources by granting landowner sources an allowance equal to their emissions. This is closest to the implicit assumption that landowners would enter a programme voluntarily and have no burden to reduce unless they chose to do so. Thus, landowner 2 could do nothing to change his or her land use emissions, and still would not have to acquire additional allowances. While on the face of it, this is close to current policy approaches to include sinks via a credit system, capping landowner 2 is actually far better. While he or she gets allowances equal to expected emissions requiring them to do nothing, emissions cannot increase without acquisition of permits to cover them. Thus, it prevents spatial leakage, at least within the countries that follow this policy. Of course, the more allowances one grants to landowners, the more the burden shifts to fossil sources. Allocation rules 4 and 5 used 1990 conditions as the basis for establishing allocations. The same principles could be applied to reference 2010 emissions as in allocation principle 2. We are not proposing that one or the other of these allocation rules is preferable, but use these examples to illustrate that there are a number of ways to extend simple allocation principles that might be used for fossil sources to terrestrial carbon sources and sinks with very different implications for burden-sharing.

The problem Kyoto negotiators ran into was that they agreed to the burden on fossil emission reductions first. They then needed to produce language and processes to make sure that sinks credits would really be reductions beyond a baseline; otherwise the situation in which 'hot air' from sinks credits might cover all emissions increases would have been a distinct possibility. As the negotiations occurred in the run-up to signing the Kyoto Protocol in 1997, because they had little data on sinks in 1990 or projected levels in 2010, it was impossible to adjust the allowance levels to take these into consideration. At the time, the chosen approach – caps on fossil emissions and sinks allowed in as credits against the cap – was perhaps the best that could be done. The approach, however, has left us with a legacy of poorly defined categories of land use activities.

A reading of the views of the community that usually discusses sinks and sees profit in them is that they envision allocation rules like 4 and 5. The moral premise for getting this windfall gain appears to be that sequestering carbon is virtuous and it should be rewarded. However, the main reason the uptake is now occurring is that in the past history of this land, deforestation or tillage practices occurred that released carbon. So today's virtue is only erasing yesterday's vice. Thus, most people would not automatically find allocation principles such as 4 and 5 compelling. These are issues of equity or relate to perceptions of what is fair. Potentially being forced to buy additional allowances even though a landowner is a net sink would no doubt strike some as unfair. The issue of credit for past actions is one that also affects combustion sources, whereby firms would like to get credit on the basis of having adopted less emitting practices before adopting the policy. At the start-up of a programme there is an incentive issue beyond the fairness issue: if allocations are based on actual performance in years before the start of the programme, as they have been in most cap-and-trade systems, firms would have an incentive to perform poorly up to the start of the programme or risk receiving a small allocation based on low emissions. Thus, there is some basis for giving such credits to encourage early action, but determining a baseline is difficult. If one begins applying such early action credits, it only makes sense to maintain 'policy neutrality' so that every credit given for past action is balanced by tightening the overall cap. At least one must recognize that generous crediting for prior action may mean that a cap will not achieve the reduction originally planned.

One issue that affects perceptions of fairness with regard to sinks allowances, however, is that any sink is likely to be temporary. Thus, if a landowner receives a permanent annual allocation requiring it to be a permanent sink, eventually it will not be possible to achieve uptake at that level. The landowner would thus need to purchase permits indefinitely even if carbon levels were fully restored to a natural state (or higher through permanent management). Such a permanent liability does not necessarily create an economic inefficiency. The lump sum (negative) allocation would result in a drop in the value of the land reflecting expectations of the cost of the permanent liability, just as a generous lump sum allocation would result in an upward value of the land reflecting the fact that the landowner could have permanent income from the sale of allowances. It should then not affect future production decisions. If one wishes to correct perceived unfairness, one solution is a one-time negative allocation, with an annual requirement of no emissions. The one-time allocation could be based on the difference between an estimated 'steady-state' carbon stock under 'good' practices and the current carbon stock under degraded conditions. The landowner could work off this negative allocation by following good practices, and after that would only need to maintain the stock of carbon.

In showing various principles by which a cap-and-trade system could be extended to sinks and sources related to land use, we have hoped to demonstrate that there is no reason why sink needs to be treated in a widely different manner as a fossil source. A target can be fashioned to achieve the same net effect on the atmosphere with land use sinks and sources included as when they are not. To do so requires an adjustment in the cap level to account for the net land use sink or source, and given different changes over time among countries or entities, their inclusion can have potentially large effects on burden-sharing that can be overcome through differentiation or choice of allocation rule. Blindly excluding land use emissions and sinks, or giving landowners the choice to voluntarily sell credits or not does not make these issues go away. It only eliminates or limits economic incentives to reduce emissions or increase sinks in the most cost-effective manner.

With this idealized system laid out, we turn to issues that have been the subject of considerable investigation regarding the inclusion of sinks with the goal of identifying which of these remain an issue, and which of them largely disappear when the policy architecture is better formulated.

8.5 Sinks Issues in Policy Discussions

As noted previously, a confusing array of issues related to the inclusion of sinks in a climate mitigation policy has arisen. The Intergovernmental Panel on Climate Change (IPCC) brought out a special report providing a good compendium and a full discussion of these issues (Watson *et al.*, 2000). It is structured and hamstrung, however, by the policy environment and governmental interests to which it was reporting. In trying to be comprehensive and responsive while avoiding to be policy-prescriptive, it is not as effective as it could be in sorting out reasonable approaches and strategies from those that create problems rather than solve them.

8.5.1 How much to pay for an additional tonne of sequestration compared to an avoided tonne of emissions?

Many issues have been wrapped into this question, and various solutions proposed. Some would like to pay landowners up front for prospective storage once a forestation project has been established. Worried that the carbon may not remain stored, the concept of 'discounted' tonnes has been created, whereby a fractional discount factor would be applied to account for possible return of carbon in the future – leakage. Others have proposed renting carbon storage – paying a price per tonne-year stored so that if the landowner chose to do something differently in the future, he or she could do so and would have received payment only for the time they actually stored the carbon. This is a solution to the problem of paying for a 'permanent' tonne only to have the landowner abandon the activity that is keeping it sequestered. Many of these approaches are based on solid economic analysis, recognizing that carbon storage is an investment problem, and can be analysed using the same formulas as for any investment. McCarl *et al.* (2005) and Lewandrowski *et al.* (2004) provide good reviews of different approaches.

Key to investment problems is the net present value (NPV) of the stream of returns. A landowner considering a sequestration project would compare the NPV of carbon storage to the investment cost plus the discounted stream of annual maintenance costs, just as he or she might compare the NPV of returns to installing irrigation to enhance crop production or establishing a forest for purposes of harvesting the wood. Herzog *et al.* (2003) offer one formulation of this NPV problem:

$$\text{NPV} = p(0)a(0) + \sum_{t}^{\infty} p(t)a(t)(1+r)^{-t} \quad (8.3)$$

where $p(t)$ is the price of carbon in year t, $a(t)$ is the net amount sequestered or leaked in year t and r is the interest rate. They use the formulation to estimate a discount factor for ocean sequestration, imagining that the carbon would be sequestered in year zero and would gradually return to the atmosphere over a very long time. Thus in their problem $a(0)$ is positive and $a(t)$, for $t = 1, \ldots \infty$, is negative. The same approach has been proposed for land use sinks, and for conceptual purposes the time periods could be of a length where all sequestration occurred in period zero – e.g. each period could be 10, 20 or 40 years – and leakage then might occur in later periods.

The simple and economically efficient approach for pricing carbon is to allow the market to price it once a cap has been established. A landowner who sequesters a tonne of carbon in period t may choose to sell the tonne at the full market price in time t or could hold it for future use or sales. Should the landowner at time $t + n$ emit a tonne of carbon back into the atmosphere, he or she would then be responsible for purchasing a carbon allowance at the going price in year $t + n$ or could use the banked tonne. This treatment is symmetrical to that of a fossil fuel emitter, say an electric power producer, who might be considering different power plant options that would have different streams of carbon emissions in the future. If the carbon stream were less than the allowance stream, the power producer could sell the extra allowances into the market or bank them against the possibility that it may not be of interest to continue the operation of the carbon-saving power plant indefinitely just as the landowner might decide to change his or her land use practice in such a way that carbon previously sequestered is released. The zero point on the axis, going from sink to source, has no special meaning in this trading environment. All that is important is how an entity's emissions or uptake compares with its baseline allocation of allowances so that it can determine whether it has allowances to sell or must acquire allowances.

Alternative solutions whereby there is an established rental price or an established discount for land use sinks lead to potential economic inefficiencies by asymmetrically treating fossil emitters and landowners. If we knew for certain future carbon prices and market rate of returns, and which sinks would leak at which rates, or at least the average leakage rate, one could establish an equivalency between rental rates, the carbon price and a discount factor.

Herzog *et al.* (2003) calculate the discount factor by calculating the NPV as in Eq. 8.3 and dividing it by the NPV of permanent storage (i.e. when $a(t)$, for $t = 1, \ldots \infty$, is zero). Lewandrowski *et al.* calculated a rental payment as

$$a = rP \quad (8.4)$$

where r is here the discount rate and P is the price of a tonne of permanently sequestered carbon. This result is derivable from a formulation like Eq. 3 under some highly simplified assumptions, namely that the price of carbon is constant over time. As Herzog *et al.* (2003) show, if the price of carbon rises at the rate of discount, the value of temporary storage is zero, and there are conditions under which we might reasonably expect the carbon price to rise at that rate. In particular, with a stabilization target and no backstop, efficient allocation of the reduction through time would require a constant discounted price – i.e. the actual price rises at the discount rate. We would not press the case that actual carbon price will necessarily rise at the discount rate but use this example to illustrate that the rental rate for carbon depends on what you assume about the future carbon price path – and, under some not implausible assumptions, the right rental rate could be zero.

The various formulations of: (i) sell or buy permits as you go, (ii) discounted tonnes, or (iii) renting carbon are all derived from the same basic formulation and so it would seem that any of these options could be used. Although the mathematics can be manipulated to derive one formulation from the other, problems arise because:

1. Calculating the discount factor or the rental price requires someone to know or estimate future carbon prices and the appropriate discount rate. If a public agency is to compute the discount factor or the rental value, they must make some projection of these.
2. Whether and when leakage occurs is not purely a phenomenon of nature that occurs with a known (or knowable) frequency, but rather is at least partly under the control of the landowner.

Problem 1 indicates that the public agency bears the risk of being wrong with rental calculations or with the discounted tonnes calculation, whereas when the fossil emitter's investment decisions require forecasting, the risk is on the private entity. One can make a case that the public agency should take steps to limit risk to private entities, but there is no good reason to have some segment of mitigators (fossil emitters) bearing the risk, and another segment (land use sequesters or emitters) not bearing the risk. Problem 2 indicates that an upfront discounted payment with no requirement to be responsible for the future of the carbon creates no incentive for the landowner to take actions that would prevent return of the carbon to the atmosphere. The rental formulation partly avoids this by only paying as you go, but because it produces incentives for sequestering but not avoiding emissions, it leaves land use emissions uncapped.

The 'disconnected tonnes' makes carbon sequestration less attractive – those landowners who might be willing to assure that the carbon had been permanently stored will be less willing to sequester at a discounted payment. If leakage were a purely natural and random phenomenon with no ability to know what its rate was for a specific parcel or to control it, the discount approach would on average credit the right amount. Since with these assumptions the landowner had no control over leakage, the lack of incentive to control it has no effect on leakage. However, these are unreasonable assumptions. The landowners who, a few years after accepting the payment, decide to do something else face no penalty for releasing the carbon. Realistically a programme of upfront payment would likely include conditions that would limit the landowner's actions, or penalize him or her for actions that led to sequestered carbon being emitted. But the efficient penalty is for the landowner to purchase carbon permits at the going price at the time the carbon is emitted. The notion of a penalty – that a wrong was committed – is mischaracterizing the decision. Simply allowing the landowner to essentially buy out of the commitment to store the carbon by purchasing credits preserves the option to use the land in another way if it is more economic. From a broader economic standpoint, preserving this option makes a lot of sense. If for some reason food is short and agricultural commodity prices rise, the landowner can switch to crop production. As long as carbon allowances are purchased to cover the emissions, the country will continue to be in compliance with its GHG mitigation targets; yet it allows land to be used to solve another pressing problem, food supply. There is no net leakage that is not covered by a reduction in emissions (or more uptake) elsewhere, and so there is no need to apply a discount to sequestered tonnes in the first place.

We have been careful to identify problems with tonne-years and discounting as a problem of a public agency implementing these formulas. All of the market approaches we see in capital and investment markets are likely to develop in a carbon market if it is set up as we propose – selling when sequestering at the then current price, and requiring allowances to cover emissions if at some point the carbon is released back to the atmosphere. In particular, landowners who wanted an upfront payment would probably find intermediaries prepared to pay some amount for the future stream of sequestration. The payment would reflect the intermediary's expectations of future prices of carbon, and a contract would need to be structured to describe who would bear the risk if the landowner was later found not to have sequestered the carbon. For this system to work, this requires that the sequestration agreement is legally enforced and the sequestration is monitored over time by a public agency. Landowners might simply bank credits they have created through sequestration, speculating that the price might increase and leave them in a difficult financial position if they wanted to do something that would release the carbon. Future prices and future contracts would likely develop, and intermediaries may be willing to rent carbon based on their speculation of what such temporary storage was worth – i.e. speculating on how carbon prices would change. Contracts and agreements between landowners and such intermediaries could be negotiated or might vary depending on the interests of the landowner, and the risks the intermediaries were willing to accept. In short, the market would quickly invent solutions to illiquidity or the need for upfront payments to cover investment, at a price, just as it has for other investments. Many concerns about the ability of landowners and markets to deal with carbon pricing over time have been expressed in the literature. However, investing in a forest for the sake of receiving payment in the future for the carbon stored is no different than the problem of investing in a forest with the goal of selling the timber in the future.

8.5.2 What should be done about the possibility of catastrophic release of carbon or the high variability of ecosystem uptake?

The amount of carbon taken up by plants varies dramatically from year to year depending on the weather. Rapid growth in one year may produce a lot of litter subject to rapid decomposition, and if followed by a year of poor growth, the result may be net emissions that year, with decomposition release greater than the carbon taken up by new growth. Wildfires might lead to large net emissions that would destroy well-meaning efforts to sequester carbon, and are the most dramatic example of catastrophic release evidently beyond the control of the landowner. However, these natural phenomena that lead to variability again would seem to be no different from the normal situation landowners face. Bad weather that leads to little carbon uptake, and possibly net emissions, is no different to the situation where bad weather leads to crop failure and financial loss because there is no revenue to cover the cost of planting and other costs of farming. Similarly, the forest manager who had planted a forest in anticipation of harvesting the timber faces potentially catastrophic loss if there is a forest fire that wipes out the young forest. Limiting financial liability for these risks in the case of carbon storage would limit the incentives landowners would have to take actions to limit the effect of these events. The prudent landowner would enter into carbon sequestration with the same set of risk calculations he or she would use in cropping or timber management, taking into account an estimate of the variability over time of carbon uptake. This might include carrying a bank of credits from good years to cover bad years, the use of various financial instruments to cover the risk (saving, insurance, forward options on purchase of allowances to cover potential risks) or fire prevention and weather amelioration strategies (irrigation) that would limit the effects of these natural conditions.

One element of the variability issue deserves some consideration. Public monitoring and enforcement will need to create a periodicity to inventory requirements. It is not likely that land use carbon would be 'continuously monitored' and the concept is almost nonsensical given that carbon is exchanged continuously through the day and seasons with periods of net uptake and net release. The preferred method is likely to be to estimate a stock at time *t*, re-estimate the stock at some later time, and the difference is the net uptake or release. So there is a decision to be made as to how often that inventory must be updated and reported.

The Kyoto Protocol set a 5-year commitment period for countries, essentially allowing unlimited borrowing and banking within that 5-year period. Countries may require fossil emitters to provide inventories more frequently – e.g. annually. Because of the variability of land use carbon, the difficulty and cost of accurate measurement, as well as the likely approach of measuring the stock instead of the continuous flow, it likely makes sense to have a longer rather than shorter required inventory periodicity for land use carbon. This would automatically allow borrowing and banking over the established period by the landowner. For example, if an annual inventory period were required and the landowner had to be in compliance with the target each year, and if the first year was a bad weather year (or unluckily a forest fire struck), the landowner might be required to make a big purchase of allowances, only to have large net sequestration in subsequent years. This is not insurmountable; explicit banking and borrowing provisions could be created such that this variability could be evened out. However, inventory methods are likely to involve some cost, and may not be accurate enough to reliably measure year-to-year changes. This suggests that the goal may be to set the periodicity of the inventory at least every 5 years and possibly as much as every 10 or 20 years.

It does not seem essential that the periodicity be the same as either a national target period such as in the Kyoto Protocol or the same as for fossil emitters. If, however, one requires that allowances can only be used once the carbon is actually sequestered, and has been certified as such, this could mean that no sequestration could be credited until the second inventory was taken, perhaps 20 years later if that was the official reporting period. This does not present any fundamental problems, but for those hoping to use sequestration in early periods this would prevent it. Not to make too much of this constraint, it would not necessarily mean that landowners could not find intermediaries who would pay them early, on an intermediate assessment of carbon sequestered, and on an expectation of future carbon prices. However, one way to add flexibility without necessarily requiring frequent and costly inventories would be for the reporting rules to allow landowners to inventory more frequently. If they followed established principles, sequestered carbon could be credited in the current period. For example, a landowner might choose to inventory and report after the 5th year, even though only required to do so once every 20 years.

8.5.3 How to resolve the problem of determining direct human responsibility for sequestration?

The Kyoto Protocol limits sinks credits against targets to those due to 'direct human-induced . . . change'. In retrospect, this may be among the most problematic passages on sinks in the agreement. As we reviewed in Section 8.3, strong interactions of nature and management mean clearly that separating carbon uptake into these two categories is impossible. Felzer *et al.* (2005) estimate the tropospheric ozone damage effect to be substantial, and while the extent remains controversial, CO_2 fertilization as usually modelled strongly enhances vegetation growth and carbon uptake. Climate change itself will affect plant growth. These are probably what the framers of the Protocol considered 'indirect' effects and thus meant to exclude. However, it does not seem as easy to dismiss the US interpretation, where simply protecting property rights is a direct human action that might lead to carbon sequestration, or at least prevent deforestation and carbon release.

Even if one were to take a very narrow definition of actions – a specific forest established with the express intent of sequestering carbon – and one could somehow assess 'intent', the 'direct human-induced' language would seem to require the ability to attribute some part of the carbon sequestered to the direct human action. It would mean subtracting out that due to indirect actions (nitrogen deposition, CO_2 fertilization), or even giving credit for more than was sequestered if some indirect action (e.g. tropospheric ozone) damaged vegetation which would have otherwise taken up carbon. One would need to at some level confront the reality that once trees are planted, it is mostly nature that takes over and grows them, and so the distinction of what is due to direct human action and what to nature or indirect action is necessarily fuzzy. In reality, vegetation growth is a collaborative effort of humans and nature, where the result is not uniquely attributable to either collaborator. Trying to create rules and then measure and attribute carbon uptake to different causes would seem to be a distraction, adding to the cost of monitoring and, if anything, creating cost inefficiency. With clear property rights for land, and the ability of the landowner to sell allowances for anything sequestered on it above the established baseline, or pay for emissions above (or sequestration less than) the baseline, that landowner (or the country in the case of national targets) has an incentive to fix the problems that lead to damage. Again drawing the comparison with forest and agricultural products, products harvested by the landowner can be sold and are not subject to a test of whether the products were 'human-induced'. The public good nature of the 'indirect' human effects such as air pollution requires collective action to solve, and including all carbon sequestration or emission within an incentive structure would not automatically solve these problems. But, with landowners losing or gaining depending on whether these other environmental problems are solved, it would at least provide a motivation for them to support collective action on pollution.

As previously noted, the concern of limiting sinks to direct human action as in Kyoto would appear to arise from the fact that negotiators focused first on emissions and reduction goals, and having agreed to those, tried to bring sinks into the format. With that approach, making sure sinks credits were for uptake beyond 'business as usual' was a necessary consideration. In retrospect, however, this gave rise to language that has proved nearly impossibly to implement. The problematic language could be avoided if the caps are reformulated as caps on total emissions from fossil and land use net of sinks. This will mean, however, rethinking the targets because, as shown in Table 8.1, a given percentage below 1990 emissions will have very different implications if applied to all emissions and sinks than if only applied to fossil emissions.

8.5.4 Broad cap or sinks as credits?

In the experience with emissions trading systems, two types of approaches to creating tradable emissions reductions are identified (e.g. Ellerman *et al.*, 2000). A cap-and-trade system distributes allowances that must then be used by entities under the cap to cover their emissions. Trading is among these allowances. However they are originally distributed, entities may purchase more if they need them or sell extras they do not need, but they must hold allowances to match their emissions. The second type of system is a credit system. In a credit system, credits are earned by reducing emissions below an established baseline. Typically, entering the credit system is voluntary: there is a market for the credits and it is in the economic interest of an entity to produce credits at the going price if they can do so, but other entities may choose not to enter the credit system and so they are not required to make any reductions. A credit system is often an add-on to an allowance cap-and-trade system. The cap-and-trade system forces the entities under the cap to reduce emissions whether or not it is economically desirable, and thus allowances have a positive market price. Those entities outside the cap but allowed to produce credits can sell credits in the market if they want to. Since producing credits is voluntary, no entity covered under the credit system should bear net costs unless they have miscalculated their own cost of producing the credits. Those under the cap can be shown to gain from trading (as compared with trying to meet the allocation without trading), but in most cases they are bearing costs compared with not having the policy at all. Trading is beneficial because it reduces costs. Of course, a generous allowance allocation can mean that even under the cap there may be some entities that benefit from the policy compared to the case with no policy, but if the cap is binding, the entities under the cap on average bear a cost.

The Bush Intensity target is voluntary, and its main aspect is a credit system. At present there is not much of a market for these credits, but producing and registering credits may be worth it if entities anticipate that there will be a cap-and-trade system in the future. The McCain–Lieberman Bill was a cap-and-trade, but sinks were allowed in as credits. The language of the Kyoto Protocol is one that would allow sinks in as credits at least in terms of a country meeting its target. It is not clear that this would foreclose sinks or some amount of land area entering under a cap within a domestic system of a party under the Protocol, but whatever the result of that broader cap, it would have to be squared with the sinks language in the Protocol, making them credits against the national cap. As already noted, a problem with a credit system is the 'real reduction' problem. A baseline for emissions, the reference against which credits can be earned, is hard to establish. If very loose, many entities may have an interest in entering the credit market as sellers but many of the credits may be unrelated to real reductions. If very tight, few will have an incentive to sell credits. As a result, much effort must be expended to determine the baseline for each entity with potential credits. In contrast, if the national allowance target can be established, the integrity of the overall target is not compromised even if the allocation provides 'hot air' allowances to some participants.

Both spatial leakage (landowners not voluntarily entering the credit system) and temporal leakage (landowners selling credits this period with the sequestered carbon being emitted in later periods) are a problem with credit systems. A forest landowner, who forgoes harvesting to sequester carbon, reduces the supply of lumber. But the demand for lumber remains, and so other lumber suppliers produce more lumber, thereby offsetting most of the sequestered carbon by higher emissions from forests not in the credit system. Leakage will potentially occur anytime the policy is incomplete spatially or temporally. Cap-and-trade systems that are not geographically comprehensive also suffer leakage, and if a cap-and-trade system were only going to be in place for a few years, one might expect temporal leakage in such a system as well. A well-structured policy that covers all potential emitters and sinks across space and over time eliminates the problem of leakage. A credit system in which coverage is voluntary does not assure this, whereas a cap-and-trade system can be easily structured to do so.

8.5.5 Permanence and leakage: a special problem for carbon sequestration?

Leakage is a concern for climate change as the cap that is set, presumably based on a solid assessment of acceptable emissions of carbon to the atmosphere, is not met because reductions taken by some entities are offset by an increase in emissions by entities not under the cap. As permanence is analogous to the spatial leakage problem, it is useful to refer to it as temporal leakage. Spatial leakage occurs because, at a given time, some emitters are not covered by the cap. Temporal leakage occurs when entities are induced to make reductions or sequester carbon in one period, but are outside the incentive system in a later period. Land use emissions face a special problem with permanence and leakage only because land use has been envisioned as entering voluntarily and as a credit rather than under a cap.

8.5.6 Ancillary benefits, pre-existing distortions

Equating marginal costs of carbon reduction and sequestration across the economy is an economically efficient solution in an idealized economy where all other prices appropriately reflect the real marginal cost of goods. Taxes, subsidies and unregulated externalities (positive or negative) result in prices not reflecting the full marginal cost of all inputs, and therefore an idealized policy that results in equating marginal costs of carbon reduction among countries or across sectors may not be the most cost-effective policy (Babiker *et al.*, 2004; Paltsev *et al.*, 2005). Ancillary benefits of both carbon sequestration and emissions reductions are often cited. Emissions reductions by fuel switching may reduce the emissions of many other air pollutants (Matus *et al.*, 2006). Carbon sequestration may reduce soil erosion and leaching of agricultural chemicals, thereby reducing water pollution (e.g. Marland *et al.*, 2005). Some fuels are taxed heavily in some countries (Paltsev *et al.*, 2005); many countries have significant agricultural subsidies. All of these externalities and distortions mean that equating carbon prices across sectors and economies is unlikely to result in the economic efficiency the simple textbook story suggests. The 'first-best' solution in economics literature in these cases is to work to get rid of the other distortions by appropriately pricing other externalities or to reduce the distorting effects of taxation. Where these distortions are and how to get rid of them needs research and policy attention.

We would argue, however, that we need to avoid the often first impulse of adding a mark-up or mark-down on carbon prices from activities with different pre-existing distortions or ancillary benefits. The danger of such mark-ups or mark-downs is that the ancillary benefits or extra costs are likely to vary by fuel (in the case of fossil emissions) and by particular site and sequestration option for land use activities. The correct mark-up or mark-down will also likely change over time. Thus, it seems preferable to work towards fixing the other problems directly, and pricing the 'partial interest' in the carbon mitigation options for its climate benefit. Recognizing and pricing 'partial interests' is not a new concept (see Wiebe *et al.*, 1996 for a careful discussion of pricing partial interests in land related to environmental goals). There may be reasons to make exceptions, but it seems preferable to keep the climate policy instrument clearly focused on climate policy rather than to use it to jointly solve a myriad other problems for which it may be a relatively poor instrument. Again the existence of ancillary benefits or costs is not unique to either land use sources or sinks.

8.5.7 Measurement, monitoring and enforcement

Much scientific attention is directed at developing and improving the reliability of techniques to estimate the stock of soil carbon at a particular time. This is important and essential work, and more progress is needed. There will, however, always be uncertainty and inaccuracy in these measurements. Measurement error need not be fatal to including carbon sequestration in a cap-and-trade system. A trading system can operate as long as the measurement process is accepted as defining an authoritative measurement – it need not be accurate with certainty. The process might include not only a technical approach to measurement but also the ability to challenge a measurement and a process for resolving questions or challenges, and final certification. While error in measurement can be tolerated, it would be hard to create legal authority if errors were so large and random as to appear to lack any scientific foundation.

A more subtle problem, however, is a compromise of the effectiveness of the system if there is a bias in the measurement process. If on average the measurement process systematically underestimates carbon stored, the system will provide too little incentive to sequester carbon, whereas if it systematically overestimates carbon, the cap will be met legally, given that the measurement system is legally accepted, but the effect on the atmosphere will be less than expected. This can be remedied by further tightening the cap to meet the atmospheric target, but carbon sequestration will be overused compared to emissions reductions because pricing does not reflect the actual carbon sequestered.

An even subtler problem of bias arises when a measurement process has been constructed to be unbiased based on experimental measurement, but the incentives to participate lead to bias in the actual application. Consider the following situation in which a practice is extensively evaluated through experiment and an average sequestration level is attached to that practice. The average sequestration achieved by the practice becomes a part of the measurement method – the part of the model by which carbon inventory is estimated. Under real conditions, the vigour with which the practice is implemented may be subject to variation. If it costs more to vigorously implement the practice, actual landowners may have an incentive to minimally implement the practice. This, in turn, results in less carbon stored on average under real conditions than the average experimental result. Another way this may happen is if the cost of implementing the option is correlated with an environmental condition that also affects the amount of carbon stored. If the correlation is such that lower costs are associated with lower carbon uptake, again the average uptake under real conditions will be less than the experimental average, because the activity will be implemented at the low-cost sites but it may not be economic to implement at the higher-cost sites. Reliance on practice rather than actual measurements tends to increase the chance of these incentive effects to create bias in the estimates.

Finally, enforcement is a necessary element of a successful system and must be part of the design. One of the surprising aspects of a cap-and-trade system as described by Ellerman *et al.* (2000) is that enforcement has been much more successful. There is no direct reason in economics for this, but rather it appears that regulators are more willing to enforce a cap when they can point to allowances that can be purchased to meet it. Since all entities have opportunities to purchase in the same market, the claim of some entities facing special hardships that prevent them from complying is less compelling. Hardship is more compelling in systems where entities must comply with an individual limit, and experience shows that exclusions are often granted, and so the environmental target is rarely met. Consider the case of wildfire that resulted in carbon emissions. If the landowner were required to meet some level of sequestration, and keep the carbon sequestered for some minimum period of time, enforcement in the face of fire becomes problematic. The landowner may have little ability to actually comply. The enforcement agency can levy a fine, but this can appear unreasonable given that the landowner could not prevent the fire. This would likely give rise to hardship exclusions. In any case, the carbon is in the atmosphere, and levying a fine would not remove it. With cap-and-trade, where the landowner can purchase or sell allowances, a fire is a hardship but the landowner can still comply by purchasing allowances.

Again, homeowners and businesses that choose to locate in areas prone to disasters mostly face the economic consequences of these disasters, and therefore presumably try to limit their exposure to, and the effects of, these disasters. The same principles should be applied to carbon sequestration. To the extent emergency aid or disaster assistance would apply to carbon losses, care is required to structure the aid so as not to undermine the incentives to reduce the chance of losing the carbon. Completely exempting the landholder from any responsibility to cover these emissions with credits would certainly undermine these incentives. One alternative worth further consideration that could provide some relief would be that if the landowner demonstrated effort to re-establish the forest and restore the carbon, he or she could borrow against that planned future replacement of carbon to cover the catastrophic loss. Such borrowing automatically occurs within the inventory period, and so a long period such as 20 years automatically gives the landowner a chance to restore carbon catastrophically released in, for example, year 3 into the inventory period. However, the fixed period still creates the possibility that catastrophe in year 19 or 20 would leave the landowner short. Additional borrowing provisions could ease this problem.

8.5.8 Carbon stored in products

Harvested material from forests and farms end up in a variety of product streams. Some are relatively short-lived such as food or pulp and paper. Others may remain 'stored' for decades or centuries such as lumber used in buildings or furniture. Schlamadinger and Marland (1999) provide some estimates and discuss issues related to carbon in the product stream. This raises the question that in harvesting a forest, should harvested product be tracked until it actually decomposes, and only then be counted as an emission of carbon requiring an allowance? In principle, the answer is 'yes' because this would provide an incentive to not finally dispose of these products if they can be salvaged or reused, and would accurately account for the time between harvest and decomposition when the carbon remained out of the atmosphere. In practice, this would require a complex tracking system both of the product and of the owners of the product to ensure that they were liable for emissions if and when they dispose of the product in such a way that the carbon was released to the atmosphere.

Simpler approaches would be to ignore this storage and assume the carbon will return to the atmosphere sooner or later. Another approach is to try to apply an average discounted tonne factor as an offset to the total harvest. Neither approach creates an incentive to prolong the life of carbon stored by not destroying structures or by recycling used lumber, assuming there is value to temporary storage. Crediting via a discounted tonne approach gets us back to the problem of estimating this discount factor, which we rejected earlier.

Bioenergy is a carbon-containing product of vegetation, and it has the potential to become more important as a 'carbon-free' energy source in a carbon-constrained world. It can be carbon-free if the biomass used is from areas where the crops are continually regrown, the carbon in the soil is not being depleted and fossil energy is not used in its production. Given the potential importance of bioenergy as a solution to climate change, getting the incentives on bioenergy right is particularly important. If one simply exempted biomass energy producers from the cap, ignoring emissions that occur in processing and combustion on the basis that these are being taken up by next year's biomass crop, it would provide no incentive to regrow the biomass.[3] Schlamadinger and Marland (1999) find that in cases of clear-cutting forest stands with large amounts of biomass the loss of carbon may never return to the predisturbance level even when accounting for energy and long-lived stocks. Similarly, disturbed cropland, as shown in Section 8.3, often has significantly less carbon than in its predisturbed state. To correctly account for such land conversion losses of carbon or non-sustainable management of land, land used to produce biomass would need to come under a cap to provide correct incentives to maintain carbon stocks in soils or in standing vegetation or detritus. Because the bioenergy would be combusted relatively quickly (weeks, months, a few years at most) after production, one could exempt emissions from combustion of the fuel (e.g. at power plant or by vehicles using a liquid fuel) completely. This approach could be applied to other product streams that are short-lived, reducing the monitoring problem to the land parcel without the need to follow the product stream.

The long-lived product streams create a more severe problem of tracking and monitoring. More investigation is needed to determine the importance of long-lived product streams. An important question is whether this carbon pool would be substantially affected by creating proper incentives to manage it. Any gain should then be balanced against the cost of establishing the necessary monitoring and tracking system for the carbon.

8.6 Conclusions

The role of sinks in climate policy has been controversial and confused. Different parties had very different motivations that led to the existing 'compromise' design of climate policy as it relates to sinks; moreover, it appears that the poor design in the Kyoto Protocol stemmed from the fact that sinks were added relatively late in the negotiation process. In addition, there was a relative vacuum of information on how big these sinks were, and how they might change over time for the parties involved. The sinks issue was relatively new and not much thought had been given on how to include them. Unfortunately, the Kyoto model for sinks, designed in a rush, has been borrowed in other proposals such as in the McCain–Lieberman Bill in the USA. The crediting approach described in these policies and proposals has led to nearly unsolvable problems. Rethinking how land use activities could be brought within climate mitigation efforts seems worthwhile.

We argue that many of the problems and concerns that analysts and policymakers have spent enormous effort trying to solve are mostly the result of the faulty architecture for sinks in the Kyoto Protocol. Like legislation to close tax loopholes that mostly creates a more complex tax code and more loopholes, attempts to patch the Kyoto approach to sinks have only led to more problems. It has set scientists and policymakers to consider imponderables such as what part of a forest is due to direct human-induced change and how much is due to nature or indirect human inducement. Hundreds of pages have been written attempting to define how many trees make a forest, and what is the difference between reforestation and afforestation.

These issues mostly disappear if one brings land use fully under a cap-and-trade system. This creates incentives both to control land use emissions and to enhance land use sinks. Whether the area is defined as a forest or not is irrelevant – all that matters is changes in the stock of carbon. The problem of leakage has been raised as a special problem related to sinks, but it can best be seen as a problem of incomplete policy, either in space or time. Bringing land use fully under the cap eliminates the problem of leakage. Instead, landowners can exercise the option to maintain or not maintain the storage, by purchasing allowances to cover emissions. This keeps the atmospheric carbon goal intact, and preserves an important flexibility in how land can be used in the future should economic condition change in different ways than we now expect. Coverage under a cap allows landowners to sell current allowances at current prices but requires them to cover future emissions with allowances when the emissions occur.

The variability in land use storage due to climate or events like forest fires is often seen as a unique problem for sinks inclusion in a carbon market. Land use carbon sinks and sources are subject to much variability but landowners regularly face much variability with regard to current uses of land. Farmers and foresters face risks of natural disasters that damage their crops or their forest stands. They make investments in the face of these uncertainties. Increasingly market intermediaries have come into being to bear or pool risk, or to allow hedging against these uncertainties. There is every reason to believe that these same types of intermediaries would come into being if there were a robust market in carbon allowances. Cost-effectiveness in carbon mitigation actions requires not only an equal carbon price across sectors but also that the risk of estimating future conditions be borne equally across sectors. Proposals that shield landowners from these risks would create an asymmetry between fossil emitters and sinks, and lead to economic inefficiency.

The literature on climate policy often portrays the management of terrestrial sinks as a very different issue than management of carbon emissions from fossil fuels, and that this difference requires special provisions in policy design. There are important biophysical aspects of sinks that make them different in some regard from fossil fuel emissions of carbon. How much sink one gets from specific management practices is highly variable across different sites, and over time. Moreover, sink storage is a combined result of direct management and earth system feedbacks. We conclude that these issues generally do not present insurmountable barriers to inclusion of terrestrial sources and sinks in a cap-and-trade system on equal terms with carbon emissions.

Rethinking the inclusion of land use activities in mitigation activities will require re-evaluating targeted levels of net emissions. The 7% below 1990 fossil emissions in the Kyoto Protocol or a return to 1990 emissions as in the McCain–Lieberman Bill that was under consideration in the US Senate has very different implications if applied to the total of fossil and land use emissions (net of sinks). The benefits of rethinking these targets, in terms of eliminating needless terminology and improving cost-effectiveness of mitigation policy, seems well worth it. There remain some ways in which land use should be treated differently, and some important issues that need further investigation. We argue that the inventory period for land use should be longer than for carbon emissions – a reporting requirement of every 10 or 20 years may be appropriate with the flexibility to produce an inventory more frequently if so desired by the landowner. Measurement, monitoring and enforcement remain important issues. Measurement need not be exact but the measurement process needs to be unbiased, and more accurate measurements will be more broadly accepted. The measurement process needs to include the ability to challenge results and processes to resolve those challenges.

There are also ancillary benefits or costs related to sinks but these also exist for mitigation of CO_2 from fossil energy emissions. These are potentially important issues that can lead to an idealized method such as a cap-and-trade system that strives for equal marginal cost abatement across sectors and countries to not be cost-effective. The first best solution in these circumstances is to fix these other problems with instruments designed specifically to address them. Adding mark-ups or mark-downs for different types of mitigation actions would require consideration of how they would likely vary by site and over time. This seems to recommend against such an approach unless a very strong case can be made. An important issue is how to deal with carbon in products harvested from vegetation. Here it is useful to distinguish between short-lived and long-lived products. Emissions with short-lived products should be exempted from a carbon charge, and instead the land from which they are produced should be under a cap so that long-term changes in carbon storage are monitored and incentives are in place to maintain or increase storage as economics dictate. Longer-lived products could require a very involved system to track their fate, as well as their owners. Whether correct incentives in this regard would substantially increase these pools compared to the case where they were simply ignored (and all carbon assumed to return to the atmosphere in relatively short order) needs further investigation.

Inclusion of land use and land-use change in climate mitigation policy has been made impossibly complex, because the architecture contained in current policies for including them is flawed. Solutions and compromises that were pragmatic or were deemed necessary to make progress on a broader agreement appear to have led us to the current climate policy architecture for land use and land-use change. Looking back now at the tangle these compromises have created makes it clear that much could be gained by reconsidering the architecture of sinks in climate policy. To do so will require some very fundamental re-evaluation of goals and targets, but the cost of not doing so means that we may leave a major source of GHGs uncontrolled, and fail to effectively use low-cost sequestration and bioenergy options that will be needed to limit atmospheric concentrations of warming substances.

Notes

[1]TEM tracks RSOLC, the amount of soil organic carbon that might decompose in the time frame of decades to centuries. Total soil organic carbon (TSOLC) would also include inert soil carbon.

[2]Depending on the merit criteria, the inclusion of sinks could lead to a different optimal level of reduction or split among countries. This example is meant to indicate that through adjusting the allowance level, any reduction amount can be achieved, including the exact level one would have achieved without sinks.

[3]Ethanol as currently produced often results in significant CO_2 emissions from fossil fuels because they are used in various parts of the processing cycle such as in distillation. If fossil fuel use is fully under a cap, including that potentially used in biomass,

this is not a problem because those emissions will be controlled under the cap covering energy use emissions. In other policies where this carbon is not priced properly, the net CO_2 emissions from ethanol production could render it worse than using petroleum products in the first place. So, this is an important concern in many policy contexts, but here we are assuming that carbon from fossil emissions are priced appropriately.

References

Babiker, M., Reilly, J. and Viguuier, L. (2004) Is emissions trading always beneficial? *Energy Journal* 25(2), 33–56.

Babiker, M.H., Jacoby, H.D., Reilly, J.M. and Reiner, D.M. (2002) The evolution of a climate regime: Kyoto to Marrakech. *Environmental Science and Policy* 5(3), 195–206.

Betz, R., Eichhammer, W. and Schleich, J. (2004) Designing national allocation plans for EU-emissions trading: a first analysis of outcomes. *Energy and Environment* 15(3), 375–425.

Birdsey, R., Cannell, M., Galinski, W., Gintings, A., Hamburg, S., Jallown, B., Kirschbaum, M., Krug, T., Kurz, W., Prisley, S., Schulze, D., Singh, K.D., Singh, T.P., Solomon, A.M., Viller, S.L. and Yamagata, Y. (2000) Afforestation, reforestation, and deforestation (ARD) activities. In: Watson, R.T., Noble, I.R., Bolin, B., Ravindranath, N.H., Verardo, D.J. and Dokken, D.J. (eds) *Land Use, Land-Use Change, and Forestry*. Intergovernmental Panel on Climate Change, Cambridge University Press, Cambridge, pp. 125–179.

Bohringer, C. (2001) *Climate Politics from Kyoto to Bonn: From Little to Nothing?!?* Working Paper, Center for European Economic Research, Mannheim, Germany.

EC (2003) *Directive 2003/87/EC Establishing a Scheme for Greenhouse Emission Allowance Trading within the Community and Amending Council Directive 96/61/EC*. European Commission, Brussels, Belgium.

EC (2005) *EU Emission Trading. An Open Scheme Promoting Global Innovation to Combat Climate Change*. European Commission, Brussels, Belgium.

EIA (2003) *Analysis of S. 139, the Climate Stewardship Act of 2003*. Energy Information Administration, Office of Integrated Analysis and Forecasting, US Department of Energy, Washington, DC.

Ellerman, A.D. (2001) Le defi europeen: Issues in the implementation of greenhouse gas emissions trading in Europe. *Revue de l'Energie* 524, 105–111.

Ellerman, A.D., Joskow, P.L., Schmalensee, R., Montero, J.P. and Bailey, E.M. (2000) *Markets for Clean Air: The U.S. Acid Rain Program*. Cambridge University Press, Cambridge.

Federal Register (2002) *Department of Energy, Voluntary Reporting of Greenhouse Gas Emissions, Reductions, and Carbon Sequestration*. 67(87), 30370–30373 (May 6).

Felzer, B., Kicklighter, D.W., Melillo, J.M., Wang, C., Zhuang, Q. and Prinn, R. (2004) Effects of ozone on net primary production and carbon sequestration in the conterminous United States using a biogeochemistry model. *Tellus B* 56(3), 230–248.

Felzer, B., Reilly, J., Melillo, J., Kicklighter, D., Sarofim, M., Wang, C., Prinn, R. and Zhuang, Q. (2005) Future effects of ozone on carbon sequestration and climate change policy using a global biogeochemical model. *Climatic Change* 73, 345–373.

Herzog, H., Caldeira, K. and Reilly, J. (2003) An issue of permanence: assessing the effectiveness of ocean carbon sequestration. *Climatic Change* 59, 293–310.

Kyoto Ratification Advisory Group (2003) Report of the Kyoto Ratification Advisory Group: a risk assessment. The Cabinet Office of New South Wales Government Printer.

Lewandrowski, J., Peters, M., Jones, C., House, R., Sperow, M., Eve, M. and Paustian, K. (2004) Economics of sequestering carbon in the U.S. agricultural sector. *Economic Research Service Technical Bulletin* No. (TB1909), USDA, Washington, DC, March.

Manne, A. and Richels, R. (2001) *US Rejection of the Kyoto Protocol: The Impact on Compliance Cost and CO2 Emissions*. Working Paper, AEI-Brookings Joint Center for Regulatory Studies, 1–12 October.

Marland, G., McCarl, B.A. and Schneider, U. (2007) Soil carbon: policy and economics. *Climatic Change* (Forthcoming).

Matus, K., Yang, T., Paltsev, S., Reilly, J. and Nam, K. (2007) Toward integrated assessment of environmental change: air pollution health effects in the USA. *Climatic Change* (in press).

McCarl, B.A., Murray, B.C. and Schneider, U.A. (2005) The comparative value of biological carbon sequestration. Available at: http://agecon2.tamu.edu/people/faculty/mccarl-bruce/papers/915.pdf

Melillo, J.M., McGuire, A.D., Kicklighter, D.W., Moore III, B., Vorosmarty, C.J. and Schloss, A.L. (1993) Global climate change and terrestrial net primary production. *Nature* 363, 234–240.

Paltsev, S., Reilly, J.M., Jacoby, H.D., Ellerman, A.D. and Tay, K.H. (2003) Emissions trading to reduce greenhouse gas emissions in the United States: the McCain–Lieberman proposal. *MIT Joint Program for the Policy and Science of Global Change*. Report No. 97. Cambridge, Massachusetts, June.

Paltsev, S., Reilly, J., Jacoby, H. and Tay, K. (2007) How (and why) do climate policy costs differ among countries? In: Schlesinger, M. (ed.) *Integrated Assessment of Human-induced Climate Change*. Cambridge University Press, Cambridge (in press).

Pew Center (2005) *The European Union Emissions Trading Scheme (EU-ETS): Insights and Opportunities*. Pew Center, Alexandria, Virginia.

Point Carbon (2005a) *Carbon Market Europe*, 23 September 2005. A Point Carbon Publication. Available at: http://www.pointcarbon.com

Point Carbon (2005b) *Carbon Market Europe*, 30 September 2005. A Point Carbon Publication. Available at: http://www.pointcarbon.com

Reilly, J. (2002) *MIT EPPA Model Projections and the U.S. Administration's Proposal, Joint Program on the Science and Policy of Global Change Technical Note No. 2*. Massachusetts Institute of Technology, Cambridge, Massachusetts. Available at: http://web.mit.edu/globalchange/www/technote3.html

Reilly, J. (2003) Reconstructing climate policy: beyond Kyoto. *Colorado Journal of International Environmental Law and Policy, 2003 Yearbook*, pp. 117–124.

Reiner, D.M. (2001) Climate impasse: how The Hague negotiations failed, *Environment* 43(2), 36–43.

Schlamadinger, B. and Marland, G. (1999) Net effect of forest harvest on CO_2 emissions to the atmosphere: a sensitivity analysis on the influence of time. *Tellus* 51B, 314–325.

Tian, H., Melillo, J.M., Kicklighter, D.W., McGuire, A.D. and Helfrich III, J.V.K. (1999) The sensitivity of terrestrial carbon storage to historical climate variability and atmospheric CO_2 in the United States. *Tellus* 51B, 414–452.

Tian, H., Melillo, J.M., Kicklighter, D.W., Pan, S., Liu, J., McGuire, A.D. and Moore III, B. (2003) Regional carbon dynamics in monsoon Asia and its implications for the global carbon cycle. *Global and Planetary Change* 37, 201–217; doi:10.1016/S0921-8181(02)00205-9.

UNFCCC (United Nations Framework Convention on Climate Change) (1997) *The Kyoto Protocol*. Climate Change Secretariat, Bonn.

UNFCCC (United Nations Framework Convention on Climate Change) (2000) *Methodological Issues: Land Use, Land-Use Change, and Forestry*. FCCC/SBSTA/2000/9, 25 August.

UNFCCC (United Nations Framework Convention on Climate Change) (2005) *Kyoto Protocol: Status of Ratification*. Updated 20 January 2005. Available at: http://unfccc.int/essential_background/kyoto_protocol/status_of_ratification/items/2613.php

Watson, R.T., Noble, I.R., Bolin, B., Ravindranath, N.H., Verardo, D.J. and Dokken, D.J. (eds) (2000) *Land Use, Land-Use Change, and Forestry*. Intergovernmental Panel on Climate Change, Cambridge University Press, Cambridge.

White House (2002) *U.S. Climate Strategy: A New Approach*. Policy Briefing Book, Washington, DC, February.

Wiebe, K., Tegene, A. and Kuhn, B. (1996) *Partial Interests in Land: Policy Tools for Resource Use and Conservation*. Economic Research Service, USDA Agricultural Economic Report No. AER744, Washington, DC, November.

Xiao, X., Kicklighter, D.W., Melillo, J.M., McGuire, A.D., Stone, P.H. and Sokolov, A.P. (1997) Linking a global terrestrial biogeochemical model and a 2-dimensional climate model: implications for the carbon budget. *Tellus* 49(B), 18–37.

Xiao, X., Melillo, J.M., Kicklighter, D.W., McGuire, A.D., Prinn, R.G., Wang, C., Stone, P.H. and Sokolov A.P. (1998) Transient climate change and net ecosystem production of the terrestrial biosphere. *Global Biogeochemical Cycles* 12(2), 345–360.

9 Methane: Importance, Sources and Sinks

David S. Reay[1], Keith A. Smith[1] and C. Nick Hewitt[2]
[1]*School of GeoSciences, University of Edinburgh, Edinburgh,UK;* [2]*Lancaster Environment Centre, University of Lancaster, Lancaster, UK*

9.1 Introduction

In 1776 the Italian physicist Alessandro Volta noticed bubbles rising from the bottom of a pond. He collected some of these bubbles and found that the collected air was inflammable. Volta had discovered the powerful greenhouse gas (GHG) methane (CH_4). Today CH_4 is used throughout the world both as an industrial and a domestic fuel source, but in the past few decades there has been increasing concern about the effect CH_4 may be having on our climate.

As with the other two main anthropogenic GHGs, carbon dioxide (CO_2) and nitrous oxide (N_2O), the concentration of CH_4 in the atmosphere has been increasing rapidly in the last century or so. We now have records of CH_4 concentrations in our atmosphere going back more than 650,000 years. These show that, though we have had many peaks and troughs in that time, levels of CH_4 have never been as high as they are today (Cicerone and Oremland, 1988). Peak concentrations were previously ~700 parts per billion (ppb), but since the beginning of the industrial era levels have more than doubled to their current high of about 1800 ppb. Although such concentrations are much lower than those of CO_2 (currently about 380 ppm), CH_4 molecules are much more effective at trapping the infrared radiation (heat) reflected from the earth's surface. Indeed, the global warming potential (GWP) of CH_4 on a mass basis is 23 times that of CO_2 over a 100-year time horizon. CH_4 concentrations are rising at around 0.6%/year. Consequently, CH_4 is now a key target in many GHG reduction strategies (Hogan *et al.*, 1991).

The GWP compares the potential warming effects of different trace gases relative to that of CO_2. It is also adjusted for any production of secondary GHGs formed when the primary gas is destroyed. The GWP combines the capacity of a gas to absorb infrared radiation, its lifetime in the atmosphere and the length of time over which its effects on the earth's climate need to be quantified (the time horizon). As it has an atmospheric lifetime of only about 10 years, CH_4 is 62 times more effective than CO_2 on a 20-year time horizon and drops to being 23 times more effective on a 100-year time horizon.

9.2 Natural Sources

Global CH_4 emissions add up to ~600 Tg CH_4/year, with about half of this coming from natural sources and half from human activity (IPCC, 2001) (Fig. 9.1). Natural

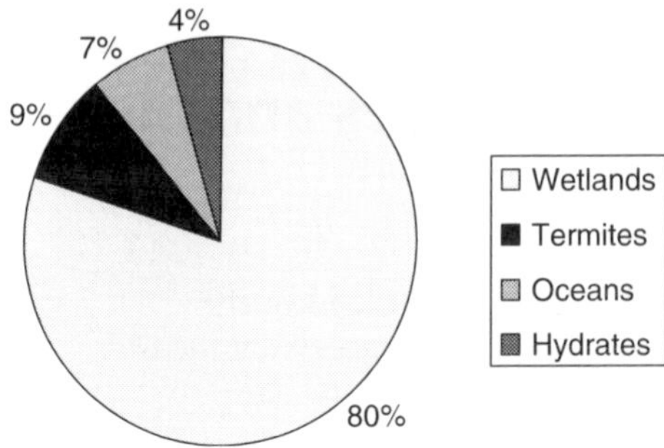

Fig. 9.1. Natural sources of methane (CH_4).

sources are dominated by microbial activity, in particular the process known as methanogenesis, common in waterlogged soils around the world.

9.2.1 Wetlands

Wetland emissions dominate the natural sources of CH_4. The amounts of CH_4 produced vary greatly from area to area, with changes in temperature, water level and organic carbon content all being important controlling factors. Current estimates of global CH_4 emissions from wetlands range between 100 and 250 Tg CH_4/year.

The process of CH_4 production in wetland soils (methanogenesis) involves the microbial mineralization of organic carbon under anaerobic conditions in the waterlogged soil. In the absence of oxygen, the organic carbon (usually simple carbon compounds such as acetate) is used as an alternative terminal electron acceptor and so provides a source of energy for the methanogens in these anoxic soils. When water levels fall, as they often do in the summer months, CH_4 emissions from wetlands can be greatly reduced or even cease completely as oxygen concentrations rise in the soil.

Through widespread land drainage and changes in land use, humans have had a huge impact on wetland CH_4 emissions in many areas of the world. Additionally, increased nitrogen and sulphate deposition from the atmosphere, again resulting from human activity, may also greatly affect net CH_4 emissions from wetland areas. Some studies have shown an increase in CH_4 consumption (CH_4 oxidation) as a result of increased nitrogen deposition, while others have reported inhibition. Sulphate deposition, which can occur due to emissions of sulphur dioxide (SO_2) from fossil fuel burning, can favour sulphate-reducing bacteria over methanogenic bacteria in waterlogged soils, thus actually reducing CH_4 production.

The potential for control of CH_4 emission from wetlands lies largely in land use policy. Under certain circumstances draining of wetlands could greatly reduce emissions. However, certain wetland environments such as peat bogs have been shown to vastly increase their emissions of CO_2 in response to such draining – so offsetting any net reduction in CH_4 emissions. In addition there is the problem of habitat destruction, with many wetland animal, plant and insect species already being endangered. Addition of sulphate-based fertilizers to wetlands may also limit methanogenesis by favouring sulphate-reducing bacteria, but again this may have negative impacts on ecosystem function and biodiversity through acidification of the soils, not to mention the significant financial costs that would be involved.

9.2.2 Termites

Each termite produces, on average, about 0.5 μg CH_4/day – a seemingly insignificant amount. However, when this is multiplied by the global population of termites, CH_4 emissions from this source are estimated to be about 20 Tg/year.

There are more than 2000 different species of termites and the amounts of CH_4 produced vary considerably between species, with some producing no CH_4 at all. CH_4 is produced in termite guts, by symbiotic bacteria and protozoa, during food digestion. This CH_4 does not always end up going straight into the atmosphere. Many species are subterranean or live in aboveground earth mounds where much of the CH_4 can be used up by soil methanotrophs before it gets out into the atmosphere – soil-mediated

CH_4 oxidation, as discussed in subsequent chapters.

9.2.3 Oceans

Globally, oceans are thought to add around 15 Tg CH_4/year to the atmosphere, with parts of the surface of the world's oceans having relatively high concentrations of dissolved CH_4. CH_4 is not very soluble in water and so is often emitted as bubbles rising up from aquatic and oceanic sediments (like those seen by Volta), rather than via diffusion between the water and air. As in wetlands, oceanic CH_4 is largely produced by methanogenic bacteria, which, because they need anaerobic conditions, are usually found to be producing CH_4 either within sinking particles or in sediments.

Humans have an impact on oceanic CH_4 emissions, primarily through the effect on oceanic nutrient inputs via rivers and estuaries. The high nutrient loads of many rivers, produced largely by sewage input and agricultural runoff, lead to eutrophic conditions in estuaries and coastal waters. Such nutrient-rich waters and sediments are ideal for methanogenesis, with oxygen levels in the water often very low and with plenty of organic carbon available that the methanogenic bacteria are able to utilize.

9.2.4 Hydrates

CH_4 hydrates (also known as clathrates) are thought to be responsible for emissions of 5–10 Tg CH_4/year to the atmosphere. Globally, there are huge amounts of CH_4 stored as hydrates – it is estimated that there is about 3000 times more CH_4 locked up as hydrates (~10^7 t) than is currently found in our atmosphere.

CH_4 hydrates occur as solid deposits in deep marine sediments and in some polar regions. They are made up of a mixture of CH_4 and water (~70% CH_4) that can quickly break down due to changes in temperature and pressure, to release the trapped CH_4.

Large-scale decomposition of CH_4 hydrate deposits has been blamed by some researchers for big surges in atmospheric CH_4 concentrations during the last 500 years. As such, CH_4 hydrates represent a potentially huge positive feedback mechanism to global warming. With increasing ocean temperatures, large CH_4 hydrate deposits may become unstable and thus lead to very rapid CH_4 emissions.

The hard-to-predict, but potentially catastrophic, consequences of global warming on the deposits of CH_4 hydrate underline the potential threat posed by feedbacks within the global climate system. Limiting future global warming may be crucial to prevent such a runaway scenario of CH_4 release leading to even greater warming.

9.3 Anthropogenic Sources

Energy-related and ruminant emissions dominate anthropogenic CH_4 sources (Fig. 9.2). Globally, these are estimated to total ~320 Tg CH_4/year.

9.3.1 Energy-related

The bulk of energy-related CH_4 emissions arises from their release during fossil fuel extraction and transportation. Some CH_4 is

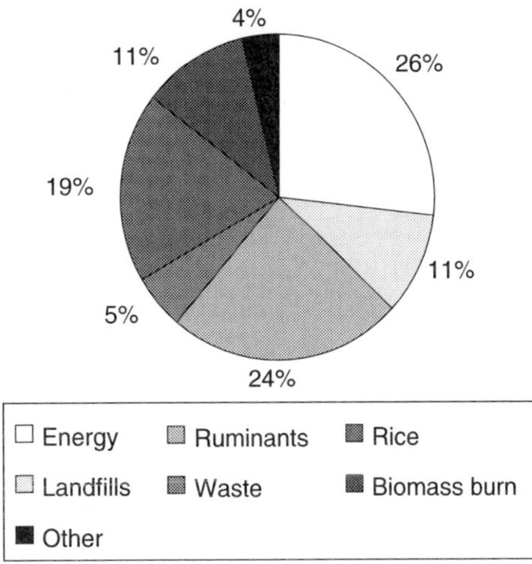

Fig. 9.2. Anthropogenic sources of methane (CH_4).

also produced during fossil fuel combustion, with sources such as fossil-fuelled power stations, transport and domestic heating all being significant contributors to atmospheric CH_4 concentrations. The total CH_4 emissions estimate from energy-related sources is ~100 Tg/year.

At 30–50 Tg CH_4/year, coal mining constitutes a large source of anthropogenic CH_4 (despite the decline of coal mining in the UK). The bulk of global CH_4 emissions due to coal mining actually comes from just a few, relatively deep mines around the world. During the geological process of coal formation, CH_4 is formed and some of this remains trapped until it is released by mining operations. Generally, the deeper the coal seam, the greater is the amount of CH_4 that is trapped. In shallow and opencast mines, trapped CH_4 is often released directly to the atmosphere during mining, whereas in deeper mines it is often released deliberately via ventilation shafts, so as to prevent potentially explosive concentrations building up.

Strategies to reduce CH_4 emissions from coal mines include recovery of CH_4 during mine construction, during coal extraction and even after the mine is closed down. Such recovery can be economically viable, particularly in deep mines where high concentrations mean that the recovered CH_4 can be used as a fuel source, and can substantially reduce emissions to the atmosphere.

As with coal, the geological formation of oil can result in large CH_4 accumulations. During oil drilling and extraction, this trapped CH_4 can be released to the atmosphere. Targeted collection of the CH_4 associated with oil can vastly reduce emissions from this source, with the collected CH_4 being either flared off (as CO_2 and water) or used as an additional fuel source.

The loss of CH_4 during natural gas extraction is obviously something that has both a direct GHG and an economic cost. Nevertheless, significant amounts are lost during its extraction and aboveground transfer. It is estimated that in the 1990s ~6% of the CH_4 piped across Russia was lost due to leaks. More efficient collection techniques, improved targeting of buried CH_4 deposits and better-maintained transfer pipelines could all help reduce incidental CH_4 emissions from this source. Leaks at the end of the supply network – in homes and businesses – also represent a potentially significant source of emissions.

9.3.2 Landfills

Global CH_4 emissions from landfill are estimated to be between 30 and 70 Tg/year. Most of this landfill CH_4 currently comes from developed countries, where the levels of waste tend to be highest. Landfills provide ideal conditions for methanogenesis, with lots of organic material and with anaerobic conditions usually prevalent. The huge amounts of waste that are buried in landfill sites can mean that CH_4 is produced for years after the site is closed, due to the waste slowly decaying under the ground.

CH_4 escapes from landfills either directly to the atmosphere or by diffusion through the cover soil. Highly active communities of CH_4-consuming bacteria (methanotrophs) can develop in these overlying soils and these can greatly reduce the amounts of CH_4 emitted, as discussed by De Visscher *et al.* (Chapter 12, this volume).

Our so-called throwaway society in developed countries has led to a large increase in the amounts of organic waste entering landfill sites. Not only does this lead to greater GHG emissions in the form of landfill CH_4, but the waste incurs further GHG costs in terms of waste transport. In recent years, landfill CH_4 emissions have been identified by several world governments as a GHG source that can be both easily defined and reduced. In addition to the practice of covering landfills with a thick soil layer – to promote CH_4 uptake by soil methanotrophs – more proactive strategies are also available.

CH_4 recovery systems are now commonly installed at landfill sites and these can reduce emissions to the atmosphere by more than half. Sometimes the recovered CH_4 is simply flared off, but often the recovery systems can provide an economically viable energy source, where the collected CH_4 is used in electricity generation.

9.3.3 Rice cultivation

At 25–60 Tg CH_4/year, rice agriculture is a substantial anthropogenic source of atmospheric CH_4, possibly the biggest of all anthropogenic sources. The warm, waterlogged soil of rice paddies provides ideal conditions for methanogenesis and, though some of the CH_4 produced is usually oxidized by methanotrophs in the shallow overlying water, much is released into the atmosphere.

Rice is grown very widely and rates of CH_4 emissions may vary greatly between different areas. Differences in average temperature, water depth and the length of time during which the rice paddy soil is waterlogged can all result in big regional and seasonal variations. However, CH_4 emissions from worldwide rice agriculture have been well studied in recent years and fairly reliable estimates of global emissions now exist.

On average, rice paddy soils are only fully waterlogged for about 4 months each year. For the rest of the time methanogenesis is generally much reduced and, where the soil dries out sufficiently, rice paddy soil can become a temporary sink for atmospheric CH_4.

With an increasing world population, reductions in rice agriculture remain largely untenable as a CH_4 emissions reduction strategy. However, through a more integrated approach to rice paddy irrigation and fertilizer application, substantial reductions remain possible. Many rice varieties can be grown under much drier conditions than those traditionally employed, with large reductions in CH_4 emissions without any loss in yield. Additionally, there is the potential for improved varieties of rice to produce a much larger crop per area of rice paddy, and so allow for a cut in the area of rice paddies, without a cut in rice production. Finally, the addition of compounds such as ammonium sulphate, which favour activity of other microbial groups over that of the methanogens, has proved successful under some conditions.

9.3.4 Ruminants

CH_4 emissions from ruminant livestock are currently estimated to be ~100 Tg/year and represent the biggest anthropogenic source. The loss of CH_4 from ruminant livestock is a problem not only with respect to GHG emissions but also to farmers, in that feed converted into, and released as, CH_4 is feed not being converted into meat and/or milk.

CH_4 is produced in the guts of ruminant livestock as a result of methanogenic bacteria. The composition of the animal feed is a crucial factor in controlling the amounts of CH_4 produced. A single sheep can produce ~30 l CH_4/day while a dairy cow can produce up to 200 l/day.

As with rice agriculture, CH_4 emissions arising from ruminant livestock are dependent on human demand. With a continuing expansion of meat and dairy product consumption around the world, the demand for ruminant livestock, and thus the size of this CH_4 source, has grown rapidly. Intensive rearing methods, developed to provide large amounts of meat and dairy products at low prices and to a wide consumer base, have led to very high densities of ruminant livestock and strong local CH_4 sources.

The best-studied and applied CH_4 reduction strategy in ruminants has been that of altering the feed composition, either to reduce the percentage that is converted into CH_4 or to improve the meat and milk yield. Improvements in the overall quality of animal feed may allow meat and dairy production to be maintained at the same level with fewer animals, and so result in less total CH_4 emissions. Recent ruminant CH_4 reduction strategies have included the introduction of methanogen inhibitors, both biological and chemical, with the animal feed, to kill off or at least reduce the activity of the methanogenic bacteria or the protozoa that they are associated within the gut.

9.3.5 Agricultural and municipal waste

Agricultural waste can represent a significant source of CH_4. The anaerobic decomposition of livestock and poultry manure, common to manure heaps and slurry tanks, often leads to large amounts of CH_4 production due to its large organic carbon content. Similarly, the processing of industrial and domestic wastewater and sewage can produce significant amounts of CH_4.

In total, such waste accounts for 14–25 Tg CH_4/year globally. Historically, emissions from this source are likely to have been much lower due to lower livestock numbers and densities. Where traditionally animal manure would have been spread over a wide area and decomposed aerobically, intensive livestock rearing methods mean high concentrations of manure build-up in relatively small areas. In a similar way, greater human numbers and population densities have led to larger concentrations of wastewater and sewage in collection areas such as sewage works.

The trapping of CH_4 from strong sources, such as slurry tanks, has already proved a very successful way of reducing emissions to the atmosphere. The recovered CH_4, often called 'biogas', can be simply flared off or can be used as a fuel. Pilot studies have shown that such biogas capture may provide an alternative to petroleum as fuel for transport.

Other CH_4 mitigation options include reducing demand for meat and dairy products and a move away from intensive rearing methods, with an increase in grazing time for animals and so a greater dispersal of their manure. For human wastewater and sewage, ensuring aerobic decomposition using aeration methods is an oft-employed strategy, though CH_4 capture and subsequent flaring is practised at some sewage treatment sites.

9.3.6 Biomass burning

Biomass burning accounts for 20–40 Tg CH_4/year. CH_4 emissions arising from biomass burning are a result of incomplete combustion and huge amounts can be produced during large-scale burning of woodlands, savannah and agricultural waste. In savannah regions of the world, burning is often carried out every few years to promote regeneration of the vegetation. The importance of CH_4 emissions from biomass burning can be overshadowed by the large amounts of CO_2 that are also produced, but in many cases the subsequent regrowth, and CO_2 uptake, of previously burned woodland and savannah areas means that CH_4 emissions have a disproportionately greater climate-forcing effect.

Burning of agricultural waste also produces significant amounts of CH_4 due to its generally high water content. Additionally, wood burning as a domestic fuel source and for charcoal production can release significant amounts of CH_4.

9.4 Methane Sinks

Balancing CH_4 emissions to our atmosphere are three main sinks: tropospheric destruction of CH_4 by hydroxyl (OH) radicals is the dominant one, with stratospheric destruction and oxidation by soil bacteria also being important (Fig. 9.3). The total sink size is estimated to be 500–600 Tg CH_4/year.

9.4.1 Soil methane sinks

Although some soils can actually be sources of CH_4, e.g. wetlands and rice paddy soils, they can act as effective sinks for CH_4 as described by Dunfield (Chapter 10, this volume). The Intergovernmental Panel on Climate Change (IPCC) estimates that soils represent a sink for CH_4 of ~30 Tg/year. The CH_4 is predominantly used by aerobic bacteria in the soil

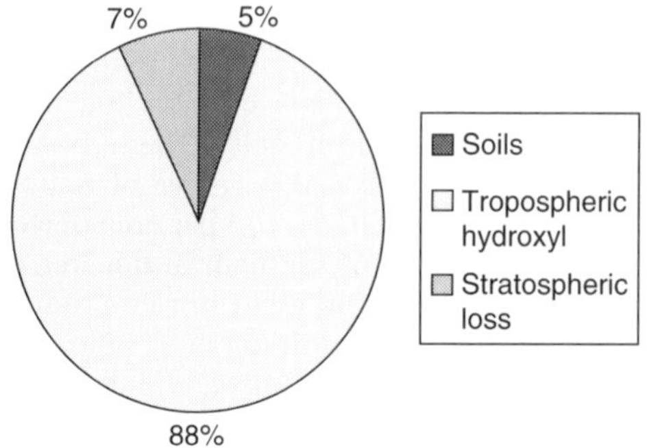

Fig. 9.3. Sinks for methane (CH_4).

(methanotrophs), which use the CH_4 as a source of carbon in the process of biological CH_4 oxidation. Methanotrophs are fairly ubiquitous in soil, but two distinct types have been reported.

The first are commonly described as high-capacity–low-affinity methanotrophs and are adapted for growth at high CH_4 concentrations (several thousand parts per million in air). Such high CH_4 concentrations arise from waterlogged soils and sediments. The second are known as low-capacity–high-affinity methanotrophs and are able to make use of the trace amounts of CH_4 in the atmosphere (~1800 ppb). Although several of the 'low-affinity' methanotrophs have been identified and cultured in laboratories, the 'high-affinity' methanotrophs remain poorly understood.

The key as to whether a soil acts as a sink or source of CH_4 tends to be water. Forest soils are often strong sinks for CH_4, because transpiration by the trees keeps the soil water content from becoming too high. In such soils the aerobic conditions required by methanotrophs tend to predominate, giving rise to a net CH_4 sink. When soils do become waterlogged, such as often happens in winter, the availability of oxygen falls and the balance shifts from methanotrophs to anaerobic CH_4-producing bacteria (methanogens), with the soil becoming a CH_4 source.

It is not uncommon for both high-affinity and low-affinity methanotrophs to be active in the same area, but at different depths. The high-affinity methanotrophs make use of the atmospheric CH_4 diffusing into the soil from the surface, and so tend to be most active in the top few centimetres of the soil. Active populations of the low-affinity methanotrophs, on the other hand, can often be found in deeper soil layers, making use of the high concentrations of CH_4 arising from soil methanogenesis. Indeed, the bulk of low-affinity methanotroph activity can be confined to a very narrow band in the soil. This band, occurring where the balance between CH_4 diffusing up through the soil and oxygen diffusing down through it, is just right. Changes in the water table and rainfall throughout the year mean that such bands of activity can move up and down through the soil following the conditions most suitable for CH_4 oxidation.

In addition to water content of the soil, factors such as pH, soil temperature and concentration of inorganic nitrogen (e.g. nitrate and ammonium) can be crucial in determining whether a particular soil will act as a sink for CH_4 or not (Hütsch *et al.*, 1994). For instance, though the soil under alder trees may be well aerated and seemingly ideal for methanotrophs, the high concentrations of nitrate that arise from nitrogen fixation in the alder's roots mean that methanotroph activity is slowed or completely stopped (Reay *et al.*, 2001).

Changes in human land use can have a huge impact on the capacity of soils to act as a CH_4 sink. Conversion of wooded and fallow land for agricultural use tends to result in increased nitrogen concentrations in the soil which then inhibit CH_4 oxidation (Steudler *et al.*, 1989). Similarly, the increased deposition of nitrogen from the atmosphere due to human activities can also reduce, or completely inhibit, CH_4 oxidation in soil.

Rapid, and often very significant, changes in CH_4 flux arise from alterations in soil drainage and soil structure caused by land-use change. Deforestation can result in higher water tables and soil water contents, due to the loss of transpiration by the trees. Soil compaction, such as that caused by agricultural machinery, can also result in anaerobic conditions developing in the soil and consequently turn the soil from a CH_4 sink to a CH_4 source.

Our potential for control of the soil CH_4 sink lies primarily in our ability to alter land use practices. The better targeting of fertilizer application, avoidance of soil compaction, and more thoughtful land conversion could all help to avoid the destruction of soil CH_4 sinks. Likewise, reducing the amount of atmospheric nitrogen pollution we produce could also help to maintain levels of CH_4 oxidation in existing soil CH_4 sinks around the world.

9.4.2 Atmospheric methane sinks

Shallcross *et al.* (Chapter 11, this volume) describe how CH_4 is removed from the Earth's atmosphere.

The dominant route of removal is by oxidation by OH. This reaction is very important in the atmosphere as it not only limits the radiative forcing potential of CH_4 but also contributes to the production of peroxy radicals, which can subsequently lead to the formation of ozone. CH_4 oxidation therefore affects the oxidizing capacity of the lower atmosphere.

The reaction of CH_4 with OH is:

$$CH_4 + OH \quad CH_3 + H_2O$$

Although this initial reaction produces methyl (CH_3) radicals and water (H_2O), the complete reaction involves a number of further reactions and leads both to increased water vapour and to production of CO_2. Another indirect effect of the reaction between CH_4 and OH is that it can magnify the effects of other pollutants, with increased CH_4 in the atmosphere meaning fewer OH radicals and so less oxidizing power in the atmosphere as a whole.

Some additional CH_4, ~40 Tg/year, is accounted for in the same way, but in our stratosphere. There are also several relatively minor sinks for CH_4, including chemical oxidation by chlorine in the atmosphere and in the surface waters of our seas.

Overall the direct human impact on the atmospheric destruction of CH_4 is relatively minor. However, our emissions of other atmospheric pollutants, such as nitrogen oxide (NO_x) gases, may reduce the levels of OH radicals in our atmosphere, thereby prolonging the lifetime of CH_4 in our atmosphere.

Because of the difficulty of directly measuring OH radical concentrations in the atmosphere (it is not yet possible to measure OH on a global scale, hence it is not possible to know with certainty what is the global average OH), models are used to calculate global OH concentrations. In constructing such models it becomes apparent that the atmospheric chemistry and physics leading to the formation and removal of OH are very complicated and include many feedbacks. Hence, understanding and predicting the rate of removal of CH_4 from the atmosphere is not as simple as it might first appear.

Shallcross *et al.* (Chapter 11, this volume) describe how changes in the isotopic ratios of $^{13}CH_4/^{12}CH_4$ and $^{12}CH_3D/^{12}CH_4$ allow quantification of the various possible routes of removal and explain the fragility of the Earth's greenhouse: changes in the abundance of OH will necessarily result in changes to the radiative forcing due to CH_4.

9.4.3 Artificial methane sinks

Biological CH_4 oxidation is hugely important in reducing CH_4 emissions from 'source' areas such as landfills, marshland and lakes. Although vast amounts of CH_4 may be produced in these areas, methanotrophs can often limit the actual release of CH_4 to the atmosphere to less than 10%. Indeed, biological oxidation of CH_4 is probably greater than total chemical oxidation in the atmosphere if the full CH_4 cycle is considered (King, 1992). De Visscher *et al.* (Chapter 12, this volume) describe the important CH_4 sinks in human-made environments. In the soils above landfill sites, for instance, large and very active populations of high-capacity methanotrophs can develop, using much of the CH_4 that diffuses up through the soil from the decomposing rubbish (Whalen *et al.*, 1990). Although the focus of their chapter is on landfill cover soils, which are the clearest examples of such artificial CH_4 sinks, De Visscеher *et al.* also discuss the more general importance of such 'artificially induced' CH_4 oxidation.

References

Cicerone, R.J. and Oremland, R.S. (1988) Biogeochemical aspects of atmospheric methane. *Global Biogeochemical Cycles* 2(4), 299–327.

Hogan, K.B., Hoffman, J.S. and Thompson, A.M. (1991) Methane on the greenhouse gas agenda. *Nature* 354, 181–182.

Hütsch, B.W., Webster, C.P. and Powlson, D.S. (1994) Methane oxidation in soil as affected by land use, soil pH and N fertilization. *Soil Biology and Biochemistry* 26(12), 1613–1622.

IPCC (2001) *Climate Change 2001: The Scientific Basis*. Contribution of working group I to the third assessment report of the Intergovernmental Panel on Climate Change. Houghton, J.T., Ding, Y., Griggs, D.J., Noguer, M., van der Linden, P.J., Dai, X., Maskell, K. and Johnson, C.A. (eds) Cambridge University Press, Cambridge.

King, G.M. (1992) Ecological aspects of methane oxidation, a key determinant of global methane dynamics. In: Marshall, K.C. (ed.) *Advances in Microbial Ecology*. Plenum Press, New York, pp. 431–468.

Reay, D.S., Radajewski, S., Murrell, J.C., McNamara, N. and Nedwell, D.B. (2001) Effects of land-use on the activity and diversity of methane oxidizing bacteria in forest soils. *Soil Biology and Biochemistry* 33(12–13), 1613–1623.

Steudler, P.A., Bowden, R.D., Melillo, J.M. and Aber, J.D. (1989) Influence of nitrogen fertilization on methane uptake in temperate forest soils. *Nature* 341, 314–316.

Whalen, S.C., Reeburgh, W.S. and Sandbeck, K.A. (1990) Rapid methane oxidation in a landfill cover soil. *Applied and Environmental Microbiology* 56, 3405–3411.

10 The Soil Methane Sink

Peter F. Dunfield
Institute of Geological and Nuclear Sciences, Wairakei Research Station, Taupo, New Zealand

10.1 Introduction

Methane is produced in flooded, anaerobic soil by methanogenic archaea, and oxidized in aerobic soil by methanotrophic bacteria. Both groups are active in wetland environments such as peat bogs, rice paddies and lake sediments, but these ecosystems are usually net methane sources. On the other hand, comparatively dry, aerobic soils (frequently termed 'upland soils' in the literature) are net sinks for atmospheric methane. Methanotrophic bacteria in upland soils oxidize methane that diffuses into the soil from the overlying atmosphere, where it is present at a trace mixing ratio of ~1.75 ppmv. The term 'atmospheric methane' will be used throughout this chapter to denote methane at this mixing ratio.

The first observation of a soil acting as a net methane sink was made in the Great Dismal Swamp, in south-eastern Virginia and north-eastern North Carolina, USA, when the water table was low and there was a thick aerobic surface soil layer (Harriss *et al.*, 1982). Since then, atmospheric methane uptake has been observed in a number of soil ecosystems, including forests (Keller *et al.*, 1983; Steudler *et al.*, 1989), savannah (Seiler *et al.*, 1984), grasslands and meadows (Mosier *et al.*, 1991; Neff *et al.*, 1994), landfill cover soils (Whalen *et al.*, 1990), desert (Striegl *et al.*, 1992), tundra (Whalen and Reeburgh, 1990), heathland (Kruse and Iversen, 1995), dryland rice soils (Singh *et al.*, 1998) and other agricultural soils (Mosier *et al.*, 1991; Hütsch *et al.*, 1993). This sink is microbial, as autoclaved soil has no methane oxidation activity (Whalen *et al.*, 1990; Yavitt *et al.*, 1990; Bender and Conrad, 1992), the kinetics of the process is typical of an enzymatic reaction (Bender and Conrad, 1992) and $^{14}CH_4$ is incorporated into microbial biomass (Roslev *et al.*, 1997).

Although aerobic upland soils are generally considered methane sinks alone, methanogenesis also occurs. Periodic net methane emissions, or elevated mixing ratios of methane in soil matrix air, have been observed in many upland soils (e.g. Yavitt *et al.*, 1990; Keller et al., 1993; Mosier *et al.*, 1993; Savage *et al.*, 1997; Castro *et al.*, 2000; Maljanen *et al.*, 2003). In some soils, methane production was localized in organic horizons (Sexstone and Mains, 1990; Adamsen and King, 1993; Yavitt *et al.*, 1995; Amaral and Knowles, 1997; Saari *et al.*, 1997); in others, methane production was located in water-saturated zones (Klemedtsson and Klemedtsson, 1997; Kammann *et al.*, 2001). Methane emissions may be particularly strong during spring thaw in the temperate zone (Dunfield *et al.*, 1995; Wang and Bettany, 1995). Isotope dilution techniques indicate that methanogenesis occurs simultaneously with methane oxidation in some soils, even when only net methane uptake is measurable using flux techniques (Andersen *et al.*, 1998; von Fischer and Hedin, 2002). Methane is also produced in the anaerobic guts of soil macrofauna, especially soil-feeding termites (Seiler *et al.*, 1984). The soil methane sink should therefore be considered a net effect of methane consumption and production processes, which may occur either simultaneously or separated in time and space. Periodic methane emission rates in localized areas may be extremely high compared to methane uptake rates, and effectively negate the sink strength of a large area (Simpson *et al.*, 1997), or cause a site that is usually a net sink to be a net source when considered over an entire year (Dunfield *et al.*, 1995).

The magnitude of the soil methane sink is usually estimated to be 30–60 Tg/year. However, this estimate is based on extrapolations of data-sets that are geographically and temporally incomplete. In a review of published flux measurements, Smith *et al.* (2000) place a potential range of 7–100 Tg/year on the sink strength. Considerable variability exists among different soil types, and many ecosystems are poorly studied. The highest methane oxidation rates have been measured in pristine forests, and the record is 13.7 $mg/m^2/day$ measured in tropical forests of India (Singh *et al.*, 1997).

The microbial soil methane sink is much less in magnitude than the atmospheric sink mediated by OH radicals (see Shallcross *et al.*, Chapter 11, this volume). However, there are several good reasons, besides scientific curiosity, to investigate it. Its magnitude is close to the present source–sink imbalance of the methane budget, and it is very sensitive to many types of anthropogenic disturbance, including conversion of forests to agriculture, fertilization, soil compaction, acidification and nitrogen deposition. Cultivated soils support very low rates of methane oxidation compared to native ecosystems. It is worthwhile to examine the mechanisms of disturbance, and determine whether the decline of the soil methane sink can be limited or reversed by proper management. The soil methane sink might also be better understood, and more amenable to preservation, if we knew exactly which microorganisms were responsible for it. Although much is known about methanotrophic bacteria in general, evidence indicates that atmospheric methane oxidizers in upland soils are species that have not yet been isolated into pure culture. The following sections will provide a review of the biogeochemical controls of the soil methane sink, and summarize what is known about the bacteria responsible.

10.2 Biogeochemical Considerations

Environmental factors affecting the soil methane sink are essentially of two types: those that have purely physical effects (primarily on diffusion), and those that influence methanotroph populations. Some factors, such as water content, have both physical and microbiological effects. The most important are described in more detail.

10.2.1 Temperature

Temperature influences methane uptake, but the effect is usually weak, with measured Q_{10}s of only 1–2 and little seasonal variation (Born *et al.*, 1990; King and Adamsen, 1992; Crill *et al.*, 1994; Priemé and Christensen, 1997; Price *et al.*, 2004). Methane uptake can continue throughout winter in snow-covered soils (Kessavalou *et al.*, 1998; Mast *et al.*, 1998). The limited temperature dependence probably indicates that the diffusion rate of atmospheric methane into the soil is the major limiting factor. Diffusion is only weakly related to temperature.

10.2.2 Diffusion and water content

Methane diffuses from the atmosphere to methanotrophic bacteria contained within a semi-porous soil matrix. A gradient of decreasing methane with depth is usually evident, declining from a mixing ratio of about 1.75 ppmv at the soil surface often to <0.5 ppmv at 10–30 cm depth (Born *et al.*, 1990; Whalen *et al.*, 1992; Adamsen and King, 1993; Dörr *et al.*, 1993; Koschorreck and Conrad, 1993; Dunfield *et al.*, 1995). Methane oxidation must therefore be diffusion-limited, because if the movement of methane from the atmosphere were much faster than microbial oxidation, no gradient would be evident. Net diffusion along a linear gradient (i.e. downward into a soil) can be described by Fick's law:

$$= {}_{CH_4soil}\ d[CH_4]/d \qquad (10.1)$$

where J is flux (e.g. μmol/m^2/day) and D_{CH_4soil} is the binary diffusion coefficient of CH_4 in soil matrix air (e.g. m^2/day) and z is the depth (m).

Soil is a three-phase system composed of water, air and solids. Diffusion of a gas in water is 10,000 times slower than diffusion in air, so the critical factor controlling the diffusion rate of methane into the soil column is the gas-filled porosity of the soil. D_{CH_4soil} can be adequately estimated by the following empirical formula:

$$D_{CH_4soil} = D_{CH_4air} \times a\varphi_g^b \qquad (10.2)$$

where D_{CH_4air} is the temperature-dependent diffusion rate in air (1.95 m^2/day at 22.5°C; Striegl, 1993), φ_g is fractional gas-filled soil porosity, and a and b are empirical factors to compensate for soil-dependent tortuosity. The actual values of these coefficients will vary with soil type, but estimated average values of a = 0.9 and b = 2.3 work well with many soils (Campbell, 1985; Dunfield *et al.*, 1995; Price *et al.*, 2004). The dependence of the diffusion rate on gas-filled pore space should therefore be an exponential rise to a maximum, the maximum rate occurring where all pore space is gas-filled (Fig. 10.1) (Dunfield *et al.*, 1995).

One corollary of this diffusion model is that coarse-textured soils have higher potential methane uptake rates than fine-textured soils, because there is more gas-filled pore space and less water retention capacity (Born *et al.*, 1990; Dörr *et al.*, 1993). Another is that soil methane uptake decreases with increasing water content (Whalen *et al.*, 1990; Adamsen and King, 1993; Dunfield *et al.*, 1995; Sitaula *et al.*, 1995; Kruse *et al.*, 1996). In addition, soil compaction decreases methane oxidation (Hansen *et al.*, 1993; Ruser *et al.*, 1998; Sitaula *et al.*, 2000). This is primarily a diffusion effect, but a long-term decline of methanotrophic populations may also be caused by the decreased methane supply to compacted soil (Sitaula *et al.*, 2000). Forest harvesting decreases the soil methane sink, or converts it to a methane source, probably because of reduced transpiration and the resulting increase of soil water content (Castro *et al.*, 2000; Kähkönen *et al.*, 2002).

At extremely high and extremely low water contents this physical diffusion model is complicated by biological factors. Methane oxidation, like any microbial activity, becomes limited by water stress (osmotic stress) under very dry conditions (Whalen *et al.*, 1990; Nesbit and Breitenbeck, 1992; Schnell and King, 1996; Priemé and Christensen, 1999). A hump-shaped dependence of methane uptake on water content may therefore be observed, with an optimum at 20–50% of water-holding capacity (or water-filled pore space), bordered by a zone of diffusion limitation at higher water contents, and a zone of desiccation stress at lower water contents (Torn and Harte, 1996; Gulledge and Schimel, 1998a). At high water contents diffusion of O_2 is also limited, resulting in anaerobiosis and potentially in methane production (Fig. 10.1).

10.2.3 Other natural factors

If only diffusion is considered, methanotrophs are expected to live in surface soil, where they are in closest proximity to the atmospheric methane source. However, in many soils surface organic-rich horizons lack appreciable activity, and there is a pronounced subsurface maximum for methane oxidation (Crill, 1991; Adamsen and King, 1993; Koschorreck and Conrad, 1993; Yavitt *et al.*, 1993; Bender and Conrad, 1994b; Schnell and King, 1994; Priemé and

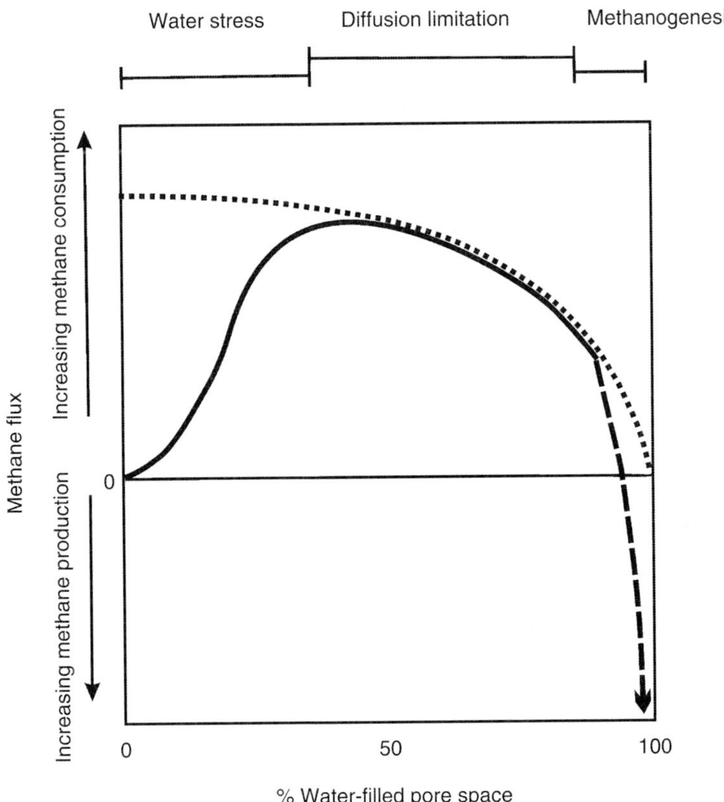

Fig. 10.1. Conceptual model of the effect of water on soil methane flux. The dotted line represents the relationship expected due to diffusion alone. The solid line is the pattern commonly observed, with a zone of water stress and a zone of diffusion limitation. The dashed line indicates that methanogenesis may occur at high water contents.

Christensen, 1997; Ishizuka *et al.*, 2000). The growth of methanotrophic bacteria in many surface soils must therefore be limited, perhaps by inhibitory substances. Potential inhibitors include ammonium (Adamsen and King, 1993; Schnell and King, 1994), ethylene (Sexstone and Mains, 1990; Jäckel *et al.*, 2004) and monoterpenes (Amaral and Knowles, 1998). Desiccation of surface soils could also prevent establishment of methanotrophs, as populations may recover very slowly from water stress (Nesbit and Breitenbeck, 1992; Whalen *et al.*, 1992).

Other natural factors affecting the soil methane sink include tree species composition and fires. Soils under different forest compositions support different methane uptake rates (Borken *et al.*, 2003; Menyailo and Hungate, 2003; Reay *et al.*, 2005). Fires may either increase or decrease the methane sink, and the effects may not be immediate (Burke *et al.*, 1997; Priemé and Christensen, 1999; Jaatinen *et al.*, 2004). Tree species and fire effects are probably mediated through soil chemistry, moisture and microbiology, but the precise mechanisms are complex and not yet clearly understood.

10.2.4 Anthropogenic factors

Conversion of natural ecosystems to agriculture usually reduces soil methane-oxidizing activity. When adjacent cultivated and pristine ecosystems are compared, methane oxidation rates are up to an order of magnitude lower in the cultivated sites (Keller *et al.*, 1990; Mosier *et al.*, 1991; Hütsch *et al.*, 1994; Jensen and Olsen, 1998; Priemé and Christensen, 1999; Knief *et al.*, 2005a). Microbial community analysis of a cultivated field and adjacent forests in Thailand indicated that dramatic changes in the methanotrophic species composition had occurred in the agricultural soil (Knief *et al.*, 2005a). Nitrogen fertilization is often postulated as the major cause of the reduced methane sink in agricultural soils (Section 2.5), but reduced methane oxidation has been noted in unfertilized arable land as well (Mosier *et al.*, 1991).

Other anthropogenic disturbances destroy or reduce the atmospheric methane sink. These include application of a variety of agrochemicals such as pesticides and herbicides (Topp, 1993; Syamsul Arif *et al.*, 1996; Boeckx *et al.*, 1998; Priemé and Ekelund, 2001). Soil acidification via acid rain can inhibit methanotrophs through direct pH effects or aluminium toxicity (Nanba and King, 2000; Bradford *et al.*, 2001b; Sitaula and Bakken, 2001). Even sample handling for scientific purposes can decrease methanotrophic activity, indicating that the bacteria are very sensitive to physical disturbance of natural chemical gradients (Whalen *et al.*, 1992; Amaral and Knowles, 1997).

10.2.5 Nitrogen

Probably the most intensely investigated factor regulating methane uptake is soil nitrogen. Nitrogen fertilization is usually inhibitory to atmospheric methane oxidation. Inhibition mechanisms are complex, and are outlined briefly in this chapter. However, in some cases nitrogen fertilization is not inhibitory, and in rare instances it even stimulates methane oxidation. Nitrogen is an essential nutrient that is frequently limiting in soils. In rice paddies, where methane concentrations are high and rice plants efficiently compete with bacteria for nitrogen, methanotrophs are nitrogen-limited and their activity is stimulated by fertilization (Bodelier *et al.*, 2000). In aerobic upland soils where less methane is available than in rice paddies, nitrogen limitation is probably less intense. However, both stimulatory and inhibitory effects of nitrogen have been observed, the balance of which is specific to the site, fertilizer dosage and fertilizer type. The vast number of studies on these effects are too numerous to list here, and the reader is referred to the review of Bodelier and Laanbroek (2004).

Fertilizers can inhibit methanotrophs through non-specific salt effects (Nesbit and Breitenbeck, 1992; Adamsen and King, 1993; Dunfield and Knowles, 1995; Gulledge and Schimel, 1998b; Saari *et al.*, 2004), but ammonia (NH_3) has a more specific effect as well. Ammonia is a structural analogue of methane and competes for the active site of the methane monooxygenase (MMO) enzyme. Typical competitive inhibition patterns have been observed in pure cultures of methanotrophic bacteria (O'Neill and Wilkinson, 1977; Carlsen *et al.*, 1991) and in soil (Dunfield and Knowles, 1995). The major inhibitory form is NH_3 rather than NH_4^+, so the degree of inhibition depends on pH (O'Neill and Wilkinson, 1977; Carlsen *et al.*, 1991). Competition for the MMO active site is obviously relieved as soon as the offending NH_3 is nitrified or removed from soil, but inhibition of methane oxidation often persists (Bodelier and Laanbroek, 2004). Active-site competition therefore does not completely explain observed patterns of inhibition. King and Schnell (1994a) proposed that the inhibitory effect of NH_3 in soils is primarily due to the production of nitrite (NO_2^-). Ammonia is oxidized by methanotrophs to NO_2^- via hydroxylamine (NH_2OH). NO_2^- is a general toxin and additions of NO_2^- inhibit CH_4 oxidation in soil and in methanotroph cultures (O'Neill and Wilkinson, 1977; King and Schnell, 1994b; Schnell and King, 1994; Dunfield and Knowles, 1995). The bacteria recover poorly from NO_2^- toxicity (King and Schnell, 1994b; Schnell and King, 1994). With this model, enzyme competition is the root but not the direct cause of inhibition by ammonium fertilizers.

As ammonium fertilizers include a counteranion to NH_4^+, the effect of the fertilizer is a combined result of competitive ammonia oxidation plus a non-competitive salt effect (Gulledge and Schimel, 1998b). Generally, because of the competitive effect, ammonia is the most inhibitory form of fertilizer nitrogen, but this is not always the case. Nitrate is a more potent inhibitor than ammonia in some soils, perhaps via NO_2^- produced from nitrate reduction (Wang and Ineson, 2003; Reay and Nedwell, 2004).

A study of the Broadbalk wheat experiment in the UK found a decreased soil methane oxidation potential after 140 years of NH_4NO_3 applications, but no significant effect of a single fertilization event (Hütsch *et al.*, 1993). In long-term experiments, organic nitrogen additions are often less inhibitory than inorganic nitrogen additions, or may even stimulate methane oxidation (Hütsch *et al.*, 1993; Willison *et al.*, 1996). These studies show that inhibition by nitrogen fertilizers is mediated not only by immediate effects on methanotrophic bacteria, but also by long-term changes in soil chemistry, microbial populations or ecological interactions.

Different experiments show different patterns: methane oxidation may be stimulated, unaffected, inhibited immediately after fertilization or only after several years of fertilization (Bodelier and Laanbroek, 2004). There are considerable study-specific and site-specific factors that determine the effects of nitrogen, including amount of fertilizer applied, type of fertilizer, degree of NH_4^+ fixation on soil minerals, degree of nitrogen limitation, rate of nitrogen uptake by plants, nitrification and denitrification rates, depth to which the fertilizer penetrates, cation exchange capacity and soil pH. Theoretically, the pattern of inhibition may also depend on the particular methanotrophic species present. Gulledge and Schimel (1998b) proposed that different inhibition patterns observed in several soils were an indication that different methanotrophic species were present. It now appears that there are indeed several different species of atmospheric methane oxidizers (Section 10.5).

10.3 Methanotrophic Bacteria

Methane oxidation in upland soils is an aerobic microbial process. Aerobic methanotrophs comprise several phylogenetically distinct clusters of proteobacteria. There are currently 13 described genera, divided by convention into two groups: type I (*Methylococcus, Methylocaldum, Methylomicrobium, Methylosphaera, Methylomonas, Methylobacter, Methylosarcina, Methylothermus* and *Methylohalobius*); and type II (*Methylocystis, Methylosinus, Methylocella* and *Methylocapsa*). These differ primarily in phylogenetic affiliation (type I are *Gammaproteobacteria*, type II are *Alphaproteobacteria*) and also in several biochemical characteristics, including carbon assimilation pathways (ribulose monophosphate versus serine pathway), the dominant phospholipid fatty acids (PLFAs 16:0, 14:0 and 16:1 versus 18:1) and the geometric arrangement of intracellular membranes (ICM) (Hanson and Hanson, 1996; Bowman, 2000).

Methanotrophs obtain energy via the oxidation of CH_4 to CO_2. The first step is the introduction of one atom of oxygen into methane. This step is catalysed by MMO, which requires three substrates: CH_4, O_2 and a reductant for the excess O atom from O_2:

$$CH_4 + O_2 + 2H^+ + 2e^- \rightarrow CH_3OH + H_2O \quad (10.3)$$

The product of MMO, methanol (CH_3OH), is sequentially oxidized to formaldehyde (HCOH), formate (HCOOH) and CO_2, reactions that contribute to the development of a proton-motive force (Hanson and Hanson, 1996). Two forms of MMO are known to exist: a particulate form (pMMO) for which reduced cytochrome-*c* is the reductant, and a soluble form (sMMO), which uses NADH + H^+. The pMMO is universal to all known methanotrophic genera except *Methylocella*, while the soluble form is present in *Methylocella* and a few other type I and type II species (Bowman, 2000; Dedysh *et al.*, 2000). All methanotrophs possessing pMMO also have an extensive ICM system, presumably in order to increase the amount of this primary metabolic enzyme. Of the two MMO forms, sMMO has a broader substrate specificity and a lower affinity for methane. Synthesis of pMMO requires high concentrations of copper (1 μmol/g cells) (Hanson and Hanson, 1996).

A major unanswered question is: Which particular methanotrophic species are responsible for the uptake of atmospheric methane in forest soils? If we knew more about the specific bacteria involved, we would certainly have a better understanding of the process and how to manage it properly. However, several lines of evidence indicate that the species most active in upland soils have yet to be isolated. The remainder of this chapter will outline what we know and do not know about methanotrophs in upland soils.

10.4 'High-affinity' Methane Oxidizers

Bender and Conrad (1992) noted that methane uptake in upland soils displayed a typical hyperbolic Michaelis–Menten response to methane concentration. This is expected for an enzymatic reaction. However, an intriguing feature was that the apparent affinity for methane was several orders of magnitude higher in upland soils (K_s~10–100 nM) than in pure cultures of methanotrophs and wetland soils (K_s~1–10 μM) (Fig. 10.2). This was interpreted to mean that the active methanotrophs in upland soils are oligotrophs and possess a high-affinity form of MMO that allows survival on the trace level of atmospheric methane. Energetic calculations support this proposal. Soil microorganisms can grow on atmospheric trace gases only if their half-saturation constants (K_s) are sufficiently low or their enzyme levels (cellular V_{max}) are sufficiently high to supply their maintenance energy requirement, which is about 4.5 kJ/(C-mol microbial biomass)/h (Tijhuis *et al.*, 1993). Given that the V_{max} measured for methanotrophs is about 250 mmol CH_4/(C-mol microbial biomass)/h, a K_s of <110 nM CH_4 would be required to allow maintenance on atmospheric methane alone (Conrad, 1999).

The findings of Bender and Conrad (1992) have been widely cited (more than 100 times to date) and the hypothesis that there are high-affinity methanotrophs has fuelled much research. However, there are difficulties with this hypothesis. It assumes that true affinity constants (K_s) are being measured in most experiments but, in fact, what is measured are kinetic constants (K_m), which are not necessarily the same as affinity constants. In addition, because concentrations of the other two reactants for MMO (O_2 and reductant) may vary, and because there may be diffusion limitation steps, only apparent kinetic coefficients are measured ($K_{m(app)}$). It is impossible to apply simple Michaelis–Menten kinetics to an enzyme like MMO that has three reactants. The $K_{m(app)}$ for one reactant will change in response to concentrations of the other reactants. Although it seems counterintuitive, a limitation of one substrate (e.g. reductant) may actually cause the $K_{m(app)}$ for another substrate (e.g. methane) to decrease (Dunfield and Conrad, 2000). This is fully consistent with certain kinetic mechanisms for multireactant enzymes (Segel, 1975). Probably due to such an effect, the $K_{m(app)}$ of a *Methylocystis* strain (LR1) was observed to decrease with starvation (Dunfield *et al.*, 1999; Dunfield and Conrad, 2000). This work did not conclude that a high-affinity enzyme was being expressed in this methanotroph during starvation, but rather suggested that there may be no such thing as a high-affinity enzyme.

There are therefore two possibilities to explain high-affinity methane oxidation: either this is an inappropriate application of a simple Michaelis–Menten model to a complex system, or there truly is a high-affinity enzyme. A final verdict probably must await isolation of the unknown atmospheric methane oxidizers from soil.

The results of Bender and Conrad (1992) are also often extrapolated too far. 'High-affinity' methane oxidizer is frequently used interchangeably with 'atmospheric' methane oxidizer, a useful generalization in many instances but not necessarily true. A high-affinity kinetic curve will show a zero-order rate at high methane

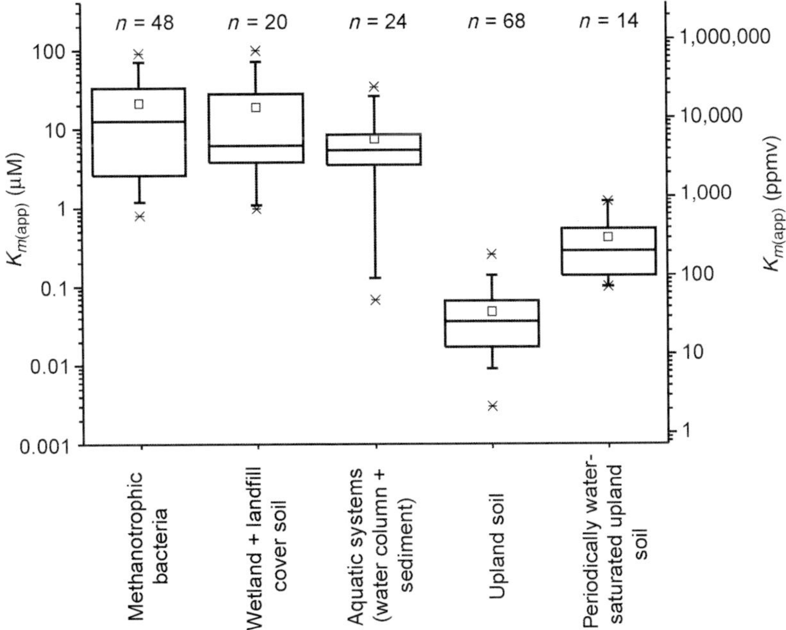

Fig. 10.2. Summary of reported $K_{m(app)}$ values for methane oxidation measured in various natural environments and in methanotroph cultures. The upper and lower box lines represent the 25th and 75th percentile values (= 50% of all values). The horizontal line in the box represents the 50th percentile (median) and the square symbol represents the mean. Error bars denote the 5th and 95th percentiles and asterisks indicate the upper and lower limits. (Data from articles cited in Knief *et al.*, 2005b. Figure from Knief *et al.*, 2005b. With permission from Blackwell Publishing.)

concentrations, and a low-affinity curve will show a first-order rate at low methane concentrations. Most low-affinity methanotrophs can oxidize atmospheric methane (Knief and Dunfield, 2005), provided they have a supply of reductant, the key co-substrate for MMO. It is not always correct to assume that high-affinity activity is responsible for methane oxidation at 1.7 ppmv and low-affinity activity is responsible for methane oxidation at higher mixing ratios. This false assumption has sometimes been used even when the relevant experimental data show a single hyperbolic curve in a soil, without any low-affinity activity (e.g. Bull *et al.*, 2000). Concurrent low-affinity and high-affinity activity has only been observed in soils after long preincubation under high (>10% v/v) methane mixing ratios (Bender and Conrad, 1992), and in many soils a low-affinity activity cannot be induced at all (Bradford *et al.*, 2001a; Reay and Nedwell, 2004).

The difference between high-affinity and low-affinity oxidation is also not clearly delineated. Forest soils oxidizing atmospheric methane show the lowest measured $K_{m(app)}$ values. Some other soils that oxidize atmospheric methane display an 'intermediate-affinity', with $K_{m(app)}$ values higher than in forests, but lower than in wetlands (Fig. 10.2). It is difficult to draw a sharp dividing line between high-affinity and low-affinity oxidation because the measured values span such a broad range. There may be many different variants of MMO with different kinetic properties. We do know that the soluble form of MMO has a lower affinity than the particulate form (Hanson and Hanson, 1996). Alternatively, the broad range may only reflect the difficulties described above in measuring apparent kinetic constants.

10.5 Oligotrophs, Flush Feeders or Facultative Methanotrophs?

There are four possible survival mechanisms for the bacteria responsible for atmospheric methane oxidation: (i) they are true oligotrophs capable of growing and surviving on atmospheric methane alone; (ii) they grow not only on atmospheric methane but also on methane produced in soil; (iii) they survive on substrates other than (or in addition to) methane; and (iv) they survive on other substrates and methane is co-oxidized in a non-energy-yielding process.

10.5.1 Oligotrophs

The hypothesis generally used to explain the low $K_{m(\text{app})}$ of soil methane uptake is that certain methanotrophs have developed a specialized enzyme to allow survival on the constant low level of atmospheric methane. However, there are several problems (described earlier) in interpreting the apparent kinetic coefficient $K_{m(\text{app})}$. A better measure of oligotrophy is specific affinity (a_s^O), defined as V_{max}/K_m, or the slope of the first-order section of a Michaelis–Menten curve. Specific affinity directly indicates how fast limiting substrate is metabolized. Measured specific affinities of cultivated methanotrophs are too low to supply maintenance energy requirements at 1.7 ppmv methane (Conrad, 1999; Knief and Dunfield, 2005). To grow on atmospheric methane alone, they would need either higher cellular $V_{max(\text{app})}$ values (i.e. more enzyme) or lower K_s values. The existence of the putative 'high-affinity' methanotroph of Bender and Conrad (1992) is therefore supported by energetic calculations, although it is probably better defined as a 'high-specific-affinity' methanotroph.

Other evidence indicates that methanotrophs obtain all or some of their maintenance energy from atmospheric methane. In forest soils, $^{14}CH_4$ is incorporated into cell material even at atmospheric concentrations (Roslev *et al.*, 1997). Potential methanotrophic activity declines when some soils are completely deprived of CH_4 (Schnell and King, 1995; Gulledge *et al.*, 1998), although this is not always the case (Benstead and King, 1997; Gulledge *et al.*, 1998).

10.5.2 Flush feeders

Methanogenesis occurs in anaerobic microsites of upland soils, especially when high water content restricts O_2 diffusion (Section 10.1). Periodic soil methane production may support growth and long-term survival of methanotrophs. Although addition of methane does not always increase atmospheric methane oxidation capacity of soils (e.g. Benstead and King, 1997), stimulated methanogenesis in a tundra soil did result in increased atmospheric methane oxidation (West and Schmidt, 2002). Some methanotrophic bacteria are able to oxidize atmospheric methane for several months after the onset of starvation, even though this does not meet their full maintenance energy requirement (Knief and Dunfield, 2005). The reducing equivalents needed to support methane oxidation in these cells presumably come from storage compounds such as poly-β-hydroxybutyrate (Dawes and Senior, 1973). Methanotrophic populations can therefore potentially grow and build up storage reserves during periods of soil methanogenesis, and then continue to consume atmospheric methane afterwards as partial support of their maintenance requirement.

10.5.3 Facultative methanotrophs

Another possibility is that atmospheric methane oxidizers do not survive on methane alone, but also consume other substrates. Until recently, all methanotrophs were thought to be obligately methylotrophic. They grow on methane, methanol and, in some cases, formate, formaldehyde and methylamines, but not on any compounds containing carbon–carbon bonds (Bowman, 2000). Therefore, potential alternative substrates for methanotrophs are limited. One of these substrates, methanol, has been shown to increase the atmospheric methane uptake rate when added to soil and methanotrophic cultures (Benstead *et al.*, 1998; Jensen *et al.*, 1998). The reducing power obtained from methanol oxidation presumably fuels MMO.

Dedysh *et al.* (2005) recently proved that several strains of the genus *Methylocella* are able to use the multicarbon compounds acetate, pyruvate, succinate, malate and ethanol in addition to one-carbon compounds. *Methylocella* is the first clear example of a facultative methanotroph. With this discovery it now seems more plausible that methanotrophs in soil can survive on alternative compounds in addition to methane and methanol. Acetate stimulated atmospheric methane oxidation in a tundra soil (West and Schmidt, 1999). This could be a secondary effect via enhanced methanogenesis, but the existence of facultative methanotrophs suggests that direct stimulation of the methanotroph population by acetate is also possible. Unfortunately, *Methylocella* contain only sMMO, which has a lower affinity for methane than pMMO has, and is therefore less likely to be used for atmospheric methane oxidation. The methane oxidation threshold for *Methylocella* is much higher than 1.7 ppmv (Dedysh *et al.*, 2005). It remains to be seen whether there are facultative methane oxidizers with a pMMO.

10.5.4 Co-oxidation

The only organisms known to co-oxidize methane are ammonia-oxidizing bacteria such as *Nitrosomonas* and *Nitrosospira*. The ammonia monooxygenase enzyme is evolutionarily related to pMMO (Holmes *et al.*, 1995) and both enzymes will oxidize both substrates, which are structural analogues of each other (Bédard and Knowles, 1989). However, there is substantial evidence that nitrifiers are not involved in atmospheric methane consumption, and no evidence to the contrary. The maximum specific CH_4 oxidation rate in nitrifier cultures is lower than the minimum rate in any methanotroph (Bédard and Knowles, 1989). Extremely high nitrifier populations would be needed to account for observed methane oxidation rates in soil – much higher populations than are actually present (Jiang and Bakken, 1999; Kolb *et al.*, 2005). Stimulation of nitrification via NH_4^+ fertilization usually does not stimulate methane oxidation rates (e.g. Bender and Conrad, 1994a). Soil nitrification rates and NH_4^+ and NO_3^- pools are often highest in surface organic soil, while the maximum methane oxidation activity occurs lower down the profile (Castro *et al.*, 1994; Schnell and King, 1994; Kruse and Iversen, 1995). A negative relationship was observed between nitrification rates and methane oxidation rates in several soils (Reay *et al.*, 2005).

10.6 Which Species are Responsible?

Cultivation-independent studies have provided direct evidence that unknown species of methanotrophs are abundant and active in many upland soils. At present, there is evidence implicating four major phylogenetic clusters of methanotrophs in atmospheric methane oxidation (Fig. 10.3). These are dealt with in the following sections.

10.6.1 USCα Cluster

Holmes *et al.* (1999) characterized atmospheric methane-oxidizing bacteria using a dual approach of retrieving *pmoA* genes from soil and radiolabelling microbial phospholipid fatty acids (PLFAs) by incubation of soil under $^{14}CH_4$. The *pmoA* gene codes for the active site-containing subunit of pMMO, which is universal to all known methanotrophs except *Methylocella* spp. Phylogenies constructed based on *pmoA* gene sequences closely reflect phylogenies based on 16S rRNA sequences, and therefore retrieval and sequencing of *pmoA* from environmental DNA extracts allows identification of the methanotrophs present (Holmes *et al.*, 1995; Heyer *et al.*, 2002).

In the study of Holmes *et al.* (1999), a novel group of *pmoA* sequences was predominantly recovered from several forest

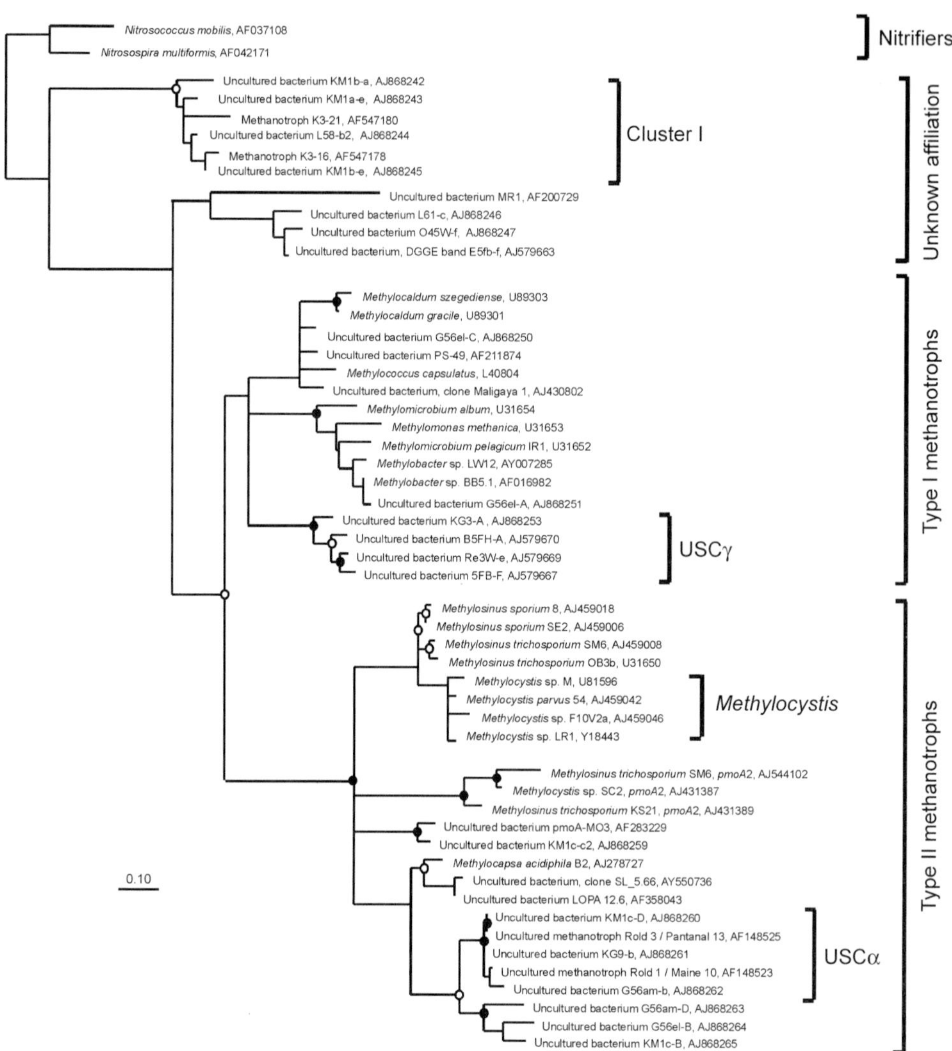

Fig. 10.3. Phylogenetic tree based on sequences of derived amino acid sequences of partial *pmoA* or *amoA* genes of methanotrophs and ammonia oxidizers. Groups that are believed to be involved in the consumption of atmospheric methane (cluster I, *Methylocystis*, USCα, USCγ) are indicated. (Modified from Knief *et al.*, 2005b. With permission of Blackwell Publishing.) Tree construction with the TREE-PUZZLE algorithm and the Jones–Taylor–Thornton evolutionary model was based on 147 derived amino acid residues. Black circles indicate branches that were recovered in 90% of 25,000 reconstructed TREE-PUZZLE trees; white circles indicate branches with 80% recovery. The *AmoA* sequences of nitrifiers were set as the outgroup. The bar represents 0.10 change per amino acid position.

soils that oxidized atmospheric methane. These sequences were not closely related to the *pmoA* of any known methanotroph. However, they were more similar to *pmoA* genes of type II (*Alphaproteobacteria*) than type I (*Gammaproteobacteria*) methanotrophs. The most closely related *pmoA* sequence from a cultivated methanotroph is that of *Methylocapsa acidiphila*, an alphaproteobacterium isolated from acidic peat (Dedysh *et al.*, 2002). The *pmoA* data were supported by incubating soils under a low mixing ratio

(<50 ppmv) of ^{14}C-labelled CH_4. The resulting ^{14}C-labelled PFLA profiles were somewhat related to profiles of known type II methanotrophs, especially *Methylocapsa* (Holmes *et al.*, 1999; Roslev and Iversen, 1999).

This unknown group of *pmoA* sequences, called unidentified soil cluster *Alphaproteobacteria* (USCα) after Knief *et al.* (2003), has been recovered from many different upland soils in subsequent studies (Henckel *et al.*, 2000; Jensen *et al.*, 2000; Bourne *et al.*, 2001; Knief *et al.*, 2003). Labelling of PLFAs and bacteriohopanoids by incubation with ^{13}C-labelled methane has supported the theory that unknown species of type II methanotrophs are responsible for atmospheric methane uptake in many forest soils (Bull *et al.*, 2000; Knief *et al.*, 2003; Crossman *et al.*, 2005).

Recent quantification of *pmoA* genes in upland soils has provided a clue that the USCα cluster represents a high-specific-affinity methane oxidizer. Kolb *et al.* (2005) used quantitative real-time polymerase chain reaction (PCR) assays targeting specific phylogenetic groups of *pmoA* genes, including USCα, to estimate methanotroph population densities in soils. In one of two upland soils tested, USCα was the most abundant *pmoA* sequence recovered, with a population density of $\sim 2 \times 10^6$ cells/g soil. According to this population, and the measured methane oxidation rates in the soil, the per-cell oxidation rates of USCα methanotrophs at 1.7 ppmv CH_4 (i.e. the specific affinities) were estimated to be much higher than rates in known methanotrophs. This estimate assumed that all abundant methanotrophic species were detectable, and that populations were not underestimated due to methodological limitations. These are problematic assumptions, but if they are correct, USCα methanotrophs might be the high-affinity species originally postulated by Bender and Conrad (1992).

10.6.2 USCγ Cluster

Using an analogous approach to that of Holmes *et al.* (1999), but on a much larger set of soils, Knief *et al.* (2003) demonstrated that different methanotrophs were present and active in different soils that oxidized atmospheric methane. In acidic soils, the USCα group was the predominant *pmoA* sequence detected. However, a novel group of *pmoA* sequences distantly related (<75% derived amino acid identity) to sequences of type I methanotrophs (*Gammaproteobacteria*) was predominantly detected in soils with pH values above 6.0. This group was named 'upland soil cluster γ' (USCγ). Soils were incubated under low mixing ratios of $^{13}CH_4$, and incorporation of ^{13}C into PLFAs characteristic of methanotrophic *Gammaproteobacteria* was observed in soils where sequences of the USCγ were detected. The results indicate that the atmospheric methane oxidizers in neutral to alkaline soils are novel type I species. The prevalence of USCα sequences in acid (pH < 6) soils and the close phylogenetic relationship to the acidophile *Methylocapsa* suggest that this group is acidophilic, whereas the USCγ group is neutrophilic (Knief *et al.*, 2003).

10.6.3 *Methylocystis* Spp.

Most cultivated methanotrophs can oxidize atmospheric methane. However, most species lose this ability within a few days when exposed to only 1.7 ppmv methane (Roslev and King, 1994; Schnell and King, 1995; Knief and Dunfield, 2005). Presumably, as long as cells have recently seen high methane (or methanol) concentrations, they have sufficient reductant to use as a co-substrate for MMO function. However, because atmospheric methane does not supply enough maintenance energy, the amount of reducing power decreases with time and cells become inactive. The minimum methane mixing ratio needed for maintenance ranges from 10 to more than 1000 ppmv, but there are differences among species (Knief and Dunfield, 2005). The most oligotrophic methanotrophs known are some strains of *Methylocystis*. These have the highest specific affinity, need only about 10–100 ppmv methane for growth and oxidize atmospheric methane for several months without loss of activity (Knief and Dunfield, 2005). Presumably the reductant for

MMO during atmospheric methane oxidation comes from storage compounds. *Methylocystis* therefore fit the 'flush feeder' profile described in Section 10.5.2.

Some cultivated *Methylocystis* can therefore theoretically contribute to atmospheric methane oxidation in soils. These bacteria are most likely to be important in soils such as wet gleysols where ephemeral methanogenesis occurs. In some gleysols, population sizes of *Methylocystis* are sufficient to account for atmospheric methane oxidation rates (Horz *et al.*, 2002; Knief *et al.*, 2005b). *Methylocystis* can also display a pseudo high-affinity activity similar to that observed *in situ* in these soils (Dunfield and Conrad, 2000). However, *Methylocystis* need additional energy sources for long-term survival, and populations of this bacterium in many forest soils are too low to account for measured activity (Knief and Dunfield, 2005; Kolb *et al.*, 2005). They may be important in some sites, but in forest soils with high methane oxidation rates uncultivated methanotrophs such as USCα and USCγ are probably more important.

10.6.4 Cluster I

Recent quantitative analysis of a soil that oxidized atmospheric methane found that the most abundant *pmoA* type detected was a cluster named 'Cluster I' (Kolb *et al.*, 2005; Ricke *et al.*, 2005). This *pmoA* cluster had been infrequently detected in other upland soils as well (Knief *et al.*, 2003) and may represent type II methanotrophs, despite its odd phylogenetic placement (Pacheco-Oliver *et al.*, 2002). The quantitative evidence suggests that this cluster may be responsible for atmospheric methane uptake in some soils, although data are as yet limited (based on a single soil), and supporting evidence such as analysis of $^{13}CH_4$-labelling patterns is lacking.

10.6.5 Others

Both molecular and cultivation techniques are subject to biases. The primer systems used to detect *pmoA* are not necessarily universal, so there may be other, undetected groups of methanotrophs in soils. Many unidentified phylogenetic groups of *pmoA* that are distant from any cultivated species have already been detected in various soils (Holmes *et al.*, 1999; Henckel *et al.*, 2000; Knief *et al.*, 2003; Knief *et al.*, 2005b). Because of primer biases, these may be more abundant than expected based on their rare detection. As an example of the importance of primer bias, an alternative *pmoA* gene present in most type II methanotrophs was only recently discovered, because it was less easily PCR-amplified than the known *pmoA* gene (Dunfield *et al.*, 2002).

It has recently been discovered that some crenarchaea are capable of autotrophic ammonia oxidation, and possess an *amoA* gene (Könneke *et al.*, 2005; Treusch *et al.*, 2005). Perhaps discoveries of yet-unknown methane oxidizers will also be made. These could possess pMMO or sMMO, or evolutionarily unrelated methane monooxygenases. It cannot be ruled out that completely unexpected species are involved in atmospheric methane oxidation.

10.7 Conclusions

The major environmental factors influencing the soil methane sink are water and nitrogen, but a variety of other factors are also important. Most anthropogenic influences decrease the soil methane sink, but interpretation of the mechanisms behind the inhibition is complicated by our limited knowledge of the microorganisms responsible.

The accumulated molecular and cultivation evidence suggests that there is no single atmospheric methane oxidizer. The process of atmospheric methane oxidation is carried out by different species in different soils, and at least four taxonomic groups appear to be important so far. There are several potential ecological strategies these bacteria may use, including oligotrophy mediated by a high-affinity enzyme, growth on alternative substrates as well as methane and flush-feeding. These strategies may differ across species and environments. The species responsible for methane oxidation in many forest soils have only been studied indirectly, and are not yet available in culture. If and when this is achieved, a better understanding of the soil methane sink might be achieved.

Acknowledgement

The author would like to thank Ralf Conrad for his critical reading of this chapter.

References

Adamsen, A.P.S. and King, G.M. (1993) Methane consumption in temperate and subarctic forest soils: rates, vertical zonation, and responses to water and nitrogen. *Applied and Environmental Microbiology* 59, 485–490.

Amaral, J.A. and Knowles, R. (1997) Localization of methane consumption and nitrification activities in some boreal forest soils and the stability of methane consumption on storage and disturbance. *Journal of Geophysical Research* 102, 29255–29260.

Amaral, J.A. and Knowles, R. (1998) Inhibition of methane consumption in forest soils by monoterpenes. *Journal of Chemical Ecology* 24, 723–734.

Andersen, B.L., Bidoglio, G., Leip, A. and Rembges, D. (1998) A new method to study simultaneous methane oxidation and methane production in soils. *Global Biogeochemical Cycles* 12, 587–594.

Bédard, C. and Knowles, R. (1989) Physiology, biochemistry, and specific inhibitors of CH_4, NH_4^+, and CO oxidation by methanotrophs and nitrifiers. *Microbiological Reviews* 53, 68–84.

Bender, M. and Conrad, R. (1992) Kinetics of CH_4 oxidation in oxic soils exposed to ambient air or high CH_4 mixing ratios. *FEMS Microbiology Ecology* 101, 261–270.

Bender, M. and Conrad, R. (1994a) Microbial oxidation of methane, ammonium and carbon monoxide, and turnover of nitrous oxide and nitric oxide in soils. *Biogeochemistry* 27, 97–112.

Bender, M. and Conrad, R. (1994b) Methane oxidation activity in various soils and freshwater sediments: occurrence, characteristics, vertical profiles, and distribution on grain size fractions. *Journal of Geophysical Research* 99, 16531–16540.

Benstead, J. and King, G.M. (1997) Response of methanotrophic activity in forest soil to methane availability. *FEMS Microbiology Ecology* 23, 333–340.

Benstead, J., King, G.M. and Williams, H.G. (1998) Methanol promotes atmospheric methane oxidation by methanotrophic cultures and soils. *Applied and Environmental Microbiology* 64, 1091–1098.

Bodelier, P.L.E. and Laanbroek, H.J. (2004) Nitrogen as a regulatory factor of methane oxidation in soils and sediments. *FEMS Microbiology Ecology* 47, 265–277.

Bodelier, P.L.E., Roslev, P., Henckel, T. and Frenzel, P. (2000) Stimulation by ammonium-based fertilisers of methane oxidation in soil around rice roots. *Nature* 403, 421–424.

Boeckx, P., Van Cleemput, O. and Meyer, T. (1998) The influence of land use and pesticides on methane oxidation in some Belgian soils. *Biology and Fertility of Soils* 27, 293–298.

Borken, W., Xu, Y.J. and Beese, F. (2003) Conversion of hardwood forests to spruce and pine plantations strongly reduced soil methane sink in Germany. *Global Change Biology* 9, 956–966.

Born, M., Dörr, H. and Levin, I. (1990) Methane consumption in aerated soils of the temperate zone. *Tellus* 42B, 2–8.

Bourne, D.G., McDonald, I.R. and Murrell, J.C. (2001) Comparison of *pmoA* PCR primer sets as tools for investigating methanotroph diversity in three Danish soils. *Applied and Environmental Microbiology* 67, 3802–3809.

Bowman, J. (2000) The methanotrophs: the families Methylococcaceae and Methylocystaceae. In: Dworkin, M. (ed.) *The Prokaryotes*. Available at: http://link.springer-ny.com/link/service/books/10125/

Bradford, M.A., Ineson, P., Wookey, P.A. and Lappin-Scott, H.M. (2001a) Role of CH_4 oxidation, production, and transport in forest soil CH_4 flux. *Soil Biology and Biochemistry* 33,1625–1631.

Bradford, M.A., Ineson, P., Wookey, P.A. and Lappin-Scott, H.M. (2001b) The effects of acid nitrogen and acid sulphur deposition on CH_4 oxidation in a forest soil: a laboratory study. *Soil Biology and Biochemistry* 33, 1695–1702.

Bull, I.D., Parekh, N.R., Hall, G.H., Ineson, P. and Evershed, R.P. (2000) Detection and classification of atmospheric methane oxidizing bacteria in soil. *Nature* 405, 175–178.

Burke, R.A., Zepp, R.G., Tarr, M.A., Miller, W.L. and Stocks, B.J. (1997) Effect of fire on soil–atmosphere exchange of methane and carbon dioxide in Canadian boreal forest sites. *Journal of Geophysical Research D: Atmospheres* 102, 29289–29300.

Campbell, G.S. (1985) *Soil Physics with BASIC Transport Models for Soil–Plant Systems*. Elsevier, Amsterdam, The Netherlands.

Carlsen, H.N., Joergensen, L. and Degn, H. (1991) Inhibition by ammonia of methane utilization in *Methylococcus capsulatus* (Bath). *Applied Microbiology and Biotechnology* 35, 124–127.

Castro, M.S., Peterjohn, W.T., Melillo, J.M., Steudler, P.A., Gholz, H.L. and Lewis, D. (1994) Effects of nitrogen fertilization on the fluxes of N_2O, CH_4, and CO_2 from soils in a Florida slash pine plantation. *Canadian Journal of Forest Research* 24, 9–13.

Castro, M.S., Gholz, H.L., Clark, K.L. and Steudler, P.A. (2000) Effects of forest harvesting on soil methane fluxes in Florida slash pine plantations. *Canadian Journal of Forest Research* 30, 1534–1542.

Conrad, R. (1999) Soil microorganisms oxidizing atmospheric trace gases (CH_4, CO, H_2, NO). *Indian Journal of Microbiology* 39, 193–203.

Crill, P.M. (1991) Seasonal patterns of methane uptake and carbon dioxide release by a temperate woodland soil. *Global Biogeochemical Cycles* 5, 319–334.

Crill, P.M., Martikainen, P.J., Nykänen, H. and Silvola, J. (1994) Temperature and N fertilization effects on methane oxidation in a drained peatland soil. *Soil Biology and Biochemistry* 26, 1331–1339.

Crossman, Z.M., Ineson, P. and Evershed, R.P. (2005) The use of ^{13}C labelling of bacterial lipids in the characterisation of ambient methane-oxidising bacteria in soils. *Organic Geochemistry* 36, 769–778.

Dawes, E.A. and Senior, P.J. (1973) The role and regulation of energy reserve polymers in microorganisms. *Advances in Microbial Physiology* 10, 135–266.

Dedysh, S.N., Liesack, W., Khmelenina, V.N., Suzina, N.E., Trotsenko, Y.A., Semrau, J.D., Bares, A.M., Panikov, N.S. and Tiedje, J.M. (2000) *Methylocella palustris* gen. nov., sp. nov., a new methane-oxidizing acidophilic bacterium from peat bogs, representing a novel subtype of serine-pathway methanotrophs. *International Journal of Systematic and Evolutionary Microbiology* 50, 955–969.

Dedysh, S.N., Khmelenina, V.N., Suzina, N.E., Trotsenko, Y.A., Semrau, J.D., Liesack, W. and Tiedje, J.M. (2002) *Methylocapsa acidiphila* gen. nov., sp. nov., a novel methane-oxidizing and dinitrogen-fixing acidophilic bacterium from *Sphagnum* bog. *International Journal of Systematic and Evolutionary Microbiology* 52, 251–261.

Dedysh, S.N., Knief, C. and Dunfield, P.F. (2005) *Methylocella* species are facultatively methanotrophic. *Journal of Bacteriology* 187, 4665–4670.

Dörr, H., Katruff, L. and Levin, I. (1993) Soil texture parameterization of the methane uptake in aerated soils. *Chemosphere* 26, 697–713.

Dunfield, P.F. and Conrad, R. (2000) Starvation alters the apparent half-saturation constant for methane in the type II methanotroph *Methylocystis* strain LR1. *Applied and Environmental Microbiology* 66, 4136–4138.

Dunfield, P.F. and Knowles, R. (1995) Kinetics of inhibition of methane oxidation by nitrite, nitrate and ammonium in a humisol. *Applied and Environmental Microbiology* 61, 3129–3135.

Dunfield, P.F., Topp, E., Archambault, C. and Knowles, R. (1995) Effect of nitrogen fertilizers and moisture content on CH_4 and N_2O fluxes in a humisol: measurements in the field and intact soil cores. *Biogeochemistry* 29, 199–222.

Dunfield, P.F., Liesack, W., Henckel, T., Knowles, R. and Conrad, R. (1999) High-affinity methane oxidation by a soil enrichment culture containing a type II methanotroph. *Applied and Environmental Microbiology* 65, 1009–1014.

Dunfield, P.F., Tchawa Yimga, M., Dedysh, S.N., Liesack, W., Berger, U. and Heyer, J. (2002) Isolation of a *Methylocystis* strain containing a novel *pmoA*-like gene. *FEMS Microbiology Ecology* 41, 17–26.

Gulledge, J. and Schimel, J.P. (1998a) Moisture control over atmospheric CH_4 consumption and CO_2 production in diverse Alaskan soils. *Soil Biology and Biochemistry* 30, 1127–1132.

Gulledge, J. and Schimel, J.P. (1998b) Low-concentration kinetics of atmospheric CH_4 oxidation in soil and mechanism of NH_4^+ inhibition. *Applied and Environmental Microbiology* 64, 4291–4298.

Gulledge, J., Steudler, P.A. and Schimel, J.P. (1998) Effect of CH_4-starvation on atmospheric CH_4 oxidizers in taiga and temperate forest soils. *Soil Biology and Biochemistry* 30, 1463–1467.

Hanson, R.S. and Hanson, T.E. (1996) Methanotrophic bacteria. *Microbiological Reviews* 60, 439–471.

Hansen, S., Mæhlum, J.E. and Bakken, L.R. (1993) N_2O and CH_4 fluxes in soil influenced by fertilization and tractor traffic. *Soil Biology and Biochemistry* 25, 621–630.

Harriss, R.C., Sebacher, D.I. and Day, J.F.P. (1982) Methane flux in the Great Dismal Swamp. *Nature* 297, 673–674.

Henckel, T., Jäckel, U., Schnell, S. and Conrad, R. (2000) Molecular analyses of novel methanotrophic communities in forest soil that oxidize atmospheric methane. *Applied and Environmental Microbiology* 66, 1801–1808.

Heyer, J., Galchenko, V.F. and Dunfield, P.F. (2002) Molecular phylogeny of type II methane-oxidizing bacteria isolated from various environments. *Microbiology* 148, 2831–2846.

Holmes, A.J., Costello, A., Lidstrom, M.E. and Murrell, J.C. (1995) Evidence that particulate methane monooxygenase and ammonia monooxygenase may be evolutionarily related. *FEMS Microbiology Letters* 132, 203–208.

Holmes, A.J., Roslev, P., McDonald, I.R., Iversen, N., Henriksen, K. and Murrell, J.C. (1999) Characterization of methanotrophic bacterial populations in soils showing atmospheric methane uptake. *Applied and Environmental Microbiology* 65, 3312–3318.

Horz, H.P., Raghubanshi, A.S., Heyer, J., Kammann, C., Conrad, R. and Dunfield, P.F. (2002) Activity and community structure of methane-oxidising bacteria in a wet meadow soil. *FEMS Microbiology Ecology* 41, 247–257.

Hütsch, B.W., Webster, C.P. and Powlson, D.S. (1993) Long-term effects of nitrogen fertilization on methane oxidation in soil of the Broadbalk wheat experiment. *Soil Biology and Biochemistry* 25, 1307–1315.

Hütsch, B.W., Webster, C.P. and Powlson, D.S. (1994) Methane oxidation in soil as affected by land use, soil pH and N fertilization. *Soil Biology and Biochemistry* 26, 1613–1622.

Ishizuka, S., Sakata, T. and Ishizuka, K. (2000) Methane oxidation in Japanese forest soils. *Soil Biology and Biochemistry* 32, 769–777.

Jaatinen, K., Knief, C., Dunfield, P.F., Yrjälä, K. and Fritze, H. (2004) Methanotrophic bacteria in boreal forest soil: long-term effects of prescribed burning and wood ash fertilization. *FEMS Microbiology Ecology* 50, 195–202.

Jäckel, U., Schnell, S. and Conrad, R. (2004) Microbial ethylene production and inhibition of methanotrophic activity in a deciduous forest soil. *Soil Biology and Biochemistry* 36, 835–840.

Jensen, S. and Olsen, R.A. (1998) Atmospheric methane consumption in adjacent arable and forest soil systems. *Soil Biology and Biochemistry* 30, 1187–1193.

Jensen, S., Priemé, A. and Bakken, L. (1998) Methanol improves methane uptake in starved methanotrophic microorganisms. *Applied and Environmental Microbiology* 64, 1143–1146.

Jensen, S., Holmes, A.J., Olsen, R.A. and Murrell, J.C. (2000) Detection of methane oxidizing bacteria in forest soil by monooxygenase PCR amplification. *Microbial Ecology* 39, 282–289.

Jiang, Q.Q. and Bakken, L.R. (1999) Nitrous oxide production and methane oxidation by different ammonia-oxidizing bacteria. *Applied and Environmental Microbiology* 65, 2679–2684.

Kähkönen, M.A., Wittmann, C., Ilvesniemi, H., Westman, C.J. and Salkinoja-Salonen, M.S. (2002) Mineralization of detritus and oxidation of methane in acid boreal coniferous forest soils: seasonal and vertical distribution and effects of clear-cut. *Soil Biology and Biochemistry* 34, 1191–1200.

Kammann, C., Grünhage, L., Jäger, H.J. and Wachinger, G. (2001) Methane fluxes from differentially managed grassland study plots: the important role of CH_4 oxidation in grassland with a high potential for CH_4 production. *Environmental Pollution* 115, 261–273.

Keller, M., Goreau, T.J., Wofsy, S.C., Kaplan, W.A. and McElroy, M.B. (1983) Production of nitrous oxide and consumption of methane by forest soils. *Geophysical Research Letters* 10, 1156–1159.

Keller, M., Mitre, M.E. and Stallard, R.F. (1990) Consumption of atmospheric methane in soils of central Panama: effects of agricultural development. *Global Biogeochemical Cycles* 4, 21–27.

Keller, M., Veldkamp, E., Weitz, A.M. and Reiners, W.A. (1993) Effect of pasture age on soil trace-gas emissions from a deforested area of Costa Rica. *Nature* 365, 244–246.

Kessavalou, A., Mosier, A.R., Doran, J.W., Drijber, R.A., Lyon, D.J. and Heinemeyer, O. (1998) Fluxes of carbon dioxide, nitrous oxide, and methane in grass sod and winter wheat-fallow tillage management. *Journal of Environmental Quality* 27, 1094–1104.

King, G.M. and Adamsen, A.P.S. (1992) Effects of temperature on methane consumption in a forest soil and in pure cultures of the methanotroph *Methylomonas rubra*. *Applied and Environmental Microbiology* 58, 2758–2763.

King, G.M. and Schnell, S. (1994a) Effect of increasing atmospheric methane concentration on ammonium inhibition of soil methane consumption. *Nature* 370, 282–284.

King, G.M. and Schnell, S. (1994b) Ammonium and nitrite inhibition of methane oxidation by *Methylobacter albus* BG8 and *Methylosinus trichosporium* OB3b at low methane concentrations. *Applied and Environmental Microbiology* 60, 3508–3513.

Klemedtsson, A.K. and Klemedtsson, L. (1997) Methane uptake in Swedish forest soil in relation to liming and extra N-deposition. *Biology and Fertility of Soils* 25, 296–301.

Koschorreck, M. and Conrad, R. (1993) Oxidation of atmospheric methane in soil: measurements in the field, in soil cores and in soil samples. *Global Biogeochemical Cycles* 7, 109–121.

Knief, C. and Dunfield, P.F. (2005) Response and adaptation of methanotrophic bacteria to low methane concentrations. *Environmental Microbiology* 7, 1307–1317.

Knief, C., Lipski, A. and Dunfield, P.F. (2003) Diversity and activity of methanotrophic bacteria in different upland soils. *Applied and Environmental Microbiology* 69, 6703–6714.

Knief, C., Vanitchung, S., Harvey, N.W., Conrad, R., Dunfield, P.F. and Chidthaisong, A. (2005a) Diversity of methanotrophic bacteria in tropical upland soils under different land use. *Applied and Environmental Microbiology* 71, 3826–3831.

Knief, C., Kolb, S., Bodelier, P.L.E., Lipski, A. and Dunfield, P.F. (2005b) The active methanotrophic community in hydromorphic soils changes in response to changing methane concentration. *Environmental Microbiology*; doi:10.1111/j.1462-2920.2005.00898.x.

Kolb, S., Knief, C., Dunfield, P.F. and Conrad, R. (2005) Abundance and activity of uncultured methanotrophic bacteria involved in the consumption of atmospheric methane in two forest soils. *Environmental Microbiology* 7, 1150–1161.

Könneke, M., Bernhard, A.E., de la Torre, J.R., Walker, C.B., Waterbury, J.B. and Stahl, D.A. (2005) Isolation of an autotrophic ammonia-oxidizing marine archaeon. *Nature* 437, 543–546.

Kruse, C.W. and Iversen, N. (1995) Effect of plant succession, ploughing, and fertilization on the microbiological oxidation of atmospheric methane in a heathland soil. *FEMS Microbiology Ecology* 18, 121–128.

Kruse, C.W., Moldrup, P. and Iversen, N. (1996) Modeling diffusion and reaction in soils: II. Atmospheric methane diffusion and consumption in a forest soil. *Soil Science* 161, 355–365.

Maljanen, M., Liikanen, A., Silvola, J. and Martikainen, P.J. (2003) Methane fluxes on agricultural and forested boreal organic soils. *Soil Use and Management* 19, 73–79.

Mast, M.A., Wickland, K.P., Striegl, R.T. and Clow, D.W. (1998) Winter fluxes of CO_2 and CH_4 from subalpine soils in Rocky Mountain National Park, Colorado. *Global Biogeochemical Cycles* 12, 607–620.

Menyailo, O.V. and Hungate, B.A. (2003) Interactive effects of tree species and soil moisture on methane consumption. *Soil Biology and Biochemistry* 35, 625–628.

Mosier, A., Schimel, D., Valentine, D., Bronson, K. and Parton, W. (1991) Methane and nitrous oxide fluxes in native, fertilized and cultivated grasslands. *Nature* 350, 330–332.

Mosier, A.R., Klemedtsson, A.K., Sommerfeld, R.A. and Musselman, R.C. (1993) Methane and nitrous oxide flux in a Wyoming subalpine meadow. *Global Biogeochemical Cycles* 7, 771–784.

Nanba, K. and King, G.M. (2000) Response of atmospheric methane consumption by Maine forest soils to exogenous aluminum salts. *Applied and Environmental Microbiology* 66, 3674–3679.

Neff, J.C., Bowman, W.D., Holland, E.A., Fisk, M.C. and Schmidt, S.K. (1994) Fluxes of nitrous oxide and methane in nitrogen-amended soils in a Colorado alpine ecosystem. *Biogeochemistry* 27, 23–33.

Nesbit, S.P. and Breitenbeck, G.A. (1992) A laboratory study of factors affecting methane uptake by soils. *Agriculture, Ecosystems and Environment* 41, 39–54.

O'Neill, J.G. and Wilkinson, J.F. (1977) Oxidation of ammonia by methane-oxidizing bacteria and the effects of ammonia on methane oxidation. *Journal of General Microbiology* 100, 407–412.

Pacheco-Oliver, M., McDonald, I.R., Groleau, D., Murell, J.C. and Miguez, C.B. (2002) Detection of methanotrophs with highly divergent *pmoA* genes from Arctic soils. *FEMS Microbiology Letters* 209, 313–319.

Price, S.J., Sherlock, R.R., Kelliher, F.M., McSeveny, T.M., Tate, K.R. and Condron, L.M. (2004) Pristine New Zealand forest soil is a strong methane sink. *Global Change Biology* 10, 16–26.

Priemé, A. and Christensen, S. (1997) Seasonal and spatial variation of methane oxidation in a Danish spruce forest. *Soil Biology and Biochemistry* 29, 1165–1172.

Priemé, A. and Christensen, S. (1999) Methane uptake by a selection of soils in Ghana with different land use. *Journal of Geophysical Research D: Atmospheres* 104, 23617–23622.

Priemé, A. and Ekelund, F. (2001) Five pesticides decreased oxidation of atmospheric methane in a forest soil. *Soil Biology and Biochemistry* 33, 831–835.

Reay, D.S. and Nedwell, D.B. (2004) Methane oxidation in temperate soils: effects of inorganic N. *Soil Biology and Biochemistry* 36, 2059–2065.

Reay, D.S., Nedwell, D.B., McNamara, N. and Ineson, P. (2005) Effect of tree species on methane and ammonium oxidation capacity in forest soils. *Soil Biology and Biochemistry* 37, 719–730.

Ricke, P., Kolb, S. and Braker, G. (2005) Application of a newly developed ARB software-integrated tool for in silico terminal restriction fragment length polymorphism analysis reveals the dominance of a novel *pmoA* cluster in a forest soil. *Applied and Environmental Microbiology* 71, 1671–1673.

Roslev, P. and Iversen, N. (1999) Radioactive fingerprinting of microorganisms that oxidize atmospheric methane in different soils. *Applied and Environmental Microbiology* 65, 4064–4070.

Roslev, P. and King, G.M. (1994) Survival and recovery of methanotrophic bacteria starved under oxic and anoxic conditions. *Applied and Environmental Microbiology* 60, 2602–2608.

Roslev, P., Iversen, N. and Henriksen, K. (1997) Oxidation and assimilation of atmospheric methane by soil methane oxidizers. *Applied and Environmental Microbiology* 63, 874–880.

Ruser, R., Flessa, H., Schilling, R., Steindl, H. and Beese, F. (1998) Soil compaction and fertilization effects on nitrous oxide and methane fluxes in potato fields. *Soil Science Society of America Journal* 62, 1587–1595.

Saari, A., Martikainen, P.J., Ferm, A., Ruuskanen, J., Boer, W.D., Troelstra, S.R. and Laanbroek, H.J. (1997) Methane oxidation in soil profiles of Dutch and Finnish coniferous forests with different soil texture and atmospheric nitrogen deposition. *Soil Biology and Biochemistry* 29, 1625–1632.

Saari, A., Rinnan, R. and Martikainen, P.J. (2004) Methane oxidation in boreal forest soils: kinetics and sensitivity to pH and ammonium. *Soil Biology and Biochemistry* 36, 1037–1046.

Savage, K., Moore, T.R. and Crill, P.M. (1997) Methane and carbon dioxide exchanges between the atmosphere and northern boreal forest soils. *Journal of Geophysical Research D: Atmospheres* 102, 29279–29288.

Schnell, S. and King, G.M. (1994) Mechanistic analysis of ammonium inhibition of atmospheric methane consumption in forest soils. *Applied and Environmental Microbiology* 60, 3514–3521.

Schnell, S. and King, G.M. (1995) Stability of methane oxidation capacity to variations in methane and nutrient concentrations. *FEMS Microbiology Ecology* 17, 285–294.

Schnell, S. and King, G.M. (1996) Responses of methanotrophic activity in soils and cultures to water stress. *Applied and Environmental Microbiology* 62, 3203–3209.

Segel, I.H. (1975) *Enzyme Kinetics: Behaviour and Analysis of Rapid Equilibrium and Steady-state Enzyme Systems*. Wiley, New York.

Seiler, W., Conrad, R. and Scharffe, D. (1984) Field studies of methane emission from termite nests to the atmosphere and measurements of methane uptake by tropical soils. *Journal of Atmospheric Chemistry* 1, 171–186.

Sexstone, A.J. and Mains, C.N. (1990) Production of methane and ethylene in organic horizons of spruce forest soils. *Soil Biology and Biochemistry* 22, 135–139.

Simpson, I.J., Edwards, G.C., Thurtell, G.W., Den Hartog, G., Neumann, H.H. and Staebler, R.M. (1997) Micrometeorological measurements of methane and nitrous oxide exchange above a boreal aspen forest. *Journal of Geophysical Research D: Atmospheres* 102, 29331–29341.

Singh, J.S., Singh, S., Raghubanshi, A.S., Singh, S., Kashyap, A.K. and Reddy, V.S. (1997) Effect of soil nitrogen, carbon and moisture on methane uptake by dry tropical forest soils. *Plant and Soil* 196, 115–121.

Singh, J.S., Raghubanshi, A.S., Reddy, V.S., Singh, S. and Kashyap, A.K. (1998) Methane flux from irrigated paddy and dryland rice fields, and from seasonally dry tropical forest and savanna soils of India. *Soil Biology and Biochemistry* 30, 135–139.

Sitaula, B.K. and Bakken, L.R. (2001) Nitrification and methane oxidation in forest soil: acid deposition, nitrogen input and plant effects. *Water, Air, and Soil Pollution* 130, 1061–1066.

Sitaula, B.K., Bakken, L.R. and Abrahamsen, G. (1995) CH_4 uptake by temperate forest soil: effect of N input and soil acidification. *Soil Biology and Biochemistry* 27, 871–880.

Sitaula, B.K., Hansen, S., Sitaula, J.I.B. and Bakken, L.R. (2000) Methane oxidation potentials and fluxes in agricultural soil: effects of fertilisation and soil compaction. *Biogeochemistry* 48, 323–339.

Smith, K.A., Dobbie, K.E., Ball, B.C., Bakken, L.R., Sitaula, B.K., Hansen, S., Brumme, R., Borken, W., Christensen, S., Priemé, A., Fowler, D., Macdonald, J.A., Skiba, U., Klemedtsson, L., Kasimir-Klemedtsson, A., Degorska, A. and Orlanski, P. (2000) Oxidation of atmospheric methane in Northern European soils, comparison with other ecosystems, and uncertainties in the global terrestrial sink. *Global Change Biology* 6, 791–803.

Steudler, P.A., Bowden, R.D., Melillo, J.M. and Aber, J.D. (1989) Influence of nitrogen fertilization on methane uptake in temperate forest soils. *Nature* 341, 314–316.

Striegl, R.G. (1993) Diffusional limits to the consumption of atmospheric methane by soils. *Chemosphere* 26, 715–720.

Striegl, R.G., McConnaughey, T.A., Thorstenson, D.C., Weeks, E.P. and Woodward, J.C. (1992) Consumption of atmospheric methane by desert soils. *Nature* 357, 145–147.

Syamsul Arif, M.A., Houwen, F. and Verstraete, W. (1996) Agricultural factors affecting methane oxidation in arable soil. *Biology and Fertility of Soils* 21, 95–102.

Tijhuis, L., Van Loosdrecht, M.C.M. and Heijnen, J.J. (1993) A thermodynamically based correlation for maintenance Gibbs energy-requirements in aerobic and anaerobic chemotrophic growth. *Biotechnology and Bioengineering* 42, 509–519.

Topp, E. (1993) Effects of selected agrochemicals on methane oxidation by an organic agricultural soil. *Canadian Journal of Soil Science* 73, 287–291.

Torn, M.S. and Harte, J. (1996) Methane consumption by montane soils: implications for positive and negative feedback with climatic change. *Biogeochemistry* 32, 53–67.

Treusch, A.H., Leininger, S., Kletzin, A., Schuster, S.C., Klenk, H.P. and Schleper, C. (2005) Novel genes for nitrite reductase and Amo-related proteins indicate a role of uncultivated mesophilic crenarchaeota in nitrogen cycling. *Environmental Microbiology*; doi:10.1111/j.1462-2920.2005.00906.x.

von Fischer, J.C. and Hedin, L.O. (2002) Separating methane production and consumption with a field-based isotope pool dilution technique. *Global Biogeochemical Cycles* 16, 8.1–8.13.

Wang, F.L. and Bettany, J.R. (1995) Methane emission from a usually well-drained prairie soil after snowmelt and precipitation. *Canadian Journal of Soil Science* 75, 239–241.

Wang, Z.P. and Ineson, P. (2003) Methane oxidation in a temperate coniferous forest soil: effects of inorganic N. *Soil Biology and Biochemistry* 35, 427–433.

West, A.E. and Schmidt, S.K. (1999) Acetate stimulates atmospheric CH_4 oxidation by an alpine tundra soil. *Soil Biology and Biochemistry* 31, 1649–1655.

West, A.E. and Schmidt, S.K. (2002) Endogenous methanogenesis stimulates oxidation of atmospheric CH_4 in alpine tundra soil. *Microbial Ecology* 43, 408–415.

Whalen, S.C. and Reeburgh, W.S. (1990) Consumption of atmospheric methane by tundra soils. *Nature* 346, 160–162.

Whalen, S.C., Reeburgh, W.S. and Sandbeck, K.A. (1990) Rapid methane oxidation in a landfill cover soil. *Applied and Environmental Microbiology* 56, 3405–3411.

Whalen, S.C., Reeburgh, W.S. and Barber, V.A. (1992) Oxidation of methane in boreal forest soils: a comparison of seven measures. *Biogeochemistry* 16, 181–211.

Willison, T.W., Cook, R., Müller, A. and Powlson, D.S. (1996) CH_4 oxidation in soils fertilized with organic and inorganic-N: differential effects. *Soil Biology and Biochemistry* 28, 135–136.

Yavitt, J.B., Downey, D.M., Lang, G.E. and Sexstone, A.J. (1990) Methane consumption in two temperate forest soils. *Biogeochemistry* 9, 39–52.

Yavitt, J.B., Simmons, J.A. and Fahey, T.J. (1993) Methane fluxes in a northern hardwood forest ecosystem in relation to acid precipitation. *Chemosphere* 26, 721–730.

Yavitt, J.B., Fahey, T.J. and Simmons, J.A. (1995) Methane and carbon dioxide dynamics in a northern hardwood ecosystem. *Soil Science Society of America Journal* 59, 796–804.

11 The Atmospheric Methane Sink

Dudley E. Shallcross[1], M. Aslam K. Khalil[2] and Christopher L. Butenhoff[2]

[1]*Biogeochemistry Research Centre, School of Chemistry, University of Bristol, Bristol, UK;* [2]*Department of Physics, Portland State University, Portland, Oregon, USA*

11.1 Introduction

Methane (CH_4) is the most abundant organic species in the atmosphere. Like CO_2, CH_4 levels have risen dramatically over the last 200 years. Pre-industrial levels, which were ~750 ppb, now stand at ~1760 ppb in the northern hemisphere and ~1630 ppb in the southern hemisphere. CH_4 has been increasing at 1.3% per year for most of the 20th century until the early 1990s (Blake and Rowland, 1988) as illustrated in Fig. 11.1. Since then the rate of growth has slowed to 0.6% per year (Steele *et al.*, 1992). There are various reasons for the slowdown in the growth rate. Bekki *et al.* (1994) have argued that stratospheric ozone depletion was so severe in 1991–1992 that hydroxyl (OH) levels in the troposphere rose due to the increase in the short-wavelength radiation penetrating into the lower atmosphere. Their two-dimensional (2D) model calculations suggest that up to half of the decrease in CH_4 growth rate in 1991 and 1992 could be attributed to this rise in OH levels. Wang *et al.* (2004) used a three-dimensional (3D) atmospheric model to carry out an inverse analysis of the CH_4 budget. They concluded that the economic downturn of the former Soviet Union and east-European countries in the early 1990s, together with more efficient industrial practices, has been an important factor in the slowdown of CH_4, in addition to potential changes in OH levels.

These model results suggest that the current slowdown may be temporary. Quay *et al.* (1999) and Fletcher *et al.* (2004) have used measurements of $^{13}CH_4$, $^{12}CH_4$ and $^{12}CH_3D$ to determine changes in source strengths for CH_4 over the last 10–15 years. They concluded that there have been decreases in northern hemisphere sources such as bogs and landfills but increases in tropical and southern hemisphere sources such as swamps and biomass burning. Finally, Warwick *et al.* (2002) have shown that year-to-year fluctuations in meteorology can cause significant interannual fluctuations in the levels of atmospheric CH_4 measured at the surface. Therefore, there are several potential reasons why CH_4 trends have been fluctuating in recent years.

CH_4 is a far more effective greenhouse gas than CO_2, being 62 times more effective on a 20-year time horizon and 23 times more effective on a 100-year time horizon. The changes in effectiveness with time reflect the fact that CH_4 is much shorter-lived in the atmosphere than CO_2. The sources of CH_4 are discussed in detail by Reay *et al.* (Chapter 9, this volume) and

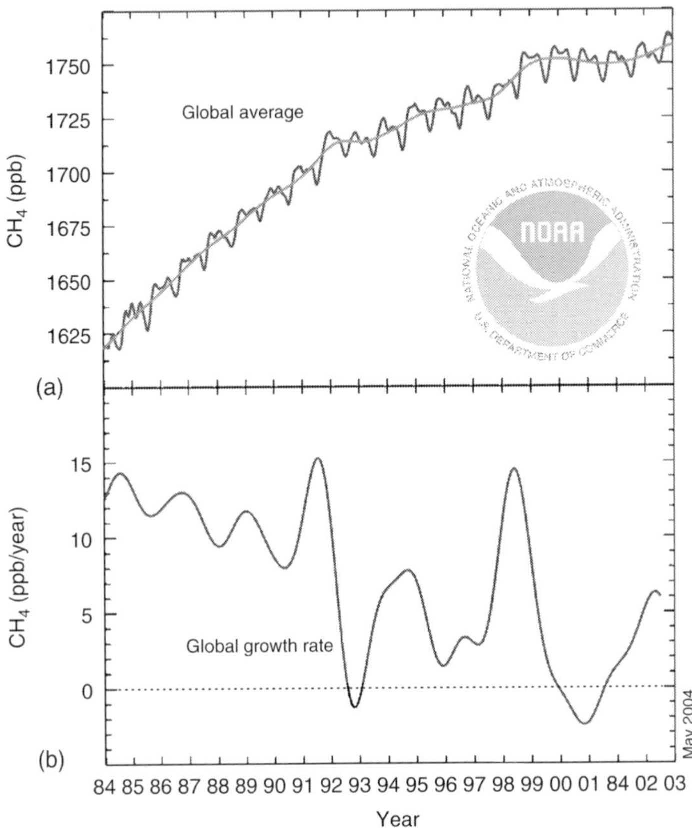

Fig. 11.1. Changes in global methane (CH_4) surface concentrations over the last 20 years. (a) Global average atmospheric CH_4 mixing ratios (wavy line) determined using measurements from the NOAA/CMDL cooperative air sampling network. The smooth line represents the long-term trend. (b) Global average growth rate for CH_4. (From GLOBALVIEW, 2005.)

are summarized in Table 11.1. It is apparent that anthropogenic sources dominate, being more than double the total from natural sources. The three main anthropogenic sources are evenly spread between fossil fuel-related (27%), waste management (24%) and enteric fermentation of cattle (23%), with biomass burning (11%) and rice paddies (15%) making a non-negligible contribution to the total. Natural sources are dominated by wetland emissions, particularly in the tropics. CH_4 can be generated by the decay of organic matter under oxygen-deficient conditions, and the microbes that initiate this decomposition are most active where temperatures are highest.

11.2 Removal of Methane

11.2.1 Physical removal (wet and dry deposition)

Henry's law constant, $K_H(T_0)$, for CH_4 is very small and it is therefore unlikely that CH_4 will partition into the aqueous phase. Hence, physical removal by wet deposition (i.e. removal from the atmosphere by uptake into rain or aerosol species) will not be a significant sink for CH_4. Table 11.2 summarizes some Henry's law constants for a range of C_1 compounds found in the atmosphere for comparison, from which it

Table 11.1. Sources of methane (CH_4) in the atmosphere. (From IPCC, 1996.)

	Likely (Tg/year)	Range (Tg/year)
Natural sources	–	–
Wetlands	–	–
Tropics	65	30–80
Northern latitudes	40	20–60
Termites	20	10–50
Ocean	10	5–50
Freshwater	5	1–25
Geological	10	5–15
Others	10	5–15
Total	160	–
Anthropogenic sources	–	–
Fossil fuel-related	–	–
Coal mines	30	15–45
Natural gas	40	25–50
Petroleum industry	15	5–30
Coal combustion	15	5–30
Waste management system		–
Landfills	40	20–70
Animal waste	25	20–30
Domestic waste treatment	25	15–80
Enteric fermentation	85	65–100
Biomass burning	40	20–80
Rice paddies	60	20–100
Total	375	–
Total sources	535	–

Table 11.2. Henry's Law constants, $K_H(T_0)$, at 298 K for some C_1 compounds. (From Sander, 1999.)

Gas	Symbol	$K_H(T_0)$/mol/atm
Methane	CH_4	1.4×10^{-3}
Methanol	CH_3OH	2.2×10^{-2}
Methanal	$CH_2(O)$	1.3
Methanoic acid	CH(O)OH	5.5×10^{3}
Hydrogen peroxide	HOOH	8.3×10^{4}
Nitric oxide	NO	1.9×10^{-3}

emerges that methanoic acid (formic acid) is very soluble and partitions preferentially into the aqueous phase, but as one progresses from carbonyl to alcohol to alkane, the partitioning into the aqueous phase becomes smaller. For comparison, hydrogen peroxide, a highly soluble species, and nitric oxide, a highly insoluble species, are included in the list.

The dry deposition velocities for hydrocarbons have not been measured extensively, but data suggest that they are rather small (PORG, 1997). Table 11.3 summarizes the data from a series of chamber experiments covering a range of hydrocarbons that span molecular masses and molecule type. Surface types were chosen to give a variation in plant canopy type and leaf area index. Although it is not measured explicitly in this study, owing to the particularly non-polar nature of CH_4, removal by dry deposition to the surface is assumed to be small as well. As discussed by Dunfield (Chapter 10, this volume), the soil sink for CH_4 is not insignificant, and is estimated to be between 10 and 44 Tg CH_4/year. CH_4 is consumed in soils by methanotrophic bacteria and archaea (e.g. Pancost *et al.*, 2000; Aloisi *et al.*, 2002). Hence, it is concluded that wet and dry depositions are not significant sinks for CH_4 from the atmosphere, but consumption of CH_4 within the soil is indeed a significant atmospheric sink.

11.2.2 Chemical removal

Preliminary survey

The Earth's atmosphere is oxidizing in nature (Ravishankara, 1988) and therefore CH_4 will be photo-oxidized via reaction with free radicals and possibly some closed-shell species in the atmosphere. There are a number of potential species in the atmosphere that may initiate this photo-oxidation, but it emerges that very few of these are important sinks for CH_4. Although the dissociation energy for a C–H bond in CH_4 is 4.55 eV (~440 kJ/mol), which equates to a wavelength of 272.2 nm, there are no electronic states of CH_4 in this region of the spectrum (Yung and DeMore, 1999). Consequently, CH_4 can only be broken down by high-energy radiation (100–140 nm) and this is found in the mesosphere (equivalent to an altitude of 50–80 km) and upper regions of the atmosphere. Whilst direct photolysis of CH_4 is important in the atmospheres of Jupiter, Saturn, Uranus and Neptune, it is

Table 11.3. Empirical deposition velocities (μm/s) for selected hydrocarbons to various surfaces. (From PORG, 1997.)

	Ethene	Ethyne	1,3-Butadiene	n-Hexane	Benzene
Spinach	74	2	78	43	48
Grass	150	65	82	49	56
Soil	27	16	63	22	27
Salt water	1.6	1.6	1.5	3	1.4

not an important loss process in the Earth's atmosphere. Table 11.4 collects data for the enthalpy of reaction for the most likely species that will initiate oxidation of CH_4 in the Earth's atmosphere.

This table shows that for the majority of species (both radicals and closed-shell species) the enthalpy of reaction is large and positive, i.e. the reaction is endothermic at temperatures found in the Earth's atmosphere (~200–300K) and will either proceed very slowly or not proceed at all. The reactions that can be ignored on thermodynamic grounds have been shaded in Table 11.4. Table 11.5 collects the rate coefficient data, where available, for the list of reactions compiled in Table 11.4. A rate coefficient of 1×10^{-10} cm³/molecule/s would be typical of a gas-phase reaction occurring at the kinetic limit (i.e. the reaction occurs every time the reactants collide). A rate coefficient of 1×10^{-11} cm³/molecule/s means that one in ten collisions lead to reaction and so on. All the reactions that were discarded on thermodynamic grounds have rate coefficients between 5×10^{-18} and 1×10^{-52} cm³/molecule/s, which effectively means that no reaction occurs. Of the reactions that are thermodynamically feasible, only those with excited oxygen atoms, O(¹D), and fluorine atoms, F, are very fast. In Table 11.6 the loss rate for CH_4 for each of the 11 unshaded gas-phase reactions from Table 11.4 is estimated for the troposphere (the lowest 10–15 km of the atmosphere) and the stratosphere (the region encompassing the altitude range of 10–50 km). In each case the loss rate R is determined using Eq. 11.1:

$$R = k(T)[X][CH_4] \text{ molecule/cm}^3\text{/s} \quad (11.1)$$

where [X] is the concentration of the co-reactant with CH_4 and $k(T)$ is the rate coefficient at a temperature T, taken nominally to be 300 K for the troposphere (surface) and 250 K for the stratosphere (30 km). $[CH_4]$ is the concentration of CH_4. The concentration of X has been derived from the modelling studies of Lary and Toumi (1997) and Lary and Shallcross (2000), at the Earth's surface for the troposphere and at 30 km for

Table 11.4. Enthalpy of reaction data for the reaction of major atmospheric constituents with methane (CH_4). The unshaded data (exothermic or near thermo-neutral enthalpies of reaction) indicate reactions that may take place in the atmosphere, shaded data are considered to be too endothermic to be of importance. (Thermodynamic data from DeMore et al., 1997 and Chase, 1998.)

Reaction		$\Delta_r H_{298K}$ (kJ/mol)
$O(^1D) + CH_4$	$\rightarrow OH + CH_3$	−178.7
$F + CH_4$	$\rightarrow HF + CH_3$	−130.6
$CN + CH_4$	$\rightarrow HCN + CH_3$	−78.7
$CF_3O + CH_4$	$\rightarrow CF_3OH + CH_3$	−63.3
$OH + CH_4$	$\rightarrow H_2O + CH_3$	−59.5
$H + CH_4$	$\rightarrow H_2 + CH_3$	+3.3
$CH_3O + CH_4$	$\rightarrow CH_3OH + CH_3$	+3.7
$Cl + CH_4$	$\rightarrow HCl + CH_3$	+8.0
$O(^3P) + CH_4$	$\rightarrow OH + CH_3$	+10.9
$NO_3 + CH_4$	$\rightarrow HNO_3 + CH_3$	+12.5
$FO + CH_4$	$\rightarrow HOF + CH_3$	+14.6
$BrO + CH_4$	$\rightarrow HOBr + CH_3$	+37.2
$ClO + CH_4$	$\rightarrow HOCl + CH_3$	+43.8
$HO_2 + CH_4$	$\rightarrow H_2O_2 + CH_3$	+73.1
$Br + CH_4$	$\rightarrow HBr + CH_3$	+73.3
$CH_3O_2 + CH_4$	$\rightarrow CH_3OOH + CH_3$	+73.6
$N + CH_4$	$\rightarrow NH + CH_3$	+105.4
$N_2O_5 + CH_4$	$\rightarrow HNO_3 + NO_2 + CH_3$	+107.9
$NO_2 + CH_4$	$\rightarrow HONO + CH_3$	+108.7
$O_3 + CH_4$	$\rightarrow OH + O_2 + CH_3$	+117.5
$I + CH_4$	$\rightarrow HI + CH_3$	+140.8
$O_2 + CH_4$	$\rightarrow CH_3 + HO_2$	+233.0
$NO + CH_4$	$\rightarrow HNO + CH_3$	+238.1
$CO + CH_4$	$\rightarrow HCO + CH_3$	+373.6

Table 11.5. Gas-phase rate coefficients for the reaction of major atmospheric constituents with methane (CH_4). $k(T) = A\exp(\ \ Ea/RT)(T/300)^n$.

Reaction	Arrhenius A factor (cm^3/molecule/s)	Ea (kJ/mol)	$(T/298)^n$	k (300 K) (cm^3/molecule/s)	Reference
$O(^1D) + CH_4 \rightarrow OH + CH_3$	1.5×10^{-10}	0.0	0.0	1.5×10^{-10}	Atkinson *et al.* (2001)
$F + CH_4 \rightarrow HF + CH_3$	1.3×10^{-10}	1.8	0.0	6.4×10^{-11}	Atkinson *et al.* (2001)
$CN + CH_4 \rightarrow HCN + CH_3$	5.1×10^{-13}	2.64	0.0	8.6×10^{-13}	Baulch *et al.* (1994)
$CF_3O + CH_4 \rightarrow CF_3OH + CH_3$	2.5×10^{-12}	11.8	0.0	2.2×10^{-14}	Atkinson *et al.* (2001)
$OH + CH_4 \rightarrow H_2O + CH_3$	2.5×10^{-12}	14.8	0.0	6.5×10^{-15}	DeMore *et al.* (1997)
$H + CH_4 \rightarrow H_2 + CH_3$	5.8×10^{-13}	33.6	3.0	8.4×10^{-19}	Baulch *et al.* (1992)
$CH_3O + CH_4 \rightarrow CH_3OH + CH_3$	2.6×10^{-13}	37.0	0.0	9.4×10^{-20}	Tsang and Hampson (1986)
$Cl + CH_4 \rightarrow HCl + CH_3$	6.6×10^{-12}	10.3	0.0	1.1×10^{-13}	Atkinson *et al.* (2001)
$O(^3P) + CH_4 \rightarrow OH + CH_3$	8.3×10^{-12}	35.5	1.56	5.5×10^{-18}	Baulch *et al.* (1992)
$NO_3 + CH_4 \rightarrow HNO_3 + CH_3$	$<8.0 \times 10^{-19}$	–	–	–	Boyd *et al.* (1991)
$ClO + CH_4 \rightarrow HOCl + CH_3$	1.0×10^{-12}	30.8	0.0	4.4×10^{-18}	DeMore *et al.* (1997)
$HO_2 + CH_4 \rightarrow H_2O_2 + CH_3$	1.5×10^{-11}	103.1	0.0	1.7×10^{-29}	Baulch *et al.* (1992)
$Br + CH_4 \rightarrow HBr + CH_3$	50×10^{-10}	73.9	0.0	6.8×10^{-23}	Russell *et al.* (1988)
$CH_3O_2 + CH_4 \rightarrow CH_3O_2H + CH_3$	3.0×10^{-13}	77.3	0.0	1.0×10^{-26}	Tsang and Hampson (1986)
$N + CH_4 \rightarrow NH + CH_3$	–	–	–	$<3.0 \times 10^{-18}$	Aleksandrov *et al.* (1990)
$N_2O_5 + CH_4 \rightarrow HNO_3$ Cantrell *et al.* $+ NO_2 + CH_3$	–	–	–		$<2.0 \times 10^{-23}$ (1987)
$NO_2 + CH_4 \rightarrow HONO + CH_3$	2.0×10^{-11}	125.6	0.0	2.7×10^{-33}	Slack *et al.* (1981)
$O_3 + CH_4 \rightarrow OH + O_2 + CH_3$	–	–		$<1.2 \times 10^{-21}$	Stedman and Niki (1973)
$I + CH_4 \rightarrow HI + CH_3$	2.5×10^{-10}	138.0	0.0	2.3×10^{-34}	Pardini and Martin (1983)
$O_2 + CH_4 \rightarrow CH_3 + HO_2$	6.6×10^{-11}	237.8	0.0	2.6×10^{-52}	Baulch *et al.* (1992)
$NO + CH_4 \rightarrow HNO + CH_3$	–	–		$<4.3 \times 10^{-30}$	Yamaguchi *et al.* (1999)

Table 11.6. Gas-phase rate coefficients for the reaction of major atmospheric constituents with methane (CH_4) under tropospheric and stratospheric conditions, see text for more details. (From Lary and Toumi,1997 and Lary and Shallcross, 2000.)

Reaction	k (300 K) (cm^3/molecule/s)	Tropospheric concentration (molecule/cm^3)	CH_4 loss rate troposphere (molecule/cm^3/s)	Relative CH_4 loss rate
$O(^1D) + CH_4 \rightarrow OH + CH_3$	1.5×10^{-10}	2.5×10^{-1}	1.7×10^{6}	5.8×10^{-1}
$F + CH_4 \rightarrow HF + CH_3$	6.4×10^{-11}	2.5×10^{-10}	7.2×10^{-4}	2.5×10^{-10}
$CN + CH_4 \rightarrow HCN + CH_3$	8.6×10^{-13}	2.5×10^{-13}	9.7×10^{-9}	3.3×10^{-16}
$CF_3O + CH_4 \rightarrow CF_3OH + CH_3$	2.2×10^{-14}	2.5×10^{-8}	2.5×10^{-5}	8.5×10^{-13}
$OH + CH_4 \rightarrow H_2O + CH_3$	6.5×10^{-15}	1.0×10^{6}	2.9×10^{11}	100.0
$H + CH_4 \rightarrow H_2 + CH_3$	8.4×10^{-19}	2.5×10^{-2}	9.5×10^{0}	3.2×10^{-6}
$CH_3O + CH_4 \rightarrow CH_3OH + CH_3$	9.4×10^{-20}	2.5×10^{2}	1.1×10^{0}	3.6×10^{-7}
$Cl + CH_4 \rightarrow HCl + CH_3$	1.1×10^{-13}	2.5×10^{3}	1.2×10^{7}	4.2
$O(^3P) + CH_4 \rightarrow OH + CH_3$	5.5×10^{-18}	2.5×10^{3}	6.2×10^{2}	2.1×10^{-4}
$NO_3 + CH_4 \rightarrow HNO_3 + CH_3$	$<8.0 \times 10^{-19}$	2.5×10^{8}	9.0×10^{6}	3.1
Reaction	**k (250 K) (cm^3/molecule/s)**	**Stratospheric concentration (molecule/ cm^3) (/s)**	**CH_4 loss rate stratosphere**	**Relative CH_4 loss rate**
$O(^1D) + CH_4 \rightarrow OH + CH_3$	1.5×10^{-10}	2.9×10^{1}	1.9×10^{6}	37.0
$F + CH_4 \rightarrow HF + CH_3$	5.5×10^{-11}	2.5×10^{-10}	6.0×10^{-6}	1.2×10^{-10}
$CN + CH_4 \rightarrow HCN + CH_3$	3.2×10^{-13}	2.9×10^{-4}	4.1×10^{-2}	7.9×10^{-7}
$CF_3O + CH_4 \rightarrow CF_3OH + CH_3$	8.6×10^{-15}	2.5×10^{-8}	9.5×10^{-8}	1.8×10^{-12}
$OH + CH_4 \rightarrow H_2O + CH_3$	2.0×10^{-15}	5.9×10^{6}	5.2×10^{6}	100.0
$H + CH_4 \rightarrow H_2 + CH_3$	3.3×10^{-20}	3.3×10^{0}	4.8×10^{-5}	9.2×10^{-10}
$CH_3O + CH_4 \rightarrow CH_3OH + CH_3$	4.8×10^{-21}	2.5×10^{2}	5.3×10^{-4}	1.0×10^{-8}
$Cl + CH_4 \rightarrow HCl + CH_3$	4.6×10^{-14}	1.7×10^{5}	3.4×10^{6}	66.1
$O(^3P) + CH_4 \rightarrow OH + CH_3$	2.4×10^{-19}	3.3×10^{9}	3.5×10^{5}	6.7
$NO_3 + CH_4 \rightarrow HNO_3 + CH_3$	$<8.0 \times 10^{-19}$	2.5×10^{8}	8.8×10^{4}	1.7

the stratosphere; the level of CH_4 assumed is 1800 ppb and 1500 ppb for the troposphere and stratosphere, respectively. The combination of rate coefficient and concentration of species leads to the conclusion that in the troposphere the OH radical is the major sink for CH_4, and chlorine atoms and possibly nitrate radicals also play a minor role. In the stratosphere, the OH radical is still the dominant sink, whilst the role of $O(^1D)$ as a sink becomes more prominent with altitude, and at ~30 km is a significant sink, as are chlorine atoms.

The accurate measurement of gas-phase rate coefficients is non-trivial and a source of error in any estimation of CH_4 atmospheric sinks. However, for many of the species proposed in Table 11.4, the atmospheric concentration can only be inferred from the models. The concentration of radical (open-shell) species in the troposphere and stratosphere is usually very small and poses a significant measurement challenge. In the troposphere there have been several measurements of the OH radical and, whilst these have not been extensive in time and space, there is reasonable agreement between models and measurements (e.g. Heard and Pilling, 2003). Several measurements have been made of the nitrate radical (NO_3) in the troposphere (e.g. Brown *et al.*, 2003; Stutz *et al.*, 2004) that are in keeping with the model estimates used. In contrast, the measurements of Aliwell and Jones (1998) suggest that if NO_3 concentrations in urban environments could be as much as 40 times larger than those used in this assessment, such a level of NO_3 would become an important sink for CH_4 locally. However, since the rate coefficient measured for the reaction between NO_3 and CH_4 is an upper limit, it would be judicious to view NO_3 as a potentially minor sink only for CH_4 in and near urban areas, and unimportant elsewhere. There are no direct measurements of chlorine atom concentrations in the atmosphere, but they have been inferred from measurements of other species (e.g. Wingenter *et al.*, 2001; Hopkins *et al.*, 2002). These studies suggest that in marine environments the model-derived concentration used in Table 11.6 for the troposphere may be up to a factor of ten or lower. Therefore, chlorine atoms (generated from the oxidation of sea salt) could be an important sink for CH_4 in the marine boundary layer. In conclusion to this preliminary survey, reaction with OH radicals is the major sink for CH_4 in the lower atmosphere (stratosphere and troposphere), with chlorine atoms making a minor contribution globally in the troposphere but being potentially important in marine environments. In the stratosphere both chlorine atoms and excited oxygen atoms are non-negligible sinks for CH_4.

Isotopic analysis of the chemical sinks for methane

An emerging field in atmospheric science is the measurement of isotopic abundances. Such analyses have been used frequently in Earth sciences but have only recently been possible in atmospheric studies due to improvements in sensitivity of instrumentation and pre-concentration techniques. Isotopic analysis is proving to be a useful technique to further refine an assessment of atmospheric CH_4 sinks. Table 11.7 collects gas-phase rate coefficients for the reactions of OH radicals and chlorine atoms with $^{12}CH_4$, $^{12}CH_3D$, $^{12}CH_2D_2$, $^{12}CHD_3$ and $^{12}CD_4$. The kinetic isotope effect is particularly pronounced when comparing hydrogen with deuterium, and Table 11.7 exemplifies this, where substitution of the heavier isotope slows down the rate of reaction dramatically. Less pronounced is the effect of substituting ^{12}C with ^{13}C. Recently, measurement of $^{12}CH_4$ (98.8% abundant), $^{13}CH_4$ (1.1% abundant) and $^{12}CH_3D$ (0.06% abundant) in the atmosphere has become possible. Using inverse modelling it is possible to use these data to study the atmospheric sinks for CH_4. Bergamaschi *et al.* (2000) have made measurements of these three species at Izaña on Tenerife over a period of 2 years. Interestingly, the seasonal cycle observed for $^{12}CH_3D$ is shifted by several months with respect to that observed for $^{12}CH_4$. By using an inverse model they were able to account for the seasonal changes in CH_4 with just two main sinks in the troposphere: (i) loss by reaction with OH radicals;

Table 11.7. Gas-phase rate coefficients for the reaction between OH and various isotopomers of methane (CH_4) (from Gierczak *et al.*, 1997) and chlorine (Cl) atoms and various isotopomers of CH_4 (from Boone *et al.*, 2001).

OH	A (cm³/molecule/s)	Ea (kJ/mol)	k_{298} (cm³/molecule/s)
$^{12}CH_4$	2.50×10^{-12}	14.8	6.50×10^{-15}
$^{12}CH_3D$	3.11×10^{-12}	15.9	5.12×10^{-15}
$^{12}CH_2D_2$	2.30×10^{-12}	16.0	3.54×10^{-15}
$^{12}CHD_3$	1.46×10^{-12}	16.4	1.94×10^{-15}
$^{12}CD_4$	1.00×10^{-12}	17.5	0.87×10^{-15}
Cl			k_{298} (cm³/molecule/s)
$^{12}CH_4$	–	–	1.10×10^{-13}
$^{12}CH_3D$	–	–	0.65×10^{-13}
$^{12}CH_2D_2$	–	–	0.42×10^{-13}
$^{12}CHD_3$	–	–	0.19×10^{-13}
$^{12}CD_4$	–	–	0.05×10^{-13}

and (ii) loss due to soil uptake. Under certain wind directions, characterized by air that had spent many days over the sea, an additional loss due to chlorine atoms was consistent with these observations.

Combining models and measurements in this way allowed these researchers to estimate that the OH sink was responsible for ~95% of the atmospheric loss and that chlorine atoms contributed up to 5% of the total loss. More extensive measurements of isotope data across the globe will be of considerable benefit in determining more accurately the role of chlorine atoms as a CH_4 sink in the troposphere, particularly in remote marine environments.

Saueressig *et al.* (2001) have investigated the kinetic isotope effects in the reaction between $O(^1D)$ atoms and the three CH_4 isotopomers discussed earlier. In addition to studies on the reaction of OH and chlorine with these species, they have used a 2D time-dependent chemical transport model to estimate the strengths of these three main sinks in the stratosphere. In the middle and upper stratosphere ($p < 10$ hPa or an altitude above 30 km) the sink, due to reaction with chlorine atoms, approaches 20% of the total CH_4 sink at low and middle latitudes. At high latitudes, ~30% of CH_4 is destroyed by chlorine atoms above 30 km. $O(^1D)$ atoms also contribute to the total CH_4 sink, accounting for 30% of the loss in the tropics and middle latitudes at ~30 km. The OH radical is the remaining (and dominant) sink for CH_4.

In conclusion, a very simple preliminary analysis shows that OH radicals, chlorine atoms and $O(^1D)$ atoms are the only significant atmospheric sinks for CH_4 in the stratosphere and that OH radicals (with a minor contribution from chlorine atoms) comprise the main sink in the troposphere.

11.3 Oxidation Mechanism for Methane

In addition to limiting the radiative forcing potential of CH_4, its oxidation in the atmosphere is very important, and contributes to the production of radicals, thereby affecting the oxidizing capacity. Under clean air conditions, CH_4 oxidation leads to the formation of hydrogen peroxide (H_2O_2) and methyl hydroperoxide (CH_3O_2H). Equation 11.2 describes the formation of methyl radicals (CH_3), which instantly add to oxygen (O_2) to form a methyl peroxy radical (CH_3O_2) in Eq. 11.3:

$$OH + CH_4 \rightarrow H_2O + CH_3 \quad (11.2)$$

$$CH_3 + O_2 + M \rightarrow CH_3O_2 + M \quad (11.3)$$

CH_3O_2 can react with HO_2 radicals (Eq. 11.4) to form CH_3O_2H, or can react with other peroxy radicals (RO_2) (Eq. 11.5) to form the methoxyl radical (CH_3O), which is a source of HO_2 (Eq. 11.6) and hence of H_2O_2 (Eq. 11.7):

$$CH_3O_2 + HO_2 \rightarrow CH_3O_2H + O_2 \quad (11.4)$$

$$CH_3O_2 + RO_2 \rightarrow CH_3O + RO + O_2 \quad (11.5)$$

$$CH_3O + O_2 \rightarrow HCHO + HO_2 \quad (11.6)$$

$$HO_2 + HO_2 + M \rightarrow H_2O_2 + M \quad (11.7)$$

Both H_2O_2 and CH_3O_2H are susceptible to wet and dry depositions and H_2O_2 is particularly important as an oxidant in aqueous media. Formaldehyde (HCHO) formed in Eq. 11.6 can be photolysed in the troposphere:

$$HCHO + h\upsilon \rightarrow H_2 + CO \quad (11.8a)$$

$$HCHO + h\upsilon \rightarrow H + HCO \quad (11.8b)$$

Equation 11.8a shows molecular products H_2 and CO and is in fact a major source of H_2 in the atmosphere. Equation 11.8b yields the radical products H and HCO that ultimately form HO_2 radicals and is therefore a source of HO_x (OH and HO_2) in the atmosphere.

Under polluted conditions, CH_3O_2 radicals react with NO to yield CH_3O and NO_2. NO_2 is rapidly photolysed to yield NO and O atoms, which in turn rapidly form ozone (O_3):

$$CH_3O_2 + NO \rightarrow CH_3O + NO_2 \quad (11.9)$$

$$CH_3O + O_2 \rightarrow HCHO + HO_2 \quad (11.6)$$

$$NO_2 + h\upsilon \rightarrow NO + O \quad (11.10)$$

$$O + O_2 + M \rightarrow O_3 + M \quad (11.11)$$

11.4 Trends in Atmospheric Methane Sinks

11.4.1 Atmospheric OH trends

From Section 11.3 it is apparent that the atmospheric concentrations of CH_4 can be affected by changes in the oxidizing capacity of the atmosphere, particularly changes in global or regional OH concentrations. Earlier theories on increases of CH_4 were based on the hypothesis that as more carbon monoxide is put into the atmosphere it will deplete OH and hence cause CH_4 concentrations to rise, even if CH_4 emission rates do not increase (Chameides *et al.*, 1977; Sze, 1977; Thompson and Cicerone, 1986). More detailed studies, including the atmospheric changes that also increase the production of OH, have suggested that long-term changes of OH are not as large as may be expected by considering the increases of carbon monoxide alone. Hence, the concentrations of CH_4 currently observed, which are more than double pre-industrial levels, are accepted to be driven primarily by increased emissions of CH_4 itself (Khalil and Rasmussen, 1985). If the pre-industrial OH concentration is taken as a standard, the present values are ~5–10% below this level, and during the ice ages OH would have been ~15–30% above this level (Lu and Khalil, 1991; Pinto and Khalil, 1991). Yet the question remains whether there is an important effect on the concentrations of CH_4 over time and accumulation in the atmosphere caused by changes of OH, especially in recent decades. For instance, one proposed theory of the recent slowdown of the CH_4 trend is that OH is increasing as a feedback from stratospheric ozone depletion, which stimulates its production by increasing ultraviolet (UV) radiation in the troposphere (Madronich and Granier, 1992).

It is not yet possible to measure OH on a global scale; hence, it is not possible to know with certainty the global average of OH. Models are used to calculate global OH concentrations. In constructing such models it becomes apparent that the atmospheric chemistry and physics are very complicated and include many feedbacks. As the completeness of the models is difficult to establish, predicted trends are not reliable, as already discussed. An empirical method has been used to determine global or hemispherical mean OH concentrations and long-term trends. In this method a tracer with known atmospheric concentrations, sources and OH reactivity is used to determine average OH levels. The tracer of choice is the solvent 1-1-1 trichloroethane, commonly known as methyl chloroform (CH_3CCl_3), because it has a number of characteristics that make it a reliable indicator of OH such as: (i) accurate global measurements over nearly 30 years; (ii) well-known emissions based on industrial production; (iii) removal mostly by reacting with OH; and (iv) a relatively short lifetime of ~5 years that reduces the uncertainties in the OH estimates (Lovelock, 1977).

A global mass balance for any atmospheric constituent can be written by equating the rate of change in concentration with time (dC/dt), with the emissions (S) less than the losses due to reactions. For gases such as CH_3CCl_3 the losses due to reactions with OH are proportional to the concentrations in the atmosphere or $k[OH][CH_3CCl_3]$ where k is the reaction rate coefficient. Then, on a global scale we can calculate OH as equal to $(S \sim d[CH_3CCl_3]/dt)/(k[CH_3CCl_3])$ where S is determined from industrial production records; the usages of CH_3CCl_3, $d[CH_3CCl_3]/dt$ and $[CH_3CCl_3]$ are determined from direct atmospheric measurements; and k is determined from laboratory studies.

Actual calculations use more complex spatially and temporally resolved models, a more detailed mass balance taking into account other sinks, potential effects of transport processes and more sophisticated techniques to obtain a de-convolution by which OH is calculated. The results of two such calculations are shown in Fig. 11.2. The mean levels are about 9.4×10^5 and 1.2×10^6 molecules/cm^3 in Prinn *et al.* (2001)

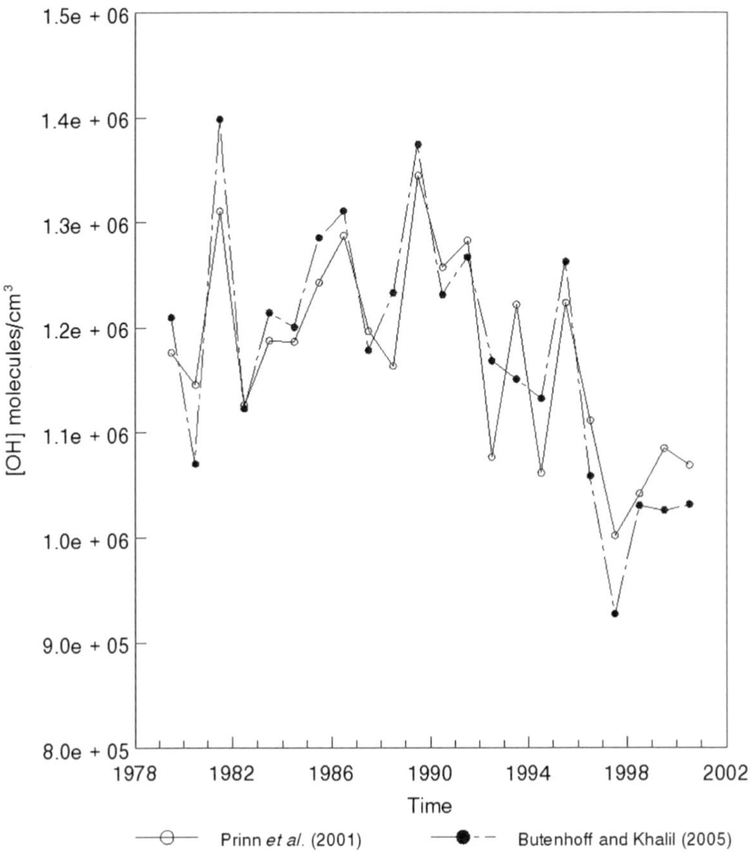

Fig. 11.2. Calculated concentrations and trends of OH in the atmosphere using methyl chloroform as a tracer. (From Prinn *et al.*, 2001 and Butenhoff and Khalil, 2005, personal communication.)

and Butenhoff and Khalil (2005, personal communication) respectively, which are consistent with atmospheric chemistry models mentioned earlier. In Fig. 11.2 the results of Prinn *et al.* (2001) are adjusted by a constant multiplicative factor of 1.25. The additional benefit of these calculations is the ability to detect trends in OH concentration over time. According to these calculations there is a suggestion of small increases during 1979–1990 of ~0.7% ± 0.8% per year and 0.7% ± 1% per year that are not statistically significant, and decreases in 1990–2000 of 1.8% ± 1.1% per year and 2.4% ± 1.2% per year respectively for the studies reported by Prinn *et al.* (2001) and Butenhoff and Khalil (in preparation). Note that these two studies use different data-sets and models. In general the year-to-year changes are variable, but this cannot be said to represent the changes in the real atmosphere as the variabilities introduced by the model and measurements are significant. The calculations are also uncertain for the recent years since the inventory of CH_3CCl_3 has become increasingly unreliable.

11.4.2 Changes in concentrations of atmospheric chlorine and $O(^1D)$

Chlorine loading of the stratosphere as a result of the use of chlorofluorocarbons (CFCs) and related halons has increased during the last century and led to the formation of the 'ozone hole' (WMO, 2003). As a result of several international agreements the use of these compounds is now banned and levels of chlorine precursors are predicted to stabilize in the atmosphere between 2010 and 2020. Hence, the role of chlorine as a CH_4 sink in the stratosphere is expected to decrease in the future. In the troposphere it is very difficult to predict future chlorine impacts. For example, a change in wind speed and wind direction patterns will change the loading of sea salt into the atmosphere and hence the production of chlorine. Changes in ocean pH and circulation will affect marine organism populations that produce organochlorine species, which are transferred from the ocean to the atmosphere. Therefore, it is impossible to gauge the future trend with any certainty.

Predictions from climate models suggest that a warming of the lower atmosphere will lead to a cooling of the mid and upper stratosphere, which will in turn decrease the efficiency of natural catalytic cycles that destroy ozone. Under these conditions, stratospheric ozone levels will increase in the mid and upper stratosphere and hence the level of $O(^1D)$ too will increase. There are many other factors that must be taken into consideration before a definitive prediction can be made on the future impact of $O(^1D)$ on stratospheric CH_4, but an increase in $[O(^1D)]$ will decrease the CH_4 lifetime.

11.5 Summary

The major sink for CH_4 is the OH radical throughout the troposphere and stratosphere (lowest 25 km of the atmosphere). Chlorine and $O(^1D)$ atoms play an important role in the removal of CH_4 in the mid and upper stratosphere. A simple loss rate analysis confirms that this is the case qualitatively, and inspection of the ratios of $^{13}CH_4/^{12}CH_4$ and $^{12}CH_3D/^{12}CH_4$ allows one to quantify the contributions more precisely. Changes in the hydroxyl radical concentration over the last 20 years are estimated and future trends in the three major sinks are discussed.

Acknowledgements

Dudley E. Shallcross thanks the Royal Society and Natural Environment Research Council (NERC) under whose auspices various aspects of this work has been carried out. M. Aslam K. Khalil and Christopher L. Butenhoff thank the US Department of Energy. We thank the National Oceanic and Atmospheric Administration/Climate Monitoring and Diagnostics Laboratory (NOAA/CMDL) cooperative network for making data available to us.

References

Aleksandrov, E.N., Vedeneev, V.I., Kozlov, S.N., Obvivalneva, A.A. and Pryakhin, G.A. (1990) Insertion of $N(^4S)$ atoms into the C-I bond in the reaction of $N(^4S)$ and CH_3I. *Bulletin of Academy of Sciences USSR Division of Chemical Sciences* (Part 2) 39, 625–626.

Aliwell, S.R. and Jones, R.L. (1998) Measurements of tropospheric NO_3 at midlatitude. *Journal of Geophysical Research* 103, 5719–5727.

Aloisi, G., Bouloubassi, I., Heijs, S.K., Pancost, R.D., Pierre, C., Damste, J.S.S., Gottschal, J.C., Forney, L.J. and Rouchy, J.M. (2002) CH_4 consuming microorganisms and the formation of carbonate crusts at cold seeps. *Earth and Planetary Science Letters* 203, 195–203.

Atkinson, R., Bauch, D.L., Cox, R.A., Crowley, J.N., Hampson, R.F. Jr, Kerr, J.A., Rossi, M.J. and Troe, J. (2001) *Summary of Evaluated Kinetic and Photochemical Data for Atmospheric Chemistry IUPAC Subcommittee on Gas Kinetic Data Evaluation for Atmospheric Chemistry*. Web Version December 2001, 1–56.

Baulch, D.L., Cobos, C.J., Cox, R.A., Esser, C., Frank, P., Just, Th., Kerr, J.A., Pilling, M.J., Troe, J., Walker, R.W. and Warnatz, J. (1992) Evaluated kinetic data for combustion modelling. *Journal of Physical and Chemical Reference Data* 21, 411–429.

Baulch, D.L., Cobos, C.J., Cox, R.A., Frank, P., Hayman, G., Just, Th., Kerr, J.A., Murrells, T., Pilling, M.J., Troe, J., Walker, R.W. and Warnatz, J. (1994) Evaluated kinetic data for combustion modelling. *Journal of Physical and Chemical Reference Data* 23 (Suppl. 1), 847–1033.

Bekki, S., Law, K.S. and Pyle, J.A. (1994) Effect of ozone depletion on atmospheric CH_4 and CO concentrations. *Nature* 371, 595–597.

Bergamaschi, P., Bräunlich, M., Marik, T. and Brenninkmeijer, C.A.M. (2000) Measurements of the carbon and hydrogen isotopes of atmospheric methane at Izaña, Tenerife: seasonal cycles and synoptic-scale variations. *Journal of Geophysical Research* 105, 14531–14546.

Blake, D.R. and Rowland, F.R. (1988) Continuing world-wide increase in tropospheric methane, 1978–1987. *Science* 239, 1129–1131.

Boone, G.D., Agyin, F., Robichaud, D.J., Fu-Ming, T. and Hewitt, S.A. (2001) Rate constants for the reactions of chlorine with deuterated methanes: experiment and theory. *Journal of Physical Chemistry A* 105, 1456–1464.

Boyd, A.A., Conosa-Mas, C.E., King, A.D., Wayne, R.P. and Wilson, M.R. (1991) Use of a stopped-flow technique to measure the rate constants at room temperature for reactions between the nitrate radical and various organic species. *Journal of Chemical Society, Faraday Transactions* 87, 2913–2919.

Brown, S.S., Stark, H., Ryerson, T.B., Williams, E.J., Nicks, D.K., Trainer, M., Fehsenfeld, F.C. and Ravishankara, A.R. (2003) Nitrogen oxides in the nocturnal boundary layer: simultaneous *in situ* measurements of NO_3, N_2O_5, NO_2, NO and O_3. *Journal of Geophysical Research* 108, Art. No. 4299.

Cantrell, C.A., Davidson, J.A., Shetter, R.E., Anderson, B.A. and Calvert, J.G. (1987) Reactions of NO_3 and N_2O_5 with molecular species of possible atmospheric interest. *Journal of Physical Chemistry* 91, 6017–6021.

Chameides, W.L., Liu, S.C. and Cicerone, R.J. (1977) Possible variations in atmospheric methane. *Journal of Geophysical Research* 82, 1795–1798.

Chase, M.W. Jr (1998) NIST–JANAF Thermochemical Tables, 4th edn. *Journal of Physical and Chemical Reference Data*, Monograph 9, 1–1951.

DeMore, W.B., Sander, S.P., Golden, D.M., Hampson, R.F., Kurylo, M.J., Howard, C.J., Ravishankara, A.R., Kolb, C.E. and Molina, M.J. (1997) *Chemical Kinetics and Photochemical Data for Use in Stratospherics Modelling*. Evaluation number 12, JPL Publications, pp. 1–266.

Fletcher, S.E.M., Tans, P.P., Bruhwiler, L.M., Miller, J.B. and Heimann, M. (2004) CH_4 sources estimated from atmospheric observations of CH_4 and its C-13/C-12 isotopic ratios: Part 1. Inverse modelling of source processes. *Global Biogeochemical Cycles* 18, Art. No. GB4004.

Gierczak, T., Talukdar, R.K., Herndon, S.C., Vaghjiani, G.L. and Ravishankara, A.R. (1997) Rate coefficients for the reactions of hydroxyl radicals with methane and deuterated methanes. *Journal of Physical Chemistry A* 101, 3125–3134.

GLOBALVIEW (2005) *Cooperative Atmospheric Data Integration Project – Methane*, Chapter 4. CD-ROM, NOAA/CMDL, Boulder, Colorado. Available via anonymous FTP to ftp.cmdl.noaa.gov at: ccg/ch4/GLOBALVIEW

Heard, D.E. and Pilling, M.J. (2003) Measurement of OH and HO_2 in the troposphere. *Chemical Reviews* 103, 5163–5198.

Hopkins, J.R., Jones, I.D., Lewis, A.C., McQuaid, J.B. and Seakins, P.W. (2002) Non-methane hydrocarbons in the Arctic boundary layer. *Atmospheric Environment* 36, 3217–3229.

IPCC (1996) *Climate Change 1995: The Science of Climate Change*. Cambridge University Press, Cambridge.

Khalil, M.A.K. and Rasmussen, R.A. (1985) Causes of increasing atmospheric methane: depletion of hydroxyl radicals and the rise of emissions. *Atmospheric Environment* 19, 397–407.

Lary, D.J. and Shallcross, D.E. (2000) The central role of carbonyl compounds in atmospheric chemistry. *Journal of Geophysical Research* 105, 19771–19778.

Lary, D.J. and Toumi, R. (1997) Halogen catalysed methane oxidation. *Journal of Geophysical Research* 102, 23421–23428.

Lovelock, J.E. (1977) Methyl chloroform in the troposphere as an indicator of OH radical abundance. *Nature* 267, 32–33.

Lu, Y. and Khalil, M.A.K. (1991) Tropospheric OH: model calculations of spatial, temporal and secular variations. *Chemosphere* 23, 397–444.

Madronich, S, and Granier, C. (1992) Impact of recent total ozone changes on tropospheric ozone photoemission, hydroxyl radicals and methane trends. *Geophysical Research Letters* 19, 465–467.

Pancost, R.D., Damste, J.S.S., de Lint, S., van der Maarel, M.J.E.C. and Gottschal, J.C. (2000) Biomarker evidence for widespread anaerobic methane oxidation in Mediterranean sediments by a consortium of methanogenic archea and bacteria. *Applied and Environmental Microbiology* 66, 1126–1132.

Pardini, S.P. and Martin, D.S. (1983) Kinetics of the reaction between methane and iodine from 830 to 1150K in the presence and absence of oxygen. *International Journal of Chemical Kinetics* 15, 1031–1043.

Pinto, J.P. and Khalil, M.A.K. (1991) The stability of tropospheric OH during ice ages, inter-glacial epochs and modern times. *Tellus* 43B, 347–352.

PORG (1997) *Ozone in the United Kingdom*. Fourth Report of the UK Photochemical Oxidants Review Group, Department of the Environment, Transport and the Regions, London.

Prinn, R.G., Huang. J., Weiss, R.F., Cunnold, D.M., Fraser, P.J., Simmonds, P.G., McCulloch, A., Harth, C., Salameh, P., O'Doherty, S., Wang, R.H.J., Porter, L. and Miller, B.R. (2001) Evidence for substantial variations of atmospheric hydroxyl radicals in the past two decades. *Science* 292, 1882–1888.

Quay, P., Stutsman, J., Wilbur, D., Snover, A., Dlugokencky, E. and Brown, T. (1999) The isotopic composition of atmospheric methane. *Global Biogeochemical Cycles* 13, 445–461.

Ravishankara, A.R. (1988) Kinetics of radical reactions in the atmospheric oxidation of CH_4. *Annual Review of Physical Chemistry* 39, 367–394.

Russell, J.J., Seetula, J.A. and Gutman, D. (1988) Kinetics and thermochemistry of CH_3, C_2H_5, and *i*-C_3H_7. Study of the equilibrium R + HBr → R–H + Br. *Journal of American Chemical Society* 110, 3092–3099.

Sander, R. (1999) Modeling atmospheric chemistry: interactions between gas-phase species and liquid cloud/aerosol particles. *Surveys in Geophysics* 20, 1–31.

Saueressig, G., Crowley, J.N., Bergamaschi, P., Brühl, C., Brenninkmeijer, C.A.M. and Fischer, H. (2001) Carbon 13 and D kinetic isotope effects in the reactions of CH_4 with $O(^1D)$ and OH: new laboratory measurements and their implications for the isotopic composition of stratospheric methane. *Journal of Geophysical Research* 106, 23127–23138.

Slack, M.W. and Grillo, A.R. (1981) Shock tube investigations of methane – oxygen ignition sensitized by NO_2. *Combustion and Flame* 40, 155–172.

Stedman, D.H. and Niki, H. (1973) Ozonolysis rates of some atmospheric gases. *Environmental Letters* 14, 303–310.

Steele, L.P., Dlugokencky, E.L., Lang, P.M., Tans, P.P., Martin, R.C. and Masarie, K.A. (1992) Slowing down of the global accumulation of atmospheric methane during the 1980s. *Nature* 358, 313–316.

Stutz, J., Alicke, B. and Ackermann, R. (2004) Vertical profiles of NO_3, N_2O_5, O_3, and NO_x in the nocturnal boundary layer: Part 1. Observations during the Texas Air Quality Study 2000. *Journal of Geophysical Research* 109, D12306.

Sze, N.D. (1977) Atmospheric CO emissions: implications for the atmospheric CO–OH–CH_4 cycle. *Nature* 195, 673–675.

Thompson, A.M. and Cicerone, R.J. (1986) Possible perturbations to atmospheric CO, CH_4 and OH. *Journal of Geophysical Research* 91, 10853–10864.

Tsang, W. and Hampson, R.F. (1986) Chemical kinetic data base for combustion chemistry: Part 1. Methane and related compounds. *Journal of Physical and Chemical Reference Data* 15, 1087–1279.

Wang, J.S., Logan, J.A., McElroy, M.B., Duncan, B.N., Megretskaia, I.A. and Yantosca, R.M. (2004) A 3-D model analysis of the slowdown and interannual variability in the methane growth rate from 1988 to 1997. *Global Biogeochemical Cycles* 18, Art. No. GB3011.

Warwick, N.J., Bekki, S., Law, K.S., Nisbet, E.G. and Pyle, J.A. (2002) The impact of meteorology on the interannual growth rate of atmospheric methane. *Geophysical Research Letters* 29, Art. No. 1947.

Wingenter, O.W., Sive, B.C., Blake, N.J., Blake, D.R. and Rowland, F.S. (2001) *Abstracts of Papers of the American Chemical Society* 222: U25.

World Meteorological Organization (2003) Scientific assessment of ozone depletion: 2002. Global Ozone Research and Monitoring Project. Report No. 47, Geneva, p. 498.

Yamaguchi, Y., Teng, Y., Shimomura, S., Tabata, K. and Suzuki, E. (1999) Ab initio study for selective oxidation of methane with NO_x (x = 1,2). *Journal of Physical Chemistry A* 103, 8272–8278.

Yung, Y.L. and DeMore, W.B. (1999) *Photochemistry of Planetary Atmospheres*. Oxford University Press, Oxford.

12 Artificial Methane Sinks

Alex De Visscher[1], Pascal Boeckx[2] and Oswald Van Cleemput[2]
[1]*Department of Chemical and Petroleum Engineering, Schulich School of Engineering, University of Calgary, Calgary, Alberta, Canada;* [2]*Laboratory of Applied Physical Chemistry (ISOFYS), Ghent University, Ghent, Belgium*

12.1 Introduction

Methane (CH_4) is produced in every environment that contains biodegradable organic matter in anaerobic conditions. Examples are wetland soils and sediments, the digestive tract of many animals and landfills. Microbial CH_4 oxidation (methanotrophy) occurs in any ecosystem that contains an oxic zone permeated by CH_4 from a source of sufficient strength to support microbial life. A typical example is a rice field, where the aerenchyma of rice plants brings oxygen to the roots, thus forming a thin methanotrophic zone in the rhizosphere. A second example is a landfill cover soil, which absorbs part of the CH_4 produced in the landfill waste mass (Fig. 12.1).

This chapter refers to CH_4 sinks in human-made environments. The primary focus is on landfill cover soils, which are the clearest examples of such artificial CH_4 sinks. Rice fields, which are more natural environments similar to swamps, are covered, but to a lesser extent, as less quantitative information is available on these ecosystems. Due to the strong interaction between CH_4 production and CH_4 oxidation in rice fields, it is difficult to distinguish the two processes. However, in both these ecosystems the process is mediated by the same microorganisms, so many observations compiled here are applicable to rice fields as well.

CH_4 production in landfills is a complex process involving different types of microorganisms that are responsible for hydrolysis of complex organic molecules into smaller ones, fermentation into H_2, CO_2 and formic acid (HCOOH), and CH_4 production from either HCOOH or CO_2. The first global estimate of CH_4 production from landfills was made by Bingemer and Crutzen (1987). They arrived at a global emission of 50 ± 20 Tg CH_4/year, with the assumption that 1 kg of degradable carbon in waste leads to 0.5 kg of CH_4 production. Recent estimates indicate lower emissions. Bogner and Matthews (2003), for instance, indicated that 20.7 Tg CH_4/year was a more realistic emission estimate for 1996. An estimated 3.8 Tg CH_4/year is recovered from landfills by combustion or flaring. In their estimate, Bogner and Matthews assumed that 10% of the CH_4 was oxidized biologically before entering the atmosphere. This leads to an *in situ* CH_4 production estimate of 26.8 Tg CH_4/year, of which 2.3 Tg CH_4/year is oxidized biologically.

However, estimates of oxidation efficiencies of landfill cover soils vary widely. Depending on climate and operation conditions, efficiency estimates range from 10% to 100%. Table 12.1 shows an overview of

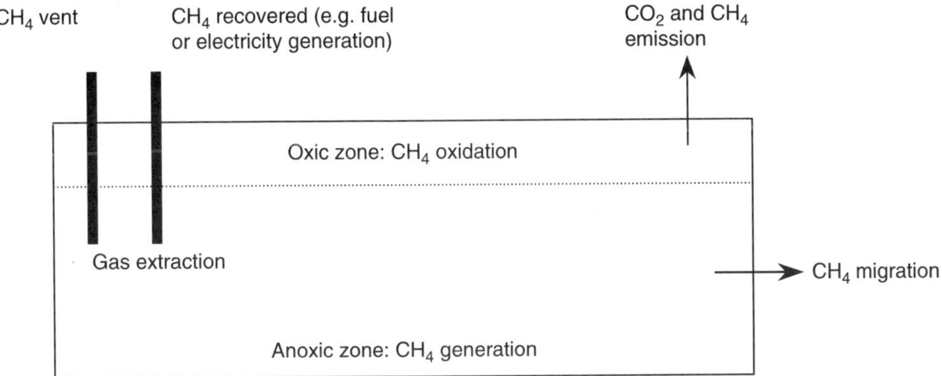

Fig. 12.1. Conceptual methane (CH_4) balance in a landfill.

Table 12.1. Efficiency estimates of CH_4 oxidation in landfill cover soils.

Source	Efficiency (%)	Landfill status
Whalen *et al.* (1990)	50	All landfills in USA
Jones and Nedwell (1993)	10–40[a]	8–9 years closed
Oonk and Boom (1995)	10–20[b]	Active
Boeckx *et al.* (1996)	40–100	>5 years closed
Czepiel *et al.* (1996a)	10	Active
Kjeldsen *et al.* (1997)	up to 80	4 years closed
Bergamaschi *et al.* (1998)	53	Active
Liptay *et al.* (1998)	30[c]	Not specified
Chanton *et al.* (1999)	12	Active
Chanton and Liptay (2000)	20[d]	5 years closed
Börjesson *et al.* (2001)	0/41–50[e] 0/60–94[e]	<1 year closed 17 years closed
De Visscher (2001)	30	Model result
Barlaz *et al.* (2004)	21; 55[f]	Active

[a]Conservative estimate.
[b]Estimate depends on stoichiometry.
[c]In summer.
[d]Average of two cover soils: mulch/topsoil (26%) and clay soil (14%); each are annual mean estimates.
[e]Winter/summer.
[f]Soil cover and biocover, respectively. Measurements do not include winter.

efficiency estimates found in the literature. If an estimate of 30% oxidation is applied to the data of Bogner and Matthews (2003), a global biological oxidation of 6.9 Tg CH_4/year and a residual emission of 16.1 Tg CH_4/year are obtained. Clearly, biological oxidation is an important factor affecting net CH_4 emissions from landfills, and could hence be the main artificial CH_4 sink.

12.2 Methanotrophy in Landfill Cover Soils

Microbial CH_4 oxidation shows two types of kinetics. Soils exposed to ambient CH_4 concentrations display high-affinity (K_m = 22–37 μl/l), low-capacity (V_{max} = 0.7–3.6 nmol/g dry weight/h) CH_4 oxidation kinetics. In contrast, soils exposed to high CH_4 mixing ratios show low-affinity (K_m = 1290–20680 μl/l), high-capacity (V_{max} = 270–3690 nmol/g dry weight/h) CH_4 oxidation kinetics (Bender and Conrad, 1992). These researchers found that CH_4 mixing ratios of 100–1000 μl/l are required to increase the CH_4 oxidizing activity of soils. Schnell and King (1995) found that this threshold is between 170 and 1000 μl/l. The CH_4 concentrations prevalent in landfill cover soils (1–60%) are sufficiently high for induction of low-affinity, high-capacity CH_4 oxidation. This is illustrated in Fig. 12.2, which shows the relationship between the CH_4 mixing ratio in landfill cover soils, and V_{max} values of samples taken from these soils (Czepiel *et al.*,

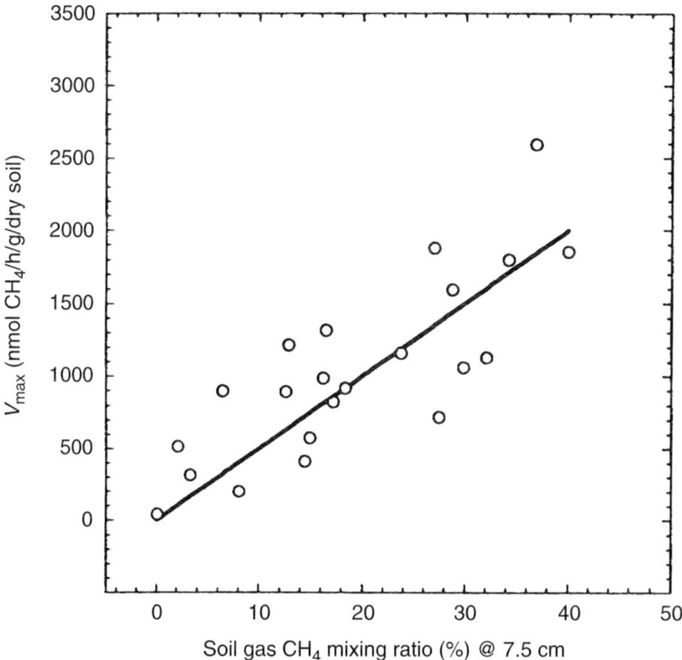

Fig. 12.2. V_{max} versus methane (CH_4) mixing ratios in landfill cover soils. (From Czepiel *et al.*, 1996a.)

1996a). Pure cultures of known methanotrophic bacteria also show a low-affinity, high-capacity pattern. Indeed, Wise *et al.* (1999) and Stralis-Pavese *et al.* (2004) showed that a wide variety of methanotrophic bacteria are present in landfill cover soils. For these reasons, it is useful to consider some characteristics of methanotrophs.

12.2.1 Methanotrophic bacteria

The properties of methanotrophic bacteria have been reviewed extensively by Hanson and Hanson (1996). This section summarizes the aspects most relevant for understanding functional methanotrophic ecology in landfill cover soils (see also Chapter 10).

Methanotrophs are Gram-negative bacteria that are able to use CH_4 as the sole carbon and energy source. They are a subset of the methylotrophs, which are able to use one-carbon compounds more reduced than HCOOH as sources of carbon and energy. What distinguishes methanotrophs from the other methylotrophs is the use of the methane monooxygenase (MMO) enzyme, which catalyzes the oxidation of CH_4 to methanol.

The current taxonomy of methanotrophs was first established by Whittenbury *et al.* (1970), based on morphological and biochemical characteristics. A more comprehensive study including numerical taxonomy and DNA–DNA hybridization has necessitated some revisions (Bowman *et al.*, 1993, 1994, 1995). Two types of methanotrophs, labelled type I and type II, were distinguished. Bowman and his co-workers consistently refer to these types as 'group I' and 'group II'. As of 1995, type I included the genera *Methylococcus*, *Methylomicrobium*, *Methylobacter* and *Methylomonas*, and formed the family *Methylococcaceae*. Type II included the genera *Methylosinus* and *Methylocystis*.

Type I methanotrophs form a distinct branch within the gamma subdivision of the *Proteobacteria*. They use a particulate MMO (pMMO) to oxidize CH_4. Formaldehyde, a CH_4 oxidation product, is assimilated using the ribulose monophosphate pathway (RuMP). Most type I methanotrophs cannot fix N_2 (absence of nitrogenase activity). Exceptions are some *Methylomonas* and *Methylococcus* species.

Type II methanotrophs form a distinct branch within the alpha subdivision of the *Proteobacteria*. They use pMMO, but in the absence of copper a soluble enzyme (sMMO) is produced in most type II methanotrophs and in some type I methanotrophs. This enzyme has broad substrate specificity and enables these microorganisms to oxidize chlorinated hydrocarbons and aromatic hydrocarbons. Type II methanotrophs assimilate formaldehyde via the serine pathway. They are also able to fix N_2.

The RuMP pathway is more efficient than the serine pathway. Consequently, type I methanotrophs tend to outgrow their type II counterparts, unless inorganic nitrogen or copper limitation provides an advantage for type II species, expressing sMMO or nitrogenase activity.

In this taxonomy, *Methylococcus* occupies a special position within the type I genera. Species from this genus are mildly thermophilic, express sMMO and nitrogenase activity, and sometimes use the serine pathway. For these reasons this genus was formerly referred to as type X. This nomenclature became irrelevant with the discovery of other genera with unusual properties and is no longer in use.

Novel methanotrophic species and genera are continuously being discovered. New type I genera include the psychrophilic *Methylosphaera* (Bowman *et al.*, 1997), the thermophilic *Methylocaldum* (Bodrossy *et al.*, 1997) and *Methylothermus* (Tsubota *et al.*, 2005), the strongly clustering *Methylosarcina* (Wise *et al.*, 2001) and the halophilic *Methylohalobius* (Heyer *et al.*, 2005). A new type II genus is the mildly acidophilic *Methylocella* (Dedysh *et al.*, 2000, 2004; Dunfield *et al.*, 2003). Dedysh *et al.* (2002) even suggested a 'type III' classification for the new genus *Methylocapsa*, which consists of mildly acidophilic bacteria not expressing sMMO, but using the serine pathway. This genus belongs to the α-*Proteobacteria* and its closest relation is *Methylocella*.

Wise *et al.* (1999) discovered that the result of methanotrophic isolation by enrichment series depends on the strength of the nitrate mineral medium used, indicating that results of enrichments are not necessarily representative of the *in situ* ecology.

12.2.2 Processes occurring in landfill cover soils

To understand what affects the efficiency of landfill cover soils to oxidize CH_4, it is important to consider the different biogeophysical processes occurring in this environment. The main processes are gas diffusion, gas flow (advection), microbial CH_4 oxidation and growth of methanotrophic bacteria. Each of these processes can be affected by environmental factors.

Figure 12.3 shows typical concentration profiles that develop in the soil gas phase in the absence and presence of CH_4 oxidation. In the absence of methanotrophy both oxygen and N_2 penetrate the soil against the gas flow, and the entire cover soil is aerobic. CH_4 and CO_2 are produced below the soil and move upwards by advection and diffusion. The gases are diluted by oxygen and N_2; hence a concentration decrease in itself is not an indication of CH_4 oxidation. In the presence of methanotrophy, oxygen penetrates only 10–15 cm deep due to consumption by the methanotrophs. CH_4 is depleted and CO_2 accumulates in comparison with the absence of methanotrophy. CH_4 oxidation is confined to the aerobic zone, so the oxidizing capacity of the soil is mainly determined by the methanotrophic activity and the oxygen diffusion into the soil. The following section reviews the influence of environmental factors on these two processes.

12.2.3 Factors influencing CH_4 oxidation in landfill cover soils

Temperature

Methanotrophic activity is strongly influenced by temperature. The temperature

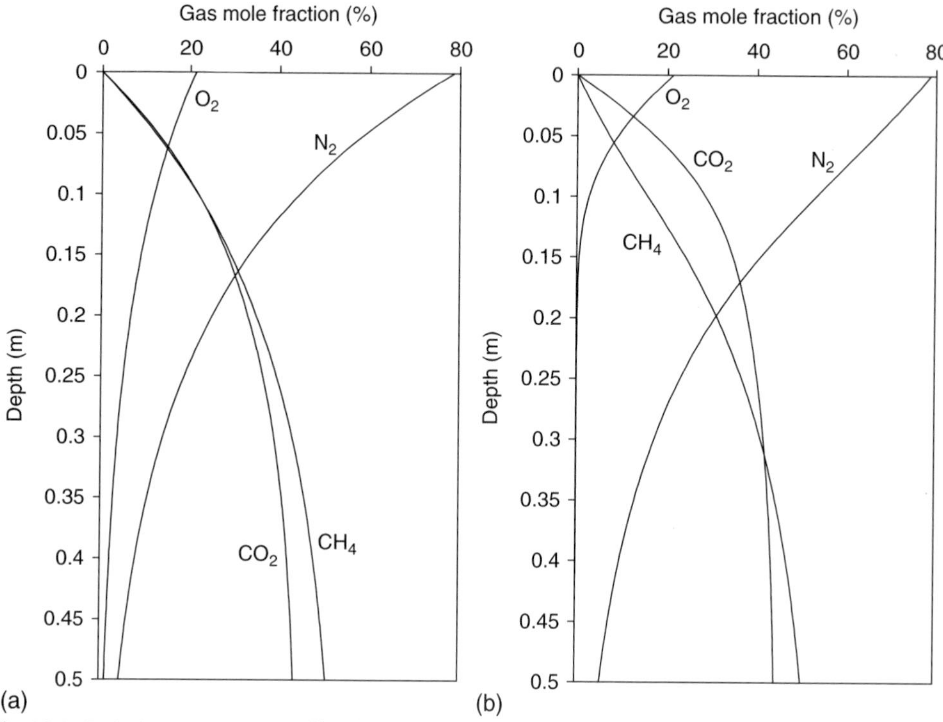

Fig. 12.3. Typical concentration profiles that develop in the gas phase of a landfill cover soil in the (a) absence and (b) presence of methane (CH_4) oxidation.

response is typical for enzyme kinetics with both high-temperature and low-temperature inactivation as described by Sharpe and DeMichele (1977).

At 10–30°C the temperature response is approximately exponential with Q_{10} values ranging from 1.7 to 4.1 (Boeckx *et al.*, 1996; Czepiel *et al.*, 1996a; De Visscher *et al.*, 2001; Börjesson *et al.*, 2004; Park *et al.*, 2005). King and Adamsen (1992) and De Visscher *et al.* (2001) found that Q_{10} is substantially larger at 1–3% CH_4 than at 0.01–0.025% CH_4.

Low-temperature inactivation is evidenced by increased Q_{10} values. This was observed by Christophersen *et al.* (2000), who found a Q_{10} value of ~5 at 2–15°C. High-temperature inactivation is evidenced by the existence of an optimum temperature for CH_4 oxidation, which ranges from 22°C to 38°C (Boeckx and Van Cleemput, 1996; Boeckx *et al.*, 1996; Czepiel *et al.*, 1996a; Whalen and Reeburgh, 1996; Gebert *et al.*, 2003; Park *et al.*, 2005). It appears that in landfill conditions type I methanotrophs tend to have a lower temperature optimum than type II methanotrophs (Gebert *et al.*, 2003). Consequently, type I methanotrophs are more dominant at 10°C than at 20°C (Börjesson *et al.*, 2004).

A 10°C temperature rise stimulates the diffusion coefficient in the gas phase by only 6–7%. Consequently, CH_4 oxidation in landfill cover soils is increasingly limited by oxygen transfer as the temperature increases. The result is a less pronounced temperature dependence of CH_4 oxidation on temperature than indicated by the Q_{10} value. This is illustrated with model calculations in Section 12.5.

Moisture content

Diffusion, advection and methanotrophy are strongly affected by soil moisture content. As a consequence, moisture content is the most important factor influencing CH_4 oxidation in landfill cover soils.

Moisture content affects diffusion at two different scales, a complication that is not always fully appreciated. The two scales are associated with gas-phase diffusion (large scale: centimetres to decimetres) and liquid-phase diffusion (micro scale or pore scale: micrometres to millimetres).

Gas-phase diffusion is four orders of magnitude faster than liquid-phase diffusion. Consequently, large-scale diffusion can only occur in the gas phase. Vertical diffusive transport as indicated in Fig. 12.3 belongs to this category. At increasing moisture content, the gas-filled pore space becomes smaller and more tortuous, which hinders diffusion and limits the amount of oxygen that can penetrate the soil. Wet soils can only support a thin methanotrophic layer below the surface and do not oxidize CH_4 efficiently. Empirical equations to describe the effect of soil moisture content on the effective diffusion coefficient of gases in soils have been developed by many researchers (e.g. Penman, 1940; Millington, 1959; Millington and Shearer, 1971; Moldrup *et al.*, 1996, 2000a,b).

At the pore scale, CH_4 usually has to diffuse through a water layer before it reaches the methanotrophic bacteria. This layer becomes thicker with increasing moisture content, which can lead to a substantial decrease of the CH_4 oxidation rate, especially in aggregated soils. Mass transport effects at the pore scale cannot usually be distinguished from physiological effects within the bacteria. For that reason the physiological effect and the pore-scale effect of moisture content on CH_4 oxidation are treated simultaneously.

Many researchers have found that CH_4 oxidation rates pass through a maximum when plotted as a function of moisture content (Boeckx and Van Cleemput, 1996; Boeckx *et al.*, 1996; Whalen and Reeburgh, 1996). Low CH_4 oxidation rates at low moisture contents are associated with water stress of the methanotrophs (Striegl *et al.*, 1992; Kruse *et al.*, 1996). Schnell and King (1995) quantified this effect by inducing an osmotic stress in soils and methanotrophic cultures. In both cases they found that the methanotrophic activity decreases with osmotic potential (increasing stress), to become zero at an osmotic potential of −3 to −4 MPa.

Low CH_4 oxidation rates at high moisture contents are associated with pore-scale diffusion limitations (Whalen *et al.*, 1990; Boeckx and Van Cleemput, 1996; De Visscher *et al.*, 2001).

Water stress in soils depends largely on the matric potential of the soil, whereas pore-scale diffusion limitation depends largely on the water-filled pore space. Consequently, the optimum moisture content for CH_4 oxidation will depend on the soil type. Some researchers have tried to overcome this difficulty by normalizing their results to the water-holding capacity. Boeckx *et al.* (1996) found an optimum at 50% of water-holding capacity, whereas Whalen and Reeburgh (1996) found an optimum at 20–50% of water-holding capacity. De Visscher and Van Cleemput (2000) developed a model that incorporates pore-scale diffusion, water stress and the influence of temperature on CH_4 oxidation.

The air permeability of soils decreases with increasing moisture content. This can potentially lead to a decrease of the advective transport of CH_4 through a landfill cover. However, the gas produced by the waste must somehow escape to the atmosphere, so high moisture contents in a cover soil will either lead to lateral transport of CH_4 and emissions adjacent to the landfill or to a pressure build-up creating the necessary driving force for advective transport through the soil. A tragic instance of the former possibility was documented by Kjeldsen and Fischer (1995) at Skelligsted landfill, Denmark, where lateral landfill gas migration led to a fatal explosion in a nearby house. The latter possibility will not affect the CH_4 transport observed in the cover soil if the soil is homogeneous. However, cracks and fissures usually occur in landfill cover soils as a result of waste settlement, so wet landfill cover soils lead to large CH_4 emissions through these 'hot spots'. Czepiel *et al.* (1996b) found that about 5% of the cover area of a landfill in New Hampshire was responsible for 50% of the CH_4 emissions. Bergamaschi *et al.* (1998) found that 70% of CH_4 emissions through cover soils of landfills in Germany and the Netherlands happened through cracks and leakages in the soil. Mice burrows have also been found to increase CH_4 flux variability in landfill cover soils (Giani *et al.*, 2002).

Figure 12.4 shows a model fit to the data on the influence of moisture and temperature on CH_4 oxidation from Boeckx *et al.* (1996) and De Visscher and Van Cleemput (2000).

2.3.3 Inorganic nitrogen

The influence of inorganic nitrogen on microbial CH_4 oxidation is exceedingly complex and not yet fully understood. This is because inorganic nitrogen can act as both nutrient and inhibitor for methanotrophy. The role of nitrogen in acting on CH_4 concentration, pH and type of methanotroph depends on its form (NH_4^+, NO_2^- or NO_3^-) and concentration. A further complicating factor is the potential inhibiting effect of Cl^- when NH_4^+ is applied as NH_4Cl (Gulledge and Schimel, 1998; De Visscher and Van Cleemput, 2003a). A review by Bodelier and Laanbroek (2004) on this inhibiting effect was published recently.

At atmospheric CH_4 concentrations, inorganic nitrogen is not likely to be the limiting substrate for CH_4 oxidation. Consequently, stimulation of CH_4 oxidation by adding inorganic nitrogen is rarely observed under such conditions. Most studies indicate that NH_4^+ and NO_2^- inhibit CH_4 oxidation at atmospheric concentrations, whereas NO_3^- has no influence other than salt effects (Boeckx and Van Cleemput, 1996; Hütsch *et al.*, 1996; Hütsch, 1998). However, some recent studies indicate that NO_3^- can be more inhibitive than NH_4^+ in very acidic soils (Wang and Ineson, 2003; Reay and Nedwell, 2004).

At high CH_4 concentrations observed in landfill cover soils, inorganic nitrogen can both stimulate and inhibit CH_4 oxidation. These inhibition mechanisms are the same as with atmospheric CH_4 concentrations: competitive inhibition by NH_4^+ (or rather NH_3; Carlsen *et al.*, 1991) on the enzyme, toxic effects of NO_2^- or salt effects (see also Dunfield, Chapter 10, this volume). Stimulation of CH_4 oxidation by inorganic nitrogen can be interpreted as a relief of nitrogen limitation of the methanotrophs (De Visscher *et al.*, 1999; Bodelier and Laanbroek, 2004).

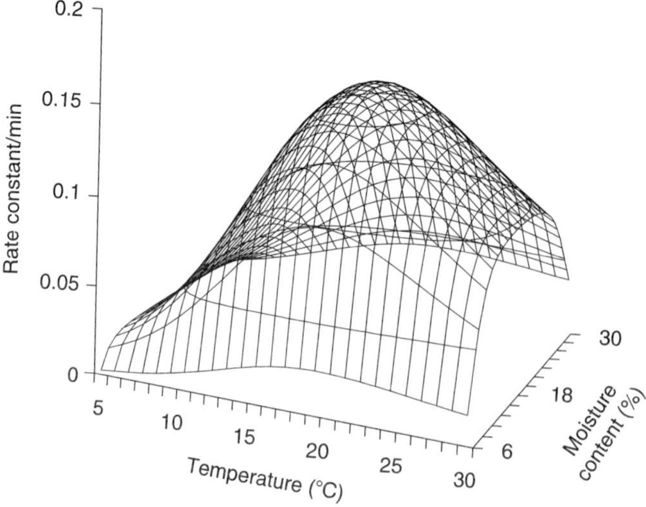

Fig. 12.4. Model fit to the data on the influence of moisture and temperature on methane (CH_4) oxidation. (From De Visscher and Van Cleemput, 2000.)

The effect of inorganic nitrogen on methanotrophy depends on exposure time to CH_4 and is closely linked to the microbial ecology of CH_4 oxidation (De Visscher and Van Cleemput, 2003a; Wilshusen *et al.*, 2004a). As explained in Section 2.1, type I methanotrophs usually grow faster than type II methanotrophs, but the former need a source of inorganic nitrogen, whereas the latter are able to fix N_2.

De Visscher *et al.* (1999) investigated the effect of adding wheat straw and sugarbeet leaves on CH_4 oxidation in microcosms imitating landfill cover soils. Adding sugarbeet leaves, with a low carbon/nitrogen ratio, leads to a net nitrogen mineralization, and this extra inorganic nitrogen source is used by type I methanotrophs for cell synthesis. The result is a stimulation of CH_4 oxidation that disappears after 1 month, when the mineralization effect subsides. Adding wheat straw, with a high carbon/nitrogen ratio, leads to immobilization of the available inorganic nitrogen and immediate nitrogen stress of the type I methanotrophs, allowing the type II methanotrophs to gain an immediate advantage as N_2-fixing bacteria. The result is a lasting stimulation of CH_4 oxidation.

De Visscher and Van Cleemput (2003a) investigated the influence of exposure time to CH_4 on the immediate effect of adding NH_4Cl and $(NH_4)_2SO_4$ to a soil. From this experiment they concluded that soil exposed to high (>1%) CH_4 mixing ratios develops methanotrophic activity in three stages. The first stage is a rapid growth of methanotrophs, probably of type I. The second stage is a decline of the methanotrophic activity, probably caused by nitrogen limitation of type I methanotrophs. After a few weeks of steady-state behaviour, a new growth phase – the third stage – is observed, probably dominated by N_2-fixing type II methanotrophs. Wilshusen *et al.* (2004a) came to similar conclusions based on phospholipid fatty acid analysis. They also found that type I/type II selective pressures are related to oxygen availability and extracellular polysaccharide (EPS or exopolymer) production, as outlined in the following section.

Exopolymer formation

After long periods of exposure to high CH_4 mixing ratios, methanotrophic activity sometimes shows a decline that cannot be attributed simply to inorganic nitrogen limitation. It has been suggested that this decline is the result of the production of EPS, which limit the availability of oxygen and CH_4 by acting as a barrier to diffusion (Hilger *et al.*, 1999, 2000a; Chiemchaisri *et al.*, 2001; Wilshusen *et al.*, 2004a,b). Figure 12.5 shows a photo of EPS deposits in soils (Hilger *et al.*, 2000a). Although serious detrimental effects of EPS on CH_4 oxidation have never been reported on a field scale, this factor is a cause for concern in the development of CH_4 biofilters and other engineered controls of microbial CH_4 oxidation. It is also becoming increasingly clear that a better insight into ecological controls of EPS production will lead to a better understanding of the ecology of CH_4 oxidation itself.

The propensity of methanotrophs to produce EPS was established by Huq *et al.* (1978). Wrangstadh *et al.* (1986) hypothesized that oxygen or inorganic nitrogen deprivation leads to the formation of EPS in methanotrophic systems. Wilshusen *et al.* (2004a) hypothesized that the key to understanding EPS production is N_2 fixation by the nitrogenase enzyme. It is well known that nitrogenase activity requires a microaerophilic environment. Whittenbury and Dalton (1981), for instance, found that oxygen concentrations below 4% were necessary to obtain N_2 fixation in pure cultures of methanotrophs. Wilshusen *et al.* (2004a) hypothesized that nitrogen-starved methanotrophs produce EPS as a nitrogen-free carbon sink, allowing them to cycle carbon without the need for inorganic nitrogen. The ensuing pore clogging caused by the EPS creates the microaerophilic environment needed to initiate nitrogenase activity. When the environment is microaerophilic from the beginning, nitrogenase activity is initiated without excessive EPS production.

The link between nitrogen limitation and oxygen limitation described here also explains why methanotrophs are known

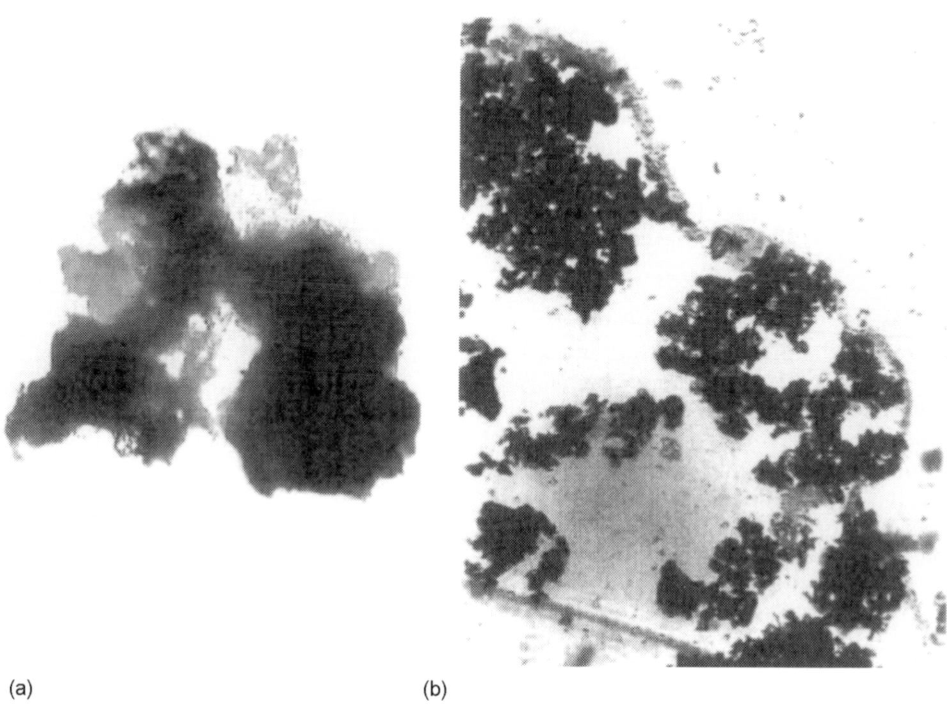

Fig. 12.5. Soils sampled from columns sparged with synthetic landfill gas. The regions stained with blue dye denote the presence of polysaccharide: (a) dense polymer coated on the soil particles; (b) polymer strands separated from soil particles. (Photo from Hilger *et al.*, 2000a.)

as microaerophilic bacteria (Stein and Hettiaratchi, 2001), in spite of the fact that kinetic studies showed a barely perceptible microaerophilic effect (Joergensen, 1985; Bender and Conrad, 1994). Microaerophilic N_2 fixation also helps to explain why nitrogen-fixing type II methanotrophs are more competitive at low oxygen concentrations (<30% of saturation in aqueous solution) than at high concentrations. Amaral and Knowles (1995) studied competitive pressure of methanotrophs in gel-stabilized systems and found type II methanotrophs at high CH_4/O_2 ratios, and type I methanotrophs at low CH_4/O_2 ratios. Ren *et al.* (1997), on the other hand, found no evidence of microaerophily in pure cultures when CH_4 and inorganic nitrogen were not limiting.

Other factors

Arif *et al.* (1996) found that 5 mg/kg soil dry weight of 2,4-dichlorophenoxyacetic acid (2,4-D) caused partial inhibition of CH_4 oxidation by soil. Adding 50 mg/kg soil dry weight caused complete inhibition. Top *et al.* (1999) used this effect as a biomarker for assessing 2,4-D biodegradation by plasmid-mediated bioaugmentation. Addition of nitrapyrin, a nitrification inhibitor, also caused a strong inhibition of CH_4 oxidation (Arif *et al.*, 1996). Boeckx *et al.* (1998) investigated the influence of various pesticides on CH_4 oxidation by arable soils. They found a general decrease of the CH_4 oxidation rate. The decrease was statistically significant for lenacil, mikado and oxadixyl in a sandy soil, as well as for mikado, atrazine and dimethenamid in a clayey soil. They also found that landfill cover soils are at least ten times less sensitive to pesticides than arable soils that are not exposed to high CH_4 mixing ratios. Börjesson (2001) found that methanethiol and carbon disulphide inhibit CH_4 oxidation in landfill cover soils.

Hilger *et al.* (2000b) found that vegetation (grass) mitigated the inhibiting effect of NH_4^+ on CH_4 oxidation in landfill cover soils, probably due to assimilation of the NH_4^+ by the grass. They also found that liming has a positive effect on CH_4 oxidation, as the oxidation of 1 mol CH_4 generates 0.1–0.12 mol H^+.

12.3 Measurement Techniques for CH_4 Oxidation

Techniques for the measurement of CH_4 oxidation in landfill cover soils and other ecosystems were recently reviewed by Nozhevnikova *et al.* (2003).

Boeckx *et al.* (1996) determined CH_4 oxidation in a cover soil by removing the cover on part of the landfill and measuring the CH_4 flux with a closed-box method directly on the waste as well as on the surrounding cover soil. Similar techniques have been used in wetlands (Nozhevnikova *et al.*, 2003). The advantage of this method is that it is a direct measurement. The disadvantage is the amount of disturbance required to perform the measurement, creating a bypass for the gas.

Czepiel *et al.* (1996a) developed a method to estimate the year-round, whole-landfill average CH_4 oxidation by combining field, laboratory and computer modelling methods. They used CH_4 concentrations at 7.5 cm depth as a proxy for the soil CH_4 oxidizing capacity, and combined this information with a typical depth profile of the activity as well as representative corrections for temperature and soil moisture contents. A year-round whole-landfill average CH_4 oxidation for a landfill in New Hampshire was 10%. The method is non-intrusive, but it is unclear if the proxy and the corrections are representative of other landfills in other climates.

Kjeldsen *et al.* (1997) estimated the CH_4 oxidizing capacity of a landfill cover soil by integrating the depth profile of the CH_4 oxidation rate of soil samples taken at different depths in the cover, as determined in the laboratory.

Oonk and Boom (1995) used a mass balance method to determine CH_4 oxidation in cover soils of several landfills in the Netherlands. They measured CH_4 and CO_2 fluxes leaving the landfill cover soil and compared the CH_4/CO_2 flux ratio with the CH_4/CO_2 production ratio inside the waste. This enabled them to deduce the CH_4 oxidation efficiency. The main advantage of this method is that it is direct and non-intrusive. The disadvantage is the potential interference of CO_2 production and consumption by plants that is sometimes more important than the landfill CO_2 production. Oonk and Boom (1995) solved this problem by taking the average of a 24 h monitoring.

A very promising method to determine CH_4 oxidation in landfill cover soils is by stable isotope measurements. Methanotrophs oxidize $^{12}CH_4$ slightly faster than $^{13}CH_4$ (Barker and Fritz, 1981; King *et al.*, 1989). A similar effect was observed for CH_3D (Coleman *et al.*, 1981). This effect can be used to quantify CH_4 oxidation by measuring the $\delta^{13}C$ or δD abundance in CH_4 emitted at the cover soil surface and comparing them with the abundance in CH_4 produced inside the landfill (Liptay *et al.*, 1998; Chanton and Liptay, 2000; Börjesson *et al.*, 2001). A simple equation to relate changes in the isotope abundance to the fate of a compound was provided by Blair *et al.* (1985):

$$f_{ox} = \frac{\delta E - \delta A}{1000(\alpha_{ox} - \alpha_{trans})}$$

where δE is the $\delta^{13}C$ abundance of the emitted CH_4 (‰), and δA the $\delta^{13}C$ abundance of the produced CH_4. α_{ox} and α_{trans} are fractionation factors of oxidation and transport, respectively. Usually, α_{trans} is assumed to be equal to 1. However, recent research has indicated that gas transport also causes isotope fractionation ($\alpha_{trans} > 1$), which leads to an underestimation of CH_4 oxidation by the isotope method (De Visscher *et al.*, 2004). Scharff *et al.* (2003) compared the mass balance method with the isotope method. They found that the results were not significantly different, but the uncertainty of the methods was very large.

More qualitative methods to test the occurrence of CH_4 oxidation include methanotrophic cell counts in the cover soil and the measurement of the N_2/O_2 ratio in the landfill cover soil (Nozhevnikova *et al.*, 1993, 2003).

Determination of *in situ* CH_4 oxidation can also be carried out using selective inhibitors for CH_4 oxidation. Krüger *et al.* (2001) tested the selectivity for difluoromethane (CH_2F_2) and showed that 1% CH_2F_2 had no effect on methanogenesis. The difference between CH_4 emission found with and without the use of CH_2F_2 is then a measure for the *in situ* CH_4 oxidation rate.

12.4 Optimizing the CH_4 Sink

12.4.1 Biological control

Techniques to optimize biological CH_4 oxidation in landfill cover soils were recently reviewed by Hilger and Humer (2003). Efforts to optimize microbial CH_4 oxidation from landfill gas are concentrating on two approaches: (i) optimizing the cover soil itself (the biocap); and (ii) sending the landfill gas through a separate biofilter. Optimization of CH_4 oxidation by using compost as a landfill biocap was studied extensively by Humer and Lechner (1999, 2001) in both the laboratory and field scales. They found that compost from municipal solid waste (MSW) and fully matured sewage sludge compost were optimal substrates for maintaining a high CH_4 oxidation efficiency in the cover. Thanks to the high porosity of MSW compost, a 60 cm layer can remove up to 350 g $CH_4/m^2/day$ in laboratory-scale experiments. Field trials confirmed that CH_4 oxidation in MSW compost covers can be 100% effective. Wilshusen *et al.* (2004b) investigated CH_4 oxidation by various composts in laboratory-scale set-ups and found the highest removal efficiency in the case of leaf compost. Barlaz *et al.* (2004) reported that compost covers oxidized more CH_4 than conventional clay covers in field trials, but warned that compost covers can also produce CH_4 if the moisture content is too high. Berger *et al.* (2005) warned that a clear boundary between two cover layers with different hydrologic properties can lead to water accumulation.

Oonk *et al.* (2004) investigated different types of cover material on two pilot-scale sites in the Netherlands. They found that spatial homogeneity is an important requirement for efficient CH_4 oxidation. When a mixture of 70% (weight) uniform sand and 30% (weight) garden waste compost is used, the CH_4 flux that can be oxidized is 30–50 g/m^2/day. It is expected that emission reductions are possible for € 3.5–5/t CO_2 equivalents, provided that sufficiently high fluxes are present.

CH_4 oxidation in separate biofilters has been investigated by Gebert *et al.* (2003, 2004) and by Streese and Stegmann (2003). Gebert *et al.* (2003) experimented with expanded clay as biofilter material, which had good moisture-holding characteristics for CH_4 oxidation at high air-filled pore space, but contained inhibiting amounts of salt. The CH_4 oxidizing capacity increased after leaching of the salts. Streese and Stegmann (2003) tested pilot-scale biofilters containing fine-grained compost and found CH_4 degradation rates of up to 63 g $CH_4/m^3/h$. These results indicate that separate biofilters can be a feasible alternative solution when regulations do not allow for biocaps.

As has been pointed out, homogeneity is a requirement for high CH_4 oxidation efficiencies, because heterogeneities such as cracks lead to bypassing of the landfill gas. This conclusion has been confirmed by modelling studies (De Visscher, 2001). For this reason, it is important that efforts be made to keep landfill cover soils homogeneous, even after settling of the waste. Patches of reduced vegetation growth are indicative of landfill gas hot spots.

12.4.2 Landfill gas extraction

In 1995 there were more than 500 commercial CH_4 recovery projects worldwide, and current estimates exceed 900 plants (Bogner and Matthews, 2003). Most of these projects involve electricity production, but some involve the production of substitute natural gas by the removal of CO_2 and trace components. At many other sites landfill gas is flared. CH_4 recovery in 1996 was estimated at 3.8 Tg/year (Bogner and Matthews, 2003).

12.5 Modelling of CH_4 Formation, Diffusion and Oxidation in Landfills and Cover Soils

Bogner *et al.* (1997) developed a simulation model describing diffusion and oxidation of CH_4 in a landfill cover soil in terms of collisions of CH_4 molecules with soil particles and biomass. The model was successfully validated using field data, but the approach is unconventional in gas transport modelling, and the conceptual validity of the model assumptions remains an issue. Simulation models based on more conventional concepts were developed by Hilger *et al.* (1999), Stein *et al.* (2001) and De Visscher and Van Cleemput (2003b).

The model of Hilger *et al.* (1999) is based on the Stefan–Maxwell equations for gas flow and diffusion. Here, a model applicable to thick biofilms was used. Oxygen is assumed to be the only limiting substrate. This is an acceptable assumption because the depth of the methanotrophically active zone is limited by oxygen penetration, but it might lead to overestimates of the CH_4 oxidation close to the soil surface, where CH_4 concentration is low.

The model of Stein *et al.* (2001) incorporates flow and Fickian diffusion with concentration-dependent diffusion coefficients. This approach is well established in chemical engineering because it combines high accuracy with ease of implementation (Froment and Bischoff, 1990). A dual-substrate CH_4 oxidation model was used to describe both CH_4 and oxygen limitation. Perera *et al.* (2002a) developed a variant with the aim of calculating the relationship between local CH_4 production and CH_4 emission. Recently, this model was combined with a geographic information systems (GIS) approach to obtain a 'pseudo-3D' model (Perera *et al.*, 2004). A two-dimensional (2D) model was developed by Perera *et al.* (2002b) with the aim of optimizing closed-box measurements on landfills.

The model of De Visscher and Van Cleemput (2003b) incorporates Stefan–Maxwell mass transport equations. They used the same dual-substrate model for CH_4 oxidation as Stein *et al.* (2001). The model also incorporates a growth model, which makes possible the calculation of a methanotrophic activity profile from minimal information. A sensitivity analysis was also performed, which indicated that the parameter $V_{max,max}$, describing the maximum value of V_{max} that can be supported by the soil, has the largest influence on the model predictions.

Figure 12.6 shows model predictions of the influence of environmental factors on CH_4 oxidation efficiencies at conditions given in Table 12.2. Temperature and moisture content are the main governing factors of the process. The temperature results suggest a Q_{10} value below 2, which is less than the physiological Q_{10}. This is due to diffusion limitation effects, as discussed in Section 12.2.3. Mahieu *et al.* (2005) extended the model of De Visscher and Van Cleemput (2003b) to include isotope fractionation effects.

In conclusion, artificial CH_4 sinks – such as those that form in landfill cover soils – represent an important component of net CH_4 exchange globally. The efficiency at which CH_4 is removed before emission to the atmosphere is highly dependent on soil water content, temperature and CH_4 supply.

Table 12.2. Environmental factors for the simulation of laboratory conditions and field conditions in Fig. 12.6. (From De Visscher and Van Cleemput, 2003b.)

Factor	Laboratory value	Field value
T (°C)	22	10
ε (m^3_{gas}/m^3_{soil})	0.412	0.25
Gas flux (mol/m²/s)	3.1×10^{-4}	1×10^{-4}
CH_4/CO_2 (mol/mol)	50/50	50/50
ϕ (m^3_{void}/m^3_{soil})	0.5878	0.5
ρ_{DB} ($kg_{soil\ DW}/m^3$)	1039	1300

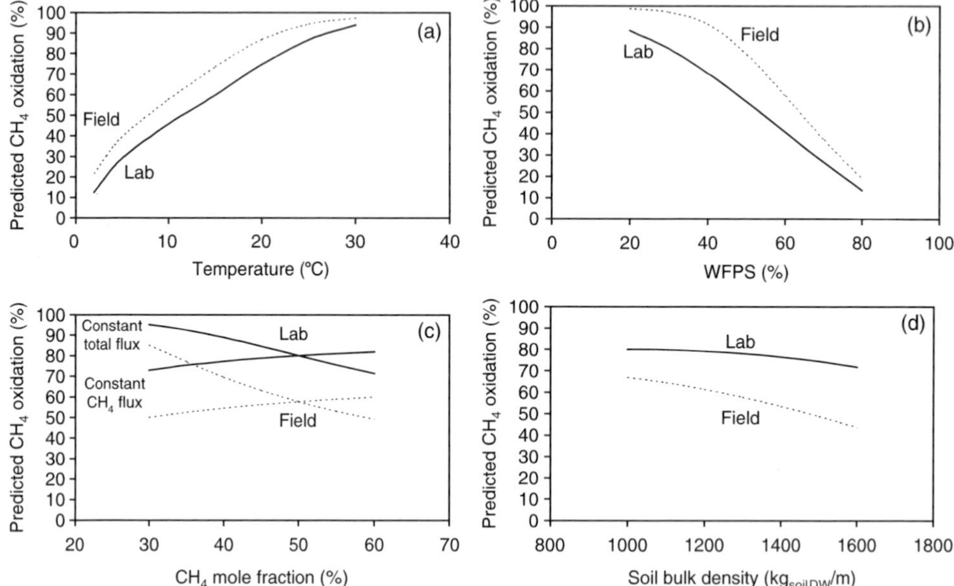

Fig. 12.6. Influence of environmental factors on methane (CH_4) oxidation predicted by the model for laboratory soil column (solid lines) and field conditions (dotted lines): (a) temperature; (b) water-filled pore space (as percentage of total void space); (c) CH_4 mole fraction in the landfill gas (at constant total flux and at constant CH_4 flux); (d) soil bulk density. (From De Visscher and Van Cleemput, 2003b.)

References

Amaral, J.A. and Knowles, R. (1995) Growth of methanotrophs in methane and oxygen counter gradients. *FEMS Microbiology Letters* 126, 215–220.

Arif, M.A.S., Houwen, F. and Verstraete, W. (1996) Agricultural factors affecting methane oxidation in arable soil. *Biology and Fertility of Soils* 21, 95–102.

Barker, J.F. and Fritz, P. (1981) Carbon isotope fractionation during microbial methane oxidation. *Nature* 293, 289–291.

Barlaz, M.A., Green, R.B., Chanton, J.P., Goldsmith, C.D. and Hater, G.R. (2004) Evaluation of a biologically active cover for mitigation of landfill gas emissions. *Environmental Science and Technology* 38, 4891–4899.

Bender, M. and Conrad, R. (1992) Kinetics of CH_4 oxidation in oxic soils exposed to ambient air or high CH_4 mixing ratios. *FEMS Microbiology Ecology* 101, 261–270.

Bender, M. and Conrad, R. (1994) Methane oxidation activity in various soils and freshwater sediments: occurrence, characteristics, vertical profiles, and distribution on grain size fractions. *Journal of Geophysical Research* 99, 16531–16540.

Bergamaschi, P., Lubina, C., Königstedt, R., Fischer, H., Veltkamp, A.C. and Zwaagstra, O. (1998) Stable isotopic signatures ($\delta^{13}C$, δD) of methane from European landfill sites. *Journal of Geophysical Research* 103, 8251–8265.

Berger, J., Fornés, L.V., Ott, C., Jager, J., Wawra, B. and Zanke, U. (2005) Methane oxidation in a landfill cover with capillary barrier. *Waste Management* 25, 369–373.

Bingemer, H.G. and Crutzen, P.J. (1987) The production of methane from solid wastes. *Journal of Geophysical Research* 92, 2181–2187.

Blair, N., Leu, A., Muñoz, E., Olsen, J., Kwong, E. and Des Marais, D. (1985) Carbon isotopic fractionation in heterotrophic microbial metabolism. *Applied and Environmental Microbiology* 50, 996–1001.

Bodelier, P.L.E. and Laanbroek, H.J. (2004) Nitrogen as a regulatory factor of methane oxidation in soils and sediments. *FEMS Microbiology Ecology* 47, 265–277.

Bodrossy, L, Holmes, E.M., Holmes, A.J., Kovacs, K.L. and Murrell, J.C. (1997) Analysis of 16S rRNA and methane monooxygenase gene sequences reveals a novel group of thermotolerant and thermophilic methanotrophs. *Methylocaldum gen. nov*. *Archives of Microbiology* 168, 493–503.

Boeckx, P. and Van Cleemput, O. (1996) Methane oxidation in a neutral landfill cover soil: influence of moisture content, temperature, and nitrogen-turnover. *Journal of Environmental Quality* 25, 178–183.

Boeckx, P., Van Cleemput, O. and Villaralvo, I. (1996) Methane emission from a landfill and the methane oxidising capacity of its covering soil. *Soil Biology and Biochemistry* 28, 1397–1405.

Boeckx, P., Van Cleemput, O. and Meyer, T. (1998) The influence of land use and pesticides on methane oxidation in some Belgian soils. *Biology and Fertility of Soils* 27, 293–298.

Bogner, J. and Matthews, E. (2003) Global methane emissions from landfills: new methodology and annual estimates 1980–1996. *Global Biogeochemical Cycles* 17(2); doi:10.1029/2002GB001913.

Bogner, J.E., Spokas, K.A. and Burton, E.A. (1997) Kinetics of methane oxidation in a landfill cover soil: temporal variations, a whole-landfill oxidation experiment, and modeling of net CH_4 emissions. *Environmental Science and Technology* 31, 2504–2514.

Börjesson, G. (2001) Inhibition of methane oxidation by volatile sulfur compounds (CH_3SH and CS_2) in landfill cover soils. *Waste Management and Research* 19, 314–319.

Börjesson, G., Chanton, J. and Svensson, B.H. (2001) Methane oxidation in two Swedish landfill covers measured with carbon-13 to carbon-12 isotope ratios. *Journal of Environmental Quality* 30, 369–376.

Börjesson, G., Sundh, I. and Svensson, B. (2004) Microbial oxidation of CH_4 at different temperatures in landfill cover soils. *FEMS Microbiology Ecology* 48, 305–312.

Bowman, J.P., Sly, L.I., Nichols, P.D. and Haywerd, A.C. (1993) Revised taxonomy of the methanotrophs: description of *Methylobacter gen. nov*., emendation of *Methylococcus*, validation of *Methylosinus* and *Methylocystis* species, and a proposal that the family *Methylococcaceae* includes only the group I methanotrophs. *International Journal of Systematic Bacteriology* 43, 735–753.

Bowman, J.P., Sly, L.I., Nichols, P.D. and Haywerd, A.C. (1994) Revised taxonomy of the methanotrophs: description of *Methylobacter gen. nov*., emendation of *Methylococcus*, validation of *Methylosinus* and *Methylocystis* species, and a proposal that the family *Methylococcaceae* includes only the group I methanotrophs. Errata. *International Journal of Systematic Bacteriology* 44, 375.

Bowman, J.P., Sly, L.I. and Stackebrandt, E. (1995) The phylogenetic position of the family *Methylococcaceae*. *International Journal of Systematic Bacteriology* 45, 182–185.

Bowman, J.P., McCammon, S.A. and Skerratt, J.H. (1997) *Methylosphaera hansonii* gen. nov., sp. nov., a psychrophilic, group I methanotroph from Antarctic marine-salinity, meromictic lakes. *Microbiology* 143, 1451–1459.

Carlsen, H.N., Joergensen, L. and Degn, H. (1991) Inhibition by ammonia of methane utilization in *Methylococcus capsulatus* (Bath). *Applied Microbiology and Biotechnology* 35, 124–127.

Chanton, J. and Liptay, K. (2000) Seasonal variation in methane oxidation in a landfill cover soil as determined by an in situ stable isotope technique. *Global Biogeochemical Cycles* 14, 51–60.

Chanton, J.P., Rutkowski, C.M. and Mosher, B. (1999) Quantifying methane oxidation from landfills using stable isotope analysis of downwind plumes. *Environmental Science and Technology* 33, 3755–3760.

Chiemchaisri, W., Wu, J.S. and Visvanathan, C. (2001) Methanotrophic production of extracellular polysaccharide in landfill cover soils. *Water Science and Technology* 43(6), 151–158.

Christophersen, M., Linderød, L., Jensen, P.E. and Kjeldsen, P. (2000) Methane oxidation at low temperatures in soil exposed to landfill gas. *Journal of Environmental Quality* 29, 1989–1997.

Coleman, D.D., Risatti, J.B. and Schoell, M. (1981) Fractionation of carbon and hydrogen isotopes by methane-oxidizing bacteria. *Geochimica et Cosmochimica Acta* 45, 1033–1037.

Czepiel, P.M., Mosher, B., Crill, P.M. and Harriss, R.C. (1996a) Quantifying the effect of oxidation on landfill methane emissions. *Journal of Geophysical Research* 101D, 16721–16729.

Czepiel, P.M., Mosher, B., Harriss, R.C., Shorter, J.H., McManus, J.B., Kolb, C.E., Allwine, E. and Lamb, B.K. (1996b) Landfill methane emissions measured by enclosure and atmospheric tracer methods. *Journal of Geophysical Research* 101D, 16711–16719.

Dedysh, S.N., Liesack, W., Khmelenina, V.N., Suzina, N.E., Trotsenko, Y.A., Semrau, J.D., Bares, A.M., Panikov, N.S. and Tiedje, J.M. (2000) *Methylocella palustris* gen. nov., sp. nov., a new methane-oxidizing acidophilic bacterium from peat bogs, representing a novel subtype of serine-pathway methanotrophs. *International Journal of Systematic and Evolutionary Microbiology* 50, 955–969.

Dedysh, S.N., Khmelenina, V.N., Suzina, N.E., Trotsenko, Y.A., Semrau, J.D., Liesack, W. and Tiedje, J.M. (2002) *Methylocapsa acidiphila* gen. nov., sp. nov., a novel methane-oxidizing and dinitrogen-fixing acidophilic bacterium from *Sphagnum* bog. *International Journal of Systematic and Evolutionary Microbiology* 52, 251–261.

Dedysh, S.N., Berestovskaya, Y.Y., Vasylieva, L.V., Belova, S.E., Khmelenina, V.N., Suzina, N.E., Trotsenko, Y.A., Liesack, W. and Zavarzin, G.A. (2004) *Methylocella tundrae* sp. nov., a novel methanotrophic bacterium from acidic tundra peatlands. *International Journal of Systematic and Evolutionary Microbiology* 54, 151–156.

De Visscher, A. (2001) Modelling of diffusion and reaction of CH_4 and N_2O in soils. PhD thesis, Ghent University, Belgium.

De Visscher, A. and Van Cleemput, O. (2000) Modelling moisture and temperature effects on methane oxidation in soils. In: Van Ham J. *et al.* (eds) *Non-CO_2 Greenhouse Gases: Scientific Understanding, Control and Implementation*, Kluwer Academic Publishers, Dordrecht, The Netherlands, pp. 137–138.

De Visscher, A. and Van Cleemput, O. (2003a) Induction of enhanced methane oxidation in soils: ammonium inhibition patterns. *Soil Biology and Biochemistry* 35, 907–913.

De Visscher, A. and Van Cleemput, O. (2003b) Simulation model for gas diffusion and methane oxidation in landfill cover soils. *Waste Management* 23, 581–591.

De Visscher, A., Thomas, D., Boeckx, P. and Van Cleemput, O. (1999) Methane oxidation in simulated landfill cover soil environments. *Environmental Science and Technology* 33, 1854–1859.

De Visscher, A., Schippers, M. and Van Cleemput, O. (2001) Short-term kinetic response of enhanced methane oxidation in landfill cover soils to environmental factors. *Biology and Fertility of Soils* 33, 231–237.

De Visscher, A., De Pourcq, I. and Chanton, J. (2004) Isotope fractionation effects by diffusion and methane oxidation in landfill cover soils. *Journal of Geophysical Research* 109, D18111; doi:10.1029/2004JD004857.

Dunfield, P.F., Khmelenina, V.N., Suzina, N.E., Trotsenko, Y.A. and Dedysh, S.N. (2003) *Methylocella silvestris* sp. nov., a novel methanotroph isolated from an acidic forest cambisol. *International Journal of Systematic and Evolutionary Microbiology* 53, 1231–1239.

Froment, G.F. and Bischoff, K.B. (1990) *Chemical Reactor Analysis and Design*. Wiley, New York.

Gebert, J., Groengroeft, A. and Miehlich, G. (2003) Kinetics of microbial methane oxidation in biofilters. *Waste Management* 23, 609–619.

Gebert, J., Groengroeft, A., Schloter, M. and Gattinger, A. (2004) Community structure in a methanotroph biofilter as revealed by phospholipid fatty acid analysis. *FEMS Microbiology Letters* 240, 61–68.

Giani, L., Bredenkamp, J. and Eden, I. (2002) Temporal and spatial variability of the CH_4 dynamics of landfill cover soils. *Journal of Plant Nutrition and Soil Science* 165, 205–210.

Gulledge, J. and Schimel, J.P. (1998) Low-concentration kinetics of atmospheric CH_4 oxidation in soil and mechanism of NH_4^+ inhibition. *Applied and Environmental Microbiology* 64, 4291–4298.

Hanson, R.S. and Hanson, T.E. (1996) Methanotrophic bacteria. *Microbiological Reviews* 60, 439–471.

Heyer, J., Berger, U., Hardt, M. and Dunfield, P.F. (2005) *Methylohalobius crimeensis* gen. nov., sp. nov., a moderately halophylic, methanotrophic bacterium isolated from hypersaline lakes of Crimea. *International Journal of Systematic and Evolutionary Microbiology* 55, 1817–1826.

Hilger, H. and Humer, M. (2003) Biotic landfill cover treatments for mitigating methane emissions. *Environmental Monitoring and Assessment* 84, 71–84.

Hilger, H.A., Liehr, S.K. and Barlaz, M.A. (1999) Exopolysaccharide control of methane oxidation in landfill cover soil. *Journal of Environmental Engineering* 125, 1113–1123.

Hilger, H.A., Cranford, D.F. and Barlaz, M.A. (2000a) Methane oxidation and microbial exopolymer production in landfill cover soil. *Soil Biology and Biochemistry* 32, 457–467.

Hilger, H.A., Wollum, A.G. and Barlaz, M.A. (2000b) Landfill methane oxidation response to vegetation, fertilization, and liming. *Journal of Environmental Quality* 29, 324–334.

Humer, M. and Lechner, P. (1999) Alternative approach to the elimination of greenhouse gases from old landfills. *Waste Management and Research* 17, 443–452.

Humer, M. and Lechner, P. (2001) Microbial methane oxidation for the reduction of landfill gas emissions. *Journal of Solid Waste Technology and Management* 27, 146–151.

Huq, M.N., Ralph, B.J. and Rickard, P.A.D. (1978) The extracellular polysaccharide of a methylotrophic culture. *Australian Journal of Biological Sciences* 31, 311–316.

Hütsch, B.W. (1998) Methane oxidation in arable soil as inhibited by ammonium, nitrite, and organic manure with respect to soil pH. *Biology and Fertility of Soils* 28, 27–35.

Hütsch, B.W., Russell, P. and Mengel, K. (1996) CH_4 oxidation in two temperate arable soils as affected by nitrate and ammonium application. *Biology and Fertility of Soils* 23, 86–92.

Joergensen, L. (1985) The methane mono-oxygenase reaction system studied in vivo by membrane-inlet mass spectrome try. *Biochemical Journal* 225, 441–448.

Jones, H.A. and Nedwell, D.B. (1993) Methane emission and methane oxidation in land-fill cover soil. *FEMS Microbiology Ecology* 102, 185–195.

King, G.M. and Adamsen, A.P.S. (1992) Effects of temperature on methane consumption in a forest soil and in pure cultures of the methanotroph *Methylomonas rubra*. *Applied and Environmental Microbiology* 58, 2758–2763.

King, S.L., Quay, P.D. and Lansdown, J.M. (1989) The $^{13}C/^{12}C$ kinetic isotope effect for soil oxidation of methane at ambient atmospheric concentrations. *Journal of Geophysical Research* 94, 18273–18277.

Kjeldsen, P. and Fischer, E.V. (1995) Landfill gas migration: field investigations at Skellingsked landfill, Denmark. *Waste Management and Research* 13, 467–484.

Kjeldsen, P., Dalager, A. and Broholm, K. (1997) Attenuation of methane and nonmethane organic compounds in landfill gas affected soils. *Journal of the Air and Waste Management Association* 47, 1268–1275.

Krüger, M., Frenzel, P. and Conrad, R. (2001) Microbial processes influencing methane emission from rice fields. *Global Change Biology* 7, 49–64.

Kruse, C.W., Moldrup, P. and Iversen, N. (1996) Modeling diffusion and reaction in soils: II. Atmospheric methane diffusion and consumption in a forest soil. *Soil Science* 161, 355–365.

Liptay, K., Chanton, J., Czepiel, P. and Mosher, B. (1998) Use of stable isotopes to determine methane oxidation in landfill cover soils. *Journal of Geophysical Research* 103, 8243–8250.

Mahieu, K., De Visscher, A., Vanrolleghem, P.A. and Van Cleemput, O. (2005) Improved quantification of methane oxidation in landfill cover soils by numerical modeling of stable isotope fractionation. In: *Proceedings of Sardinia 2005, Tenth International Waste Management and Landfill Symposium*. S. Margherita di Pula, Cagliari, Italy, 3–7 October. CISA, Environmental Sanitary Engineering Centre, Italy.

Millington, R.J. (1959) Gas diffusion in porous media. *Science* 130, 100–102.

Millington, R.J. and Shearer, R.C. (1971) Diffusion in aggregated porous media. *Soil Science* 111, 372–378.

Moldrup, P., Kruse, C.W., Rolston, D.E. and Yamaguchi, T. (1996) Modeling diffusion and reaction in soils: III. Predicting gas diffusivity from the Campbell soil-water retention model. *Soil Science* 161, 366–375.

Moldrup, P., Olesen, T., Gamst, J., Schjønning, P., Yamaguchi, T. and Rolston, D.E. (2000a) Predicting the gas diffusion coefficient in repacked soil: water-induced linear reduction model. *Soil Science Society of America Journal* 64, 1588–1594.

Moldrup, P., Olesen, T., Schjønning, P., Yamaguchi, T. and Rolston, D.E. (2000b) Predicting the gas diffusion coefficient in undisturbed soil from soil water characteristics. *Soil Science Society of America Journal* 64, 94–100.

Nozhevnikova, A., Nekrasova, V., Lebedev, V.S. and Lifshits, A.B. (1993) Microbiological processes in landfills. *Water Science and Technology* 27(2), 243–252.

Nozhevnikova, A., Glagolev, M., Nekrasova, V., Einola, J., Sormunen, K. and Rintala, J. (2003) The analysis of methods for measurement of methane oxidation in landfills. *Water Science and Technology* 48(4), 45–52.

Oonk, J. and Boom, A. (1995) *Landfill Gas Formation, Recovery and Emissions*. TNO, Apeldoorn, The Netherlands.

Oonk, H., Hensen, A., Mahieu, K., De Visscher, A., van Velthoven, F. and Woelders, H. (2004) *Verbeterde Methaanoxidatie in Toplagen van Stortplaatsen*. TNO Report R 2004/377. TNO, Apeldoorn, The Netherlands.

Park, J.R., Moon, S. Ahn, Y.M., Kim, J.Y. and Nam, K. (2005) Determination of environmental factors influencing methane oxidation in a sandy landfill cover soil. *Environmental Technology* 26, 93–102.

Penman, H.L. (1940) Gas and vapour movements in the soil: I. The diffusion of vapours through porous solids. *Journal of Agricultural Science* 30, 437–462.

Perera, L.A.K., Achari, G. and Hettiaratchi, J.P.A. (2002a) Determination of source strength of landfill gas: a numerical modeling approach. *Journal of Environmental Engineering* 128, 461–471.

Perera, M.D.N., Hettiaratchi, J.P.A. and Achari, G. (2002b) A mathematical modeling approach to improve the point estimation of landfill gas surface emissions using the flux chamber technique. *Journal of Environmental Engineering and Science* 1, 451–463.

Perera, L.A.K., Achari, G. and Hettiaratchi, J.P.A. (2004) An assessment of the spatial variability of greenhouse gas emissions from landfills: a GIS based statistical–numerical approach. *Journal of Environmental Informatics* 4, 11–30.

Reay, D.S. and Nedwell, D.B. (2004) Methane oxidation in temperate soils: effects of inorganic N. *Soil Biology and Biochemistry* 36, 2059–2065.

Ren, T., Amaral, J.A. and Knowles, R. (1997) The response of methane consumption by pure cultures of methanotrophic bacteria to oxygen. *Canadian Journal of Microbiology* 43, 925–928.

Scharff, H., Martha, A., van Rijn, D.M.M., Hensen, A., van den Bulk, W.C.M., Flechard, C., Oonk, J., Vroon, R., De Visscher, A., and Boeckx, P. (2003) *A Comparison of Measurement Methods to Determine Landfill Methane Emissions*. Novem, Utrecht, The Netherlands.

Schnell, S. and King, G.M. (1995) Stability of methane oxidation capacity to variations in methane and nutrient concentrations. *FEMS Microbiology Ecology* 17, 285–294.

Sharpe, P.J.H. and DeMichele, D.W. (1977) Reaction kinetics in poikilotherm development. *Journal of Theoretical Biology* 64, 649–670.

Stein, V.B. and Hettiaratchi, J.P.A. (2001) Methane oxidation in three Alberta soils: influence of soil parameters and methane flux rates. *Journal of Environmental Technology* 22, 1–11.

Stein, V.B., Hettiaratchi, J.P.A. and Achari, G. (2001) A numerical model for biological oxidation and migration of methane in soils. *ASCE Practice Periodical of Hazardous, Toxic, and Radioactive Waste Management* 5, 225–234.

Stralis-Pavese, N., Sessitsch, A., Weilharter, A., Reichenauer, T., Riesing, J., Csontos, J., Murrell, J.C. and Bodrossy, L. (2004) Optimization of diagnostic microarray for application in analysing landfill methanotroph communities under different plant covers. *Environmental Microbiology* 6, 347–363.

Streese, J. and Stegmann, R. (2003) Microbial oxidation of methane from old landfills in biofilters. *Waste Management* 23, 573–580.

Striegl, R.G., McConnaughey, T.A., Thorstenson, D.C., Weeks, E.P. and Woodward, J.C. (1992) Consumption of atmospheric methane by desert soils. *Nature* 357, 145–147.

Top, E.M., Maila, M.P., Clerinx, M., Goris, J., De Vos, P. and Verstraete, W. (1999) Methane oxidation as a method to evaluate the removal of 2,4-dichlorophenoxyacetic acid (2,4-D) from soil by plasmid-mediated bioaugmentation. *FEMS Microbiology Ecology* 28, 203–213.

Tsubota, J., Eshinimaev, B.T., Khmelenina, V.N. and Trotsenko, Y.A. (2005) *Methylothermus thermalis*, gen. nov., sp. nov., a novel moderately thermophilic obligate methanotroph from a hot spring in Japan. *International Journal of Systematic and Evolutionary Microbiology* 55, 1877–1884.

Wang, Z.P. and Ineson, P. (2003) Methane oxidation in a temperate coniferous forest soil: effects of inorganic N. *Soil Biology and Biochemistry* 35, 427–433.

Whalen, S.C. and Reeburgh, W.S. (1996) Moisture and temperature sensitivity of CH_4 oxidation in boreal soils. *Soil Biology and Biochemistry* 28, 1271–1281.

Whalen, S.C., Reeburgh, W.S. and Sandbeck, K.A. (1990) Rapid methane oxidation in a landfill cover soil. *Applied and Environmental Microbiology* 56, 3405–3411.

Whittenbury, R. and Dalton, H. (1981) The methylotrophic bacteria. In: Starr, M.P., Stolph, H., Trüper, H.G., Balows, A. and Schlegel, H.G. (eds) *The Prokaryotes*, Vol. I. Springer, New York, pp. 894–902.

Whittenbury, R., Phillips, K.C. and Wilkinson, J.C. (1970) Enrichment, isolation and some properties of methane utilizing bacteria. *Journal of General Microbiology* 61, 205–218.

Wilshusen, J.H., Hettiaratchi, J.P.A., De Visscher, A. and Saint-Fort, R. (2004a) Methane oxidation and EPS formation in compost: effect of oxygen concentration. *Environmental Pollution* 23, 305–314.

Wilshusen, J.H., Hettiaratchi, J.P.A. and Stein, V.B. (2004b) Long-term behavior of passively aerated compost methanotrophic biofilter columns. *Waste Management* 24, 643–653.

Wise, M.G., McArthur, J.V. and Shimkets, L.J. (1999) Methanotrophic diversity in landfill soil: isolation of novel type I and type II methanotrophs whose presence was suggested by culture-independent 16S ribosomal DNA analysis. *Applied and Environmental Microbiology* 65, 4887–4897.

Wise, M.G., McArthur, J.V. and Shimkets, L.J. (2001) *Methylosarcina fibrata* gen. nov., sp. nov. and *Methylosarcina quisquiliarum* sp. nov., novel type I methanotrophs. *International Journal of Systematic and Evolutionary Microbiology* 51, 611–621.

Wrangstadh, M., Conway, P.L. and Kjellberg, S. (1986) The production and release of an extracellular polysaccharide during starvation of a marine *Pseudomonas* sp. and the effect thereof on adhesion. *Archives of Microbiology* 145, 220–227.

13 Nitrous Oxide: Importance, Sources and Sinks

David S. Reay[1], C. Nick Hewitt[2] and Keith A. Smith[1]
[1]School of GeoSciences, University of Edinburgh, Edinburgh UK; [2]Lancaster Environment Centre, University of Lancaster, Lancaster, UK

13.1 Introduction

Nitrous oxide (N_2O) is commonly known as laughing gas and is widely used as an anaesthetic in the field of medicine. It has a slightly sweet odour and was first described by Joseph Priestley in 1772. It is a relatively inert gas and, like the other greenhouse gases (GHGs) – carbon dioxide (CO_2) and methane (CH_4) – its mixing ratio in the atmosphere has risen markedly in the last 200 years, increasing from ~0.270 ppm in the pre-industrial era to 0.315 ppm currently, with the rate of increase estimated to be ~0.25% per year. Although its concentration is very small relative to that of CO_2, N_2O has two attributes that make it a potent GHG: (i) it is chemically rather inert, with a lifetime of ~120 years in the atmosphere; and (ii) each molecule of N_2O has a much greater radiative forcing potential than a molecule of CO_2.

Together, these attributes mean that N_2O has a global warming potential (GWP) of 296 on a 100-year time horizon (see Chapter 5), relative to CO_2 on a mass basis (i.e. 1 kg of N_2O is equivalent to 296 kg of CO_2). Thus it has contributed a significant amount to the enhanced radiative forcing from GHGs since pre-industrial times – ~6% of the total. A doubling in its concentration would lead to a global temperature rise of ~0.7°C.

Before considering the major sinks for N_2O let us first examine where the N_2O that enters our atmosphere comes from. Global emissions total ~18 Tg N_2O-N/year. Of this, ~10 Tg arises from 'natural sources'.

13.2 'Natural' Nitrous Oxide Sources

As can be seen in Fig. 13.1, temperate and tropical soils dominate natural N_2O emissions on a global scale, although emissions from the world's oceans are also important. However, for all these 'natural' N_2O sources it is difficult to separate the impacts of human activities. In particular, our release of reactive nitrogen in the form of leached fertilizers, ammonia emissions from livestock and NO_x from fossil fuel combustion can be important drivers of 'natural' N_2O emissions.

13.2.1 The oceanic source

Globally, oceans add ~3 Tg N_2O/year to the atmosphere. Like CH_4, much of the N_2O in the surface waters of the oceans arises from microbial activity in and around sinking particles, such as faecal pellets. These particles provide the anaerobic conditions necessary for denitrification. Denitrification is a predominantly

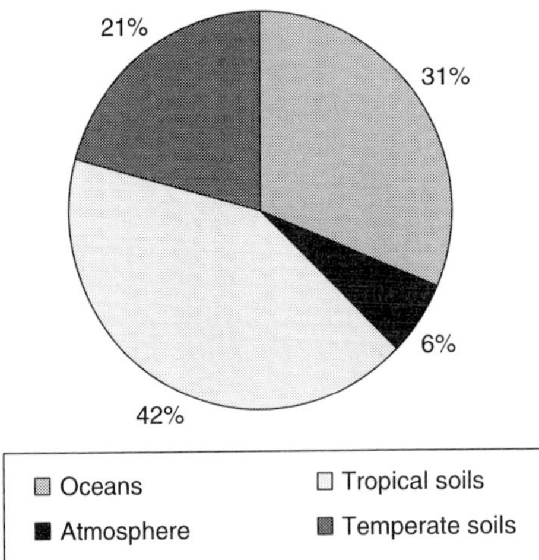

Fig. 13.1. Natural sources of nitrous oxide (N_2O).

microbially mediated process where bacteria use nitrate, rather than oxygen, as a terminal electron acceptor to respire organic matter and hence gain energy. In doing so, they reduce nitrate to nitrogen gas (N_2). N_2O is one of the intermediate steps in the process and, where the reduction of nitrate to N_2 is incomplete (as is invariably the case), significant production of N_2O can result.

In some areas of the ocean, surface waters can become oxygen-depleted allowing active denitrification and N_2O production in open water. Large amounts of oceanic N_2O can also arise from denitrification in marine sediments, particularly in nutrient-rich areas such as those of estuaries. Some N_2O is also produced as a by-product of nitrification in the world's oceans. The rate of emission of this dissolved N_2O from ocean waters is dependent on its outgassing from the surface waters to the atmosphere, which can be very rapid when wind speeds are high and the surface of the ocean is well mixed.

As with CH_4, there can be significant human impacts on 'natural' oceanic N_2O emissions. These lie primarily in our effect on oceanic nutrient inputs through rivers and estuaries. The nitrogen-rich waters of many rivers, produced by sewage input and agricultural runoff, can lead to eutrophic conditions in estuaries and coastal waters. Such nutrient-rich waters and sediments are ideal for denitrification and hence N_2O production, with oxygen levels in the water often being very low, and with plenty of nitrogen and organic carbon on which the denitrifying bacteria grow. Increased atmospheric deposition of reactive nitrogen to the surface of the world's oceans, again due to the activities of humans, may also result in increased oceanic N_2O emissions.

Elevated nitrogen concentrations in estuaries and coastal waters, through sewage and agricultural runoff, have been the subject of increasing concern in recent years. Steps have been taken in some countries to reduce key nitrogenous pollutants, such as nitrate fertilizers, through better land management practices. Similarly, efficient sewage treatment has also led to reductions in levels of coastal eutrophication in some areas. However, in many parts of the world such coastal nitrogen pollution continues to increase in line with a growing human population and the need for an ever more intensive agriculture.

13.2.2 The atmospheric source

The atmosphere acts as a source for N_2O through the oxidation of ammonia (NH_3). In total, atmospheric NH_3 oxidation is thought to be responsible for ~0.6 Tg N_2O/year. Again the line between natural and anthropogenic sources is blurred. The primary sources of atmospheric NH_3 are human-made, with the largest increases in emissions in recent decades being due to increased global livestock farming. Ammonia is emitted from both solid and liquid livestock waste through volatilization and can also induce N_2O production in soils by deposition after its initial release. Plumes of ammonia can often be detected arising from intensive livestock-rearing facilities, such as poultry and pig farms. Fertilizer application and agricultural chemical use are also significant sources of atmospheric ammonia, as is fossil fuel-powered transport.

13.2.3 The 'natural' soil sources

Soils are significant natural sources of N_2O on a global scale, with both tropical and temperate soils being important. Tropical soils are estimated to add ~4 Tg N_2O/year to our atmosphere. Of this, ~3 Tg comes from wet forest soils, with the rest being emitted from the soils of dry savannahs.

Temperate soils are estimated to add ~2 Tg N_2O to the atmosphere, with about half being emitted by forest soils, and the other half by the soils of temperate grasslands.

Tropical and temperate soils generally have different ratios of nutrients, with tropical soils often being phosphorus-limited, rather than nitrogen-limited (like many temperate soils). Because of this, extra nitrogen inputs to these phosphorus-limited tropical soils may cause N_2O emissions hundreds of times greater than that seen in nitrogen-limited temperate soils. N_2O arises from soils primarily via the two biological pathways of nitrification and denitrification.

Nitrification in soils is carried out by aerobic, ammonia-oxidizing bacteria (AOB), which produce nitrate from ammonium in the soil, but can also produce some N_2O during this process. As the nitrification process relies on good availability of oxygen, it is most important in well-drained and aerated soils. These AOB have also been shown to oxidize certain amounts of the GHG CH_4 as part of the nitrification process, though whether they have a significant impact on CH_4 emissions from soil is still open to debate.

In the wetter or more compact soils common to many temperate soils, the anaerobic conditions suitable for denitrification become more prevalent. As for oceanic N_2O emissions, denitrification involves the reduction of nitrate to gaseous nitrogen (N_2) by anaerobic bacteria. Again, N_2O can be produced during this process and, generally, denitrification produces more N_2O than nitrification. During soil denitrification, as in the oceans, the N_2O that is produced can be further reduced to N_2, but usually a proportion escapes to the atmosphere. Soil conditions, such as water content, temperature and the availability of nitrogen are key determinants of how much N_2O a particular soil will produce.

Increased atmospheric nitrogen deposition following NO_x and ammonia emissions can induce elevated rates of N_2O emission over large areas of otherwise pristine soil. Rates of N_2O from such soils at all latitudes are also likely to change in response to variations in temperature and rainfall resulting from global climate change. Ensuring that fertilizers do not end up on natural soils, whether directly or indirectly, makes sense both environmentally and economically.

13.3 Anthropogenic N_2O Sources

Global anthropogenic N_2O-N emissions total ~8 Tg/year (Fig. 13.2). Agricultural soils and livestock dominate these emissions through a combination of both direct and indirect pathways.

13.3.1 Agricultural soils

Direct sources

Agricultural soils represent a very large, and growing, global source of N_2O to the atmosphere. Current estimates for annual emissions from this source range from ~2 to ~4 Tg N_2O globally. With a rapid increase

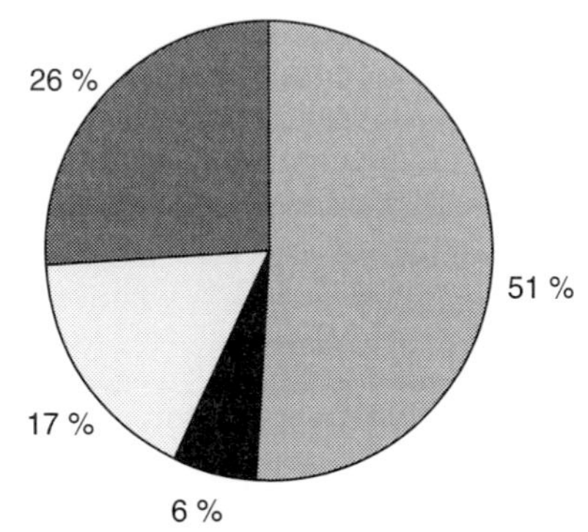

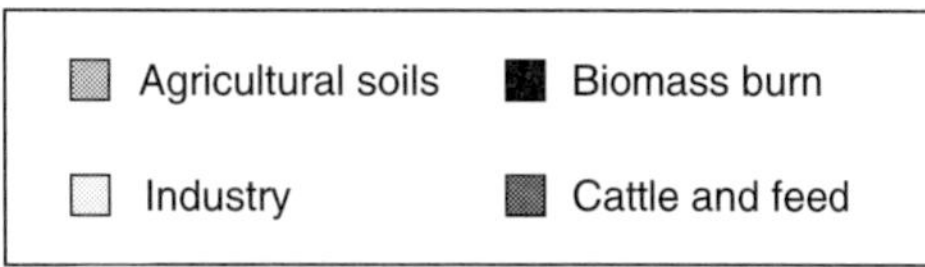

Fig. 13.2. Anthropogenic sources of nitrous oxide (N_2O).

in population growth, and the consequent need for more food production, both the area of agricultural soils and the intensity of their use are likely to continue to rise rapidly in the coming decades.

A major direct source of N_2O from agricultural soils is synthetic fertilizer use. Widespread increase in the use of such nitrogen-based fertilizers has been driven by the need for greater crop yields, and by more intensive farming practices. Where applications of nitrogen-based fertilizers are combined with soil conditions favourable to denitrification, large amounts of N_2O can be produced.

Similarly, the widespread and often poorly controlled use of animal waste as fertilizer can lead to substantial emissions of N_2O from agricultural soils. Some additional N_2O arises in agricultural soils through the introduction of nitrogen by biological fixation of atmospheric N_2 by leguminous crops (and symbiotic associations in rice fields).

The application of nitrogen-based fertilizers and manures in many areas has been excessive, with large proportions of the added fertilizer providing no benefit to crop yield, but inducing increased leaching and elevated N_2O emissions. Better targeting of fertilizer applications, both in space and time, can significantly reduce N_2O emissions from agricultural soils. Land management strategies that accurately take account of the optimum amounts of fertilizer addition necessary for maximum crop yield and minimum waste are therefore crucial.

Indirect

Indirect agricultural sources of N_2O remain poorly defined in most cases. There are several ways in which such indirect emissions occur. The most important of these is thought to be the N_2O arising from nitrogen leaching and runoff from agricultural soils.

After fertilizer application or heavy rain, large amounts of nitrogen may leach from the soil into drainage ditches, streams, rivers and eventually estuaries. Some of the N_2O produced in agricultural soils is also lost in this way, as it is emitted to the atmosphere as soon as the drainage water is exposed to the air. Still more N_2O is produced from such polluted drainage waters when the leached nitrogen undergoes the processes of nitrification or denitrification in aquatic and estuarine sediments. Other important indirect N_2O sources from agricultural soils include the volatilization of NH_3 and subsequent deposition of this nitrogen on other soils. Increased food consumption and consequent increases in municipal sewage treatment – also rich in nitrogen – have led to increased indirect N_2O emissions.

Again, it is through properly informed land management practice and fertilization campaigns that N_2O emissions from agriculture can be reduced. Much of the impetus for control of nitrogen-based fertilizers has come from concern over high nitrate levels in drinking water supplies and the threat of eutrophication in estuaries and coastal waters. Individual governments have enacted changes in policy to bring about reductions in nitrogen leaching, with areas designated as 'nitrate-vulnerable zones' (NVZs) requiring particular attention in the UK.

13.3.2 Biomass burning

Biomass burning, largely resulting from the activities of humans, accounts for ~0.5 Tg N_2O emissions each year. N_2O emissions arising from biomass burning are a result of incomplete combustion, and significant amounts can be produced during large-scale burning of forests, savannah and grasslands.

An important source of N_2O emission related to biomass burning is crop residue burning. The high nitrogen and water contents of many crop residues mean that the burning of such material can produce a relatively high percentage (around 1%) of the total emissions in the form of N_2O.

Wood burning is used for domestic heating and cooking purposes and for charcoal production, and can also release significant amounts of N_2O.

13.3.3 Industrial sources

Together, industrial sources are thought to emit ~1.3 Tg N_2O/year to the atmosphere. Of these, nitric acid production is one of the most important. Nitric acid is a key ingredient in nitrogen-based fertilizers. Nitric acid production involves the oxidation of ammonia (NH_3) using a platinum catalyst. N_2O emissions arise from this process during the catalytic oxidation and it is estimated that ~5 g of N_2O is produced in this way for every kilogram of nitric acid manufactured. Much recent progress has been made in reducing N_2O emissions from this process by using more efficient NH_3 oxidation conditions – lower reaction temperatures preventing large amounts of the NH_3 ending up as N_2O.

Nylon manufacture is responsible for N_2O emission through its requirement for adipic acid (also called hexanedioic acid). Adipic acid is a fine powder used in the manufacture of nylon fibres, as well as for some plastics, clothing, carpets and tyres. It is also used in the production of dyes and insecticides. N_2O arises from adipic acid production during the oxidation of a ketone–alcohol mixture with nitric acid. It is estimated that for each kilogram of adipic acid made, ~30 g of N_2O is also produced. However, big improvements in the removal of N_2O before its emission to the atmosphere, such as by thermal decomposition, have achieved cuts in emissions from this source of more than 90%.

Burning of fossil fuels, particularly coal, in electricity generation can lead to substantial amounts of N_2O production. Emissions arise primarily from the reaction of nitric oxide during combustion, though the high temperatures involved often mean that a large proportion of the N_2O produced in this way is destroyed before it can escape to the atmosphere. Nevertheless, increasing global energy demands have led to increased emissions from this source in recent decades, and the widespread use of coal as power station fuel means that N_2O emissions from this source should not be overlooked.

13.3.4 Livestock

Globally, livestock-related N_2O emissions are estimated to total between 1 and 2 Tg N/year. Livestock produce only relatively minor amounts of N_2O directly. Although some N_2O can be produced in ruminant guts via the reduction of nitrate in livestock feed to NH_3 and NH_4; the very anoxic conditions in the gut mean that actual N_2O production is likely to be very low.

Areas where livestock do have a large impact on N_2O emissions are through initial production of their feed and through subsequent management of their waste. Livestock feed production, like human food production, often involves large applications of nitrogen-based fertilizers to agricultural soils. N_2O production from animal wastes occurs during both storage and treatment, again by the processes of nitrification and denitrification. Additionally, N_2O is indirectly produced via the volatilization of ammonia as discussed earlier.

13.3.5 Transport

Current estimates range from 0.11 to 0.24 Tg of transport-related N_2O emission per year, though with rapidly increasing global car use, N_2O emissions from this source may rise in the future. Additional N_2O emissions arise from other forms of fossil fuel-powered transport such as planes, boats and lorries.

The increased use of catalytic convertors in cars around the world has led to concerns about the resulting increase in transport-related N_2O emissions. However, although catalytic convertors in cars can lead to an increase in N_2O emissions of ~20 times that of comparable vehicles without catalytic convertors, global emissions from this source are thought to be relatively small.

Indeed, advances in catalytic convertor technology have led to significant reductions in such N_2O emissions in recent years.

13.4 The Sinks

13.4.1 The stratosphere

The stratosphere is the layer of our atmosphere that is between ~10 and 50 km above the Earth's surface. Butenhoff and Khalil (Chapter 14, this volume) describe how most N_2O is removed from the atmosphere by direct photolysis by sunlight in the stratosphere, with ~10% being removed by chemical reaction with the excited oxygen atom $O(^1D)$. This latter reaction is also of significance since it is a major source of N_2O to the stratosphere that plays a significant role in the catalytic removal of stratospheric ozone. The photolysis of N_2O peaks at wavelengths between 195 and 205 nm and at altitudes between 25 and 35 km. Together, photolysis and the $O(^1D)$ reaction remove ~12–13 Tg N_2O/year from the atmosphere.

The isotopic composition of N_2O in the stratosphere can be used to constrain estimates of N_2O source fluxes. Photolysis preferentially removes the light molecule $^{14}N^{14}N^{16}O$ from the stratosphere, leaving the remaining N_2O enriched in the heavy isotopically substituted species $^{14}N^{15}N^{16}O$, $^{15}N^{14}N^{16}O$, $^{14}N^{14}N^{17}O$ and $^{14}N^{14}N^{18}O$. The enrichment of these species increases through the stratosphere as the air ages. Back-flux of mass from the lower stratosphere to the troposphere enriches the isotopic composition of tropospheric N_2O, which balances the flux of light N_2O from surface sources. Successful efforts at modelling the fractionation processes of N_2O reveal that the dynamical mixing of tropospheric air into the lower stratosphere decreases the fractionation constants in the lower stratosphere, though dependencies of the fractionation on temperature and wavelength also contribute to a certain extent.

Although some outstanding questions remain about proposed exotic sinks of N_2O, as well as the origin of the oxygen isotope anomaly, our understanding of the stratospheric sinks of N_2O is essentially complete. Detailed examination of the rates of these removal processes is important since, coupled with observations of the rates of change in atmospheric concentration of the gas, it allows the rates of emission of N_2O from the Earth's surface to be constrained, and hence future concentrations and the resultant radiative forcing to be estimated.

13.4.2 Soils and surface waters

Soils, as we have seen, are important sources of N_2O, but they can also act as a sink for atmospheric N_2O. Soil uptake of N_2O is driven by denitrification by bacteria, which convert N_2O into nitrogen gas (N_2). In many global budgets this bacterial reduction of N_2O is not explicitly accounted for, but simply included in the total net flux of N_2O from soils. This may, however, lead to errors in estimates of atmospheric budgets and trends in emissions.

Kroeze *et al.* (Chapter 15, this volume) analyse the processes underlying bacterial reduction of N_2O to N_2, and discuss the likeliness of N_2O uptake in different systems. They focus not only on soils, but also on aquatic systems, including groundwater systems, riparian zones and surface waters. Here a given system is considered a sink when the net N_2O uptake occurs over a relatively large area and prolonged period of time. Their results indicate that soils, surface waters and riparian zones may be potential sinks for atmospheric N_2O, while groundwater systems are not likely to be major sinks. Whether or not a system acts as a sink depends on local circumstances. The most important factors affecting N_2O uptake by soils are nitrogen availability, soil wetness and temperature. In addition, soil drainage conditions and soil pH are important. Based on these factors, Kroeze *et al.* hypothesize that areas prone to high N_2O sink activity in soils may be located in northern regions of the Earth.

14 Stratospheric Sinks of Nitrous Oxide

Christopher L. Butenhoff and M. Aslam K. Khalil
Department of Physics, Portland State University, Portland, Oregon, USA

14.1 Introduction

The importance of nitrous oxide (N_2O) in atmospheric chemistry is well known. Not only is it a potent greenhouse gas, being some 200 times more effective than carbon dioxide (CO_2) on a per molecule basis at absorbing infrared radiation (Ramaswamy, 2001), but its reaction with $O(^1D)$ in the stratosphere is the major source of the ozone-destroying catalyst NO. Its concentration in the troposphere has increased by 9.4% since pre-industrial times and continues to increase by 0.6 ppbv/year (0.2% per year) (Khalil *et al.*, 2002). It currently contributes ~6% to the total enhanced radiative forcing from greenhouse gases relative to pre-industrial times, and a doubling of its concentration would increase the global temperature by ~0.7°C (Ramaswamy, 2001). N_2O is produced primarily through the biological processes of nitrification and denitrification in soils and the ocean. The reason for its increase is the additional release of N_2O from agricultural activities, most notably the use of nitrogen fertilizers. As agriculture in the developing countries transitions from organic to synthetic nitrogen fertilizers, it is likely that the level of N_2O in the atmosphere will continue to rise. As such it is important to understand the life history of this gas for an effective management policy.

The individual flux strengths of N_2O sources are poorly constrained and contribute the largest uncertainties to N_2O's global budget. Bottom-up estimates of N_2O's global source range from ~7 to 37 Tg N/year (Prather and Ehhalt, 2001) with a likely value of ~18 Tg N/year (Kroeze *et al.*, 1999).

The global source can be constrained by using a simple one-box atmospheric model:

$$\frac{dB}{dt} = S - L \qquad (14.1)$$

where dB/dt is the rate of change of the atmospheric N_2O burden, and S and L are the respective global source and loss rates. If we use the recommended Intergovernmental Panel on Climate Change (IPCC) values of L = 12.6 Tg N/year and dB/dt = 3.8 Tg N/year (Prather and Ehhalt, 2001), the estimated source strength is 16.4 Tg N/year. Although this method can put a constraint on the aggregate source total, it does not provide information on fluxes for N_2O's individual sources. However, advances in isotopic analysis on both theoretical and instrumental grounds have opened new pathways to a better understanding of trace gas budgets. Since the various production and removal processes isotopically fractionate N_2O differently, they leave isotopic signatures in the troposphere and stratosphere, which can be used to put constraints on the source and sink fluxes.

After an initial discussion on the bulk removal processes of N_2O in the stratosphere, we continue with a review of recent work on the measurement and analysis of its isotopic composition in the atmosphere. In this chapter, we restrict our discussion to the stratosphere, where the isotopic fractionation is limited to the sink processes of photolysis and reaction with $O(^1D)$, with perhaps some minor contributions from exotic processes. The stratosphere is a supplier of heavy N_2O isotopes to the troposphere; this, combined with the surface emissions of relatively light isotopes from N_2O's major sources, determines the isotopic composition of the troposphere.

14.2 Bulk Removal of Nitrous Oxide in the Stratosphere

N_2O is inert and well mixed in the troposphere but decreases rapidly with height in the stratosphere. Its mixing ratio varies little throughout the troposphere with a pole-to-pole gradient of only ~1.2 ppbv, less than a 0.5% change from its average global surface value of 315 ppbv (Khalil *et al.*, 2002). Any significant departure from the mean indicates the presence of a local source. N_2O's long lifetime allows it to enter the stratosphere, primarily through convection at tropical latitudes, where it is transported through the strong upwelling currents. Once in the stratosphere, it is mixed vertically and horizontally, and is eventually removed through photolysis:

$$N_2O + h\nu \rightarrow N_2 + O(^1D) \qquad (14.2)$$

and reaction with excited oxygen atoms

$$N_2O + O(^1D) \rightarrow 2NO \qquad (14.3a)$$

$$\rightarrow N_2 + O_2 \qquad (14.3b)$$

Although the relative proportions vary throughout the stratosphere, Eq. 14.2 is responsible for ~90% of the total stratospheric loss, while Eq. 14.3 accounts for the remaining 10% (Garcia and Solomon, 1994). The limited vertical exchange between the troposphere and stratosphere reduces the effectiveness of these processes at removing N_2O (Bates and Hays, 1967).

14.2.1 Reaction with $O(^1D)$

Recommended rate constants for Eqs 14.3a and 14.3b are 6.7×10^{-11} and 4.9×10^{-11} cm^3/molecule/s respectively, and produce a branching ratio ($R = k_{3a}/k_3$) of 0.57 (Sander *et al.*, 2003). This ratio is slightly lower than the average branching ratio of 0.61 ± 0.06 calculated from relevant measurements since 1957 (Cantrell *et al.*, 1994).

As Eq. 14.3a is a major source of odd NO_x species to the stratosphere, there is concern that increased concentrations of N_2O will contribute to the decline of stratospheric ozone (Crützen, 1970). The production of nitric oxide (NO) from Eq. 14.3a peaks near the equator at an altitude of ~26 km (Johnston *et al.*, 1979) where it reaches a maximum value of 320 molecules/cm^3/s (Cantrell *et al.*, 1994). Production is symmetric around the equator in both hemispheres during spring and fall but is biased towards the summer hemisphere during winter and summer. The global production of NO is estimated to be 1.25 Tg N/year (Cantrell *et al.*, 1994). As ozone photolysis is the main source of $O(^1D)$, the loss of N_2O by $O(^1D)$ diminishes with increasing altitude in the stratosphere.

14.2.2 Photolysis

Photolysis by ultraviolet (UV) radiation is the largest sink of N_2O and removes ~90% of all N_2O that is lost from the atmosphere. This occurs primarily in the stratosphere. The reaction photodissociates the molecule, producing ground-state N_2 and excited oxygen $O(^1D)$. A minor pathway (<1%) also produces $N(^4S)$ and $NO(^2\Pi)$ (Greenblatt and Ravishankara, 1990). Although $O(^1D)$ can combine with oxygen to form ozone, the net impact of N_2O in the stratosphere is the destruction of ozone due to the production of NO from the reaction of N_2O with $O(^1D)$.

The rate of photolysis varies both spatially and temporally throughout the stratosphere and is given by:

$$J = [N_2O]\int \sigma(\lambda)I(\lambda)d\lambda \qquad (14.4)$$

where $\sigma(\lambda)$ is the absorption cross section at wavelength λ, $I(\lambda)$ is the spectrum of the photolysing radiation or actinic flux and $[N_2O]$ is the local concentration of N_2O. The actinic flux is a measure of the direct, scattered and reflected radiation, and varies significantly with time and location. This produces considerable variation in the photolysis rate. It is not surprising then that the N_2O photolysis rate peaks near the equator where the solar radiation is greatest. In the remainder of the section we briefly comment on the factors that determine the photolysis rate in Eq. 14.4.

Nitrous oxide absorption spectrum

The absorption spectrum of N_2O over relevant wavelengths in the stratosphere is shown in Fig. 14.1. The absorption between 174 and 320 nm peaks near 180 nm and falls to near zero at 260 nm (Johnston and Selwyn, 1975; Selwyn *et al.*, 1977). The spectrum is characterized by a featureless absorption continuum that underlies a series of structured vibrational bands whose peak-to-trough intensities increase with both energy and temperature. The continuum is produced by transitions from the ground electronic state $X(^1\Sigma^+)$ to the repulsive excited electronic state $B(^1\Delta)$. Banded structure arises from transitions from a vibrationally excited ground state to the $X(^1\Sigma^-)$ state. Although this transition would be normally

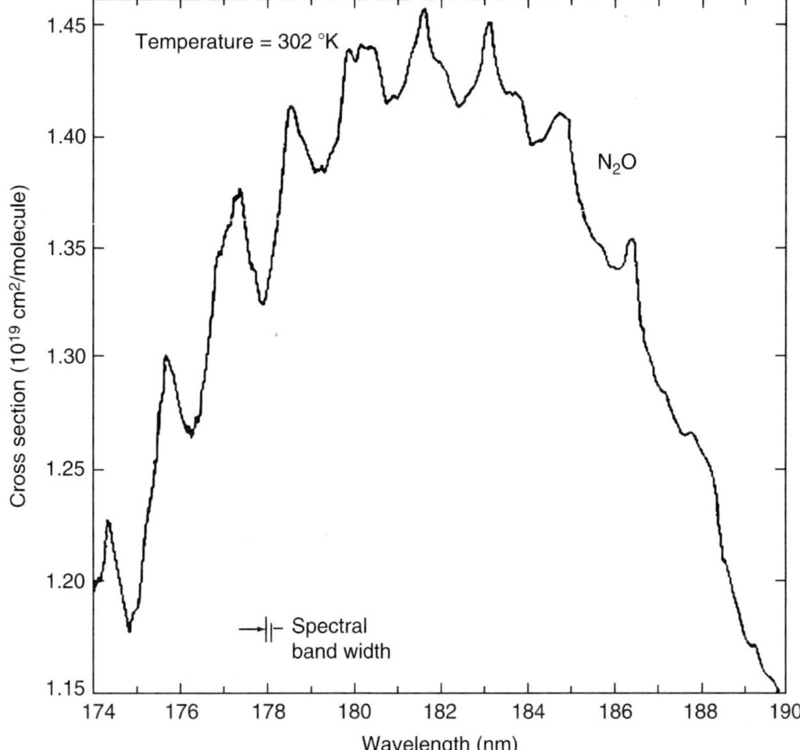

Fig. 14.1. Experimental absorption spectrum of nitrous oxide (N_2O) showing structure at high energies. (From Selwyn *et al.*, 1977. Reprinted with permission.)

weak, in both states the N_2O molecule is bent, and this enhances the transition due to quantum mechanical effects. As the temperature increases the vibrationally bent modes become more populated and the intensity of the absorption increases, especially towards the low energy side of the spectrum. Between 151 and 485 K, absorption cross sections increase by a factor of 2 at 188 nm and by a factor of 1.3 at 172 nm (Selwyn and Johnston, 1981). The temperature-dependent UV absorption spectrum can be expressed at low spectral resolution by:

$$\ln \sigma(\lambda,T) = A_1 + A_2\lambda + A_3\lambda^2 + A_4\lambda^3 + A_5\lambda^4 + (T-300)\exp(B_1 + B_2\lambda + B_3\lambda^2 + B_4\lambda^3) \quad (14.5)$$

where λ is wavelength (nm), T temperature (K), and the coefficients are:

$A_1 = 68.21023$ $\quad B_1 = 123.4014$
$A_2 = -4.071805$ $\quad B_2 = -2.116255$
$A_3 = 4.301146 \times 10^{-2}$ $\quad B_3 = 1.111572 \times 10^{-2}$
$A_4 = -1.777846 \times 10^{-4}$ $\quad B_4 = -1.881058 \times 10^{-5}$
$A_5 = 2.520672 \times 10^{-7}$ $\quad$ (Selwyn *et al.*, 1977).

The function is plotted in Fig. 14.2 over a range of temperatures spanning the stratosphere. The greatest relative changes in absorption intensity occur at the red end of the spectrum where photolysis is strongest. At 200 nm near the photolysis peak, the increase in the cross section is about 20%. Since temperatures increase with altitude in the stratosphere, absorption intensities will be larger in the upper stratosphere.

Opacity of the stratosphere

Although N_2O absorbs UV radiation most efficiently at wavelengths near 180 nm, the greatest rate of N_2O loss is near 200 nm. This is due to the opacity of the stratosphere. Molecular oxygen is an efficient absorber of radiation at wavelengths where N_2O absorption is greatest. Due to the large concentration of oxygen, the opacity of the stratosphere is high between 130 and 200 nm (Schumann–Runge system) and low between 200 and 300 nm (Herzberg continuum). Between 180 and 195 nm, the Schumann–Runge system consists of absorption bands. Between these bands the opacity is modest and radiation is transmitted. Ozone in turn is a strong absorber of radiation between 200 and 300 nm (Hartley bands) and prevents harmful radi-

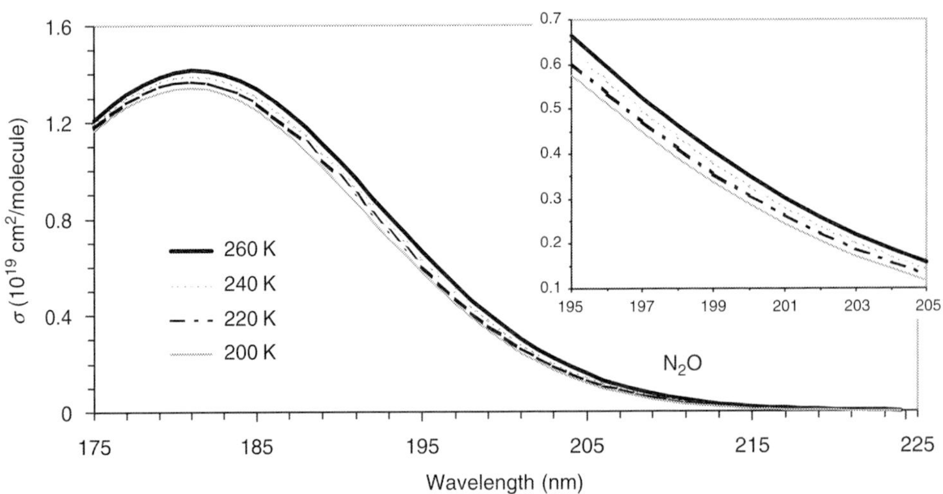

Fig. 14.2. Absorption spectra of nitrous oxide (N_2O) using polynomial fits of Selwyn and Johnston (1981) at stratospherically relevant temperatures. The intensity of the absorption increases at higher temperature representing the lower stratosphere.

ation from reaching the troposphere. This prevents the photolysation of N_2O in the troposphere.

Thus over the wavelengths that N_2O absorbs most effectively, radiation can penetrate only through a small window centred about 200nm and through some smaller windows within the Schumann–Runge band system. Hence, when the absorption spectrum of N_2O is convolved with the actinic flux of the stratosphere, the resulting spectrum peaks near 200nm (Fig. 14.3).

Photolytic loss rates

The final factor contributing to the photolysis rate is the distribution of N_2O throughout the stratosphere. The concentration of N_2O decreases with altitude above the troposphere, falling to near zero at ~50km (Toyoda *et al.*, 2004). Throughout the stratosphere, concentrations are highest in the tropics where strong tropical convection sweeps gas-rich air from the troposphere to higher altitudes (Johnston *et al.*, 1979; Minschwaner *et al.*, 1993).

Photolysis rates peak at altitudes between 30 and 35km at noon when actinic flux is highest (Fig. 14.4). Global annual loss rates are estimated by integrating local rates over the whole stratosphere and for all seasons. Estimates of the annual loss rate range from 12.2 to 13.1Tg N/year, which require an instantaneous global atmospheric lifetime of ~120 years for current N_2O burdens (Table 14.1). This is close to its steady-state lifetime since the ratio between its stratospheric and tropospheric abundances is relatively constant.

Uncertainties in the loss rate come from a variety of sources. The Schumann–Runge bands of oxygen are complex and finely structured, and thousands of overlapping lines make it difficult to accurately specify the actinic flux in the atmospheric layers where N_2O is photolyzed. High spectral resolution modelling by Minschwaner *et al.* (1992, 1993) has reduced the errors of the Schumann–Runge cross sections between 175 and 210nm to 15%. Other uncertainties include the global distribution of N_2O concentrations, absolute solar irradiance, the

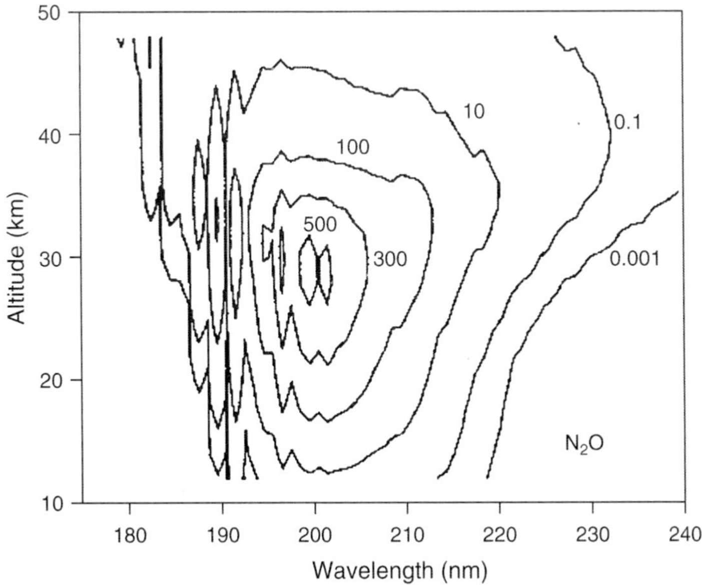

Fig. 14.3. Calculated photolysis rates of nitrous oxide (N_2O) in units of molecules/cm^3/s. The rates are from a model run at 5° latitude, at noon, during equinox. (From Minschwaner *et al.*, 1993. Reproduced by permission of American Geophysical Union.)

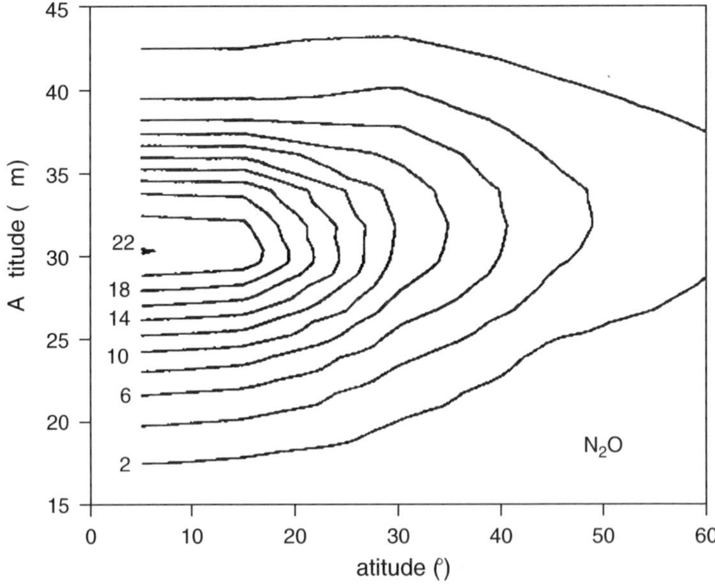

Fig. 14.4. Total loss rates for nitrous oxide (N_2O) including contributions from both photolysis and photo-oxidation. The rates are diurnal averages with units of 10^2 molecules/cm^3/s. (From Minschwaner *et al.*, 1993. Reproduced by permission of American Geophysical Union.)

cross sections in the Herzberg continuum and the absorption cross sections of N_2O. These limit the certainty of the global N_2O loss rate to about 20–30% (Minschwaner *et al.*, 1993; McLinden *et al.*, 2003).

14.3 Fractionation of Nitrous Oxide

Much has been learned about N_2O's loss processes through the analysis of its isotopic composition in the troposphere and stratosphere. The reactions and processes that produce, remove and transport N_2O throughout the atmosphere, fractionate N_2O into various ratios of heavy and light N_2O. The influence of each process on the isotopic composition depends on its contributing flux and its enrichment capacity, which can be quantified through a parameter called a fractionation constant. By measuring the fractionation constant for each production and removal process, and by sampling the atmospheric isotopic composition, we can constrain the fluxes of the relevant processes.

The most abundant naturally occurring species of N_2O is $^{14}N^{14}N^{16}O$ (446). When the major oxygen and nitrogen isotopes (^{15}N,

Table 14.1. Global loss rates and lifetime of nitrous oxide (N_2O).

	Global loss rate (Tg N/year)	Global inventory (Tg N) (reference year)	Lifetime (years)
Minschwaner *et al.* (1993)	12.2	1500 (1980)	123
Prather and Ehhalt (2001)	12.6	1519 (1998)	120
McLinden *et al.* (2003)	13.1	1410[a]	114

[a]Using 310 ppbv for N_2O concentration at lower boundary of stratosphere.

^{17}O, and ^{18}O) are isotopically substituted into the base molecule, a number of isotopologues can be created – the most abundant of these are $^{14}N^{15}N^{16}O$, $^{15}N^{14}N^{16}O$, $^{14}N^{14}N^{17}O$ and $^{14}N^{14}N^{18}O$ (abbreviated here as 456, 546, 447 and 448, respectively). Other isotopologues consisting of two or more isotopes are rare and only found in insignificant quantities. As analytic methods to separate the two isotopomers 456 and 546 are difficult, some instrumental methods fail to distinguish them, but instead measure the aggregate, abbreviated as $^{15}N^{bulk}$.

The isotopic fraction of an air sample relative to the abundant light isotope is often reported with respect to a standard. For example, a permil value for the heavy isotopologue 448 can be calculated as:

$$\delta^{448} = \left(\frac{[448]/[446]_{sm}}{[448]/[446]_{st}} - 1\right) \times 1000 = \left(\frac{R^{448}_{sm}}{R^{448}_{st}} - 1\right) \times 1000 \quad (14.6)$$

where the values within brackets [] denote concentration, sm is the sample of interest, and st is the standard. With this definition, δ will be positive when the sample is enriched in the isotopologue and negative when it is depleted. The chosen standard is typically atmospheric nitrogen and oxygen, though standard mean ocean water is often used for oxygen.

14.3.1 Source fractionation

Relative to the isotopic signature of atmospheric nitrogen and oxygen, the troposphere is enriched in heavy N_2O. Typical values for $\delta^{15}N^{bulk}$ and δ^{448} are 5–8‰ and 16–22‰ respectively (Moore, 1974; Yoshida and Matsuo, 1983; Yoshida *et al.*, 1984; Wahlen and Yoshinari, 1985; Kim and Craig, 1990; Johnston *et al.*, 1995; Cliff and Thiemens, 1997; Rahn and Wahlen 1997; Cliff *et al.*, 1999; Stein and Yung, 2003). Proximity to local sources most likely produces these variations (Cliff and Thiemens, 1997) and enrichments in the boundary layer tend to be lower than in the free troposphere.

We expect that the isotopic composition of N_2O in the troposphere will reflect the fractionation of its sources. However, we find that emissions of N_2O from its major sources are isotopically light relative to mean tropospheric air (Yoshida *et al.*, 1984; Wahlen and Yoshinari, 1985; Yoshinari and Wahlen, 1985; Kim and Craig, 1990; Kim and Craig, 1993; Cliff and Thiemens, 1997; Dore *et al.*, 1998; Naqvi *et al.*, 1998). The isotopic enrichment of $^{15}N^{bulk}$ and 448 in soil-derived N_2O ranges from –7‰ to –25‰ and 3‰ to 19‰ respectively (Kim and Craig, 1993). The composition of the other major N_2O source, the ocean, is more complex. The ocean is enriched in heavy N_2O below ~600 m ($\delta^{15}N^{bulk}$ = 7–10‰ and δ^{448} = 22–32‰), but is slightly depleted in heavy N_2O near the surface ($\delta^{15}N^{bulk}$ = 6–7‰ and δ^{448} = 18–19‰). Since gas exchange occurs mainly in the surface waters, the ocean appears to be mainly a source of light N_2O, although some heavy N_2O may escape during upwelling of deep water at the continental margins (Kim and Craig, 1993).

14.3.2 Rayleigh fractionation

Given that the emissions of N_2O are enriched in light isotopologues, the isotopic composition of N_2O in the troposphere is unexpectedly heavy. As there are no sinks of N_2O in the troposphere, there must be an additional flux of heavy N_2O to the troposphere. The source of this flux is the stratosphere. Moore (1974) first observed that the stratosphere is enriched in heavy N_2O and since then many other measurements have confirmed this. The heavy N_2O is introduced into the lower atmosphere through stratosphere–troposphere exchange processes, which balances the light N_2O from the surface sources (Yoshida and Matsuo, 1983; Kim and Craig, 1993). All of the heavy N_2O isotopologues are enriched in the stratosphere relative to the tropospheric mean and source emissions. Furthermore, the enrichment increases in altitude within the stratosphere as the N_2O concentrations decrease (Figs 14.5 and 14.6). For example, $\delta^{15}N^{bulk}$ increases from 7‰ in

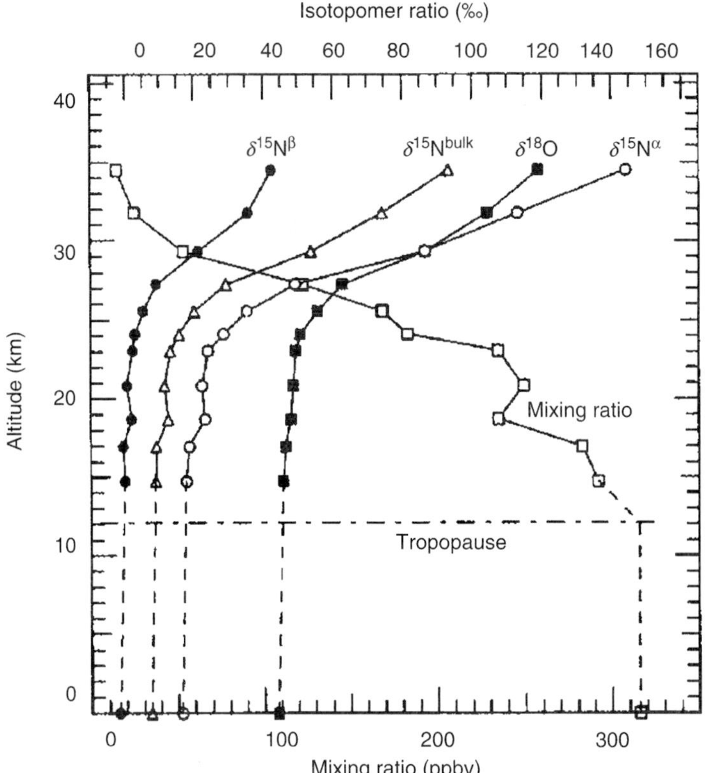

Fig. 14.5. Vertical profiles of the isotopic enrichment of nitrous oxide (N_2O) in the stratosphere over Japan on 31 May 1999. (From Toyoda *et al.*, 2001. Reproduced by permission of American Geophysical Union.)

the troposphere to more than 90‰ at 35 km in the stratosphere. Likewise, δ^{448} increases from 20‰ to ~95‰ (Toyoda *et al.*, 2001).

For some time the cause of this enrichment was unknown. Laboratory experiments suggested that neither photolysis nor the reaction with $O(^1D)$ fractionated N_2O (Johnston *et al.*, 1995). As there are no known sources of N_2O in the stratosphere, new atmospheric processes were proposed to explain the stratospheric enrichment (McElroy and Jones, 1996; Prasad, 1997; Prasad *et al.*, 1997).

There is now considerable theoretical and experimental evidence which confirms that photolysis is responsible for the enrichment of heavy N_2O in the stratosphere (Table 14.2). At wavelengths important for stratospheric photolysis, light N_2O is preferentially dissociated, leaving the remaining N_2O pool enriched in 456, 546 and 448. The evolution of the isotopic composition of the N_2O pool can be modelled as a Rayleigh fractionation process, which describes how the composition changes after some fraction of the original N_2O has been photolysed. For the Rayleigh model to be valid there can be no local N_2O sources and the sink must be an irreversible loss process. As N_2O is photolyzed, the enrichment of isotopologue X will change from an initial value of δ_o to some later value δ_i as governed by:

$$\ln\left(\frac{1+\delta_i^x}{1+\delta_o^x}\right)=(\alpha-1)\times\ln\left(\frac{c_i^x}{c_o^x}\right) \qquad (14.7)$$

where c_o is the initial concentration of bulk N_2O – usually set equal to the tropospheric

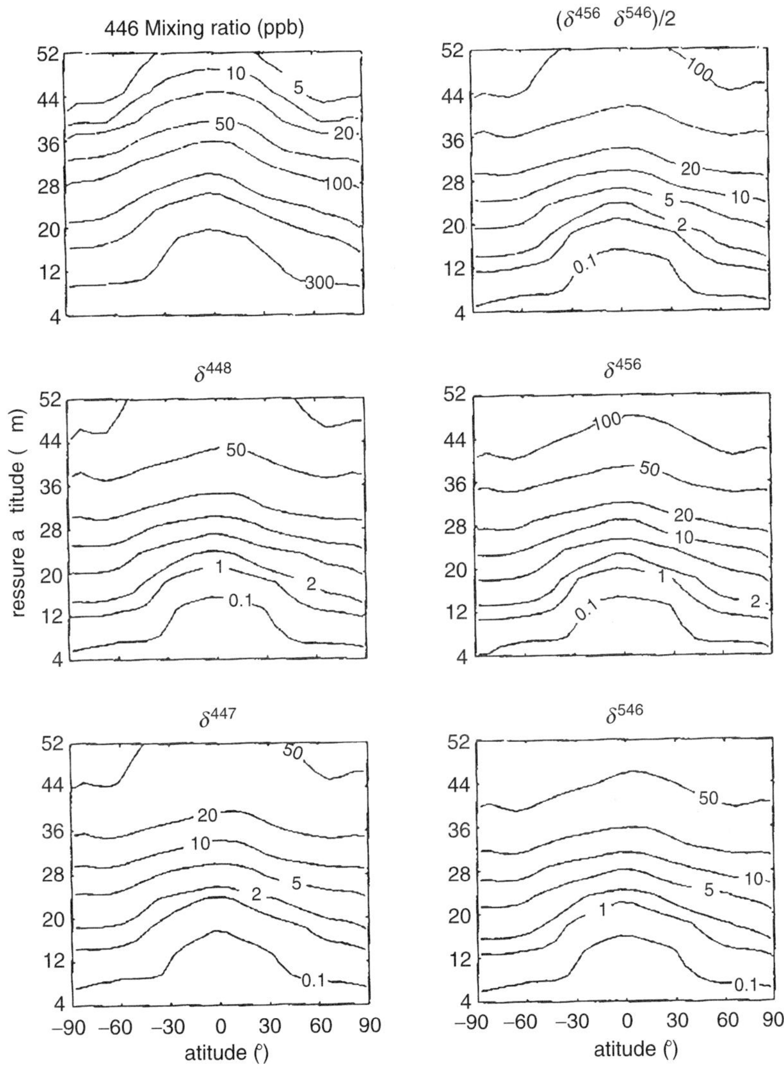

Fig. 14.6. Modelled enrichments of nitrous oxide (N_2O) and its heavy isotopologues. Units are ppbv for 446 and per mil (‰) for the isotopologues. (From McLinden *et al.*, 2003. Reproduced by permission of American Geophysical Union.)

concentration as this is the source of non-photolysed N_2O to the stratosphere, and c_i is the concentration after some loss has occurred. The constant α is the 'fractionation factor', which is the ratio of isotopologue X's photolysis (or reaction) rate coefficient to that of 446. Since α is quite small, a 'fractionation constant' is defined as $\varepsilon = 1000(1 - \alpha)$, which has units of per mil (‰). With this definition, the fractionation constant will be negative if the isotopologue is enriched by the loss process and will be positive if it is depleted.

Equation (14.7) can be linearized by making the approximation δ<<1, in which case $\ln(1+\delta) \approx \delta$ and Eq. 14.7 becomes $\delta_i = \delta_0 + \varepsilon \times \ln f$, where f is the fraction of substrate remaining. The fractionation constant can then be found from a plot of δ_i versus

Table 14.2. Fractionation constants measured in the laboratory, stratosphere and calculated in models.

	ε^{456} (‰)	ε^{546} (‰)	$^{15}\varepsilon^{bulk}$ (‰)	ε^{448} (‰)	$^{15}\varepsilon^{bulk}/\varepsilon^{448}$ (ψ)	$\varepsilon^{456}/\varepsilon^{546}$ (η)	Reference
Photolysis experiments (λ)							
185 nm	–	–	–	≤ ± 0.3	–	–	Johnston *et al.* (1995)
	−18.6 (0.5)	3.7 (0.2)	−7.5	4.5 (0.2)	−1.7	−5.0	Kaiser *et al.* (2003)
193 nm	–	–	−18.4	−14.5	1.27	–	Rahn *et al.* (1998)
	−35.7 (0.5)	−10.9 (1.7)	−23.3 (0.5)	−17.3 (0.5)	1.35	3.27	Röckmann *et al.* (2000)
	−25.7 (2)	−13.1 (2)	−19.4 (1.2)	−15.9 (3)	1.54	1.96	Turatti *et al.* (2000)
207 nm	–	–	−48.7	−46	1.06	–	Rahn *et al.* (1998)
207.6 nm	−66.5 (5)	−27.1 (6)	−46.8 (5)	−49 (10)	0.96	2.45	Turatti *et al.* (2000)
211.5 nm	−65.3 (4)	−31.4 (8)	−48.3 (5)	−46 (11)	1.05	2.08	Turatti *et al.* (2000)
213 nm	−73.5 (5)	−41 (10)	−57 (15)	–	–	1.79	Zhang *et al.* (2000)
Broadband	−54.0 (1.6)	−21.9 (1.1)	−37.9 (1.1)	34.2 (0.8)	1.11	2.47	Röckmann *et al.* (2001)
Reaction with O(^{1}D) experiments							
	–	–	–	−6	–	–	Johnston *et al.* (1995)
	−2.21	−8.79	–	−12.23	0.45	0.25	Kaiser *et al.* (2002a)
Stratospheric measurements							
14–22 km	–	–	−14.5	−12.9	1.12	–	Rahn and Wahlen (1997)
15–35 km	−57.1 (9.5)	−27.3 (13.5)	−42.3(10.1)	−42.6 (29.4)	0.99	2.09	Griffith *et al.* (2000)
<24 km	−22.9 (1.2)	−8.8 (1.4)	−15.9 (1.1)	−11.5 (1.8)	1.42	2.6	Toyoda *et al.* (2001)
>24 km	−40.9 (1.3)	−15.5 (0.4)	−28.6 (0.6)	−24.6 (0.6)	1.16	2.63	Toyoda *et al.* (2001)
10–320 ppbv	−33.4	−16.3 (0.6)	−24.9 (0.7)	−21.4 (0.6)	1.16	2.05	Röckmann *et al.* (2001a)
200–320 ppbv	−21.3 (1.5)	−12.9 (2.4)	−17.1 (1.6)	−14.0 (2.0)	1.22	1.65	Röckmann *et al.* (2001a)
<200 ppbv	−30.8 (6.4)	−12.9 (3.0)	−22.1 (4.2)	−18.9 (3.5)	1.17	2.35	Park *et al.* (2004)
>200 ppbv	−22.4 (2.5)	−7.1 (2.9)	−14.9 (1.1)	−13.3 (0.9)	1.12	2.65	Park *et al.* (2004)
Modelling studies							
193 nm	−13.1	−7.5	−10.3	−9.1	1.13	1.75	Yung and Miller (1997)
207 nm	−30.6	−17.4	−24	−21.3	1.13	1.75	Yung and Miller (1997)
6–310 ppbv	−27.1	−10.6	−19.1	−19.3	0.99	2.56	McLinden *et al.* (2003)
200–310 ppbv	−19.5	−7.4	−13.4	−14.0	0.96	2.63	McLinden *et al.* (2003)
10–170 ppbv	−30.4	−11.7	−21.1	−21.8	0.97	2.60	McLinden *et al.* (2003)

ln f. Although numerous authors use the linearized form, the approximation introduces errors of 1% for δ_i = 20‰ and 5‰ for δ_i = 100‰ (Kaiser *et al.*, 2002a; Morgan *et al.*, 2004). Therefore, it is more accurate to find ε by plotting the whole left-hand side of Eq. 14.7 against ln f.

14.3.3 Fractionation in the laboratory

Photolysis fractionation has been well studied in the laboratory and we now have a clear picture of the process (Table 14.2). The degree of fractionation depends on a number of factors, of which the most important is the isotopologue of interest. In the wavelength range 190–215 nm, where most N_2O is photolysed, the fractionation constants follow the relation $|\varepsilon^{456}| > |\varepsilon^{448}| > |\varepsilon^{546}|$, indicating that 456 is most highly enriched. At 207 nm, for example, ε^{456} = −66‰, ε^{448} = −49‰ and ε^{546} = −27‰ (Turatti *et al.*, 2000). Similar results are found at other wavelengths.

The enrichment is wavelength-dependent. While ε^{456} = −66‰ at 207 nm, it falls to −27‰ at 193 nm (Turatti *et al.*, 2000). A similar decrease is observed for the other isotopologues. These results, which are consistent throughout all published reports, indicate that the enrichment of the heavy species decreases at longer wavelengths. Because the make-up of the photolysing radiation shifts to longer wavelengths towards the top of the stratosphere, the enrichment of the heavy isotopologues should increase upwards through the stratosphere.

The above relational order of the fractionation factors is maintained at wavelengths down to 185 nm. At this wavelength, which is near the peak of the absorption spectrum, Kaiser *et al.* (2003) find that ε^{456} = −18.5‰, ε^{546} = 3.7‰ and ε^{448} = 4.5‰. Not only is the relational order between 546 and 448 reversed, but these species are also depleted relative to 446, not enriched.

The variation of the fractionation constant with isotopologue and wavelength is caused by small differences between the absorption spectra of the different N_2O species. High-resolution spectroscopy reveals that the absorption spectra of the heavy species are blue-shifted relative to 446 by ~0.1–0.2 nm (Selwyn and Johnston, 1981). The spectrum of 456 is most highly shifted, followed by 448 and 546, respectively. Because of the shifts, the cross sections of the heavy species are reduced relative to 446 at wavelengths longer than the peak of the absorption spectrum. This reduces the rate of photolysis for the heavy species and leads to their enrichment.

At wavelengths near the absorption peak, the spectrum is highly structured and banded with many peaks. This small-scale structure produces complex variations in the ratios between cross sections as one spectrum is shifted with respect to the other. At some wavelengths the cross section of the heavy species will be larger than 446 and the heavy species will be depleted due to its faster photolysis. This explains why 546 and 448 are depleted during photolysis at 185 nm.

Laboratory experiments show that the enrichment of heavy N_2O increases with decreasing temperature (Kaiser *et al.*, 2002b). During broadband photolysis ε^{456} changed from −75‰ at 190 K to −57‰ at 290 K, ε^{448} changed from −45‰ to −37‰ and ε^{546} changed from −30‰ to −25‰. Because the temperature of the stratosphere increases with altitude, this dependence will reduce the enrichment in the upper stratosphere.

In addition to photolysis, reaction with $O(^1D)$ also fractionates N_2O. Since this process is only 10% of the total N_2O sink, its overall contribution to the stratospheric enrichment is small, but may be important in the lower stratosphere where the reaction peaks. In the laboratory Kaiser *et al.* (2002a) found that this reaction fractionates according to ε^{448} = −12.4‰, ε^{546} = −8.9‰ and ε^{456} = −2.2‰. Not only are these constants lower than their respective photolysis counterparts, but the relational order is also changed: $|\varepsilon^{448}| > |\varepsilon^{546}| > |\varepsilon^{456}|$. Thus, the sink processes have their own unique signatures and these can be used to assess the relative strengths of the sink processes in the stratosphere.

14.3.4 Theoretical models of fractionation

Photolytic fractionation was put on firm theoretical footing beginning with the zero-point energy (ZPE) model of Yung and Miller (1997). During photolysis, N_2O undergoes a transition from the electronic ground state to the continuum level $B(^1\Delta)$. The larger mass of the isotopically substituted species lowers their ground-state energies, which shifts their absorption spectra to shorter wavelengths (Fig. 14.7). Since most photolysis occurs at wavelengths on the red side of the absorption peak, cross sections for 446 will be larger than for the heavy species and 446 will be photolysed preferentially, leaving the remaining pool enriched in the heavy species. As discussed earlier, the fractionation constant is the ratio of the heavy to light cross sections. The ratio, or fractionation constant, decreases with increasing wavelengths up to the absorption peak, where the model predicts that a crossover in the absorption spectra occurs, and the heavy cross sections become larger than the light ones, and the heavy species become depleted.

The model predicts the correct relational order between the heavy N_2O fractionation constants, but its quantitative estimates of the constants are about a factor of 2 smaller than laboratory-based results (Rahn *et al.*, 1998; Umemoto, 1999; Röckmann *et al.*, 2000; Turatti *et al.*, 2000; Zhang *et al.*, 2000; Röckmann *et al.*, 2001a). Improvements to the ZPE model using more sophisticated quantum mechanical calculations (Johnson *et al.*, 2001; Blake *et al.*, 2003; Liang *et al.*, 2004) produce quantitative results in good agreement with experiments (Kaiser *et al.*, 2003); though the best matches to the experimental fractionation constants are found using the high-resolution absorption spectra of Selwyn and Johnston (1981) (Kaiser *et al.*, 2003).

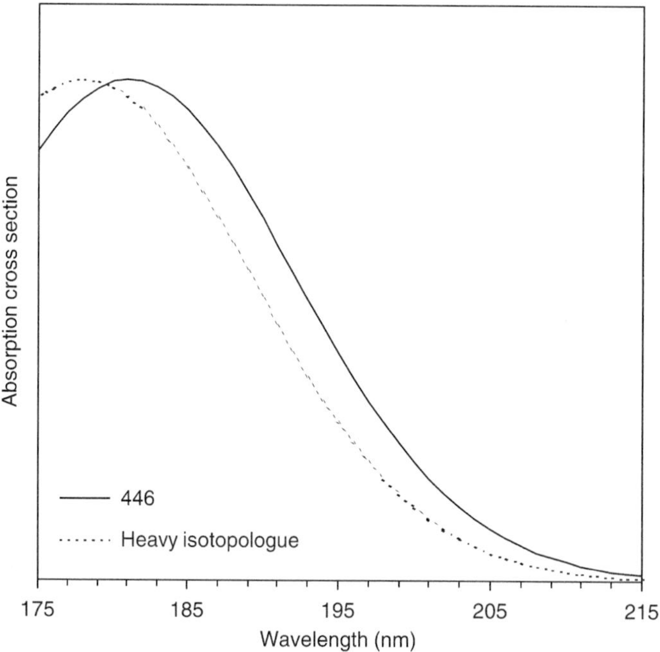

Fig. 14.7. Schematic diagram illustrating the shift in the absorption spectrum for the heavy isotopologues. In the ZPE model (Yung and Miller, 1997), the additional mass of the heavy isotopologues decreases the ZPE level and blue-shifts the absorption spectrum, causing enrichment of the isotopologues at wavelengths longer than the peak absorption.

14.4 Variation of Fractionation Constants Throughout the Stratosphere

Vertical profiles of N_2O isotopes in the stratosphere reveal that not only does the enrichment of the heavy species increase with altitude, but the fractionation constants of these species increase as well (Griffith *et al.*, 2000; Röckmann *et al.*, 2000; Toyoda *et al.*, 2001; Park *et al.*, 2004; Toyoda *et al.*, 2004). The former observation is understood by the simple recognition that N_2O undergoes continuous fractionation through photolysis and reaction with $O(^1D)$ as it moves from the lower to the upper layers of the stratosphere. As a result, air in the upper stratosphere is isotopically older and is more enriched in the heavy species relative to fresh air entering from the troposphere.

The latter observation that fractionation constants increase with altitude (i.e. they become more negative) is difficult to explain. That they do is clearly seen in the isotopic measurements in Fig. 14.8 and is evident from the summary of stratospheric measurements in Table 14.2. In general, fractionation constants are on an average approximately twice as large in the upper stratosphere as in the lower stratosphere. The altitude at which this transition occurs is not clearly delimited but rather falls in a band between 18 and 26 km (Toyoda *et al.*, 2004). In addition, fractionation constants measured in the laboratory are consistently larger than those in the lower stratosphere and agree approximately with those measured in the middle-to-upper stratosphere. In the following sections we explore a number of both physical and chemical factors that account for the observed behaviour of ε.

14.4.1 Effect of diffusion and mixing on ε

One factor that may cause the stratospheric fractionation constant to deviate from the measured laboratory values is the effect of eddy diffusion. Rahn *et al.* (1998) calculate

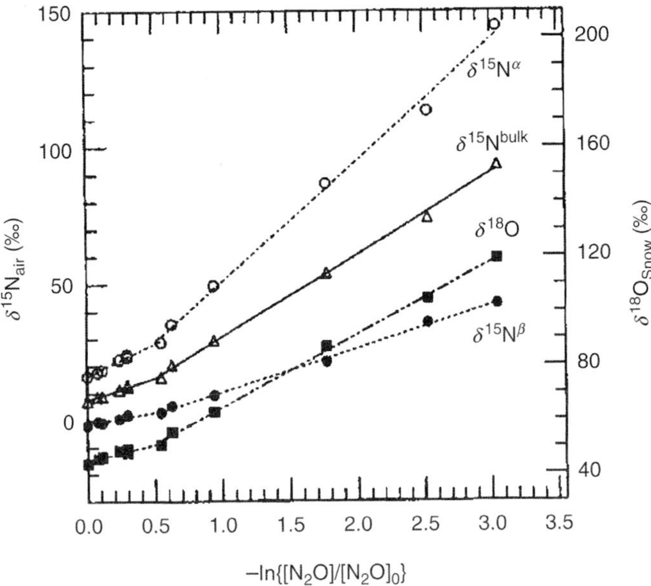

Fig. 14.8. Measurements of stratospheric $\delta^{15}N^{bulk}$ indicate that fractionation increases from the lower to the upper stratosphere. When the data are plotted as in the figure, the slope is an estimate of the fractionation constant. The transition in the slope occurs near 24 km. (From Toyoda *et al.*, 2001. Reproduced by permission of American Geophysical Union.)

that in a diffusion-limited one-dimensional (1D) stratosphere, the apparent fractionation constant (ε_{strat}) is expected to be about half the true value (ε_{lab}). This result is promising since this is approximately the relation between fractionation constants measured in the lower stratosphere and in the laboratory. However, for diffusion and photolysis rates typical of stratospheric conditions, photolysis in the lower stratosphere is not diffusion-limited, but rather rate-limited, in which case ε_{strat} would be similar to ε_{lab} (Toyoda *et al.*, 2004). In addition, it is likely that photolysis becomes diffusion-limited in the upper stratosphere, which would cause ε_{strat} to decrease with altitude. This is opposite to the trend observed. Changes in the diffusion constant may however produce small seasonal and interannual variations in ε_{strat} in the more sensitive mid-to-upper stratosphere (Toyoda *et al.*, 2004).

It appears that mixing can explain some if not most of the departure of ε_{strat} from ε_{lab} and also the increase of ε_{strat} with altitude. We expect measurements of isotopic enrichments in the stratosphere to follow Rayleigh fractionation only if they track the chemical evolution of a single air mass. If multiple air masses are mixed, the combined enrichment will reflect contributions from multiple chemical histories. This tends to dilute the overall enrichment and lower the fractionation constant.

Dynamical mixing is common throughout the lower stratosphere. Although most tropospheric air enters the stratosphere through vertical advection in the tropics, there is considerable stratosphere–troposphere exchange at extra-tropical latitudes along isentropic surfaces and during tropopause folding events. Incursion of non-enriched tropospheric air dilutes the photolysed stratospheric air mass, thus reducing ε_{strat}. Although the dynamics of the stratospheric circulation is complex, observations indicate that N_2O-depleted air is brought down from the upper stratosphere to the lower and middle stratospheres through the polar vortex at high latitudes (Park *et al.*, 2004). The mixing of these air masses with the N_2O-enriched air of the lower stratosphere creates a lighter isotopic composition and also reduces ε_{strat}. Since subsidence is greatest at high latitudes, mixing effects produce a gradient in the latitudinal variation of ε_{strat}. Air in the subtropics undergoes less mixing than in the high latitudes, which causes ε_{strat} to decrease with latitude as mixing becomes more pronounced (Park *et al.*, 2004).

The impact of mixing processes on ε_{strat} has been simulated by a number of atmospheric chemical-tracer models (McLinden *et al.*, 2003; Morgan *et al.*, 2004). These models successfully reproduce the observed increases in the N_2O fractionation constants in the middle stratosphere. Both models find that the enrichment of isotopologues in the stratosphere can be described by two fractionation constants: one for the lower stratosphere stretching from the tropopause to ~25 km, and another for the stratosphere above 25 km. This boundary is consistent with the transition altitude measured by Toyoda *et al.* (2004).

14.4.2 Chemical effects on ε

Besides physical processes, changes in temperature and actinic flux can potentially cause ε_{strat} to vary throughout the stratosphere. Photolytic experiments indicate that the magnitude of ε decreases as temperatures increase (Kaiser *et al.*, 2002b). Since the temperature of the stratosphere increases by ~50°C from the lower to upper stratosphere, this temperature effect should produce higher fractionation constants in the lower stratosphere. This is not observed, which suggests that temperature effects are overwhelmed by other processes.

Actinic flux varies with altitude in the stratosphere in large part due to the changes in ozone concentration. Ozone-mixing ratios peak around 20–30 km and fall off with altitude. Since ozone strongly absorbs UV radiation at the red end of N_2O's absorption spectrum, actinic flux will shift to longer wavelengths in the upper stratosphere. Both theoretical and experimental studies show that the enrichment of the heavy N_2O species increases with longer wavelengths as differences between their absorption cross sections and 446 s increase. Rahn *et al.* (1998)

and Turatti *et al.* (2000) found that ε^{456}, ε^{546} ($^{15}\varepsilon^{bulk}$ for Rahn *et al.*) and ε^{448} nearly triple in absolute size from 193 to 207 nm. To account for the full increase in ε_{strat} from the lower to the upper stratosphere, the effective wavelength peak of the actinic flux would need to shift by 10 nm (Toyoda *et al.*, 2001). This is outside the range of expected actinic flux shifts. In addition, there is little variation in measured fractionation constants when the photolysing actinic flux is changed appreciably in laboratory experiments (Röckmann *et al.*, 2001a).

Fractionation constants may vary if the relative contribution of photolysis and $O(^1D)$ reaction to the total nitrous sink is variable throughout the stratosphere. The fractionation of N_2O is much less for its reaction with $O(^1D)$ than for photolysis. We therefore expect ε_{strat} to reflect changes in the relative strengths of these processes. If both sink processes follow Rayleigh fractionation, for a region of the stratosphere where photolysis accounts for x fraction of the total sink, the combined sink would produce fractionation given by $\varepsilon_{sink} = x\varepsilon_{photolysis} + (1-x)\varepsilon_{oxidation}$ (Toyoda *et al.*, 2001, 2004). From Table 14.2 the average photolysis fractionations in the lower stratosphere are −22‰, −9.6‰ and −13‰ for ε^{456}, ε^{546} and ε^{448}, respectively, and in the mid-to-upper stratosphere are −35‰, −15‰ and −22‰. If we use the broadband photolysis fractionations of Röckmann *et al.* (2001a) and the $O(^1D)$ reaction fractionations of Kaiser *et al.* (2002a), the derived photolysis fraction x ranges from 4% to 40% in the lower stratosphere, and from 45% to 65% in the upper stratosphere. Although the trend is correct – photo-oxidation should be more important in the lower stratosphere – the fractions are too low compared with standard chemistry, which predicts that photolysis should be 90% of the total stratospheric sink. This indicates that photo-oxidation alone cannot explain the altitudinal increase in the fractionation.

14.4.3. Ratios of fractionation constants

Other potentially useful diagnostics in understanding the role of physical and chemical effects on the stratospheric fractionation constant are the ratios $\eta = \varepsilon^{456}/\varepsilon^{546}$ and $\psi = {}^{15}\varepsilon^{bulk}/\varepsilon^{448}$ (Röckmann *et al.*, 2001a; Kaiser *et al.*, 2002a). Although mixing processes alter the absolute magnitude of the fractionation constants, they should conserve the relative magnitudes. Therefore, any process that changes either η or ψ within the stratosphere should be chemical and not physical in nature (Röckmann *et al.*, 2001a; Kaiser *et al.*, 2002a).

These ratios are calculated for the datasets in Table 14.2. Although there are differences in η and ψ between the lower and upper regions, in a reanalysis of the data, Park *et al.* (2004) showed that only the difference in η computed from the Röckmann *et al.* (2001a) data is statistically significant. Since temperature and actinic flux contributions are shown to have minor impacts on ε_{strat}, we can investigate what this behaviour of η means for the relative contribution of the $O(^1D)$ sink.

The ratio η has the potential to track changes in the relative contributions of the photolysis and $O(^1D)$ sinks because each process fractionates N_2O differently. Specifically, ε^{546} is greater than ε^{456} in the $O(^1D)$ reaction but is larger than ε^{456} in photolysis. With the relative contributions of these reactions to the total sink, η should range from ~0.5 (only $O(^1D)$ sink) to 2.5 (only photolysis) (Kaiser *et al.*, 2002a). To explain the Röckmann *et al.* η value would require that 60% of the N_2O sink in the lower stratosphere is due to reaction with $O(^1D)$, and 30% of the overall sink is due to $O(^1D)$. This result qualitatively agrees with standard chemistry estimates, which suggest that the $O(^1D)$ sink peaks in the lower regions, but overestimates its total strength. The standard view is that the $O(^1D)$ sink is about 10% of the total. It is unlikely then that the variation in η can be explained solely by the relative strengths of N_2O sinks.

Additional analysis by Park *et al.* (2004) shows that η calculated from the Röckmann *et al.* data is larger in the mid-latitudes than the subtropics. This may suggest that the difference is due to second-order transport effects that do not cancel in the ratios.

In summary it appears that known dynamical and chemical processes can adequately explain the observed variations in the stratospheric fractionation with mixing playing the most significant role. The altitudinal dependence of the relative sink strengths likely contributes secondary effects to the variability, whereas marginal effects may be due to small shifts in the effective wavelength of the photolysing radiation.

14.4 Constraining Isotopic Fluxes from Loss Rates

Source fluxes remain the greatest uncertainty in the global budget of N_2O. Emissions from N_2O's primary sources – soil, ocean and agricultural practices – are variable over small spatial and temporal scales, and depend on many factors such as temperature, soil moisture and soil type (Prather and Ehhalt, 2001). Since these sources have unique isotopic signatures their fluxes can be constrained if the other contributions to the isotopic budget are known. One element of this budget is the isotopic flux from the stratosphere to the troposphere. The net isotopic flux reflects differences in the isotopic composition of the troposphere and stratosphere. Specifically it is the isotopic composition of the lower stratosphere that is relevant since this is the source region of the stratosphere–troposphere exchange.

Through a series of reasonable approximations and assumptions (Park *et al.*, 2004), the isotopic flux F_{ST} is related to the net loss of N_2O in the stratosphere L and the fractionation constant for the isotopologue of interest ε (McLinden *et al.*, 2003; Park *et al.*, 2004) by:

$$F_{ST}\,(\delta_S - \delta_T) \approx -\,\varepsilon L \qquad (14.8)$$

where δ_T and δ_S are representative δ values for the troposphere and lower stratosphere. Estimates of the isotopic flux from both McLinden *et al.* (2003) and Park *et al.* (2004) are shown in Table 14.3. Both analyses agree in spite of some error, and the bulk of any difference is due to the use of different fractionation constants. Those used by McLinden *et al.* are mass flux-weighted model averages, whereas Park *et al.* used values based on measurements in the lower stratosphere at high latitudes. As Kim and Craig (1993) first suggested, exchange between the stratosphere and the troposphere is a significant contributor to the isotopic composition of the troposphere. Calculations show that tropospheric $\delta^{15}N^{bulk}$ and $\delta^{18}O$ increase from 5‰ to 25‰ and from 1‰ to 17‰ respectively, if the contribution from the stratospheric isotopic flux is considered (Park *et al.*, 2004).

14.5 Oxygen Isotope Anomaly

There has been considerable debate about the role of photolysis in causing the observed oxygen isotope anomaly observed in $\delta^{17}O$ and $\delta^{18}O$ record of N_2O (Cliff and Thiemens,

Table 14.3. N_2O isotopic fluxes from the stratosphere to the troposphere.

	446	456	546	(456+546)/2	448	447	Reference
Lifetime (years)	113.8	116.0	114.3	–	115.4	114.6	McLinden *et al.* (2003)
Stratospheric loss (Tg N/year)	13.1	–	–	–	–	–	McLinden *et al.* (2003)
Fractionation constant (‰)	–	−19.1	−4.0	−15.0	−13.9	−7.3	McLinden *et al.* (2003)
		−22.5 ± 1.2		−14.9 ± 0.5	−13.3 ± 0.5		Park *et al.* (2004)
Isotopic flux (‰ Tg N/year)	–		–	196 ± 39	182 ± 37	–	McLinden *et al.* (2003)
		290 ± 74		192 ± 48	172 ± 43		Park *et al.* (2004)

1997; Cliff *et al.*, 1999; Röckmann *et al.*, 2001b; McLinden *et al.*, 2003; Kaiser *et al.*, 2004; Yung *et al.*, 2004). The oxygen anomaly is a measure of the isotopic composition departure from pure mass dependence. Processes that fractionate isotopologues unevenly based on mass are considered mass-dependent. For N_2O, these processes fractionate oxygen according to $\delta^{17}O = \beta \times \delta^{18}O$ where β has been experimentally measured to be 0.515 (Cliff and Thiemens, 1997), though it can vary over a small range (Kaiser *et al.*, 2004). Deviations from mass dependence can be quantified by defining an oxygen anomaly, which can be written in its simplest form as:

$$\Delta^{17}O = \delta^{17}O - 0.515 \times \delta^{18}O \quad (14.9)$$

Observations show that the anomaly is non-zero (~1%) in both the troposphere and stratosphere, and increases in magnitude with altitude (Cliff and Thiemens, 1997; Cliff *et al.*, 1999; Röckmann *et al.*, 2001b). This is surprising since the major sources and sinks of N_2O are generally thought to be mass-dependent and should not produce the anomaly. The exact origin of the mechanism is yet to be clearly identified, but it may be a combination of new atmospheric sources (McElroy and Jones, 1996; Röckmann *et al.*, 2001b; see McLinden *et al.*, 2003 for a discussion), chemical pathways that transfer heavy oxygen from ozone to N_2O (Röckmann *et al.*, 2001b; Yung *et al.*, 2004) and fractionation differences in mass-dependent processes (Kaiser *et al.*, 2004). While some modelling efforts have shown that photolysis contributes significantly to the oxygen anomaly (Johnson *et al.*, 2001; McLinden *et al.*, 2003), laboratory experiments indicate that photolysis is strictly mass-dependent over the relevant wavelengths of the stratosphere (Johnston *et al.*, 1995; Röckmann *et al.*, 2001b).

14.6 Non-standard Stratospheric Sinks

Within the range of the estimated source fluxes of N_2O the global budget can be considered closed (Kroeze *et al.*, 1999; Prather and Ehhalt, 2001). In addition, the observed atmospheric vertical profile of N_2O agrees well with modelled predictions based upon known sources and sinks. With our current knowledge, there is no need for additional sources or sinks, though such processes could exist if their fluxes were small (Rahn and Wahlen, 2000). A number of new reactions have been proposed in recent years to close the N_2O budget (Prasad, 1994, 1997; Prasad *et al.*, 1997).

Most relevant to this chapter are the proposed reactions of N_2O with vibrationally excited molecular oxygen:

$$O_2(\text{high v}) + N_2O \rightarrow O_2 + N_2 + O \quad (14.10a)$$

$$\rightarrow O_3 + N_2 \quad (14.10b)$$

$$\rightarrow NO + NO_2 \quad (14.10c)$$

and with electronically excited oxygen:

$$O2(b^1\Sigma) + N_2O \rightarrow O_3 + N_2 \quad (14.11a)$$

$$\rightarrow NO + NO_2 \quad (14.11b)$$

In the absence of kinetic rate coefficients, it is difficult to assess the significance of these reactions although McElroy and Jones (1996) rule out Eqs 14.10b and 14.11a on quantum mechanical grounds, and feel that Eq. 14.10a is not likely since quenching of the excited oxygen molecule by N_2 may dominate. Prasad *et al.* (1997) show that if the above reactions are considered along with other new source reactions proposed by Prasad (1997), they retain the vertical structure of N_2O as predicted by standard chemistry alone. Therefore, these reactions are consistent with the observations.

14.7 Conclusion

The stratospheric sink of N_2O has been well characterized and can be used to put constraints on the magnitude of the source fluxes. N_2O is primarily removed by photolysis in the stratosphere with secondary contributions by its reaction with $O(^1D)$. Its absorption spectrum peaks near 185 nm, though the available photolysing radiation in the stratosphere shifts the region of peak photolysis to 195–205 nm. Maximum loss of N_2O occurs in the tropics at an altitude of ~30 km and drops rapidly at higher latitudes. The global loss of N_2O is about 13 Tg N/year. According to this loss and an observed increase of 4 Tg N/year in the total atmospheric burden, the global source of N_2O must be ~17 Tg N/year.

Experimental and theoretical studies indicate that photolysis isotopically fractionates N_2O and enriches the stratosphere in heavy N_2O species. Some of these species are returned to the troposphere via stratospheric–tropospheric exchange processes, balancing the flux of light N_2O from surface sources. They combine to produce global mean enrichments of $\delta^{15}N$ ~7‰ and $\delta^{18}O$ ~20‰ in the troposphere.

The destruction of N_2O and the subsequent enrichment of the heavy isotopologues can be described as a Rayleigh fractionation process, where each isotopologue is characterized by its own unique fractionation constant ε. The fractionation follows the order $|\varepsilon^{456}| > |\varepsilon^{448}| > |\varepsilon^{546}| > |\varepsilon^{447}|$. As the fractionation constants are dependent on wavelength and temperature, we expect to see ε vary throughout the stratosphere. While measurements clearly show that ε is non-constant, increasing from the lower to the upper stratosphere, its variability can be explained primarily through the mixing of air masses with different transport histories.

The understanding of the stratospheric sinks of N_2O is essentially complete. There remain some outstanding questions regarding the variability in the ratio of fractionation constants throughout the stratosphere and the possibility of non-standard chemical sinks. These are likely to be quite small and have little influence on the vertical structure of N_2O. Finally, it appears likely that the oxygen anomaly is primarily due to an isotopic transfer of heavy oxygen from ozone to N_2O. These unknown components of the stratospheric sink are small compared with the much larger uncertainties in the individual source fluxes of N_2O. Therefore, more measurements of the enrichment of N_2O in the lower stratosphere will be useful to better constrain the isotopic flux to the troposphere, and in turn to better constrain individual source fluxes.

References

Bates, D.R. and Hays, P.B. (1967) Atmospheric nitrous oxide. *Planetary Space Science* 15, 189–197.

Blake, G.A., Liang, M.-C., Morgan, C.G. and Yung, Y.L. (2003) A Born–Oppenheimer photolysis model of N_2O fractionation. *Geophysical Research Letters* 30; doi: 10.1029/2003GL016932.

Cantrell, C.A., Shetter, R.E. and Calvert, J.G. (1994) Branching ratios for the $O(^1D) + N_2O$ reaction. *Journal of Geophysical Research* 99, 3739–3743.

Cliff, S.S. and Thiemens, M.H. (1997) The $^{18}O/^{16}O$ and $^{17}O/^{16}O$ ratios in atmospheric nitrous oxide: a mass-independent anomaly. *Science* 278, 1774–1776.

Cliff, S.S., Brenninkmeijer, C.A.M. and Thiemens, M.H. (1999) First measurement of $^{18}O/^{16}O$ and $^{17}O/^{16}O$ ratios in stratospheric nitrous oxide: a mass-independent anomaly. *Journal of Geophysical Research* 104, 16171–16175.

Crützen, P.J. (1970) The influence of nitrogen oxides on the atmospheric ozone content. *Quarterly Journal of the Royal Meteorological Society* 96, 320–325.

Dore, J.E., Popp, B.N., Karl, D.M. and Sansone, F.J. (1998) A large source of atmospheric nitrous oxide from subtropical North Pacific surface waters. *Nature* 396, 63–66.

Garcia, R.R. and Solomon, S. (1994) A new numerical model of the middle atmosphere, 2. Ozone and related species. *Journal of Geophysical Research* 99, 12937–12951.

Greenblatt, G.D. and Ravishankara, A.R. (1990) Laboratory studies on the stratospheric NO_x production rate. *Journal of Geophysical Research* 95, 3539–3547.

Griffith, D.W.T., Toon, G.C., Sen, B., Blavier, J.F. and Toth, R.A. (2000) Vertical profiles of nitrous oxide isotopomer fractionation measured in the stratosphere. *Geophysical Research Letters* 27, 2485–2488.

Johnson, M.S., Billing, G.D., Gruodis, A. and Janssen, M.H.M. (2001) Photolysis of nitrous oxide isotopomers studied by time-dependent Hermite propagation. *Journal of Physical Chemistry A* 105, 8672–8680.

Johnston, H.S. and Selwyn, G.S. (1975) New cross sections for the absorption of near ultraviolet radiation by nitrous oxide (N_2O). *Geophysical Research Letters* 2, 549–551.

Johnston, H.S., Serang, O. and Podolske, J. (1979) Instantaneous global nitrous oxide photochemical rates. *Journal of Geophysical Research* 84, 5077–5082.

Johnston, J.C., Cliff, S.S. and Thiemens, M.H. (1995) Measurement of multioxygen isotopic ($\delta^{18}O$ and $\delta^{17}O$) fractionation factors in the stratospheric sink reactions of nitrous oxide. *Journal of Geophysical Research* 100, 16801–16804.

Kaiser, J., Brenninkmeijer, C.A.M. and Röckmann, T. (2002a) Intramolecular ^{15}N and ^{18}O fractionation in the reaction of N_2O with $O(^1D)$ and its implications for the stratospheric N_2O isotope signature. *Journal of Geophysical Research* 107; doi: 10.1029/2001JD001506.

Kaiser, J., Röckmann, T. and Brenninkmeijer, C.A.M. (2002b) Temperature dependence of isotope fractionation in N_2O photolysis. *Physical Chemistry Chemical Physics* 4, 4420–4430.

Kaiser, J., Röckmann, T., Brenninkmeijer, C.A.M. and Crützen, P.J. (2003) Wavelength dependence of isotope fractionation in N_2O photolysis. *Atmospheric Chemistry and Physics* 3, 303–313.

Kaiser, J., Röckmann, T. and Brenninkmeijer, C.A.M. (2004) Contribution of mass-dependent fractionation to the oxygen isotope anomaly of atmospheric nitrous oxide. *Journal of Geophysical Research* 109, D03305; doi: 10.1029/2003JD004088.

Khalil, M.A.K., Rasmussen, R.A. and Shearer, M.J. (2002) Atmospheric nitrous oxide: pattern of global change during recent decades and centuries. *Chemosphere* 47, 807–821.

Kim, K.-R. and Craig, H. (1990) Two-isotope characterization of N_2O in the Pacific Ocean and constraints on its origin in deep water. *Nature* 347, 58–61.

Kim, K.-R. and Craig, H. (1993) Nitrogen-15 and oxygen-18 characteristics of nitrous oxide: a global perspective. *Science* 262, 1855–1857.

Kroeze, C., Mosier, A. and Bouwman, L. (1999) Closing the global N_2O budget: a retrospective analysis 1500–1994. *Global Biogeochemical Cycles* 13, 1–8.

Liang, M.-C., Blake, G.A. and Yung, Y.L. (2004) A semianalytic model for photo-induced isotopic fractionation in simple molecules. *Journal of Geophysical Research* 109, D10308; doi: 10.1029/2004JD004539.

McElroy, M.B. and Jones, D.B. (1996) Evidence for an additional source of atmospheric N_2O. *Global Biogeochemical Cycles* 10, 651–659.

McLinden, C.A., Prather, M.J. and Johnson, M.S. (2003) Global modeling of the isotopic analogues of N_2O: stratospheric distributions, budgets, and the ^{17}O–^{18}O mass-independent anomaly. *Journal of Geophysical Research* 108, D84233; doi: 10.1029/2002JD002560.

Minschwaner, K., Anderson, G.P., Hall, L.A. and Yoshino, K. (1992) Polynomial coefficients for calculation O_2 Schumann–Runge cross sections at 0.5 cm^{-1} resolution. *Journal of Geophysical Research* 97, 10103–10108.

Minschwaner, K., Salawitch, R.J. and McElroy, M.B. (1993) Absorption of solar radiation by O_2: implications for O_3 and lifetimes of N_2O, $CFCl_3$, and CF_2Cl_2. *Journal of Geophysical Research* 98, 10543–10562.

Moore, H. (1974) Isotopic measurement of atmospheric nitrogen compounds. *Tellus* 26, 169–174.

Morgan, C.G., Allen, M., Liang, M.C., Shia, R.L., Blake, G.A. and Yung, Y.L. (2004) Isotopic fractionation of nitrous oxide in the stratosphere: comparison between model and observations. *Journal of Geophysical Research* 109, D04305; doi: 10.1029/2003JD003402.

Naqvi, S.W.A., Yoshinari, T., Jayakumar, D.A., Altabet, M.A., Narvekar, P.V., Devol, A.H., Brandes, J.A. and Codispoti, L.A. (1998) Budgetary and biogeochemical implications of N_2O isotope signatures in the Arabian Sea. *Nature* 394, 462–464.

Park, S., Atlas, E.L. and Boering, K.A. (2004) Measurements of N_2O isotopologues in the stratosphere: influence of transport on the apparent enrichment factors and the isotopologue fluxes to the troposphere. *Journal of Geophysical Research* 109, D01305; doi: 10.1029/2003JD003731.

Prasad, S.S. (1994) Natural atmospheric sources and sinks of nitrous oxide 1: an evaluation based on 10 laboratory experiments. *Journal of Geophysical Research* 99, 5285–5294.

Prasad, S.S. (1997) Potential atmospheric sources and sinks of nitrous oxide 2: possibilities from excited O_2, 'embryonic' O_3, and optically pumped excited O_3. *Journal of Geophysical Research* 102, 21527–21536.

Prasad, S.S., Zipf, E.C. and Zhao, X. (1997) Potential atmospheric sources and sinks of nitrous oxide 3: consistency with the observed distributions of the mixing ratios. *Journal of Geophysical Research* 102, 21537–21541.

Prather, M. and Ehhalt, D. (2001) Atmospheric chemistry and greenhouse gases. In: Houghton, J.T., Ding, Y., Griggs, D.J., Noguer, M., van der Linden, P.J., Dai, X., Maskell, K. and Johnson, C.A. (eds) *Climate Change 2001: The Scientific Basis*. Cambridge University Press, Cambridge, pp. 241–287.

Rahn, T. and Wahlen, M. (1997) Stable isotope enrichment in stratospheric nitrous oxide. *Science* 278, 1776–1778.

Rahn, T. and Wahlen, M. (2000) A reassessment of the global isotopic budget of atmospheric nitrous oxide. *Global Biogeochemical Cycles* 14, 537–543.

Rahn, T., Zhang, H., Wahlen, M. and Blake, G.A. (1998) Stable isotope fractionation during ultraviolet photolysis of N_2O. *Geophysical Research Letters* 25, 4489–4492.

Ramaswamy, V. (2001) Radiative forcing of climate change. In: Houghton, J.T., Ding, Y., Griggs, D.J., Noguer, M., van der Linden, P.J., Dai, X., Maskell, K. and Johnson, C.A. (eds) *Climate Change 2001: The Scientific Basis*. Cambridge University Press, Cambridge, pp. 349–416.

Röckmann, T., Brenninkmeijer, C.A.M., Wollenhaupt, M., Crowley, J.N. and Crützen, P.J. (2000) Measurement of the isotopic fractionation of 15N14N16O, 14N15N16O and 14N14N18O in the UV photolysis of nitrous oxide. *Geophysical Research Letters* 27, 1399–1402.

Röckmann, T., Kaiser, J., Brenninkmeijer, C.A.M., Crowley, J.N., Borchers, R., Brand, W.A. and Crützen, P.J. (2001a) Isotopic enrichment of nitrous oxide ($^{15}N^{14}NO$, $^{14}N^{15}NO$, $^{14}N^{14}N^{18}O$) in the stratosphere and in the laboratory. *Journal of Geophysical Research* 106, 10403–10410.

Röckmann, T., Kaiser, J., Crowley, J.N., Brenninkmeijer, C.A.M. and Crützen, P.J. (2001b) The origin of the anomalous or 'mass-independent' oxygen isotope fractionation in tropospheric N_2O. *Geophysical Research Letters* 28, 503–506.

Sander, S.P., Friedl, R.R., Golden, D.M., Kurylo, M.J., Huie, R.E., Orkin, V.L., Moortgat, G.K., Ravishankara, A.R., Kolb, C.E., Molina, M.J. and Finlayson-Pitts, B.J. (2003) *Chemical Kinetics and Photochemical Data for Use in Atmospheric Studies, Evaluation Number 14*. Jet Propulsion Laboratory, California Institute of Technology, Pasadena, California.

Selwyn, G.S. and Johnston, H.S. (1981) Ultraviolet absorption spectrum of nitrous oxide as function of temperature and isotopic substitution. *Journal of Chemical Physics* 74, 3791–3803.

Selwyn, G., Podolske, J. and Johnston, H.S. (1977) Nitrous oxide ultraviolet absorption spectrum at stratospheric temperatures. *Geophysical Research Letters* 4, 427–430.

Stein, L.Y. and Yung, Y.L. (2003) Production, isotopic composition, and atmospheric fate of biologically produced nitrous oxide. *Annual Review of Earth and Planetary Science* 31, 329–356.

Toyoda, S., Yoshida, N., Urabe, T., Aoki, S., Nakazawa, T., Sugawara, S. and Honda, H. (2001) Fractionation of N_2O isotopomers in the stratosphere. *Journal of Geophysical Research* 106, 7512–7522.

Toyoda, S., Yoshida, N., Urabe, T., Nakayama, Y., Suzuki, T., Tsuji, K., Shibuya, K., Aoki, S., Nakazawa, T., Ishidoya, S., Ishijima, K., Sugawara, S., Machida, T., Hashida, G., Morimoto, S. and Jonda, H. (2004) Temporal and latitudinal distributions of stratospheric N_2O isotopomers. *Journal of Geophysical Research* 109, D08308; doi: 10.1029/2003JD004316.

Turatti, F., Griffith, D.W.T., Wilson, S.R., Esler, M.B., Rahn, T., Zhang, H. and Blake, G.A. (2000) Positionally dependent ^{15}N fractionation factors in the UV photolysis of N_2O determined by high resolution FTIR spectroscopy. *Geophysical Research Letters* 27, 2489–2492.

Umemoto, H. (1999) $^{14}N/^{15}N$ isotope effect in the UV photodissociation of N_2O. *Chemical Physics Letters* 314, 267–272.

Wahlen, M. and Yoshinari, T. (1985) Oxygen isotope ratios in N_2O from different environments. *Nature* 313, 780–782.

Yoshida, N. and Matsuo, S. (1983) Nitrogen isotope ratio of atmospheric N_2O as a key to the global cycle of N_2O. *Geochemical Journal* 17, 231–239.

Yoshida, N., Hattori, A., Toshiro, S., Matsuo, S. and Wada, E. (1984) $^{15}N/^{14}N$ ratio of dissolved N_2O in the eastern tropical Pacific Ocean. *Nature* 307, 442–444.

Yoshinari, T. and Wahlen, M. (1985) Oxygen isotope ratios in N_2O from nitrification at a wastewater treatment facility. *Nature* 317, 349–350.

Yung, Y.L. and Miller, C.E. (1997) Isotopic fractionation of stratospheric nitrous oxide. *Science* 278, 1778–1780.

Yung, Y.L., Liang, M.C. and Blake, G.A. (2004) Evidence for O-atom exchange in the $O(^1D) + N_2O$ reaction as the source of mass-independent isotopic fractionation in atmospheric N_2O. *Geophysical Research Letters* 31, L19106; doi: 10.1029/2004GL020950.

Zhang, H., Wennberg, P.O., Wu, V.H. and Blake, G.A. (2000) Fractionation of $^{14}N^{15}N^{16}O$ and $^{15}N^{14}N^{16}O$ during photolysis at 213 nm. *Geophysical Research Letters* 27, 2481–2484.

15 Sinks for Nitrous Oxide at the Earth's Surface

Carolien Kroeze[1], Lex Bouwman[2] and Caroline P. Slomp[3]
[1]*Environmental Sciences, Environmental Systems Analysis Group, Wageningen University, Wageningen, The Netherlands;* [2]*Netherlands Environmental Assessment Agency, RIVM, Bilthoven, The Netherlands;* [3]*Department of Earth Sciences – Geochemistry, Faculty of Geosciences, Utrecht University, Utrecht, The Netherlands*

15.1 Introduction

Nitrous oxide (N_2O) is one of the most important greenhouse gases. Although it is only a trace component in the atmosphere, it is the third most important contributor to the present radiative forcing. Furthermore, N_2O is involved in the catalytic destruction of stratospheric ozone (Crutzen, 1970). Atmospheric concentrations of N_2O have been increasing for at least a century from ~275 ppbv in pre-industrial times to more than 310 ppbv at present. Studies on the N_2O budget have, so far, mainly focused on the sources of N_2O in order to explain this increase. Relatively little is known about the soil sinks for N_2O. Hence, it is difficult to balance the global N_2O budget.

Initially, coal-generated power plants were considered to be the most important source of N_2O. However, we now know that fossil fuel combustion is in fact a relatively small source of N_2O and, since the late 1990s, it is widely accepted that the increase in atmospheric N_2O is largely associated with agricultural activities. This anthropogenic source of N_2O is estimated to be of the same order of magnitude as the natural release of N_2O from soils and surface waters (Kroeze *et al.*, 1999). Although the most important sources of N_2O have thus been identified, the actual rates of emissions on a global scale are subject to large uncertainties. Current emission estimates typically have large uncertainty ranges. For instance, global emissions from agriculture as calculated using the Intergovernmental Panel on Climate Change (IPCC) guidelines are estimated at ~6 Tg N/year, with an uncertainty range of 1.2–17.9 Tg N/year (Mosier *et al.*, 1998). These global budgets usually ignore the possibility of potential soil sinks or assume that the estimates for biogenic sources of N_2O are in fact net fluxes that account for possible sinks. However, this may lead to errors in estimates of atmospheric budgets and trends in emissions as pointed out by Cicerone (1989), who showed that atmospheric mass balance calculations are relatively sensitive to uncertainties in the sink strength.

Most observations of N_2O removal at the Earth's surface relate to soils. Here, both uptake and emission of N_2O may occur simultaneously. The direction of the net flux, determining whether soils are a net source or sink, depends on the environmental factors regulating consumption and production. A useful framework for measuring and modelling N_2O exchange between soils and the atmosphere is provided by the compensation point concept introduced by Conrad and Dentener (1999):

> At the N_2O compensation point the concentration of N_2O in the soil gas phase for

 Greenhouse Gas Sinks (eds D.S. Reay, C.N. Hewitt, K.A. Smith and J. Grace)

which there is no net exchange of N_2O between ambient atmosphere and soil gas phase. Sink activity can occur when the ambient N_2O concentration in the atmosphere exceeds that in the soil gas phase. The compensation point thus determines the direction of the flux at a given ambient atmospheric N_2O concentration.

The compensation concentrations of N_2O are typically significantly higher than the ambient atmospheric concentrations. This is in agreement with the observation that most soils are a source of atmospheric N_2O. An increasing number of studies report on occasional consumption by soils, suggesting that during uptake the compensation concentrations are lower than ambient concentrations, i.e. <310 ppbv. Conrad and Dentener (1999) proposed that the compensation concentration for N_2O increases along with temperature, soil moisture and nitrogen availability.

A further consideration is the fast increase in atmospheric nitrogen deposition as observed in many countries (Bouwman *et al.*, 2002b). Even small deposition rates over prolonged periods of time may lead to changes in the nitrogen cycle of sensitive ecosystems (Bobbink *et al.*, 1998). Nitrogen enrichment of ecosystems (partly) suppresses nitrogen limitation and thus may also lead to a reduced N_2O sink strength. Even when sink activity is seasonal, this reduction of sinks for N_2O may be important considering the vast areas of nitrogen-affected ecosystems and may also contribute to increasing atmospheric N_2O concentrations.

The question arises as to what extent sinks for atmospheric N_2O at the Earth's surface are quantitatively important. So far, very little attention has been paid to the sink strength, because the uptake of N_2O by soils is generally considered to be small on a global scale. This uptake is driven by bacterial denitrification, which converts N_2O into nitrogen gas (N_2). In most global N_2O budgets bacterial reduction of N_2O is simply included in the total net flux of N_2O from soils. However, its role may be important in reducing the net release of N_2O through denitrification in the soil profile.

In this study, we will analyse the processes underlying bacterial reduction of N_2O to N_2, and discuss the likeliness of N_2O uptake in different systems. We will focus not only on soils but also on aquatic systems, including groundwater systems, riparian zones and surface waters. We will first define what we consider a sink for N_2O (Section 15.2). Next, we will describe the processes involved and identify the most important factors influencing N_2O uptake (Section 15.3). We will then discuss some observed fluxes as reported in the literature (Section 15.4). Finally, we will analyse the likeliness of N_2O uptake in different systems (Section 15.5).

15.2 Defining Sinks

When do we consider a system a sink for atmospheric N_2O? This question is not easily answered.

Many terrestrial and aquatic systems exchange N_2O with the atmosphere (Fig. 15.1). Most N_2O is formed during bacterial denitrification, during which nitrate (NO_3^-) is reduced to the gaseous nitrogen compounds nitric oxide (NO) and N_2O. In some cases, the denitrification process stops here, so that the N_2O can escape to the atmosphere (fluxes F1 and F3 in Fig. 15.1).

One of the characteristics of N_2O is that it easily dissolves in water. As a result, atmospheric N_2O may enter soil pores, and if the soil is wet, N_2O may be dissolved in soil water (fluxes F2 and F4 in Fig. 15.1). Consecutively, it may be denitrified by bacteria to N_2. This way, soils and aquatic systems can remove N_2O from the atmosphere.

Since N_2O and NO_3^- may be lost from top soils through leaching and runoff (fluxes F5 and F7 in Fig. 15.1), groundwater and riparian zones can also act as potential important sinks for atmospheric N_2O. There may also be some reverse flows of N_2O and NO_3^- from subsoils to the top soils in case of lateral water flows (fluxes F6 and F8 in Fig. 15.1). However, these fluxes are usually much smaller than the leaching fluxes (F6 << F5 and F8 << F7). Through leaching (and runoff), N_2O and NO_3^- may be transported not only to subsoils but also to deeper groundwater, riparian zones and, eventually, to

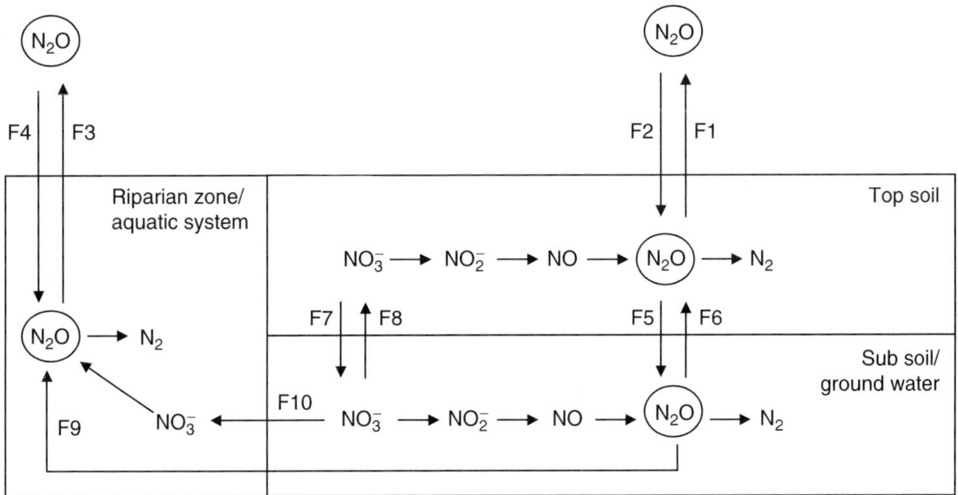

Fig. 15.1. Sources and sinks for nitrous oxide (N_2O) at the Earth's surface.

aquatic systems such as rivers and coastal waters (fluxes F9 and F10 in Fig. 15.1). In the course of this transport, denitrification may take place, during which N_2O is formed from NO_3^- and/or reduced to N_2. However, we suspect that subsoils and groundwater systems do not play an important role as a sink for atmospheric N_2O.

Whether a system is considered a source or a sink depends on the strengths of the fluxes F1, F2, F3 and F4. Net N_2O uptake may occur when N_2O uptake exceeds emission. A complicating factor is the temporal and spatial variability of all these fluxes. Both the emissions (F1 and F3) and uptake (F2 and F4) of N_2O are extremely variable in time and space. However, in most soils and aquatic systems overall emissions largely exceed overall uptake.

In this chapter, we concentrate on systems in which net N_2O uptake occurs (F2 > F1 or F4 > F3) over a relatively large area (e.g. 1 ha) and prolonged period of time (e.g. 1 year) because it is very difficult to assess the importance of temporal sink activity in systems with overall net emissions. Our aim is to deduce some general characteristics of systems that act as an N_2O sink and to identify their likely importance on a global scale.

15.3 Factors Influencing N_2O Reduction

15.3.1 The process of denitrification and N_2O reduction

In denitrification, bacteria use NO_3^- as electron acceptor, usually in the absence of oxygen. As such, denitrification could be considered as a way of breathing under low oxygen conditions. The bacteria involved first reduce NO_3^- to nitrite (NO_2^-), and then to NO, N_2O and/or N_2 (Fig. 15.1). Denitrification is the largest natural source of atmospheric N_2O. Most N_2O formed during denitrification will, however, never reach the atmosphere, because bacteria reduce it to N_2 in the last step.

Denitrification is an anaerobic, microbially mediated process that, in most aquatic and sedimentary environments, is coupled to the oxidation of organic matter. Only in groundwater systems, pyrite (FeS_2) and other Fe^{2+}-containing minerals may also act as an important electron donor (Appelo and Postma, 1993). If NO_3^- is low or absent, denitrifying organisms are able to use only N_2O as an electron acceptor (Butterbach-Bahl *et al.*, 1998).

The capacity to denitrify with reactive organic carbon as an electron donor is widespread among different types of facultative anaerobic bacteria. However, only a few genera are abundantly present in soils, sediments, marine water and freshwater such as the *Pseudomonas* and *Alcaligenes* species (Tiedje, 1988). The organisms responsible for denitrification coupled to pyrite oxidation may involve species of *Thiobacillus* and *Gallionella* (e.g. Appelo and Postma, 1993). Generally, it can be assumed that if sufficient electron donors – either as organic carbon or pyrite – and nitrogen oxides are present in a low-oxygen environment, bacteria with the appropriate metabolic capability will occupy the denitrification niche. All denitrifiers in natural environments are assumed to be capable of the complete denitrification sequence resulting in N_2 as the final product (Firestone and Davidson, 1989).

The key factors limiting rates of denitrification vary from environment to environment. Most surface soils, particularly fertilized soils, contain sufficient organic carbon and NO_3^-, and the presence of oxygen most commonly limits denitrification. The same generally holds true for the water column of lakes and the oceans. In aquatic sediments, oxygen penetration is strongly diffusion-limited and the availability of NO_3^-, NO and N_2O is generally the limiting factor (Firestone and Davidson, 1989). In contrast, electron donor availability often controls denitrification rates in many groundwater systems. As a consequence, plumes of NO_3^- and/or N_2O-containing groundwater may persist in an oxygen-depleted environment (DeSimone and Howes, 1998; Groffman *et al.*, 1998). Riparian zones are an exception because of the relatively large input of organic matter. This often allows for efficient removal of NO_3^- from groundwater, prior to discharge to surface waters (Vidon and Hill, 2004). This removal of NO_3^- may be accompanied by significant N_2O production (Hefting *et al.*, 2003).

In soils, denitrification is extremely variable in space and time. This makes it difficult to accurately measure rates of denitrification (Hofstra and Bouwman, in press; Parkin, 1990) and N_2O exchange (Lapitan *et al.*, 1999) in the field. This variability is related to the microscale (millimetre to centimetre) inhomogeneity of well-drained soils caused by the presence of, for example, earthworm castings, particles of decomposing organic matter and soil aggregates. The outside of organic carbon-rich particles and aggregates is oxic and can support high rates of nitrification while denitrification occurs in the suboxic interior (Parkin, 1987). These microsites explain why denitrification can occur even in well-drained soils.

NO_3^- production in soils (nitrification) occurs only under aerobic conditions. The oxygen status and thus coupled nitrification–denitrification in soils can change rapidly depending upon soil moisture and the consequent rate of oxygen diffusion through soils (Tiedje, 1988). An example of the rapid changes in denitrification with changing oxygen status is the pulse of soil denitrification often seen after episodic rainfall or irrigation events (Rolston *et al.*, 1982; Ryden, 1983; Sexstone *et al.*, 1985; Van Kessel *et al.*, 1993). If anoxic conditions persist until NO_3^- is depleted, soil denitrifiers may become limited by NO_3^-.

15.3.2 Identification of factors that influence N_2O reduction

Whether denitrification in a given environment acts as a net source or sink for N_2O depends on the relative abundance of oxygen, NO_3^- and suitable electron donors. Additional important factors include pH and temperature, as well as the possibility for uptake of N_2O from the atmosphere. Generally, N_2O reduction is highest when oxygen and NO_3^- are low, and when N_2O and a suitable electron donor are available (Firestone and Davidson, 1989).

N_2O reductases, the enzymes needed for the conversion of N_2O to N_2, are the most oxygen-sensitive of the enzymes involved in denitrification. A small amount of oxygen is sufficient to inhibit the conversion and will lead to accumulation instead of removal of N_2O (Betlach and Tiedje, 1981; Blicher-Matthiesen and Hoffmann, 1999). Thus environments such as waterlogged soils, sediments, groundwater and riparian zones, where diffusive transport of oxygen is limited, can potentially be sites for N_2O reduction.

High NO_3^- concentrations also commonly inhibit N_2O reductase activity (Blackmer and Bremner, 1978; Blicher-Mathiesen and Hoffmann, 1999). This explains the strong positive relationship between NO_3^- availability and N_2O accumulation generally observed for soils (e.g. Skiba *et al.*, 1998). It also explains why net uptake of N_2O is typically observed when dissolved NO_3^- concentrations in grassland and forest soils are low (e.g. <1–2 μg NO_3^--N/g soil; Ryden, 1981; Butterbach-Bahl *et al.*, 1998). NO_3^- concentrations are expected to be lowest in environments where use of N-fertilizers is limited and/or plant uptake of nitrogen is high, and where nitrification does not occur (e.g. due to a lack of oxygen and ammonia). In these environments, where electron acceptors are limiting, N_2 will be the principal gas evolved.

In most anoxic soils and natural waters, sufficient organic carbon is present to support N_2O reduction. In groundwater systems, where denitrification is frequently electron donor-limited, N_2O leached from surface soils can be transported conservatively through the system, thus making N_2O a sensitive tracer of plume movement (DeSimone and Howes, 1998). The electron donor limitation in groundwater may be alleviated when, prior to discharge, the groundwater passes through an organic carbon or pyrite-rich riparian zone. When the NO_3^- concentration in groundwater is low, such a riparian zone can act as a net sink for groundwater N_2O. When the NO_3^- concentration is high, particularly when compared to the availability of a reductant, N_2O is expected to accumulate.

Denitrification rates under field conditions are lower in acidic soils when compared to neutral or slightly alkaline soils. The amount of N_2O relative to N_2 formed during denitrification is also pH-dependent and generally decreases with increasing pH (Simek and Cooper, 2002). With these data, highest rates of N_2O reduction are expected in neutral to alkaline soils. The soil pH is also a major control of nitrification either directly or through its effect on soil cation exchange capacity (Robertson, 1989). Optimum values for nitrification range from 6.5 to 8 (Simek and Cooper, 2002). Thus, at neutral pH, enhanced NO_3^- production through nitrification could limit the possibility of a soil to act as a net sink for atmospheric N_2O.

Denitrification rates are very sensitive to temperature. Typical Q_{10} values determined for denitrification range from 5 to 16 (Ryden, 1983). At the same time, the ratio of N_2O to N_2 produced during denitrification decreases with increasing temperature (Firestone and Davidson, 1989). This indicates that the potential for removal of N_2O through denitrification is enhanced at higher temperatures.

The nitrogen cycle includes some dynamic aspects that may be important for net removal of N_2O from the atmosphere. For instance, Bakken and Bleken (1998) show that it can take decades to centuries before nitrogen is transported from soils to coastal waters depending on the route of transport (e.g. via subsoils or freshwaters). Ignoring these temporal characteristics in nitrogen budgets may lead to errors in estimates of nitrogen loads in aquatic systems, as well as in estimates of N_2O fluxes associated with nitrogen leaching and runoff.

Net N_2O uptake at the Earth's surface will occur only when *in situ* N_2O concentrations in groundwater, soil and surface waters are lower than aqueous concentrations in equilibrium with the atmosphere. Thus, we conclude that such conditions most likely occur in soils and riparian zones with low nitrogen loading. Groundwater and most surface waters are expected to be less important as a sink for N_2O because conditions are generally more conducive for N_2O production due to either abundant presence of oxygen and/or NO_3^- or electron donor limitation. As indicated in Section 15.1, the atmospheric N_2O concentration is currently increasing. Thus, the potential for N_2O uptake by low oxygen and low NO_3^- soil environments could increase in the future.

15.4 Observed Sinks for N_2O

Only a few studies are available on observed sinks for N_2O. Table 15.1 presents selected examples of observed sinks for N_2O. As explained earlier, most terrestrial and aquatic systems have the potential to act

Table 15.1. Selected examples of observed sinks for nitrous oxide (N_2O).

System/location	Observations	Period	Measurement technique	Remarks	Reference
Fertilized grassland, poorly drained loamy soil, Berkshire, UK	Annual mean sink of up to 3.2 μg N_2O-N/m^2/h	1 year	Open chamber, 3–6 gas samplings per week	N_2O uptake at moderate to high water content, low nitrate content (2–3 weeks after fertilization with 250 kg N/ha) and soil temperature >5°C	Ryden (1981, 1983)
Fertilized grassland, Siggen, south-west Germany	Some incidents of N_2O uptake, with singular large uptakes of 41 μg N_2O-N/m^2/h	1.5 years	Closed chamber, four replicates, 1–3 gas samplings per week	It is not clear whether heavily fertilized grasslands with N_2O uptake are an exception or not	Glatzel and Stahr (2001)
Unfertilized grassland, loam, poorly drained, Guelph, Ontario, Canada	Annual mean of 1.14 μg N_2O-N/m^2/h for April 1993 to March 1994	3 years	Micrometeorological flux gradient method, one measurement per month	Soil was a sink in 1993/94 possibly due to high rainfall in June; in 1992 and 1995 the site was a small source for N_2O; fluxes were often below detection limit	Wagner-Riddle *et al.* (1997)
Unfertilized grassland, clay loam, well-drained, Guelph, Ontario, Canada	Annual mean uptake of 0.8 (1996) to 1.7 (1997) μg N_2O-N/m^2/h	3 years	Micrometeorological flux gradient method, daily estimate based on hourly values	Soil was a net source in 1995	Maggiotto *et al.* (2000)
Managed grassland, Paragominas, Pará State, Brazil	Net uptake in dry season of 2 μg N_2O-N/m^2/h	2 years	Closed chamber, one measurement per month	This 'active pasture' site is a net source of N_2O with overall annual emission of 2.85 μg N_2O-N/m^2/h	Verchot *et al.* (1999)
Fertilized rice paddy, early rice, top position on slope, clay soil, Yingtan, China	Small seasonal sink of 0.63 μg/m^2/h N_2O-N in one of three plots	36 days	Closed chamber, two replicates, sampling every 3–7 days	Water management, N fertilizer (122 kg N/ha) and rice straw (7 t with high C/N) application rate influenced N_2O formation and uptake	Xu *et al.* (1997)
Spruce forest, sandy-loamy soil, Villingen, Germany	3-year mean sink of 1 μg N_2O-N/m^2/h with periods of emissions and uptake	3 years	Closed chamber, eight replicates, hourly measurements in nine campaigns of 14 days in 3-year period in spring, summer, fall and winter	Unfertilized soil in N-limited spruce forest; N application (150 kg N/ha) caused system to turn into an N_2O-N source of 0.9 μg/m^2/h	Papen (2001)
Spruce forest, drained peat soil, Scottish border, UK	Mean sink of 0.3 μg N_2O-N/m^2/h	150 days	Closed chamber, August–May with <1 measurement per week	N deposition of 48–96 kg N/ha/year caused system to turn into a source of N_2O-N of 0.5–5.7 μg/m^2/h	Skiba *et al.* (1999)

Deciduous forest, acid sand soil, Poppel, Belgium	Mean 2-year sink of 7.4 µg N_2O-N/m^2/h	21 months (631 days)	Closed chamber, six replicates, weekly to event-based measurements	High N_2O-N uptake of 1.3 kg/ha over a 2-year period was attributed to low nitrification activity at pH H_2O of 3.8; most negative fluxes were observed in wet periods (WPFS > 35%)	Goossens *et al.* (2001)
Spruce-fir forest, podzol, Mt Ascutney, Vermont, and Mt Washington, New Hampshire, USA	Mean sink of 1.12 µg N_2O-N/m^2/h (Mt Ascutney) and 0.23 µg N_2O-N/m^2/h (Mt Washington)	83 days (May–August)	Closed chamber, six replicates, one measurement per 2 weeks	Sink activity was attributed to low nitrification activity and thus low nitrate availability and high soil moisture content	Castro *et al.* (1993)
Permanently water-covered freshwater riparian fen in Denmark	In a 15 m transect from hill to fen, nitrate, oxygen and N_2O concentrations declined to very low levels, while N_2 concentrations increased	October 1993	Groundwater analysis (N compounds) and sediment slurry incubations, including ^{15}N experiments	Denitrification served as a sink for both dissolved N_2O in groundwater recharging the fen, and the N_2O produced within the riparian sediment	Blicher-Matthiesen and Hoffmann (1999)
Three riparian sites in Belgium	Closed chamber measurements show small fluxes, both negative and positive. Observed sink: −0.6 ± 0.4 mg N_2O-N/m^2/day	6 days in the period from August 2000 to Spring 2001	Closed chambers, three replicates, each box sampled five times over a total period of 72 min	In N-rich landscapes, riparian zones can act as (large) sources or small sinks for atmospheric N_2O; optimal location for denitrification, relatively low nitrate	Dhondt *et al.* (2004)
15 Pre-alpine deep lakes	Under complete anoxic conditions, water layers can be depleted of N_2O	Fall 1995	Degree of N_2O saturation based on measured concentrations in water	None of the lakes was found to act as a net sink for atmospheric N_2O, because of poor mixing of anoxic layers and surface waters	Mengis *et al.* (1997)
Antarctic surface waters (Belligausen Sea and Drake Passage)	North of the upwelling region, antarctic surface water formed from mixing of surface water and ice melt was moderately depleted of N_2O	November/ December 2001	N_2O concentration measurements in water and atmosphere, using two sampling modes	Source and sink areas were found in the studied area; overall there was a small negative flux	Rees *et al.* (1997)

as a temporary sink. In this overview, we focus only on studies in which a net sink was observed over a relatively long period of time.

The techniques and frequencies of the N_2O flux measurements presented in Table 15.1 show a large variability. As has been discussed elsewhere (Bouwman *et al.*, 2002a; Hofstra and Bouwman, 2005) this has a major effect on measured N_2O and NO emissions and denitrification. Especially in experiments with a low frequency of gas sampling the uncertainty in flux estimates is large, and important flux 'events' may have been missed.

Sinks for N_2O have been reported for terrestrial and aquatic systems, as well as for riparian zones. These systems are generally sources of N_2O. However, when nitrogen availability and oxygen concentrations are low, soils, and in some cases lakes and certain regions of the ocean, could act as a sink for atmospheric N_2O. We discuss a limited number of studies in which such N_2O sinks were reported. In theory, soils are particularly powerful sinks, as illustrated by Blackmer and Bremner (1976), who showed by laboratory experiments that under conditions that are favourable for denitrification, the capacity of soils for N_2O uptake is larger than their capacity for release.

Agricultural soils are usually fertilized, and therefore not considered to act as a sink for N_2O. Nevertheless, some studies report episodes of uptake of atmospheric N_2O in fertilized fields. For instance, uptake of N_2O was observed in fertilized grasslands in the UK (Ryden, 1981, 1983) under the following conditions: high water content, very low NO_3^- availability and soil temperature exceeding 5°C. Sink activity was observed a few weeks after fertilization. Immediately after fertilization, the soils were considerable sources of N_2O. Glatzel and Stahr (2001) also report on occasional N_2O uptake in fertilized grassland soils, but do not answer to what extent these observations are exceptions. It is likely that, in general, fertilized soils are net sources of N_2O on an annual basis. Wagner-Riddle *et al.* (1997) reported sink activity in unfertilized grassland in 1 out of 3 years. In the other 2 years, the site acted as a small source for N_2O. Maggiotto *et al.* (2000) also observed sink activity in 2 out of 3 years in a grassland site. In a tropical 'active pasture', soils acted as N_2O sinks during the dry season, and as sources during the wet season, leading to a net annual source (Verchot *et al.*, 1999).

In rice paddy soils also, N_2O uptake was observed by Xu *et al.* (1997; Table 15.1). They measured N_2O fluxes in three rice paddy fields by closed chamber techniques. Of the three, a net uptake of N_2O was measured in one field. Xu *et al.* (1997) suggest that this was related to the water management, the low fertilizer nitrogen inputs in this system and a high carbon/nitrogen ratio of rice straw return. Despite this observed sink, their results indicate that rice paddies are generally sources rather than sinks for N_2O.

Forest soils are generally considered to be sources of N_2O. However, some studies indicate that forest soils can in fact also be net sinks for N_2O. For instance, Papen *et al.* (2001) report on net sinks in German forest soils and argue that increased atmospheric nitrogen deposition may change forest soils from net sinks to net sources.

Also Skiba *et al.* (1999) show that unpolluted forest soils can occasionally be a significant sink for N_2O. In a drained peat area, they measured maximum uptake rates of -15 μg N/m^2/h. However, they do not extrapolate their hourly fluxes to longer-term fluxes, making it difficult to draw conclusions on the net uptake over a longer period of time. Also Goossens *et al.* (2001) report on relatively large N_2O uptake in a Belgian forest soil. They observed negative N_2O fluxes at low soil pH (3.8), indicating that nitrification is inhibited, during periods when the water-filled pore space exceeded 35%. Castro *et al.* (1993) observed occasional N_2O uptake in forest soils during summer in north-eastern USA and attributed this to the low nitrification activity.

There are numerous examples of measurements where fluxes are below detection (Conrad *et al.*, 1983; Slemr *et al.*, 1984; Anderson and Poth, 1989; Papen *et al.*, 1993; Jambert *et al.*, 1994; Serca *et al.*, 1994; Poth *et al.*, 1995; Clemens *et al.*, 1997; Butterbach-Bahl *et al.*, 1998; Kilian *et al.*, 1998; Petersen, 1999; Riley and Vitousek,

2000). In fact, these soils may have acted as small sinks as well as small sources. Similar to the observations of Wagner-Riddle *et al.* (1997) and Maggiotto *et al.* (2000), who measured sink activity in some years and emissions in other years, sites with small fluxes may change from sources to sinks depending on the environmental conditions.

N_2O can be denitrified during leaching and runoff or in groundwater flows. For instance, Blicher-Matthiesen and Hoffmann (1999) studied NO_3^-, oxygen and N_2O concentrations in a 15 m transect from a hill to a permanently water-covered freshwater riparian fen. They observed that NO_3^-, oxygen and N_2O concentrations in this transect declined to very low levels, while N_2 concentrations increased. This indicates that, during transport from hill to fen, NO_3^- is denitrified and N_2O reduced to N_2. However, riparian zones can also be sources of N_2O. In particular in nitrogen-rich regions, riparian zones may act as a source of N_2O (Hefting *et al.*, 2003). Dhondt *et al.* (2004) measured N_2O fluxes in three riparian zones in nitrogen-rich landscapes using closed chambers. They observed small N_2O fluxes, both positive and negative, and concluded that in these particular riparian zones the denitrification of NO_3^- was not leading to large emissions of N_2O. However, they also argue that a better understanding of the controls of the N_2O/N_2 ratio in riparian zones that are meant to reduce nitrogen inputs to surface waters is important, to avoid increased N_2O emissions as a result of water pollution control. In other words, riparian zones in eutrophic landscapes may be sources of atmospheric N_2O rather than sinks. They also concluded that denitrification is controlled by the geomorphology of the river valley, rather than by the vegetation cover.

Whether lakes can act as a sink for atmospheric N_2O is not clear. Many lakes have anoxic zones that are undersaturated with N_2O. This is generally considered to be the result of N_2O consumption by denitrifying bacteria. These zones could therefore act as a sink for N_2O that is dissolved in the water. This N_2O may be formed in the aquatic system or may have leached from terrestrial systems. It may also be of atmospheric origin and diffuse into the water from the atmosphere. In the latter case, the lake can, in theory, be a sink for atmospheric N_2O. This is why Knowles *et al.* (1981) hypothesized that eutrophic, anoxic lakes may be sinks for atmospheric N_2O. However, observations by Mengis *et al.* (1997) do not support this hypothesis. Mengis *et al.* (1997) analysed oxygen and N_2O concentrations in 15 prealpine lakes (Table 15.1). They observed that in the oxic waters, N_2O concentrations typically increase with decreasing oxygen concentrations. They also observed that anoxic water layers are N_2O-undersaturated and concluded that N_2O is consumed in completely anoxic layers. Nevertheless, their results do not support the hypothesis that eutrophic anoxic lakes may act as a sink for N_2O, because the surface layers in all the 15 lakes were supersaturated with N_2O, because of poor mixing of the deep anoxic layers with the surface waters.

In the Antarctic Ocean, both source and sink areas were observed during an ocean flux study expedition in the spring of 1992 (Rees *et al.*, 1997; Table 15.1). Whether or not an area acted as a source or sink was determined by hydrographical characteristics. Rees *et al.* (1997) hypothesize that upwelling water, supersaturated with N_2O, can accumulate under the sea ice during winter. During spring, when the ice melts, this N_2O-rich water is exposed to the atmosphere. The N_2O-rich water is, however, mixed with N_2O-depleted ice melt. As a result, a water layer can be formed that is undersaturated with N_2O, acting as a seasonal sink for atmospheric N_2O.

Further net N_2O removal could occur in the relatively small volume of the oceans in which oxygen concentrations fall to near-zero levels at intermediate water depths, where denitrification is an important process. Results for the Somali Basin, however, suggest that upwelling regions may usually be a net source of N_2O (De Wilde, 1999).

15.5 Likeliness of N_2O Sinks

From the literature review and the data in Table 15.1, we selected three major factors to distinguish areas prone to soil N_2O consumption at the global scale. These factors are nitrogen availability (or nitrogen limitation), soil wetness and temperature. These appear to be the major controls that are consistently observed in nearly all studies and are therefore most suitable to delineate global areas with possible N_2O sink activity.

We used 0.5° × 0.5° resolution maps for both inputs and outputs. Surface nitrogen inputs from atmospheric nitrogen deposition, biological N_2 fixation, nitrogen fertilizer and animal manure were taken from Bouwman *et al.* (2005) (Fig. 15.2). Surface nitrogen inputs were grouped into five classes: nitrogen loading of <5 kg/ha/year is assigned a value of 5, 5–10 kg/ha/year a value of 4, 10–15 kg/ha/year a value of 3, etc. (Fig. 15.2). This classification is largely based on the threshold of 10 kg N/ha/year, above which changes to the nitrogen cycle may occur in sensitive ecosystems (Bobbink *et al.*, 1998), and the assumption that the sink strength increases with increasing nitrogen availability.

For soil wetness, we used an index obtained from the annual net precipitation (annual precipitation minus evapotranspiration) from Van Drecht *et al.* (2003) divided by the total soil water-holding capacity for the top meter of soil (from Batjes, 1997; Batjes, 2002). The maps of classified surface nitrogen inputs and wetness index were added and areas with no month with mean temperatures >5°C were excluded. Hence, we assume N_2O uptake is negligible below 5°C, which is in accordance with the work of Ryden (1983).

By adding the numbers of both classified maps, we obtain a maximum value of 10 and a minimum of 2. These were classified into 'high' (values >8), 'moderate' (values = 7–8) and 'low' (values <7) (Fig. 15.3). As expected on the basis of the literature review, nitrogen limitation occurs in areas remote from agricultural and industrial activity, mainly in northern latitudes, and semiarid and arid regions. Ecosystems in northern latitudes are known to be nitrogen-limited due to low atmospheric nitrogen deposition and the absence or low activity of biological nitrogen fixation (Crews, 1999; Sprent, 1999). When the soil wetness index is combined with this map, the arid and semiarid areas are excluded, and the major regions prone to high N_2O sink activity are located mainly in northern regions (Fig. 15.3). According to this quick

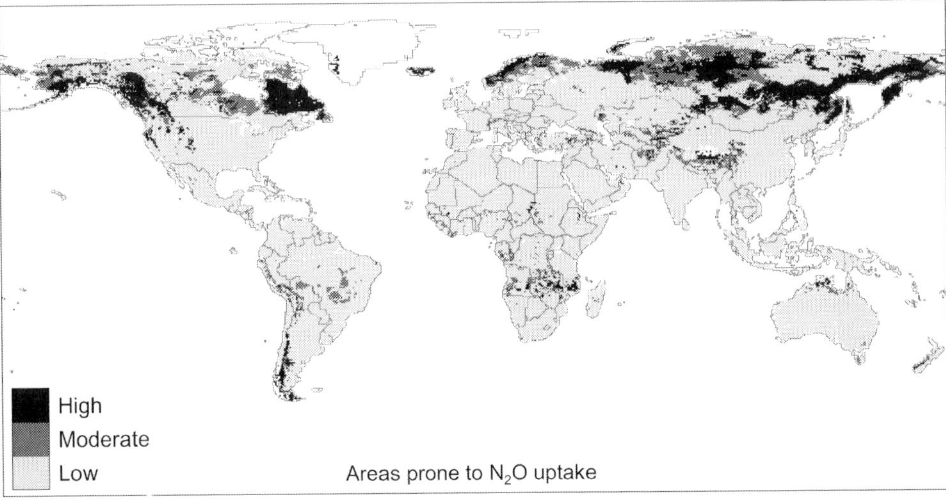

Fig. 15.2. Area in classes of probability of nitrous oxide (N_2O) sink activity calculated from data on surface nitrogen inputs and soil wetness.

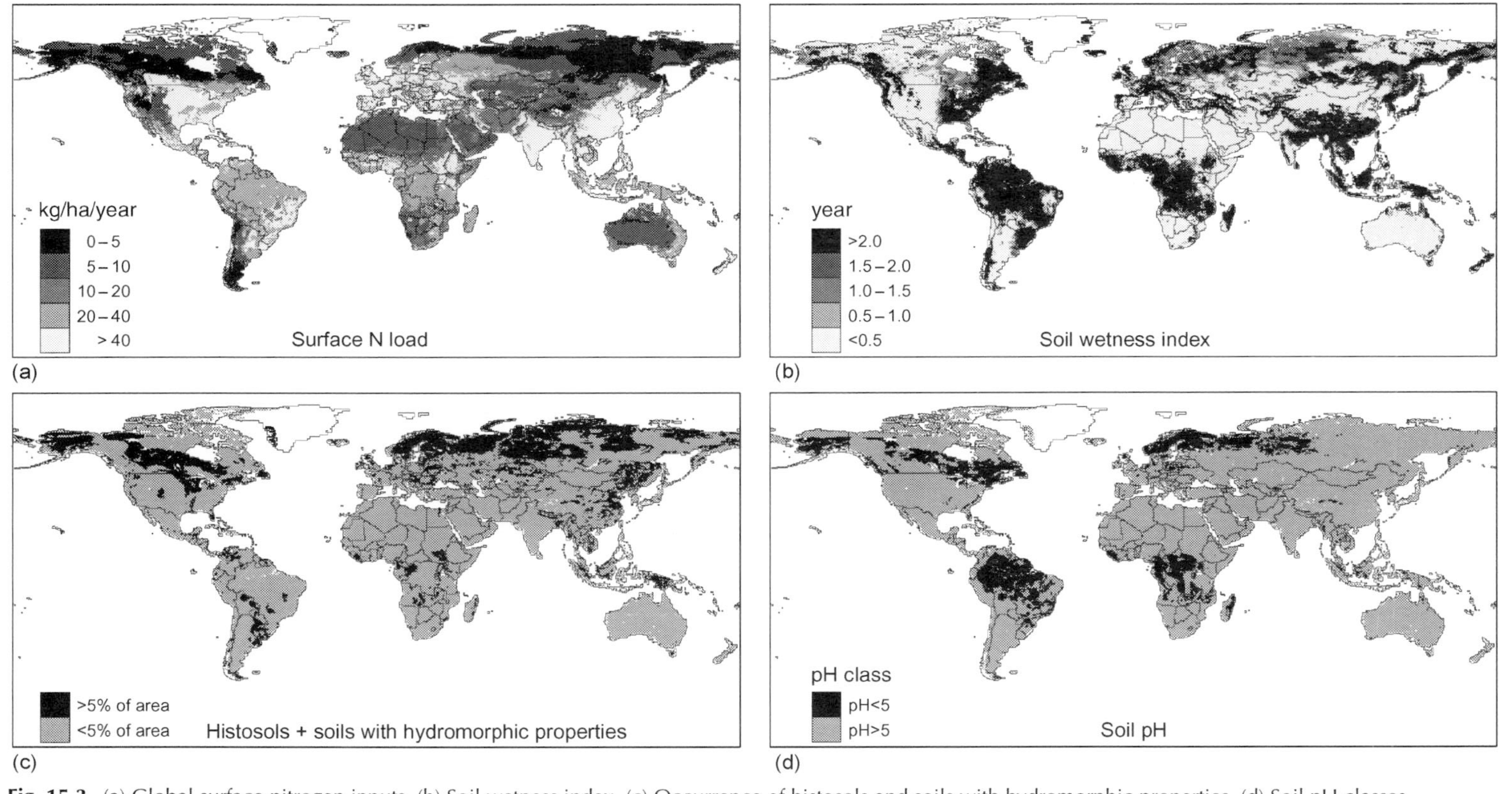

Fig. 15.3. (a) Global surface nitrogen inputs. (b) Soil wetness index. (c) Occurrence of histosols and soils with hydromorphic properties. (d) Soil pH classes.

inventory, ~8% of the global land area has a 'high', and another 8% has a 'moderate', likeliness of N_2O uptake (Table 15.2).

Our results should be interpreted as a first assessment of the likeliness of sinks for N_2O at the Earth's surface. By adding the numbers of the two classified maps, we implicitly assigned equal weights to them. In reality, however, one factor may be more important than another and the relative importance of different factors may vary from soil to soil. Moreover, we ignored other important direct or indirect factors influencing soil N_2O consumption, such as soil drainage conditions, soil fertility indicators and soil pH. These were not included in this inventory because of the difficulty in comparing the importance of these factors to surface nitrogen inputs, temperature and soil wetness. Soil wetness is also not always directly related to precipitation. This holds, for example, for histosols, which are generally poorly drained organic soils with a high water table. Also, mineral soils may have hydromorphic properties and impeded drainage because of the occurrence of impermeable layers or layers impeding air exchange.

Maps of areas with occurrence of soils with hydromorphic properties and classified soil pH (Batjes, 1997; Batjes, 2002) are presented in Fig. 15.2, but were not used in the compilation of Fig. 15.3. However, these factors could be used to modify the general patterns of Fig. 15.3.

Our inventory is a quick method to identify areas with conditions where N_2O uptake may occur during prolonged periods, during a part of the year or where there may be large interannual variability with some years showing net N_2O uptake and others indicating zero or small emissions. We recognize that such inventories are fraught with potential scaling errors. For example, for soil properties, we use those of the dominant soil within each 0.5° × 0.5° grid cell. In reality there is large heterogeneity at smaller scales that is not represented in the data used. A more sophisticated way to investigate the probability of N_2O sinks on the smaller scale is by using mechanistic models that account for all the climate, soil and hydrological factors influencing the ecosystem's carbon and nitrogen cycling (and nitrogen limitation). Our results indicate that such studies should focus on the northern regions of North America, Europe and Asia.

Table 15.2. Percentage of global land area in classes of probability of N_2O sink activity calculated from data presented in Fig. 15.3.

High	8
Moderate	8
Low	84

15.6 Conclusion

In this study, we focus on bacterial reduction of N_2O to N_2 and discuss the likeliness of N_2O uptake in different systems, including soils and aquatic systems. Our results indicate that soils are not only important sources of N_2O, but can also act as a sink for atmospheric N_2O. Global budget studies that fail to explicitly account for N_2O sinks at the Earth's surface may not be a good basis for atmospheric budget studies and analyses of trends in emissions.

First, we defined when we consider a system a sink for atmospheric N_2O, while taking into account the large variability in both N_2O production and reduction in many systems. We consider a system a sink when the net N_2O uptake occurs over a relatively large area and prolonged period of time.

Second, we described the processes involved and identified the most important factors influencing N_2O uptake. Soil uptake of N_2O is driven by denitrification by bacteria, converting N_2O into N_2. The most important factors affecting N_2O uptake by soils are nitrogen availability, soil wetness and temperature. In addition, soil drainage conditions and soil pH are important. Based on the controlling factors, we argue that soils, surface waters and riparian zones may be potential sinks for atmospheric N_2O, while groundwater systems are not likely to be major sinks.

Third, we presented an overview of a selection of studies reporting on observed

N_2O uptake. We concluded that not many experimental studies on observed sinks exist. N_2O uptake has been observed in soils, aquatic systems and riparian zones. In theory, soils are particularly powerful sinks. Whether or not a system actually acts as a sink, depends on local circumstances. From the available studies, we may conclude:

1. Although agricultural soils are usually fertilized and therefore not likely to be sinks for N_2O, some studies report on N_2O uptake in fertilized fields.
2. Several studies report on considerable N_2O uptake in forest soils, indicating that these may potentially be important sinks for atmospheric N_2O.
3. Riparian zones potentially have a large denitrification activity and depending on local conditions may be potential sinks for N_2O.
4. Whether or not lakes can act as a sink is not clear from the available studies.
5. N_2O uptake may occur in the open ocean.

In the future, the sink capacity of soils and aquatic systems may change. We showed that sink capacity of terrestrial and aquatic systems will be affected by trends in atmospheric deposition of nitrogen (in general less N_2O uptake when nitrogen deposition increases), temperature (in general more N_2O uptake at higher temperatures) and changes in annual net precipitation (in general more N_2O uptake when the soil wetness increases). Moreover, the atmospheric concentration is important (in general more N_2O uptake when the ambient concentration is higher). This indicates that the future N_2O sink strength may be reduced by increasing atmospheric nitrogen deposition caused by industrial and agricultural emissions of nitrogen compounds. What the net effects of climatic changes on the N_2O sink strength are, however, is not clear and depends on the net effect of local changes in temperature, net precipitation and atmospheric N_2O increases.

Using the controlling factors mentioned above, we identified areas prone to high N_2O sink activity in soils. Our results need to be interpreted with care because of the large uncertainties involved. Nevertheless, our analysis indicates that high sink activity may be more likely in northern regions of North America, Europe and Asia. The regions that are considered to have a 'high' or 'moderate' likeliness of N_2O uptake cover ~16% of the global land area. Clearly, further studies are needed to confirm our results. Such studies could include experimental analyses as well as mechanistic models.

References

Anderson, I.C. and Poth, M.A. (1989) Semiannual losses of nitrogen as NO and N_2O from unburned and burned chaparral. *Global Biogeochemical Cycles* 3, 121–135.

Appelo, C.A.J. and Postma, D. (1993) *Geochemistry, Groundwater and Pollution*. Balkema, Rotterdam.

Bakken, L.R. and Bleken, M. (1998) Temporal aspects of N enrichment and emission of N_2O to the atmosphere. *Nutrient Cycling in Agroecosystems* 52, 107–121.

Batjes, N.H. (1997) A world dataset of derived soil properties by FAO-UNESCO soil unit for global modelling. *Soil Use and Management* 13, 9–16.

Batjes, N.H. (2002) Revised soil parameter estimates for the soil types of the world. *Soil Use and Management* 18, 232–235.

Betlach, M.R. and Tiedje, J.M. (1981) Kinetic explanation for accumulation of nitrite, nitric oxide and nitrous oxide during bacterial denitrification. *Applied Environmental Microbiology* 42, 1074–1084.

Blackmer, A.M. and Bremner, J.M. (1976) Potential of soil as a sink for atmospheric nitrous oxide. *Geophysical Research Letters* 3, 739–742.

Blackmer, A.M. and Bremner, J.M. (1978) Inhibitory effect of nitrate on reduction of N_2O to N_2 by soil microorganisms. *Soil Biology and Biochemistry* 10, 187–191.

Blicher-Mathiesen, G. and Hoffmann, C. (1999) Denitrification as a sink for dissolved nitrous oxide in a freshwater riparian fen. *Journal of Environmental Quality* 28, 257–262.

Bobbink, R., Hornung, M. and Roelofs, J.G.M. (1998) The effects of air-borne nitrogen pollutants on species diversity in natural and semi-natural European vegetation. *Journal of Ecology* 86, 717–738.

Bouwman, A.F., Boumans, L.J.M. and Batjes, N.H. (2002a) Emissions of N_2O and NO from fertilized fields: summary of available measurement data. *Global Biogeochemical Cycles* 16(4), 1058.

Bouwman, A.F., Van Vuuren, D.P., Derwent, R.G. and Posch, M. (2002b) A global analysis of acidification and eutrophication of terrestrial ecosystems. *Water, Air and Soil Pollution* 141, 349–382.

Bouwman, A.F., Van Drecht, G. and Van der Hoek, K.W. (2005) Nitrogen surface balances in intensive agricultural production systems in different world regions for the period 1970–2030. *Pedosphere* 15(2), 137–155.

Butterbach-Bahl, K., Gasche, R., Huber, C., Kreutzer, K. and Papen, H. (1998) Impact of N-input by wet deposition on N-trace gas fluxes and CH_4-oxidation in spruce forest ecosystems of the temperate zone in Europe. *Atmospheric Environment* 32(3), 559–564.

Castro, M.S., Steudler, P.A., Melillo, J.M., Aber, J.D. and Millham, S. (1993) Exchange of N_2O and CH_4 between the atmosphere and soils in spruce-fir forests in northeastern United States. *Biogeochemistry* 18, 119–135.

Cicerone, R.J. (1989) Analysis of sources and sinks of atmospheric nitrous oxide (N_2O). *Journal of Geophysical Research* 94, 18265–18271.

Clemens, J., Vandré, R., Kaupenjohann, M. and Goldbach, H. (1997) Ammonia and nitrous oxide emissions after landpreading of slurry as influenced by application technique and dry matter reduction. II. Short-term nitrous oxide emissions. *Zeitschift für Planzenernährung und Bodenkunde* 160, 491–496.

Conrad, R. and Dentener, F.J. (1999) The application of compensation point concepts in scaling of fluxes. In: Bouwman, A.F. (ed.) *Scaling of Trace Gas Fluxes in Ecosystems.* Elsevier Science, Amsterdam, pp. 205–216.

Conrad, R., Seiler, W. and Bunse, G. (1983) Factors influencing the loss of fertilizer nitrogen in the atmosphere as N_2O. *Journal of Geophysical Research* 88, 6709–6718.

Crews, T. (1999) The presence of nitrogen fixing legumes in terrestrial communities: evolutionary vs. ecological considerations. *Biogeochemistry* 46, 233–246.

Crutzen, P.J. (1970) The influence of nitrogen oxides on the atmospheric ozone content. *Quarterly Journal of the Royal Meteorological Society* 96, 320–325.

DeSimone, L.A. and Howes, B.L. (1998) Nitrogen transport and transformations in a shallow aquifer receiving wastewater discharge: a mass balance approach. *Water Resources Research* 34, 271–285.

De Wilde, H. (1999) Nitrous oxide and methane in marine systems. PhD thesis, Groningen University, The Netherlands, ISBN 90-803470-3-5.

Dhondt, K., Boeckx, P., Hofman, G. and Van Cleemput, O. (2004) Temporal and spatial patterns of denitrification enzyme activity and nitrous oxide fluxes in three adjacent vegetated riparian buffer zones. *Biology and Fertility of Soils* 40(4), 243–251.

Firestone, M.K. and Davidson, E.A. (1989) Microbiological basis of NO and N_2O production and consumption in soil. In: Andrae, M.O. and Schimel, D.S. (eds) *Exchange of Trace Gases between Terrestrial Ecosystems and the Atmosphere*. Wiley, Chichester, UK, pp. 7–21.

Glatzel, S. and Stahr, K. (2001) Methane and nitrous oxide exchange in differently fertilised grassland in southern Germany. *Plant and Soil* 231, 21–35.

Goossens, A., Visscher, A.D., Boeckx, P. and Cleemput, O.V. (2001) Two-year study on the emission of N_2O from coarse and middle-textured Belgian soils with different land use. *Nutrient Cycling in Agroecosystems* 60, 23–34.

Groffman, P.M., Gold, A.J. and Jacinthe, P.A. (1998) Nitrous oxide production in riparian zones and groundwater. *Nutrient Cycling in Agroecosystems* 52, 179–186.

Hefting, M.M., Bobbink, R. and De Caluwe, H. (2003) Nitrous oxide emission and denitrification in chronically nitrate-loaded riparian buffer zones. *Journal of Environmental Quality* 32(4), 1194–1203.

Hofstra, N. and Bouwman, A.F. (2005) Denitrification in agricultural soils: summarizing published data and estimating global annual rates. *Nutrient Cycling in Agroecosystems* 72, 267–278.

Jambert, C., Delmas, R.A., Labroue, L. and Chassin, P. (1994) Nitrogen compound emissions from fertilizer soils in a maize field pine tree forest agroecosystem in the southwest of France. *Journal of Geophysical Research* 99, 16523–16530.

Kilian, A., Gutser, R. and Claassen, N. (1998) N_2O emissions following long-term fertilization at different levels. *Agrobiological Research* 51, 27–36.

Knowles, R., Lean, D.R.S. and Chan, Y.K. (1981) Nitrous oxide concentrations in lakes: variations with depth and time. *Limnology and Oceanography* 26, 855–866.

Kroeze, C., Mosier, A. and Bouwman, L. (1999) Closing the global N_2O budget: a retrospective analysis 1500–1994. *Global Biogeochemical Cycles* 13, 1–8.
Lapitan, R.L., Wanninkhof, R. and Mosier, A.R. (1999) Methods for stable gas flux determination in aquatic and terrestrial systems. In: Bouwman, A.F. (ed.) *Approaches to Scaling of Trace Gas Fluxes in Ecosystems*. Elsevier Science, Amsterdam, pp. 31–66.
Maggiotto, S.R., Webb, J.A., Wagner Riddle, C. and Thurtell, G.W. (2000) Nitrous and nitrogen oxide emissions from turfgrass receiving different forms of nitrogen fertilizer. *Journal of Environmental Quality* 29(2), 621–630.
Mengis, M., Gachter, R. and Wehrli, R. (1997) Sources and sinks of nitrous oxide (N_2O) in deep lakes. *Biogeochemistry* 38, 281–301.
Mosier, A., Kroeze, C., Nevison, C., Oenema, O., Seitzinger, S. and Van Cleemput, O. (1998) Closing the global atmospheric N_2O budget: nitrous oxide emissions through the agricultural nitrogen cycle. *Nutrient Cycling in Agroecosystems* 52, 225–248.
Papen, H., Hellmann, B., Papke, H. and Rennenberg, H. (1993) Emission of N-oxides from acid irrigated and limed soils of a coniferous forest in Bavaria. In: Oremland, R.S. (ed.) *The Biochemistry of Global Change: Radiatively Active Trace Gases*. Chapman & Hall, New York, pp. 245–259.
Papen, H., Daum, M., Steinkamp, R. and Butterbach-Bahl, K. (2001) N_2O and CH_4-fluxes from soils of a N-limited and N-fertilized spruce forest ecosystem of the temperate zone. *Journal of Applied Biology* 75, 159–163.
Parkin, T.B. (1987) Soil microsites as a source of denitrification variability. *Soil Science Society of America Journal* 51, 1194–1199.
Parkin, T.B. (1990) Characterizing the variability of soil denitrification. FEMS Symposium on Federal European Microbiological Society, Madison, Wisconsin. *Science Tech Publishers* 56, 213–228.
Petersen, S.O. (1999) Nitrous oxide emissions from manure and inorganic fertilizers applied to spring barley. *Journal of Environmental Quality* 28, 1610–1618.
Poth, M., Anderson, I.C., Miranda, H.S., Miranda, A.C. and Riggan, P.J. (1995) The magnitude and persistance of soil NOx, N_2O, CH_4 and CO_2 fluxes from burned tropical savanna in Brazil. *Global Biogeochemical Cycles* 9, 503–513.
Rees, A.P., Owens, N.P.J. and Upstill-Goddard, R.C. (1997) Nitrous oxide in the Bellinghausen Sea and Drake Passage. *Journal of Geophysical Research* 102, 3383–3391.
Riley, R.H. and Vitousek, P.M. (2000) Hurricane effects on nitrogen trace gas emissions in Hawaiian montane rain forest. *Biotropica* 32(4a), 751–756.
Robertson, G.P. (1989) Nitrification and denitrification in humid tropical ecosystems: potential controls on nitrogen retention. In: Proctor, J. (ed.) *Mineral Nutrients in Tropical Forest and Savanna Ecosystems*. Blackwell Scientific Publications, Oxford, pp. 55–69.
Rolston, D.E., Sharpley, A.N., Toy, D.W. and Broadbent, F.E. (1982) Field measurement of denitrification: III. Rates during irrigation cycles. *Soil Science Society of America Journal* 46, 289–296.
Ryden, J.C. (1981) N_2O exchange between a grassland soil and the atmosphere. *Nature* 292, 235–237.
Ryden, J.C. (1983) Denitrification loss from a grassland soil in the field receiving different rates of nitrogen as ammonium nitrate. *Journal of Soil Science* 1983, 355–365.
Serca, D., Delmas, R., Jambert, C. and Labroue, L. (1994) Emissions of nitrogen oxides from equatorial rain forest in central Africa. *Tellus* 46B, 243–254.
Sexstone, A.J., Parkin, T.B. and Tiedje, J.M. (1985) Temporal response of soil denitrification rates to rainfall and irrigation. *Soil Science Society of America Journal* 49, 99–103.
Simek, M. and Cooper, J.E. (2002) The influence of soil pH on denitrification: progress towards the understanding of this interaction over the last 50 years. *European Journal of Soil Science* 53, 345–354.
Skiba, U., Sheppard, L.J., MacDonald, J. and Fowler, D. (1998) Some key environmental variables controlling nitrous oxide emissions from agricultural fields and semi-natural soils in Scotland. *Atmospheric Environment* 32, 3311–3320.
Skiba, U., Sheppard, L.J., Pitcairn, C.E.R., Van Dijk, S. and Rossall, M.J. (1999) The effect of N deposition on nitrous oxide and nitric oxide emissions from temperate forest soils. *Water, Air and Soil Pollution* 116, 89–98.
Slemr, F., Conrad, R. and Seiler, W. (1984) Nitrous oxide emissions from fertilized and unfertilized soils in a subtropical region (Andalusia, Spain). *Journal of Atmospheric Chemistry* 1, 159–169.
Sprent, J.I. (1999) Nitrogen fixation and growth of non-crop legume species in diverse environments. *Perspectives in Plant Ecology, Evolution and Systematics* 2/2, 149–162.
Tiedje, J.M. (1988) Ecology of denitrification and dissimilatory nitrate reduction to ammonium. In: Zehnder, A.J.B. (ed.) *Biology of Anaerobic Microorganisms*. Wiley, New York, pp. 179–244.

Van Drecht, G., Bouwman, A.F., Knoop, J.M., Beusen, A.H.W. and Meinardi, C.R. (2003) Global modeling of the fate of nitrogen from point and nonpoint sources in soils, groundwater and surface water. *Global Biogeochemical Cycles* 17(4), 1115.

Van Kessel, C., Pennock, D.J. and Farrell, R.E. (1993) Seasonal variations in denitrification and nitrous oxide evolution at the landscape scale. *Soil Science Society of America Journal* 57, 988–995.

Verchot, L.V., Davidson, E.A., Cattanio, J.H., Ackerman, I.L., Erikson, H.E. and Keller, M. (1999) Land use change and biogeochemical controls of nitrogen oxide emissions from soils in eastern Amazonia. *Global Biogeochemical Cycles* 13, 31–46.

Vidon, P. and Hill, A.R. (2004) Denitrification and patterns of electron donors and acceptors in eight riparian zones with contrasting hydrogeology. *Biogeochemistry* 71, 259–283.

Wagner-Riddle, C., Thurtell, G.W., Kidd, G.K., Beauchamp, E.G. and Sweetman, R. (1997) Estimates of nitrous oxide emissions from agricultural fields over 28 months. *Canadian Journal of Soil Science* 77, 135–144.

Xu, H., Guangxi, X., Cai, Z.C. and Tsuruta, H. (1997) Nitrous oxide emissions from three paddy fields in China. *Nutrient Cycling in Agroecosystems* 49, 23–28.

16 Cross-cutting Issues and New Directions

David S. Reay
School of GeoSciences, University of Edinburgh, Edinburgh, UK

16.1 Introduction

In bringing together the host of expert chapters contained in this book, numerous linkages between the various sinks for carbon dioxide (CO_2), methane (CH_4) and nitrous oxide (N_2O) are apparent. The authors have attempted to identify these throughout, as well as the key cross-cutting issues that affect all of these sinks. Of particular note are the impacts of a changing global climate, driven by elevated greenhouse gas (GHG) concentrations in the atmosphere, on the capacity of these sinks. From drought-induced reductions in the terrestrial CO_2 sink to temperature-dependent stratification of the oceans and consequent reduction in the oceanic CO_2 sink, the feedbacks to the earth's climate through changes in its GHG sinks are of obvious importance and concern.

This chapter briefly reviews some of the emerging issues in GHG sink science and identifies some of the most pressing areas requiring further research to reduce uncertainties. Of particular interest is the impact of changing nitrogen deposition on GHG sinks and so this issue is covered in detail by De Vries *et al.* (Chapter 17, this volume).

16.2 Sea-level Rise and Climate Change

Globally, sea level is expected to rise between 9 and 88 cm by the end of the 21st century. While in some areas, sea defences may be able to maintain existing coastlines, in most areas of the world substantial coastal retreat is inevitable. The impacts of such retreat on the various sources and sinks of CO_2, CH_4 and N_2O may be substantial. Combined with this retreat is the recognition that wetlands creation and protection must form a central part of coastal defence strategies. As wetlands represent a strong CH_4 source, increased CH_4 emissions are likely.

Similarly, in tackling the growing problem of eutrophication of surface waters due to fertilizer runoff and leaching, interception strategies such as wetlands, buffer strips and ponds may increase N_2O and CH_4 emissions where they are used – swapping a nutrient pollution problem for a climate change issue.

Predicted climate change at high northern latitudes is expected to extend the northern ranges of forests. All other things being equal, this could be expected to increase the CO_2 sink provided by vegetation at these latitudes. However, any increase in CO_2 uptake by invading woody vegetation must be set against the reduction in albedo (land reflectance) that such northward spread would cause. While unforested tundra areas have snow cover for much of the year, and so a high albedo, forested areas have a much lower albedo and therefore absorb much more of the sun's energy as heat – increasing the warming at ground level. In addition to the direct forcing of climate from this warming, rapid rises in ground temperatures at these high latitudes may result in increased decomposition rates, and hence soil CO_2 emissions in these areas. When such warming occurs in areas of CH_4-rich permafrosts, there is the additional risk of elevated CH_4 emissions.

16.3 CO_2 Capture and Storage

CO_2 capture and storage (CCS) is a process whereby the CO_2 produced by point sources, such as industry and power stations, is separated and transported to a storage point suitable to keep the CO_2 out of the atmosphere in the long term (centuries or millennia). As a strategy, it has great potential to reduce global CO_2 emissions, though realizing this potential depends on further development of CCS technologies and the transfer of this technology to developing economies such as China and India.

Assuming fossil fuels will continue to dominate our sources of energy over the coming decades, CCS provides a way in which such continued widespread utilization of fossil fuels can occur without a burgeoning growth in global CO_2 emissions. The principle is an established one: CO_2 is collected from a strong point source like a power station, separated, compressed and then stored in geological formations (in the oceans), fixed within inorganic carbonates by reaction with metal oxides or used in industrial processes as discussed by Aresta and Dibenedetto (Chapter 7, this volume).

There are several ways in which CO_2 is separated, including postcombustion, precombustion and what is termed oxyfuel combustion. The capture of CO_2 after burning of fossil fuels already occurs in a number of power stations – CO_2 being collected from the flue gases before they are emitted to the atmosphere. Precombustion capture is used in the gas industry – separation of $CO_2$from natural gas before being used. It is also used in the manufacture of fertilizers and in the production of hydrogen. Such precombustion capture is generally easier given the higher concentrations of CO_2 involved compared to those in flue gases. The use of oxyfuel combustion is still being developed, but in essence it involves the burning of fossil fuels in high-purity oxygen to produce CO_2 concentrations in the exhaust gases that are much higher than normal, thus allowing easier CO_2 separation.

Current technology is able to capture ~85–95% of the CO_2 processed in the capture plant, though this process itself requires a significant amount of energy. A power plant using a CCS system would require 10–40% more energy than a plant without CCS, and so there is a considerable extra cost involved, in terms of both the money and the extra CO_2 produced. Overall, a power plant with CCS can be expected to reduce emissions by ~80–90% compared to a non-CCS plant. Where the captured CO_2 is stored in inorganic carbonates, rather than transported to a suitable geological or oceanic storage site, the plant would require 60–180% more energy than a non-CCS plant (IPCC, 2005).

The transport of captured CO_2 to the storage sites is usually by a pipeline, especially where the distances involved are less than 1000 km. For greater distances, the transport of captured CO_2 by ships has been suggested as a more cost-effective alternative.

Storage within geological formations can utilize depleted oil and gas fields, as well as unminable coal beds and deep saline aquifers. Much of the technology for using depleted oil and gas fields has already been developed by energy companies, who have used it as a strategy to increase yield rates from existing reservoirs – called enhanced recovery. Saline aquifers have also been shown to represent an economically viable 'sink' for captured CO_2, which is injected into subterranean aquifers where it dissolves and is held in place by impermeable rock layers above. The full potential for storage in unminable coal beds (coal beds that are too deep or too thin) is still to be proven – in theory even relatively shallow coal beds could have CO_2 pumped into them, which is then absorbed by the coal.

Clearly, a key property of any geological storage site is that it can contain a large amount of CO_2 and keep it there without leakage for centuries or millennia. In general, the existence of an impermeable rock layer above the storage site is required, with the CO_2 normally being injected to depths greater than 800 m. At these depths the pressure is great enough to compress the CO_2 to an almost liquid state, thus helping to maximize storage and reduce chances of leakage.

Industrial use of CCS is already underway, with three large-scale storage projects in operation: (i) the Sleipner project making use of an offshore saline aquifer off the coast of Norway; (ii) the In Salah project, using a gas field in Algeria; and (iii) the Weyburn project in Canada. This latter CCS project – the Weyburn–Midale CO_2 Monitoring and Storage Project (Fig. 16.1) – is located in Saskatchewan, and has been in operation since 2000. The project has successfully demonstrated that the site is suitable for long-term storage of CO_2. Results from Weyburn–Midale indicate that trapped CO_2 would not escape to groundwater or the surface in the next 5000 years.

The storage of CO_2 in the oceans can be achieved in two ways. First, it can be done by injection into the water at depths greater than 1000 m, using a fixed pipeline or ship. This method is unlikely to result in truly long-term storage, as the CO_2 will dissolve in the water (dissolution) and will eventually be released back into the atmosphere from the ocean surface. There are also concerns that such large and concentrated injections of CO_2 would cause significant reductions in water pH and hence have detrimental impacts on marine life in the vicinity. Secondly, there is the option of injection right onto the seabed at depths greater than 3000 m. Using this strategy, the CO_2 may form a lake on the seabed and this would help slow the rate at which the CO_2 dissolves into the surrounding water. Consequently, such deep injection of CO_2 may help to prolong the lifetime of this 'sink'.

The fixation of captured CO_2 with metal oxides to produce stable carbonates is an option that is still being developed. Ordinarily, the reaction rates are very slow and so pretreatment of the metal oxides is required – currently a very energy-intensive process.

Overall, the potential of CCS as a strategy to mitigate climate change through stabilization of atmospheric CO_2 concentrations is huge. In particular, geological storage appears to represent a great opportunity – many of the large point sources of CO_2 around the world are located within 300 km of geological formations, which, in theory at least, have potential for CO_2 storage. Projections indicate that by 2050 ~20–40% of global fossil fuel CO_2 emissions could be suitable for CCS, covering up to 60% of CO_2 emissions from electricity generation and up to 40% of those from industry. Given an aim of stabilizing CO_2 concentrations in the atmosphere at 450–750 ppm during the 21st century, CCS may provide 15–55% of the global mitigation effort required to meet this aim.

16.4 Manipulation of the Oceanic Carbon Sink

As we saw earlier (Sabine and Feely, Chapter 3, this volume), the oceans constitute an enormous sink for CO_2 and have buffered much of the anthropogenic CO_2 added to the atmosphere since the industrial revolution. The big uncertainties still to be addressed by scientists hinge on quantifying just how much CO_2 the oceans are actually taking up and on how they will respond to a warming planet. Currently, we can estimate the ocean uptake of CO_2 by the concentration of chlorophyll in the surface waters as recorded from satellites. This gives great coverage, but unfortunately the amount of CO_2 taken up by the phytoplankton in the surface waters of the oceans is not always directly related to the amount of chlorophyll – the phytoplankton change the ratio of carbon to chlorophyll depending on their depth and supply of nutrients. A big challenge, then, is to better link the

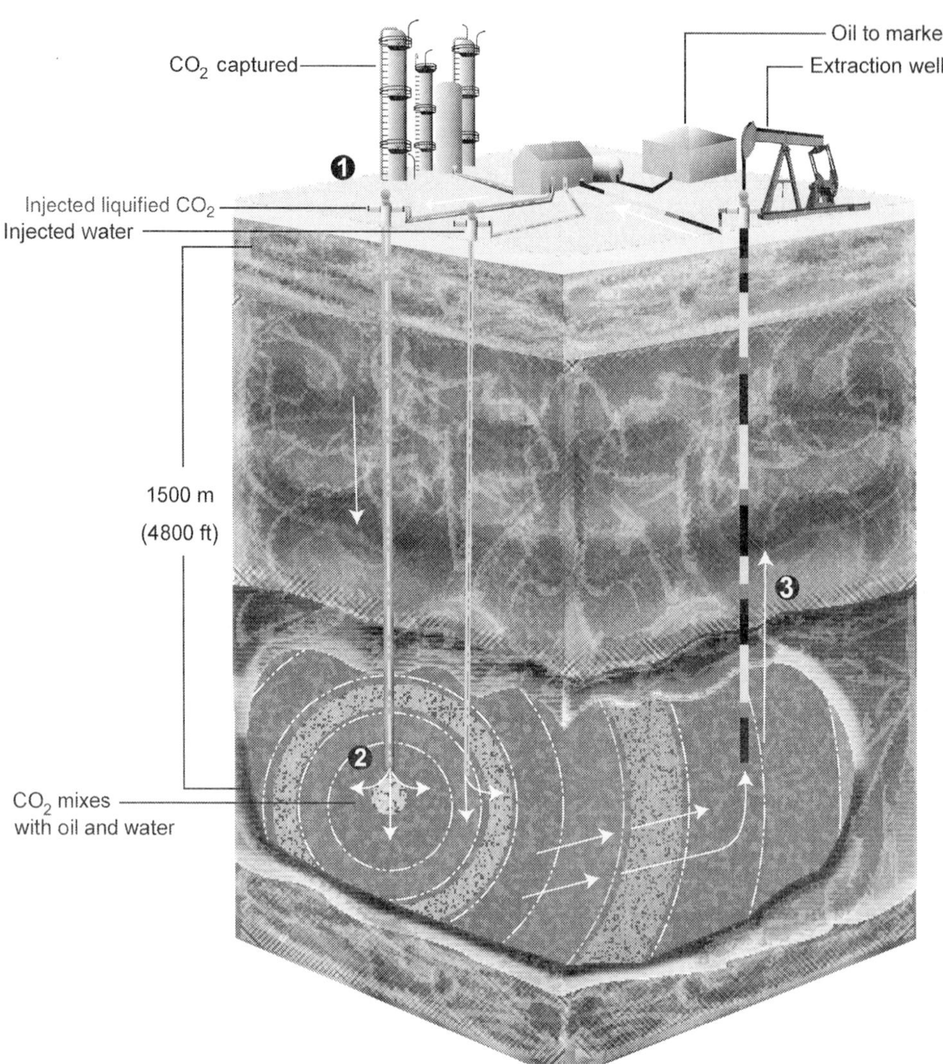

Fig. 16.1. Enhanced oil recovery by CO_2 injection. (1) Carbon dioxide (CO_2) is injected, along with water, deep underground (1500 m at the Weyburn field) into a depleted oil and gas reservoir. The CO_2 used at Weyburn and Midale comes from the Dakota Gasification Plant in Beulah, North Dakota, where the gas is captured (rather than vented to the atmosphere), liquefied by compression and pipelined 320 km north to the oilfields. (2) Oil absorbs CO_2, causing the oil to expand. Combined with water injection, CO_2 injection increases reservoir pressure and oil fluidity, enabling oil to escape from rock pores and flow more readily towards production wells. (3) Of the injected CO_2, ~20% is recycled for enhanced oil recovery and the rest is stored for thousands of years. (From Shawn Griffiths, Petroleum Technology Research Centre, Regina, SK S4S 7J7, Canada. Reproduced with thanks.)

chlorophyll seen by the satellites with the CO_2 that is taken up. Regarding the warming of the planet, the extent of increased stratification and so decreased primary production in our oceans remains the ultimate question on oceanic climate change science. With some areas likely to see an increase in storminess, and hence mixing, and others to see a greater predominance of calm high-pressure systems, the weather, as well as sea temperatures, is a key component.

Previous attempts to manipulate the oceanic sink for CO_2 have included iron fertilization in the 'high nutrient–low chlorophyll' areas of the world's oceans (e.g. the Equatorial Pacific and the Southern Ocean), where iron appears to be the limiting factor to the growth of phytoplankton. The results of these experiments have been somewhat disappointing, with iron additions resulting in greatly enhanced blooms of phytoplankton, but the bulk of the extra CO_2 taken up by these algal blooms being released back into the atmosphere within a relatively short time-span.

16.5 Warming of High-latitude Soils and Permafrosts

The permafrost soils that occur at high latitudes in the northern hemisphere represent a huge store of CH_4. For western Siberia alone, it is estimated that up to 70 billion tonnes of CH_4 are trapped as hydrates within the frozen soils. These CH_4 deposits represent a potentially very powerful positive feedback to global warming. Recent warming in these regions has been much greater than the global average, with western Siberia experiencing a warming of ~3°C in the last 40 years. If such rapid warming continues, huge amounts of CH_4 could, in theory, be released to the atmosphere. A report in 2006 by Torre Jorgensen and colleagues at Alaska Biological Research in Fairbanks suggests that such thawing of permafrost is also taking place in Alaska, with melt water pools appearing in areas thought to have been frozen for the last 300,000 years.

However, just because the thawing of CH_4-rich permafrost soils takes place does not necessarily mean that the CH_4 released will enter the atmosphere. As seen earlier (Dunfield, Chapter 10, and Visscher *et al.*, Chapter 12, this volume), methanotrophs are able to respond rapidly to an increased availability of CH_4 and so may oxidize much of the thawed CH_4 before it can leave the surface soils and enter the atmosphere. Clearly, such oxidation may still result in increased CO_2 emissions to the atmosphere, but the climate-forcing effect would be greatly reduced. A pressing 'new direction' for GHG research is to quantify the true extent of this positive feedback. We need to know how much thawing of the CH_4-rich permafrosts we are likely to see, and how the microbial communities – both methanogens and methantrophs – will respond to this thawing and warming of soils.

In addition to the release of CH_4 from permafrosts, there is the possibility of increased emissions of CO_2 and CH_4 due to elevated rates of decompostion in high-latitude soils in response to a warmer climate. A key question that has yet to be answered is to what extent will 'old' carbon in these soils – which has been stored for decades or centuries – start to decompose and be released into the atmosphere as CO_2 and CH_4 as temperatures increase?

It is hoped that research initiatives such as International Polar Year (2007–2008), and targeted research programmes such as Arctic Biosphere Atmosphere Coupling at Multiple Scales (ABACUS) will help answer such questions and thus inform climate change policy before rapid positive feedback effects on global climate occur.

16.6 CH_4 from Vegetation

In 2006, Frank Keppler and his colleagues at the Max Planck Institute in Germany reported that they had found significant CH_4 emissions from both dead and living plant material (Keppler *et al.*, 2006). This was a completely unknown source of CH_4 and therefore cast some doubt on the net climate forcing impact of the world's forests. Research is still underway into exactly how this CH_4 is produced in plants and how significant a source it is globally. Keppler *et al.* estimated that CH_4 emissions from this source could represent 10–30% of global CH_4 emissions (62–236 Tg CH_4/year for living plants, and 1–7 Tg CH_4/year from plant litter). If so, the benefits of terrestrial vegetation as sinks for CO_2 may be somewhat offset by the emissions of CH_4.

There is need for research to establish the mechanism by which such plant-derived CH_4 might be formed, the significance of this new source of CH_4 on a global scale and consequently its impact on the net climate forcing effect of terrestrial vegetation.

16.7 Nitrogen Deposition

An important feature of global change is the increased deposition of active forms of nitrogen from human activities. Nitrogen deposition has increased rapidly in recent decades and is expected to double over the next 100 years. It is intense in industrialized regions of the world but occurs to some degree everywhere. Nitrogen is a nutrient for plants and microbes and so we may expect this process to affect biological activity and consequently the fluxes of GHGs. Indeed, nitrogen supply is often a rate-limiting factor for forest growth in temperate regions, and so nitrogen deposition may be making many of our forests grow faster, hence increasing the carbon sink they represent.

De Vries *et al.* (Chapter 17, this volume) address this issue, referring to empirical and modelling studies of European forests. They examine the effects of nitrogen deposition on the emissions of the three main GHGs: CO_2, N_2O and CH_4. First, they provide an overview of the processes that lead to an increase of the vegetation CO_2 sink, an increase of N_2O emissions and a decrease of the soil CH_4 sink, in response to elevated nitrogen inputs. Secondly, they estimate the net exchange of GHGs by European forests and the impacts of nitrogen deposition on this exchange in terms of global warming potential (GWP). In performing these quantifications, they use: (i) literature data on field measurements and applications of detailed biogeochemical models; and (ii) empirical and simple process-oriented models, combined with measured and modelled nitrogen deposition data at ~500 (measurements) to 6000 (model results) forest-monitoring plots.

The authors estimate that the forest carbon sink has increased by ~10% during 1960–2000 as a result of this 'fertilization'. At the same time, there has been a stimulation of N_2O emissions and a change in the CH_4 sink. Results show that the average reduction in GWP due to CO_2 sequestration by European forests is offset by N_2O emissions by ~10%, whereas the net uptake of CH_4 is negligible compared to CO_2 sequestration.

On average, the effect of nitrogen deposition on increasing N_2O emissions is estimated at ~10–15% of the increased CO_2 sequestration. The positive effect on increasing the vegetation CO_2 sink is thus much larger than the effect on increasing the N_2O emissions. The effect of nitrogen deposition on reduced CH_4 uptake emissions is very small and highly uncertain.

De Vries *et al.* stress that, although their general conclusions are robust, the complexity of the processes involved and the large scale and diversity of the forests indicate that the impact of environmental factors on the emissions and sinks of CO_2 and specifically on N_2O and CH_4 remain highly uncertain. To improve estimates in future studies, it is deemed necessary to have an integrated approach consisting of: (i) field measurements and process studies in the laboratory; (ii) further development and testing of detailed process-oriented biogeochemical models at plot scale; and (iii) upscaling of results by further developing, as well as using empirical models and simplified process-based models.

None the less, the work with small plots discussed here is an important start. We look forward to more published work over the next few years, following the launch of a new project called 'Nitroeurope' in 2006.

17 Impact of Atmospheric Nitrogen Deposition on the Exchange of Carbon Dioxide, Nitrous Oxide and Methane from European Forests

Wim de Vries[1], Klaus Butterbach-Bahl[2], Hugo Denier van der Gon[3] and Oene Oenema[1]

[1]Wageningen University and Research Centre, Alterra, Wageningen, The Netherlands; [2]Institute for Meteorology and Climate Research, Atmospheric Environmental Research (IMK-IFU), Garmisch-Partenkirchen, Germany; [3]TNO Environment and Geosciences, Laan van Westenenk, The Netherlands

17.1 Introduction

The rapidly increasing human and animal population and the concomitant request for more food, feed and fibre have contributed to a rapid increase in available nitrogen in global agriculture during the last century (Galloway, 1998; Tilman *et al.*, 2001). Similarly, rapid industrialization and traffic development have greatly increased fossil fuel consumption, and hence the amount of reactive nitrogen in the environment (Galloway *et al.*, 2004). As a consequence, in industrialized areas with intensified agriculture, including large parts of Europe, there are increased atmospheric concentrations of nitrogen oxides (NO_x: NO and NO_2), mainly caused by traffic and industry, and of ammonia (NH_3), mainly due to emissions from (intensive) agriculture. Even though the nitrogen deposition in Europe is likely to decrease in view of NO_x emission reductions, recent estimates suggest that the global annual average nitrogen deposition over land will increase by a factor of 2.5, with the average nitrogen deposition over forests expected to increase from 10 to 20 Tg N/year in the 21st century (Lamarque *et al.*, 2005).

Increased atmospheric concentrations of NO_x and NH_3 and their subsequent deposition can have various effects on human health (Wolfe and Patz, 2002; Townsend *et al.*, 2003), for example, due to the involvement of reactive nitrogen species in the production of tropospheric O_3 (Crutzen, 1995) or the formation of secondary fine particles (e.g. Amann *et al.*, 2001), and on the (nutrient) condition and diversity of natural ecosystems like forests (e.g. De Vries *et al.*, 1995, 2003b,d; Bobbink *et al.*, 1998). Increased depositions of atmospheric NO_x and NH_3 may also influence the exchange of the three main greenhouse gases (GHGs) – carbon dioxide (CO_2), nitrous oxide (N_2O) and methane (CH_4) – between biosphere and atmosphere.

The net effect of anthropogenic nitrogen deposition on the net GHG budget of natural ecosystems is still unclear. The exchange rates of CO_2, N_2O and CH_4 between biosphere and atmosphere are the result of closely linked, but complex biologically mediated, production and consumption processes (Conrad, 1996). Increased nitrogen deposition usually increases net primary production (NPP) and the fixation of CO_2 by vegetation (e.g. Townsend *et al.*, 1996; Hungate *et al.*, 2003). Increased nitrogen availability has probably contributed to the increase in forest growth rates observed across Europe and account for some of the 'missing global carbon sink' (e.g. Hunter and Schuck, 2002). Increased productivity may in turn also increase carbon sequestration in soil due to increased litterfall (e.g. Nadelhoffer *et al.*, 1999; De Vries *et al.*, 2005b). However, N_2O emissions tend to increase when soil carbon increases, due to positive relationships between soil carbon contents, nitrogen turnover rates and N_2O production (Six *et al.*, 2004; Li *et al.*, 2005). Moreover, increased nitrogen deposition may directly increase N_2O emissions through increased availability of nitrogen for microbial processes directly involved in N_2O production (Bowden *et al.*, 1991; Butterbach-Bahl *et al.*, 1997, 2002a). On the other hand, nitrogen deposition (especially NH_4) may decrease the oxidation capacity of soils for atmospheric CH_4, thereby decreasing the net influx of CH_4 from atmosphere to biosphere (Steudler *et al.*, 1989; Sitaula *et al.*, 1995; Van den Pol-van Dasselaar *et al.*, 1999), but inverse effects have also been found (Bodelier and Laanbroek, 2004). Evidently, the net effect of anthropogenic nitrogen deposition on the net GHG budget of natural ecosystems is the resultant of complex interactions and ecosystem feedbacks, and is highly dependent on local environmental conditions.

While CO_2 exchange is measured in fairly extensive global networks (CarboEurope IP, FLUXNET), the effect of nitrogen inputs on carbon sequestration remains poorly quantified. Initiatives to model the global effect of nitrogen availability on carbon sequestration (e.g. Hudson *et al.*, 1994; Holland *et al.*, 1997; Asner *et al.*, 2001) are limited by uncertainties in the mechanistic understanding of carbon–nitrogen interactions in plant and soil and a lack of field data for verification. Although a range of nitrogen fertilization experiments have been performed at the plot level, site networks that derive both net ecosystem exchange (NEE) and nitrogen uptake are extremely limited (Nadelhoffer *et al.*, 1999). Unlike CO_2, estimates on regional and continental (global) scales of the source strength of soils for N_2O and CH_4 remain much more uncertain, both in view of the complexity of the processes involved and the lack of extensive global networks. This holds even more strongly for the impact of nitrogen deposition on these emissions.

This chapter discusses the interactions between anthropogenic nitrogen deposition and exchanges of CO_2, N_2O and CH_4 between forest ecosystems and the atmosphere in Europe. We focus on forest ecosystems in Europe because atmospheric nitrogen deposition is relatively high, while interactions between elevated anthropogenic nitrogen deposition and exchanges of CO_2, N_2O and CH_4 for these forests have been explored in various European countries. Furthermore, a Pan-European Programme for Intensive and Continuous Monitoring of Forest Ecosystems has been carried out since 1994, which includes information on nitrogen deposition and on site factors that affect GHG emissions, thus allowing the possible upscaling of relationships to the European scale. Approximately 860 permanent observation plots, with more than 500 plots with atmospheric deposition data, have been selected in 30 participating countries (level I Monitoring Programme, e.g. De Vries *et al.*, 2003d). Such upscaling is further enabled by a European Monitoring Programme on air pollution impacts since 1986, in which several forest and soil condition characteristics are monitored at a systematic 16 × 16 km grid at more than 6000 plots throughout Europe (level I Monitoring Programme, e.g. UN/ECE and EC, 2004). For these plots, relevant site and soil characteristics and modelled nitrogen deposition estimates are available (e.g. Nadelhoffer *et al.*, 1999; De Vries *et al.*, 2005b).

A quantitative assessment of this interaction requires a proper understanding of all major biotic and abiotic processes in ecosystems (Li *et al.*, 2005). Therefore, we first discuss the processes and factors involved in the interactions and the effects of increased atmospheric nitrogen deposition on them (Section 17.2). In Section 17.3, we present an overview of published emission data of CO_2, N_2O and

CH_4 from European forests and forest soils. In Section 17.4, we present the methods and data that were used to quantify the effects of nitrogen deposition on the net exchange of CO_2, N_2O and CH_4 between European forests and atmosphere, and the results of our calculations. The calculations presented in this section are new and are all based on De Vries *et al.* (2006). Finally, in Section 17.5, we discuss the net CO_2, N_2O and CH_4 emissions and the nitrogen deposition impacts on those emissions in terms of CO_2 equivalents, by using the global warming potential (GWP) approach, i.e. 1 kg N_2O is assumed to be 296 kg CO_2 equivalents and 1 kg CH_4 is 23 kg CO_2 equivalents (Ramaswamy, 2001). Furthermore, the contribution of N_2O and CH_4 emissions from forests compared to agriculture, the reliability of the assessments and the research needs to improve the quantification are discussed.

17.2 Processes and Factors Controlling Exchanges of CO_2 , N_2O and CH_4 between Forests and Atmosphere

Production and consumption of nitrogen and carbon trace gases in soils are predominantly due to the microbial processes of mineralization, nitrification, denitrification, methanogenesis and CH_4 oxidation. The interactions between carbon and nitrogen, and the exchange of CO_2, N_2O and CH_4 between biosphere and atmosphere are largely controlled by external drivers such as climate (radiation, rainfall, temperature), land use and management (forest type and its management), soil type and deposition of atmospheric nitrogen (Conrad, 1996; Groffman *et al.*, 2000). This section summarizes the effects of atmospheric nitrogen deposition on the exchanges of CO_2, N_2O and CH_4 between biosphere and atmosphere, using the relational diagram depicted in Fig. 17.1, while briefly describing the effect of other factors controlling the exchange of CO_2, N_2O and CH_4 between forests and atmosphere.

17.2.1 CO_2 exchange

Net primary production of forests greatly depends on climate, forest type, age, nutrient availability and management. Apart from changes in forest management, recent changes in the NPP of forests in Europe (Spiecker

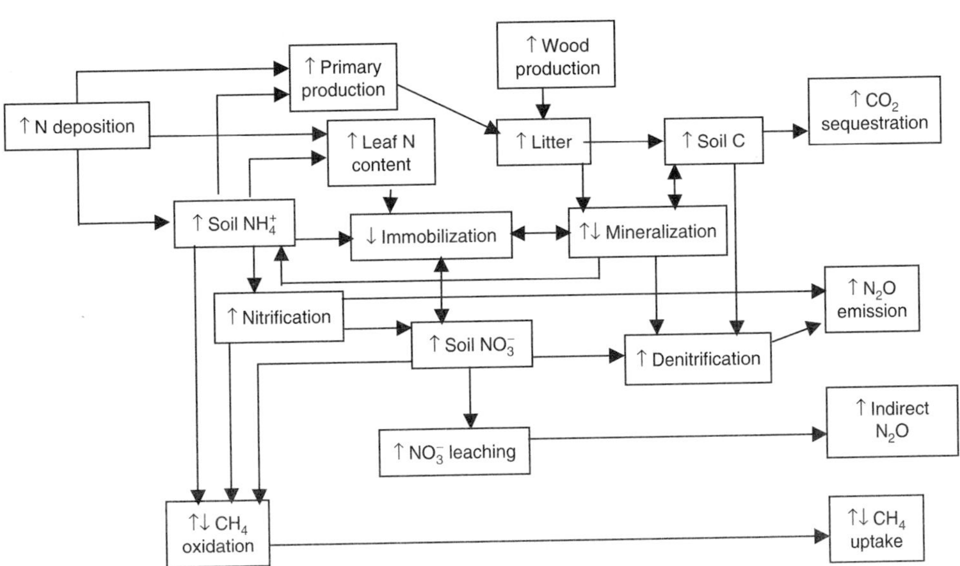

Fig. 17.1. Potential effects of increases in atmospheric nitrogen deposition on carbon dioxide (CO_2) sequestration, nitrous oxide (N_2O) emissions (direct and indirect) and methane (CH_4) uptake.

et al., 1996) have been attributed to fertilization effects caused by enhanced atmospheric CO_2 concentrations (e.g. Melillo *et al.*, 1993; Friedlingstein *et al.*, 1995) and nitrogen deposition (Holland *et al.*, 1997; Nadelhoffer *et al.*, 1999), as well as by increased temperatures or by a combination of those effects, increasing the growing season (e.g. Myneni *et al.*, 1997; Hasenauer *et al.*, 1999).

Regarding forest management, the abandonment of agricultural land and the managed re-establishment of woody vegetation is an important aspect of European forestry. The current age structure of European forests is the consequence of large-scale afforestations in the early 20th century and past management practices that centred on replanting after clear-cutting, followed by regular thinning of mono-specific even-aged stands (Nabuurs *et al.*, 2003). The young-age structure is considered the main reason for European forests actively sequestering carbon at a high rate (see also Hymus and Valentini, Chapter 2, this volume). In addition to the abandonment of agricultural land and the natural or managed re-establishment of woody vegetation, other land-use changes have played a role in changing terrestrial productivity in the recent past including: (i) conversion of forest to agricultural land; (ii) fire suppression; and (iii) expansion of woody species into grasslands and savannahs. For a more detailed discussion of these controls on past, present and future forest productivity, we refer to Hymus and Valentini (Chapter 2, this volume).

The impact of environmental variables (atmospheric CO_2 concentrations, nitrogen deposition temperature) on enhancing forest growth has been, and still is, a subject of much debate. Some authors claim that nitrogen deposition is the most important (Rehfuess *et al.*, 1999). Others claim that increases in primary production of plants are largely due to elevated atmospheric CO_2 concentrations (CO_2 fertilization effect), with related increases in water use efficiency (Osborne *et al.*, 2000). Still others indicate that such a response is rather limited, with any initial increase in the plant-based CO_2 sink disappearing within a few years (Tognetti *et al.*, 2000). Temperature effects, including advancing spring bud break and the greening up of forest canopies in mid- and high-latitude forests, lengthen the growing season which is correlated with an increase in net ecosystem production (NEP) (Baldocchi and Wilson, 2001). It is clear that a combination of changes in climate and atmospheric composition has had significant effects on terrestrial productivity over the last few decades (see also Hymus and Valentini, Chapter 2, this volume). The effect of changes in the seasonality and amount of rainfall is not yet given much attention. Regarding impacts of environmental variables on forest growth, it is clear that nitrogen deposition is an important variable, since nitrogen is often the nutrient that is mostly limiting NPP in forest ecosystems.

The increase in nitrogen deposition in forests may raise carbon sequestration by augmenting wood production and accumulation of soil organic matter (Fig. 17.1). Current hypotheses suggest that larger nitrogen deposition causes a higher rate of soil organic matter accumulation through: (i) an increased leaf/needle biomass and litter production (e.g. Schulze *et al.*, 2000); and (ii) a reduced decomposition of organic matter, depending on the stage of humus formation (Berg and Matzner, 1997; Harrison *et al.*, 2000; Hagedoorn *et al.*, 2003). Chronic nitrogen additions to temperate forest soils in the USA have been shown to cause an initial increase in soil respiration, but continued additions for more than a decade resulted in a reduction in soil respiration of more than 40% (Bowden *et al.*, 2004). The nitrogen content of forest litter and humus might thus be an important indicator of the soil carbon sequestration. However, the true potential for elevated nitrogen deposition to increase carbon accumulation in both below- and aboveground sinks appears to be much more limited than earlier thought (e.g. Holland *et al.*, 1997). A detailed 15-year-long nitrogen amendment study in Harvard Forest, USA, showed that the large increases in carbon accumulation that were predicted, using a linear relationship between leaf nitrogen content and photosynthesis, failed to materialize in the field (Bauer *et al.*, 2004).

Understanding the nitrogen cycle in semi-natural ecosystems is therefore the key to understanding the long-term source or sink strength of soils for carbon. Insight into the soil carbon sequestration is crucial since the soil is the ultimate sink or source of CO_2 for forest ecosystems in the long term. By far the largest amount of carbon stored in forests in the northern hemisphere is stored in the soil. More information on the soil CO_2 sink is given by Smith and Ineson (Chapter 4, this volume).

17.2.2 N_2O exchange

The production of N_2O in (forest) soils is predominantly due to the microbial processes of nitrification and denitrification (Granli and Bøckman, 1994; Conrad, 1996). During nitrification, ammonia oxidizers convert NH_3 to nitrite (NO_2^-) and nitrite oxidizers (NO_2^-) to nitrate (NO_3^-). N_2O develops as a by-product of ammonia oxidation or via a pathway called nitrifier denitrification, i.e. the reduction of NO_2^- via NO to N_2O and N_2 by nitrifiers (Poth and Focht, 1985; Poth, 1986; Wrage *et al.*, 2001). Denitrification is a reductive process by which denitrifiers reduce NO_3^- sequentially to NO, N_2O and finally to N_2 (Conrad, 1996), as illustrated in Fig. 17.2. Understanding the relative importance of nitrifiers and denitrifiers in producing N_2O is the key to understanding the mechanism of N_2O production and for accurately upscaling and quantifying the N_2O source strength of European forests.

The positive relationship between nitrogen deposition and N_2O emissions from forest soils has mainly been attributed to the increased availability of nitrogen (NH_4^+ and NO_3^-) for the microbial processes of nitrification and denitrification (Rennenberg *et al.*, 1998; Corre *et al.*, 1999), as illustrated in Fig. 17.1. Various studies (e.g. Brumme and Beese, 1992; Zechmeister-Boltenstern and Meger, 1997; Brumme *et al.*, 1999; Papen and Butterbach-Bahl, 1999; Butterbach-Bahl *et al.*, 2002a) showed that atmospheric nitrogen deposition is a key site parameter influencing N_2O emissions. These studies have shown that temperate forests can function as significant sources for N_2O, especially if these forests are affected over decades by high rates of atmospheric nitrogen deposition. In addition to atmospheric nitrogen deposition, many commercial forests receive applications of fertilizer nitrogen in the form of urea or ammonium nitrate. Such

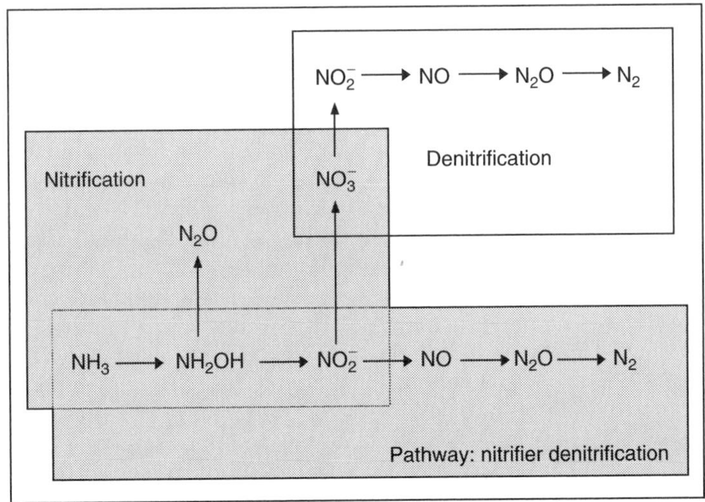

Fig. 17.2. Nitrous oxide (N_2O) production by nitrifiers and denitrifiers. (From Wrage *et al.*, 2001.)

applications have been shown to double N_2O emission rates from forest soils (Regina *et al.*, 1998). Furthermore, an increased input of nitrogen deposition affects the nitrogen leaching or runoff from forests (Dise *et al.*, 1998; Gundersen *et al.*, 1998; De Vries *et al.*, 2003a), causing an elevated indirect N_2O emission from surface waters (Fig. 17.1). Atmospheric nitrogen deposition may also increase the emission of the secondary radiatively active trace gas NO from forest soils (Gasche and Papen, 1999; Pilegaard *et al.*, 1999; Van Dijk and Duyzer, 1999).

Nitrification and denitrification can occur simultaneously in soils, though spatially separated. The rates of nitrification and denitrification increase with temperature, soil pH and substrate availability, i.e. with $[NH_4^+]$ and $[NO_3^-]$, respectively. An increased soil aeration and oxygen availability increases nitrification, but decreases denitrification. Variable but small amounts of N_2O are released during nitrification and denitrification. The ratio of N_2O produced per unit of NH_4 consumed (nitrification) tends to increase with decreased substrate pH, which is opposite to the effect on nitrification. An increase in temperature also increases the N_2O/NH_4 ratio. As with denitrification, the ratio of N_2O produced per unit of NO_3 consumed tends to increase with substrate (NO_3^-). However, in contrast to denitrification, the N_2O/NO_3 ratio decreases with an increase in temperature and soil pH and a decrease in oxygen availability (Granli and Bøckman, 1994). The fraction of N_2O released from the nitrification of NH_4^+ is usually much less than 1%, while the fraction of N_2O released from denitrification may range from 0.1% to 100%. More information on the role of the various factors is given by Lemke and Janzen (Chapter 5, this volume).

The reduction of N_2O to N_2 by denitrification is the only natural process by which N_2O is removed from the biosphere (see Kroeze *et al.*, Chapter 15, this volume), apart from its destruction in the stratosphere (see Butenhoff and Khalil, Chapter 14, this volume). The rate of N_2O consumption in soils, the importance of which is highly uncertain on a global perspective, is related to the diffusivity of the soil, which influences oxygen pressure (pO_2) and the residence time of nitrogen gases in the soil matrix. The soil diffusivity depends on the soil moisture and other soil properties such as texture and organic carbon content. Hence, the production and consumption of N_2O is controlled by a complex of biotic and abiotic factors (e.g. Brumme *et al.*, 1999; Papen and Butterbach-Bahl, 1999; Butterbach-Bahl *et al.*, 2002c). Increases in carbon sequestration in soils may increase N_2O emissions because of an increased denitrification potential (Six *et al.*, 2004; Li *et al.*, 2005). These increases in N_2O emissions, converted into CO_2 equivalent emissions, may offset the carbon sequestered completely (Li *et al.*, 2005). More information on the impact of soil carbon store on N_2O emissions and on GHG emissions is given by Lemke and Janzen (Chapter 5, this volume).

17.2.3 CH_4 exchange

Woodland soils can act as effective sinks for both atmospheric CH_4 and CH_4 produced in deeper soil layers. CH_4 is predominantly used by bacteria in the soil (methanotrophs), which use it as a source of carbon in a process called CH_4 oxidation. The 'high capacity–low affinity' methanotrophs are adapted for growth at high CH_4 concentrations (>1000 ppm in air), occurring, for example, in wetlands and in waterlogged soil layers. The 'low capacity–high affinity' methanotrophs are able to make use of the trace amounts of CH_4 in the atmosphere (~1.8 ppm in air) (Bender and Conrad, 1992; see also Dunfield, Chapter 10, this volume).

The production and consumption of CH_4 within the soil profile is controlled by several factors, of which pO_2, CH_4 partial pressure (pCH_4) and temperature are the most important. The key as to whether a soil acts as a sink or source of CH_4 is pO_2. The availability of oxygen in (parts of) the soil profile regulates whether microbial decomposition of substrates follows the aerobic pathway where CH_4 is consumed and the end product is CO_2, or the anaerobic pathway where the end products are CH_4 and CO_2. Again, pO_2 is regulated through soil diffusivity, which is a function of the water-filled pore space, rainfall and the depth of the groundwater table. In general, forest soils tend to be sinks for CH_4 because trees keep the water table well below the surface and allow the methanotrophs to grow (e.g. Keller *et al.*, 1983; Yavitt *et al.*, 1990a,b; Adamsen and King, 1993; Castro *et al.*, 1995). The CH_4 uptake by well-aerated forested soils has been related to soil texture (Striegl, 1993), soil pH (Brumme and Borken, 1999), atmospheric nitrogen deposition (Butterbach-Bahl and Papen, 2002), soil acidification (Brumme and Borken, 1999) and forest management (Borken and Brumme, 1997; Priemé and Christensen, 1997). The balance shifts from a net CH_4 sink (uptake exceeds production) to a net CH_4 source (production by methanogens exceeds uptake by methanotrophs) in water-saturated forest soils, as may sometimes happen during winter in drained soils, and more frequently in gley soils, riparian forest soils or forested bogs and fens (e.g. Crill *et al.*, 1988; Bartlett *et al.*, 1990; Roulet *et al.*, 1992).

Increased atmospheric nitrogen deposition increases [NH_4^+] in the soil and usually decreases CH_4 uptake by well-drained soils (Steudler *et al.*, 1989; Gulledge and Schimel, 1998; Van den Pol-van Dasselaar *et al.*, 1999; Le Mer and Roger, 2001). Three mechanisms have been postulated for the partial inhibition (slowdown) of CH_4 uptake by well-drained soils in response to increased nitrogen input: (i) competitive inhibition of the CH_4 monooxygenase by ammonia; (ii) inhibition of CH_4 consumption by toxic intermediates and end products of methanotrophic ammonia oxidation such as hydroxylamine and nitrite; and (iii) osmotic stress due to high concentrations of nitrate and/or ammonium (Schnell and King, 1994; Bradford *et al.*, 2001; Bodelier and Laanbroek, 2004; Reay and Nedwell, 2004). Bradford *et al.* (2001) suggested that the nitrate effect on CH_4 oxidation may be mediated through aluminium toxicity. However, Whalen and Reeburgh (2000) concluded that increased nitrogen deposition did not decrease CH_4 oxidation in boreal forests soils. Similarly, Borken *et al.* (2002) studied the long-term effect of reduced atmospheric nitrogen inputs on a coniferous forest and concluded that CH_4 oxidation did not increase. In contrast, positive effects of increased nitrogen availability on atmospheric CH_4 uptake have even been reported for severely nitrogen-limited forests (Goldman *et al.*, 1995; Börjesson and Nohrsted, 2000; Steinkamp *et al.*, 2000). Bodelier and Laanbroek (2004) hypothesized that such an increase of the methanotrophic activity of a soil due to nitrogen additions may be directly linked to nitrogen limitation of the methanotrophic bacterial community or of the biosynthesis of enzymes involved in CH_4 oxidation. An increase in the size and activity of the nitrifying population, co-oxidizing atmospheric CH_4, has also been proposed as a possible mechanism. However, the potential rate of such co-oxidation of CH_4 by soil nitrifiers has been found to be insignificant in forest soils in the UK (Reay *et al.*, 2005).

In addition to the potential for chronic atmospheric nitrogen deposition to alter the size of this sink, other sources of nitrogen input to soils have been found to have a significant effect on the forest soil CH_4 sink. For instance, Reay *et al.* (2005) reported vastly differing CH_4 oxidation potentials in soils under different vegetation types, with soils under alder having almost no capacity for CH_4 oxidation even under optimal conditions. This was apparently due to inhibition of CH_4 oxidation by the elevated nitrogen concentrations in the soils that result from the nitrogen-fixing *Frankia* spp. in the alder root nodules. Similarly, the practice of nitrogen fertilization, as used in many commercial forests to increase productivity, has been shown to significantly reduce CH_4 consumption rates (Castro *et al.*, 1994; Chan *et al.*, 2005).

In summary, the mechanisms of increased nitrogen availability through increased atmospheric nitrogen deposition on CH_4 oxidation are still poorly understood. They may differ from site to site, tipping the balance from inhibition to no effect or even an increase in CH_4 oxidation (see Fig. 17.1).

17.3 Estimates of the Net Greenhouse Gas Exchange by European Forests and Forest Soils

In this section, we present an overview of published emission data of CO_2, N_2O and CH_4 from European forests and forest soils. The quantification is based on various approaches, including: (i) literature data or reviews on measured GHG exchange fluxes from and to forests; and (ii) upscaling approaches by making use of process-based model approaches based on current understanding of processes and controlling factors (in the case of N_2O) or empirical models based on either field measurements or results from process-based modelling factors (in the case of CO_2 and N_2O).

17.3.1 CO_2 exchange

An overview of estimates of the carbon sequestration in European forests is given in Table 17.1. Apart from a distinction in the type of flux and forest compartment, a differentiation has been made in the quality of the upscaling methods, going from individual sites to the European scale. A direct comparison of the data in Table 17.1 is hampered, because of the differences in the NEP, NEE and net biome production (NBP). The NEP and NEE stand for the total uptake of CO_2 by photosynthesis, corrected for plant and soil respiration. The NBP is the NEP corrected for CO_2 emissions due to harvest and forest fires. The latter term is critical with respect to long-term carbon storage, since an aggrading forest may sequester large amounts of carbon, but most of it is emitted again to the atmosphere after logging. For a more detailed discussion about productivity terminology (gross primary productivity (GPP), NPP, NEP, NBP, etc.) we refer to Hymus and Valentini (Chapter 2, this volume). A systematic discussion related to the various approaches and results is given in De Vries *et al.* (2001, 2005b).

De Vries *et al.* (2005b) estimated the carbon sequestration of European forests during 1960–2000 by using data from the 6000 level I plots, which are assumed to be representative of 162 million hectares of forests. Data for the stand age and site quality were used to estimate the actual forest growth, using yield tables from the early 1960s, when nitrogen deposition was still low (reference deposition) (Klap *et al.*, 1997). The NEP was calculated by using a carbon content of 50%. The impact of elevated nitrogen deposition during 1960–2000 was accounted for by a method described in Section 17.4. The carbon pool change in stem wood due to forest growth thus derived equalled 0.281 Pg/year, which is comparable to the estimates given in Table 17.1. Dividing this estimate by 162 million hectares of European forests leads to a mean carbon pool change in tree stem wood of ~1730 kg/ha/year. The net carbon sequestration rate or carbon sink in stem wood was calculated by assuming that NBP equals 33% of the NEP. This percentage is based on an estimated average NBP/NEP ratio for Europe, implying a net increase in standing forest biomass of 33% of the growth since 67% is removed by harvesting or forest fires (Nabuurs and Schelhaas, 2003). This translates to an NBP of 0.094 Pg/year, which is comparable to estimates listed in Table 17.1, based on repeated forest inventories using country inventory data (Kauppi *et al.*, 1992; Nabuurs *et al.*, 1997) and by modelling forest growth (Liski *et al.*, 2002). Dividing this estimate by 162 million heatares of European forests leads to a mean net carbon sequestration rate of ~575 kg/ha/year.

Apart from sequestration in the trees, carbon can also be sequestered in the soil. The variation in soil carbon sequestration estimates is larger than for trees due to the difficulty of its assessment. For example, changes in the carbon pool in forest soils from repeated soil inventories are hard to detect within a short period of time, considering the large size of the soil carbon pools, with the possible exception of the organic layer (see also De Vries *et al.*, 2000). An estimate of the net carbon sequestration in the soil from direct measurements of the carbon input to the soil by litterfall and root decay, as well as by carbon release by mineralization (e.g. Schulze *et al.*, 2000), is hampered by the fact that the result is based

Table 17.1. Overview of different estimates of carbon sequestration on a European wide scale.

Type of carbon flux	Compartment	Method	Estimated sink (Pg/year)	Upscaling method	Reference
NBP landscape					
NBP	Landscape	Inversion modelling	0.30	Good	Bousquet *et al.* (1999)
NEE/NEP whole forest/trees					
NEE	Whole forest	CO_2 net flux measurements	0.47	Neural networks	Papale and Valentini (2003)
			0.25	Forest maps	Martin *et al.* (1998)
NEP	Total (aboveground biomass)	Tree growth measurements	0.42[a]	Multiply with forested area	Schulze *et al.* (2000)
NBP whole forest/trees					
NBP	Trees (stem wood)	Repeated forest inventories	0.10	Country inventory data	Kauppi *et al.* (1992) Nabuurs *et al.* (1997)
NBP	Trees (stem wood)	Modelling forest growth	0.06–0.10[b]	Country inventory data	Liski *et al.* (2002)
NEP contribution	Trees (aboveground biomass)	Nitrogen retention	0.039[c]	World average values	Nadelhoffer *et al.* (1999)
NBP forest soil					
NBP	Forest soil (belowground biomass)	Carbon soil input minus carbon mineralization	0.14[a]	Multiply with forested area	Schulze *et al.* (2000)
NBP	Forest soil (belowground biomass)	Modelling forest growth and decomposition	0.031–0.049[b]	Country inventory data	Liski *et al.* (2002)
NBP	Forest soil (belowground biomass)	Nitrogen retention	0.034[c]	World average values	Nadelhoffer *et al.* (1999)

NBP = net biome production; NEP = net ecosystem production; NEE = net ecosystem exchange.
[a]The estimates derived by Schulze *et al.* (2000) were slightly lower based on a forested area in Europe of 149 million hectares, but the estimates were scaled to an area of 162 million hectares, used in this study.
[b]These estimates were originally limited to the EU + Norway and Switzerland (~138 million hectares) but results were scaled to the European forested area, excluding most of Russia (~162 million hectares).
[c]These estimates were originally global but were scaled to the European nitrogen deposition on forests of 1.54 Mt/year. Actually, the estimate by Nadelhoffer *et al.* (1999) for carbon sequestration in trees refers to the contribution of nitrogen deposition to NEP by trees and not to the total NEP by forest growth.

on subtracting large numbers with relative high uncertainties. In this context a modelling exercise seems most reasonable (e.g. Liski *et al.*, 2002).

De Vries *et al.* (2005b) estimated the long-term net soil carbon sequestration in European forest soils by calculating the nitrogen immobilization (sequestration) in soils at ~6000 level I plots, multiplied by the carbon/nitrogen ratio of the forest soils. Nitrogen immobilization (sequestration) was calculated as a fraction of the nitrogen deposition corrected for nitrogen uptake, according to

$$\text{N immobilization} = \text{frN}_{\text{immobilization}} \cdot (\text{N deposition} - \text{net N uptake}) \quad (17.1)$$

The fraction $\text{frN}_{\text{immobilization}}$ was calculated as a function of the carbon/nitrogen ratio of the forest soil using available results on this relationship (Gundersen *et al.*, 1998) and those obtained from budgets for 121 intensive monitoring plots (De Vries *et al.*, 2001). By multiplying the net nitrogen immobilization with the carbon/nitrogen ratio, the variation of the carbon/nitrogen ratio with the depth of the soil profile was accounted for, according to

$$\text{C sequestration} = \text{N immobilization} \cdot (\text{fret}_{\text{ff}} \cdot \text{C/N}_{\text{ff}} + (1 - \text{fret}_{\text{ff}}) \cdot \text{C/N}_{\text{ms}}) \quad (17.2)$$

where C/N_{ff} and C/N_{ms} are the carbon/nitrogen ratios of the forest floor and the mineral soil (up to a depth of 20 cm), respectively, and fret_{ff} is the nitrogen retention fraction in the forest floor, as it is the ratio of the nitrogen retention in the forest floor to the nitrogen retention in the complete soil profile (forest floor and mineral soil). The carbon sequestration in forest soils was calculated for the period 1960–2000 using site-specific estimates for more than 6000 level I forest plots in a systematic grid of 16 × 16 km, according to:

- Nitrogen (NH_4,NO_3) deposition: European Monitoring Evaluation Programme (EMEP) model estimates for the period 1960–2000 (data at 5-year intervals that were linearly interpolated);
- Net nitrogen uptake: yield estimates as a function of stand age and site quality as described by Klap *et al.* (1997), multiplied by deposition-dependent nitrogen contents in biomass;
- Fret_{ff}: related to measured carbon/nitrogen ratios in forest soil and modelled fraction NH_4 in deposition, based on results of ^{15}N tracer experiments (Tietema *et al.*, 1998; Nadelhoffer *et al.*, 1999) as given by De Vries *et al.* (2005b);
- Carbon/nitrogen ratios for forest soils: measurements, partly extrapolations.

The estimated carbon sequestration using Eq. 17.2 equalled 0.023 Pg/year, which is 50% less than the global estimate derived by Nadelhoffer *et al.* (1999), scaled to Europe (0.034 Pg/year; De Vries *et al.*, 2005b). This is to be expected since these authors assumed a constant high retention fraction of 0.70 in the forest soil (organic layer with a carbon/nitrogen ratio of 30), whereas we used a carbon/nitrogen ratio-dependent fraction also occurring below the humus layer with lower carbon/nitrogen ratios. With the assumption that all the net incoming nitrogen is retained (total immobilization, no leaching) we got an estimate (0.042 Pg/year) that is slightly higher than that of Nadelhoffer *et al.* (1999). The calculated net carbon sequestration in the soil during 1960–2000 of ~0.023 Pg/year translates to an average accumulation of 143 kg/ha/year. This estimate is seven times lower than the value derived by Schulze *et al.* (2000), based on the estimated carbon retention in 11 sites (0.13–0.17 Pg C/year), but this is likely to be an overestimate, as it would imply that the carbon/nitrogen ratio of European forest soils is strongly increasing and there are no indications that this is the case. Furthermore, the calculated net soil carbon sequestration is in line with a calculated average value of 110 kg/ha/year for 16 typical forest types across Europe, derived by Nabuurs and Schelhaas (2002), and of 190 kg/ha/year, based on a modelling exercise for a large part of Europe by Liski *et al.* (2002).

The geographic variation in carbon sequestration in trees and soils is illustrated in Fig. 17.3. Carbon sequestration is small in northern Europe, where nitrogen input is low and nearly all incoming nitrogen is

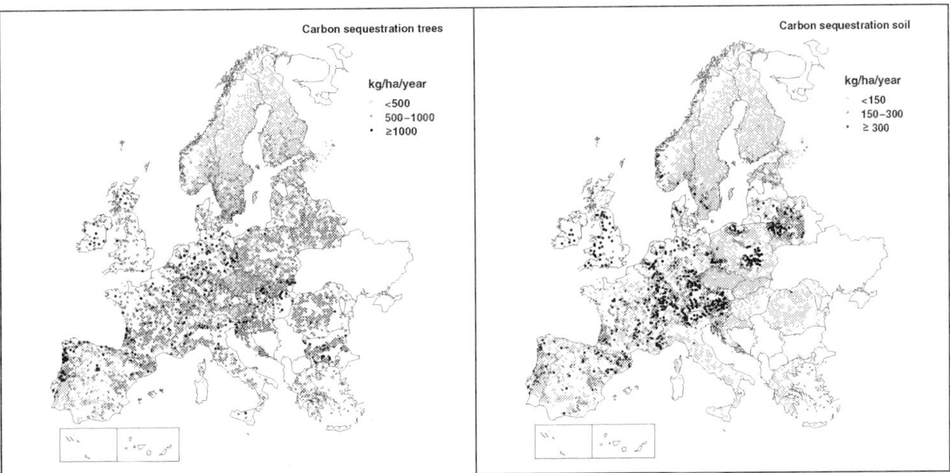

Fig. 17.3. Geographic variation of calculated carbon sequestration in trees and soil over Europe.

retained by the vegetation, and higher in central and eastern Europe where nitrogen input is higher. The finding that carbon sequestration is negligible in the northern boreal forest is in line with results from Martin *et al.* (1998) based on flux measurements for CO_2.

17.3.2 N_2O exchange

Field estimates

According to literature reviews, Papen and Butterbach-Bahl (1999), Borken and Brumme (1997), Schulte-Bisping *et al.* (2003) and Denier van der Gon and Bleeker (2005) reported that broadleaved forests show a tendency towards re-emitting a higher fraction of the nitrogen input as N_2O than coniferous forests due to species-induced differences in litter quality and soil moisture (Butterbach-Bahl *et al.*, 2002c). Papen and Butterbach-Bahl (1999), for example, observed that a beech site in Germany emitted about 10% of the nitrogen input as N_2O, whereas a spruce site emitted only 0.5% of the nitrogen input as N_2O. Differences between deciduous and coniferous forests are even more striking, because nitrogen deposition increases with surface roughness and is significantly higher in coniferous than in deciduous forests (Bleeker and Draaijers, 2001).

Generally, N_2O emissions increase when going from boreal to temperate forest ecosystems because of the increase in primary production and the increase in nitrogen deposition. For boreal forests, measured fluxes range from 0.18 to 0.27 kg N_2O-N/ha/year (Brumme *et al.*, 2004). Recently, Denier van der Gon *et al.* (2004) compiled the literature on N_2O emissions from forests in Europe (mainly Germany, but also Finland, Sweden, Denmark, UK, Belgium and Austria) and in North America (USA and Canada). Their results show that measured annual N_2O fluxes range from 0.02 to 4.5 kg N_2O-N/ha/year (Table 17.2). Some of this site variation can be explained by the effect of forest type on carbon and nitrogen cycling. However, there is usually a huge interannual variability in total annual N_2O losses. For example, the annual N_2O emissions at the Höglwald Forest ranged from 0.4 to 3.1 kg N_2O-N/ha/year within a 4-year observation period (Butterbach-Bahl *et al.*, 2002b). Evidently, it is difficult to assess long-term average fluxes from short-term measurements, and some of the observed differences between sites and forests types (see Table 17.2) may simply be a result of interannual variability. Table 17.2 also includes information on the nitrogen deposition and the N_2O emission fraction,

Table 17.2. Summary of published nitrous oxide (N_2O) emission data for deciduous forests and coniferous forests and derived emission factor as function of nitrogen input. (After Denier van der Gon and Bleeker, 2005.)

Number of observations[a]	Average nitrogen input[b] (kg N/ha/year)	N_2O emission (kg N/ha/year)	Emission factor[c]	Reference
Deciduous forest				
3	22	0.49	0.023	Ambus *et al.* (2001), Beier *et al.* (2001)
1	10	0.23	0.023	Bowden *et al.* (2000)
8	29	1.67	0.053	Brumme *et al.* (1999)
2	26	2.98	0.111	Butterbach-Bahl *et al.* (2001)
2	15	0.02	0.001	Corre *et al.* (1999)
1	20	1.45	0.072	Butterbach-Bahl *et al.* (1997)
2	46	2.65	0.044	Mogge *et al.* (1998)
1	16	0.20	0.013	Oura *et al.* (2001)
2	20	2.66	0.133[d]	Papen and Butterbach-Bahl (1999)
6	27	0.65	0.035	Skiba *et al.* (1998)
3	35	4.03	0.115	Zechmeister-Boltenstern *et al.* (2002)
Coniferous forest				
3	26	0.31	0.016	Borken *et al.* (2002)
4	26	0.58	0.034	Brumme *et al.* (1999)
2	35	0.70	0.020	Butterbach-Bahl *et al.* (2001)
4	28	1.59	0.073	Butterbach-Bahl *et al.* (1997)
3	16	0.90	0.056	Butterbach-Bahl *et al.* (2002a)
1	12	1.86	0.155	Jungkunst *et al.* (2004)
6	24	0.11	0.005	Klemedtsson *et al.* (1997)
4	105	0.36	0.005	Matson *et al.* (1992)
2	31	0.39	0.013	Oura *et al.* (2001)
2	100	3.20	0.032	Ineson *et al.* (1998)
6	30	0.60	0.020[d]	Papen and Butterbach-Bahl (1999)
14	37	0.37	0.016	Skiba *et al.* (1998)

[a]Various years and plots with different nitrogen treatments and/or tree species within the deciduous forest or coniferous forest class (e.g. an experiment where a beech and alder plot was monitored for two consecutive years would result in $n = 4$).
[b]Only indicative, as in some experiments various nitrogen input levels were studied and are averaged.
[c]Calculated from the original data of N_2O emission and nitrogen input for individual plots and then averaged. The emission factor gives the fraction of nitrogen input that is re-emitted as N_2O.
[d]Results presented in Papen and Butterbach-Bahl (1999) were corrected by excluding the extreme 1996 emissions. Results are thus based on the data for 1995 and 1997 for beech site (value of 0.133 instead of 0.222) and for 1994, 1995 and 1997 for the spruce site (value of 0.020 instead of 0.028).

assuming that nitrogen deposition is linearly related to N_2O emission, and this aspect is further discussed in Section 17.4.

Process-based model approaches

Process-oriented models simulate the biosphere–atmosphere exchange of N_2O-N based on individual production, consumption and emission processes, and their interactions. Nitrification rate ($d[NH_4^+]/dt$) is either calculated from a mechanistic description of the growth and development of nitrifying populations or as a function of substrate concentration $[NH_4^+]$, oxygen partial pressure (pO_2), temperature (T) and pH, according to:

$$d[NH_4^+]/dt = k_1 * f([NH_4^+]) * f(pO_2) * f(T)\, f(pH) \quad (17.3)$$

where k_1 is the first-order nitrification coefficient under optimal conditions and $f([NH_4^+])$, $f(pO_2)$, $f(T)$ and f(pH) are dimensionless reduction functions for $[NH_4^+]$, molecular oxygen, temperature and pH, respectively. Often, nitrifying activity is related to $[NH_4^+]$ via a Michaelis–Menten type of relationship, i.e. $f([NH_4^+]) = [NH_4^+]/\{k_2 + [NH_4^+]\}$. In this case, $[NH_4^+]$ limits nitrifying activity (cf. first-order process) at low concentrations and does not limit nitrifying activity (zero-order) at high concentrations. Constant k_2 is the Michaelis–Menten half-saturation constant or the concentration at which $f([NH_4^+]) = 0.5$. It should be noted that the meaning of k_1 changes to 'potential nitrification activity' when a Michaelis–Menten type of relationship is used for substrate dependence. For the dependence of nitrification rate on $f(pO_2)$, $f(T)$ and f(pH), various descriptions have been formulated, from simple to complex, but discussion of these is beyond the scope of this chapter (e.g. Shaffer *et al.*, 2001; Heinen, 2005).

Like nitrification, denitrification rate is either based on: (i) a mechanistic description of the growth and development of denitrifying populations; (ii) fractional or structural models; or (iii) substrate dependence using first-order kinetics. In the latter case, denitrification activity ($d[NO_3^-]/dt$) is described as a function of the estimated potential denitrification rate, which is often proportional to organic matter decomposition rate, corrected for nitrate concentration ($[NO_3^-]$), pO_2, temperature (T), and pH according to:

$$(d[NO_3^-])/dt) = k_D * f([NO_3^-]) * f(pO_2) * f(T) * f(pH) \quad (17.4)$$

where k_D is the estimated potential denitrification rate or organic matter decomposition rate, and $f([NO_3^-])$, $f(pO_2)$, $f(T)$ and f(pH) are dimensionless reduction functions for $[NO_3^-]$, molecular O_2, temperature and pH, respectively. Again, the description of these reduction functions is beyond the scope of this chapter (e.g. Shaffer *et al.*, 2001).

A well-known process-based model is the PnET-N-DNDC model (Li *et al.*, 2000), which integrates the interactions among primary drivers (climate, soil characteristics, forest type and management), soil environmental factors (e.g. temperature, moisture, pH, redox potential and substrate concentration gradients) and various biogeochemical reactions. These interactions control transformation and transport of carbon and nitrogen in the ecosystem and thereby the emission of N_2O from forest soils into the atmosphere. Extensive validation of the model for different forest sites worldwide has shown that the model is capable of predicting the emission of N_2O (Stange *et al.*, 2000; Butterbach-Bahl *et al.*, 2001).

Butterbach-Bahl *et al.* (2004) made regional inventories of N_2O emissions from forest soils for Saxony, Germany, using a coupled PnET-N-DNDC–GIS database system. Mean N_2O emissions for German forest soils were estimated at 1.4 kg N_2O-N/ha/year. The PnET-N-DNDC model has also been used to make an inventory of N_2O-N emissions from European forest soils, covering the whole of Europe excluding Cypress, former Yugoslavia and Albania, by coupling it to a Geographic information system (GIS) with a geographic resolution of 2527 grids of 50 × 50 km representing a forested area of 141 million hectares. The estimated total annual N_2O emissions were 81.6 Kt N_2O-N with an uncertainty range of 50.7–97.0 Kt N_2O-N for 2000 (Kesik *et al.*, 2005). Average N_2O emissions from forest soils across Europe were estimated at 0.58 kg N_2O-N/ha/year, which is less than half of the estimate for forests in Saxony.

Empirical approaches

On the basis of field measurements, Brumme *et al.* (1999) suggested that forest soils can be stratified according to their N_2O emission characteristics into seasonal, background and event emission pattern (SEP, BEP and EEP, respectively) types. Broadleaved forests with thick oxygen horizons on acid soils can be characterized as a large and seasonal source of N_2O (classified as SEP), which is assumed to be primarily controlled by temperature and moisture. Broadleaved forests, with thin oxygen horizons, and all coniferous forests have been identified as a low and non-seasonal source of N_2O (classified as BEP) since the oxygen horizon hardly inhibits diffusion. Estimated mean annual N_2O emissions from German forest soils for these forest types are 2.05 and 0.37 kg N/ha/year for broadleaved forests with thick and thin oxygen horizons, respectively and 0.17 kg N/ha/year for coniferous forests (Schulte-Bisping *et al.*, 2003). EEP occurs primarily during freeze–thaw processes in SEP and BEP types under beech, spruce, alder and oak forests (Brumme *et al.*, 1999; Papen and Butterbach-Bahl, 1999; Teepe *et al.*, 2000). Additional emissions by freeze–thaw processes may occasionally contribute to the BEP by up to 70%, whereas the contribution is much smaller in forests with SEP (up to 40%), but its overall impact is likely to be low (Brumme *et al.*, 1999). Schulte-Bisping *et al.* (2003) used the SEP/BEP stratification to calculate the regional emission strength of forest soils in Germany, although they used the pH of the top soil and not the thickness of the oxygen horizon as diagnostic criterion for separating forests in SEP- and BEP-type forests. Using the stratification approach and climate data for a period of 30 years (1961–1990), they calculated the average source strength of forest soils in Germany to be 0.32 kg N/ha/year, which is four times smaller than the estimate derived by Butterbach-Bahl *et al.* (2004) for Saxony with the PnET-N-DNDC model, but close to the European average estimate. However, Schulte-Bisping *et al.* (2003) did not include effects of freezing and thawing, as well as nitrogen deposition, suggesting that the estimates may not be too different from each other. Additional factors explaining the difference between the two estimates are forest typology and upscaling procedures.

17.3.3 CH_4 exchange

Although process-based approaches are becoming available (Borken and Brumme, 1997; Del Grosso *et al.*, 2000), current estimates of the exchange of CH_4 between forest soils and the atmosphere still rely on simple upscaling approaches. Using a relationship between CH_4 fluxes, soil pH and water content, Schulte-Bisping *et al.* (2001) estimated the CH_4 uptake by forest soils at 1.84 kg (CH_4)/ha/year. Brumme *et al.* (2004) used a stepwise regression technique to explore the relationship between CH_4 uptake and biome, annual rainfall, mean annual temperature, soil texture and latitude. The analysis showed that biome and soil texture class accounted for 18.1% of the variation in the global data-set of annual fluxes. The mean uptake rate was 6.6 kg CH_4/ha/year for coarse textured soils, 2.3 kg CH_4/ha/year for medium and fine-textured soils and the average for all soil types was 3 kg CH_4/ha/year (Brumme *et al.*, 2004).

An overview of published CH_4 flux rates between soils and the atmosphere for forest ecosystems as a function of climate zone, forest type, soil type and texture is given in Table 17.3. Lindner *et al.* (2004) estimated the CH_4 budget of European forest soils by linking these data to a GIS with spatial information about the forest area distribution in Europe (Schuck *et al.*, 2002), the European soils database of the European Soils Bureau (http://eusoils.jrc.it/) and the climatic zonation mentioned in Table 17.3. The total forest area in this analysis was 320 million hectares. According to these calculations, European forests have a mean net CH_4 uptake of 0.2 kg CH_4/ha/year, but with a large uncertainty range. Schulte-Bisping *et al.* (2001) and Brumme *et al.* (2004) estimated a higher net uptake of CH_4 by European forests. However, they did not consider water-saturated soils, which can be a rather large source of CH_4 (Roulet *et al.*, 1992). The current global soil sink for atmospheric CH_4 is estimated to be 30–40 Tg/year (Mosier *et al.*, 1994; Houweling *et al.*, 1999), with much of this sink occurring in well-aerated forest soils.

17.4 Estimates of Nitrogen Deposition Impacts on the Net Greenhouse Gas Exchange of European Forests and Forest Soils

This section presents our assessment of the effects of atmospheric nitrogen deposition

Table 17.3. Methane (CH_4) flux rates in depending on soil type, climate and species. (After Lindner *et al.*, 2004 and references therein.)

			CH_4 flux rates (kg CH_4-C/ha/year)									
			Boreal				Temperate					
			Atlantic		Continental		Atlantic		Continental		Mediterranean	
Soil type	Texture	Range	Broad-leaved	Coniferous forests	Broad-leaved	Coniferous forests	Broad-leaved	Coniferous forests	Broad-leaved	Coniferous forests	Broad-leaved	Coniferous forests
Lithosol	No differen-tiation	Mean	−0.2	−0.2	−0.2	−0.2	−0.35	−0.35	−0.35	−0.35	−0.35	−0.35
		Min	−0.5	−0.5	−0.5	−0.5	−0.6	−0.6	−0.6	−0.6	−0.6	−0.6
		Max	0	0	0	0	0	0	0	0	0	0
Ranker + Rendzina	No differen-tiation	Mean	6.57	6.57	6.57	12.62	−1.1	−1.1	−1.1	−1.1	−1.5	−1.5
		Min	−2.63	−2.63	−2.63	−0.79	−2.54	−2.54	−2.54	−2.54	−3	−3
		Max	23.00	23.00	23.00	18.4	−0.05	−0.05	−0.05	−0.05	−0.5	−0.5
Other soils	Coarse	Mean	−4.11	−4.11	−4.11	−1.533	−1.4	−5.82	3.09	−5.82	−3.5	−3.5
		Min	−4.3	−4.3	−4.3	−5.847	−0.7	−0.95	−5.48	−0.95	−5	−5
		Max	−0.74	−0.74	−0.74	12.51	−2.5	−10.85	21.85	−10.9	−2	−2
	Medium	Mean	−1.05	−1.05	−2.3	−1.7	−1.56	−1.7	−1.7	−1.7	−4.7	−4.7
		Min	−2.63	−2.63	−5.3	−5.4	−6.5	−5.4	−5.4	−5.4	−10.5	−10.5
		Max	−0.79	−0.79	1.3	1.3	0.12	1.3	1.3	1.3	0	0
	Fine	Mean	−1.05	−1.05	−2.3	−1.13	−0.52	−1.13	−1.13	−1.13	−2.5	−2.5
		Min	−2.63	−2.63	−5.3	−2.8	−0.7	−2.8	−2.8	−2.8	−4	−4
		Max	−0.79	−0.79	1.3	−0.36	0	−0.36	−0.36	−0.36	−0.5	−0.5
Gleysol	No differen-tiation	Mean	3	3	3	3	6.45	6.45	6.45	6.45	1	1
		Min	−0.5	−0.5	−0.5	−0.5	−0.7	−0.7	−0.7	−0.7	−1	−1
		Max	7	7	7	7	20.3	20.3	20.3	20.3	4	4
Histosol	Organic soil	Mean	−1.5	−0.06	0.8	4.50	77.3	77.3	77.3	77.3	40	40
		Min	0	0.3	−5.8	0.1	−1	−1	−1	−1	10	10
		Max	−6.5	−0.98	28	11.4	318.4	318.4	318.4	318.4	70	70

Values in shaded boxes without references are expert estimates. Negative values = deposition; positive values = emission.

on the net exchange of CO_2, N_2O and CH_4 from European forests and forest soils. The quantification is based on data for precipitation, temperature, atmospheric nitrogen deposition and soil chemistry (carbon/nitrogen and pH data) at: (i) ~500 intensive forest monitoring plots (De Vries *et al.*, 2003b) (level II Monitoring Programme, with all measured data except for temperature, which is partly estimated); and (ii) ~6000 plots (Bleeker *et al.*, 2004) at a systematic grid throughout the whole of Europe (level I Monitoring Programme, with interpolated precipitation and temperature data, modelled nitrogen deposition data and measured soil chemistry, i.e. carbon/nitrogen and pH data). The calculations presented below are all based on De Vries *et al.* (2006).

17.4.1 Impact of nitrogen deposition on CO_2 exchange

The methodology used to calculate the impact of increased nitrogen deposition on carbon sequestration by European forests is inspired by the approach of Holland *et al.* (1997) and Nadelhoffer *et al.* (1999). These authors assessed the additional carbon sequestration on a global scale from additional nitrogen uptake by trees and nitrogen immobilization in soils in response to nitrogen deposition, according to

$$\Delta C\ \text{sequestration} = \Delta N\ \text{deposition} \times (frN_{uptake} \times C/N_{stem\ wood} + frN_{immobilization} \times C/N_{soil}) \quad (17.5)$$

The basic assumption in this approach is that the additional nitrogen uptake or immobilization is reflected in carbon pool changes due to tree growth and organic matter accumulation according to the carbon/nitrogen ratios of the tree and the soil, respectively (Nadelhoffer *et al.*, 1999; De Vries *et al.*, 2003c). For the global scale, Holland *et al.* (1997) suggested that the increased carbon sequestration in forests due to increased atmospheric nitrogen deposition could take up one-third of the global CO_2 emission from fossil fuel, which is equivalent to 2 Pg C/year, if most of the nitrogen deposited were taken up by trees and used to form new woody biomass. However, recent data of Nadelhoffer *et al.* (1999) suggest that a large part of the nitrogen deposited accumulates in the soil at low carbon/nitrogen ratio (10:40) and not in the trees at high carbon/nitrogen ratio (200:500). Nadelhoffer *et al.* (1999) calculated the carbon sequestration due to increased nitrogen uptake by trees and nitrogen immobilization in soils on a global scale, assuming: (i) a constant nitrogen uptake fraction of 0.05 and a constant nitrogen immobilization fraction of 0.70, based on short-term (1–3 year) ^{15}N-labelled tracer experiments in nine temperate forests; and (ii) an average carbon/nitrogen ratio in stem wood of 500 and in forest soils of 30. Using this approach, an additional deposition of 1 kg N/ha/year leads to a sequestration of 46 kg C/ha/year, of which 25 kg C/ha/year is retained in stem wood (0.05 × 500) and 21 kg C/ha/year in soil (0.7 × 30). These results suggest that carbon sequestration in forest trees and in forest soils in response to additional nitrogen deposition is of equal magnitude, and that the impact of nitrogen deposition on carbon sequestration in forests is much lower than the estimate made earlier by Holland *et al.* (1997).

De Vries *et al.* (2005b) adapted the approach by Nadelhoffer *et al.* (1999) as summarized in Table 17.4, using measured and estimated data at the 6000 level I plots. They used 1960 as the reference year for nitrogen deposition (this leads to 'reference' or background growth) and calculated the additional nitrogen deposited during 1961–2000 relative to the reference year 1960, to assess the contribution of increased nitrogen deposition on the increase in carbon pools in trees and soil. They included the spatial differences in nitrogen deposition for the individual plots (EMEP model estimates). The nitrogen uptake fraction ranged between 0.05 and 0.10, with high values in low deposition areas because of nitrogen deficiencies, and low values in high deposition areas. Similarly, the carbon/nitrogen ratios in trees were assumed to range from 250 to 500, with low values in high deposition areas because of the assumed luxury consumption. The nitrogen immobilization fraction was assumed to be a function of the carbon/

Table 17.4. Overview of differences between the approach used by Nadelhoffer *et al.* (1999) and our study to calculate the impacts of nitrogen deposition on carbon sequestration.

Nadelhoffer *et al.* (1999)	Our approach
Reference nitrogen deposition is negligible	Reference nitrogen deposition is 1960
Constant average nitrogen deposition	Spatially distributed and time-dependent nitrogen deposition
Nitrogen uptake fraction is constant	Nitrogen uptake fraction is f(nitrogen deposition)
Nitrogen immobilization is constant	Nitrogen immobilization fraction is f(carbon/nitrogen ratio humus layer/soil, NH_4/NO_3 in deposition)
Carbon/nitrogen ratio tree is constant	Carbon/nitrogen ratio tree varies in space and time as f(nitrogen $deposition_{x,t}$)[a]
Carbon/nitrogen ratio soil is constant in space and time	Carbon/nitrogen ratio organic and mineral layer varies in space[b]

[a]Based on calculated EMEP nitrogen deposition.
[b]Based on the measured carbon/nitrogen ratio data from ~6000 forested plots.

nitrogen ratio of the soil organic matter layer, as described earlier, and not a constant of 70%. For the carbon/nitrogen ratio in the organic layer and mineral layer, we used the measured values at all level I plots instead of using a constant value of 30 (De Vries *et al.*, 2003c, 2005b).

Results for the period 1960–2000 indicate an average additional carbon sequestration in stem wood of 15 million tonnes per year in response to an additional nitrogen input (above the reference nitrogen deposition of 1960) of 0.45 million tonnes per year, which translates to 33.3kg C/ha/year per kilogram of nitrogen deposited (De Vries *et al.*, 2005b). This is close to the 25kg C/ha/year per kilogram of nitrogen used by Nadelhoffer *et al.* (1999). Assuming that the long-term carbon sequestration, corrected for CO_2 emissions due to harvest and forest fires, is 33% of the carbon pool changes – an average ratio for Europe (Nabuurs and Schelhaas, 2003) – the net carbon sequestration in stem wood is 11.1kg C/ha/year per kilogram nitrogen. For soil the sequestration is larger, namely 6.7 million tonnes per year, which translates to 14.8kg C/ha/year per kilogram of nitrogen deposited. Overall the impact of nitrogen deposition on the total carbon sequestration by trees and soils is estimated at ~26kg C/ha/year per kilogram of nitrogen. During 1960–2000, the average additional nitrogen deposition was 2.8kg N/ha/year, causing an additional overall carbon sequestration of 73kg C/ha/year. The total average carbon sequestration was estimated at 719kg C/ha/year (576 in trees and 143 in soil), implying a nitrogen deposition impact of ~10%. On a European scale, using a forested area of 162 million hectares, the impact of nitrogen deposition on carbon sequestration in trees and soil is 11.7 million tonnes per year on a total carbon sequestration of 117 million tonnes per year (0.117Gt/year with 0.094Gt/year in trees and 0.023Gt/year in soil).

17.4.2 Impact of nitrogen deposition on N_2O exchange

There are a large number of controlling variables and complex interactions that influence the net N_2O emission, which would suggest applying a detailed mechanistic model for calculating the effect of atmospheric nitrogen deposition on N_2O emissions from European forests. The problem, however, is that the application of such a model on a European scale is limited by the large number of data requirements. Consequently, we used simple, transparent and empirical approaches, with process-based models and empirical data-sets, because these are currently the most feasible to quantify the effect of anthropogenic nitrogen deposition on N_2O emissions.

Regression model based on process-based model

We derived a regression model, predicting N_2O emissions as a function of stand and site characteristics and environmental factors, including total nitrogen deposition, based on the predicted N_2O emissions (in kgN_2O-N/ha/year) for European forest soils with the PnET-N-DNDC model for a geographic resolution of 50 × 50km (Kesik *et al.*, 2005). Total nitrogen deposition was calculated from the wet deposition data, used by Kesik *et al.* (2005) according to the procedure used in PnET-N-DNDC (Li *et al.*, 2000):

$$[N]_{total} = 0.27 + 2.7\,[N]_{rainfall} \quad \text{(coniferous forests)} \qquad (17.6a)$$

$$[N]_{total} = 0.20 + 1.6\,[N]_{rainfall} \quad \text{(deciduous forests)} \qquad (17.6b)$$

where $[N]_{rainfall}$ is the nitrogen concentration in rainfall, which is equal to the wet nitrogen deposition divided by the precipitation. These concentrations were multiplied by the throughfall, which was calculated as a percentage of the precipitation depending on tree species, using precipitation and throughfall data from ~500 level II plots. On average, the throughfall fraction was ~0.7 for conifers (pine, spruce, fir and evergreen oak) and ~0.8 for deciduous trees (oak, beech, birch and hardwoods). The estimated total deposition was on average comparable to the EMEP total nitrogen deposition, but this value was not used, since it was not included in the PnET-N-DNDC calculation. The best result obtained from the regression analyses, while distinguishing between tree species, was:

$$N_2O[\text{kg N/ha/year}] = 1.3211 + a \cdot \textit{tree species} + 0.019925\,\textit{clay} - 0.01329 \cdot C_{min} - 0.05877 \cdot T + 0.0006640 \cdot C_{min} \cdot T - 0.0004592 \cdot Pr + 0.000002804 \cdot C_{min} \cdot Pr - 0.01006 \cdot pH + 0.018029 \cdot N_{dep.tot} \qquad (17.7)$$

where clay = clay content (%), C_{min} = mean value for the carbon pool in the mineral topsoil (0–30cm; t C/ha), pH = pH-H_2O, Pr = annual precipitation (mm/year), T = mean annual temperature (°C) and $N_{dep,\,tot}$ = total nitrogen deposition (kg N/ha/year). The percentage variance accounted for by this model (r^2_{adj}) is 0.42 and the standard error of observations is estimated to be 0.280kg N_2O-N/ha/year.

In the regression model, the impact of tree species was considered in different intercepts, where a is 0 for pine (reference tree), −0.2232 for larch, −0.1604 for fir, −0.0020 for evergreen oak, −0.0727 for spruce, 0.0276 for oak, 0.0396 for hardwoods, 0.1792 for birch and 0.3964 for beech. In general, the results for the deciduous trees were larger than for conifers, comparable to the measurements (Table 17.2). This could be due to differences in canopy structure, acidity of the forest floor and differences in soil moisture, which favour nitrification rather than denitrification activity in the soil of coniferous forests. The negative relationship with pH is according to expectations since the relative loss of N_2O during nitrification and denitrification increases with decreasing pH, and this overrides the stimulating effect of increasing pH on nitrification and denitrification itself. The negative relationship between N_2O emissions and carbon content, precipitation and temperature seems opposite to expectations. This is, however, due to the application of denitrification and decomposition (DNDC) on the whole of Europe and the correlations between explaining variables. For example, carbon-rich soils in Scandinavia and the Baltic states have relatively low emissions due to temperature restrictions. The positive impact of carbon is illustrated in the interaction terms of carbon with temperature and precipitation. The estimated average annual N_2O emissions calculated with Eq. 17.7 was 0.59kg N_2O-N/ha/year, which is very close to the average value of 0.58kg N_2O-N/ha/year derived by Kesik *et al.* (2005), while using the original PnET-N-DNDC model.

Figure 17.4 presents a graph showing the predicted N_2O-N emissions with PnET-N-DNDC versus the regression model, showing that the regression model deviates from PnET-N-DNDC predictions for higher annual N_2O emissions (>1.0kg N_2O-N). The regression model (Eq. 17.7) shows that a change in nitrogen deposition of 1kg N/ha/year leads to an increase of ~0.018 kg N_2O-N/ha/year. This is 1.8% of the nitrogen input, which is almost a factor of 2 higher than the default N_2O emissions factor of 1% used by the Intergovernmental Panel on Climate Change (IPCC) (e.g. IPCC, 1996; Mosier *et al.*, 1998). The value of 1.8% is close to the value derived by Denier van der Gon and Bleeker (2005) for coniferous forests (see also Tables 17.2 and 17.5 and the discussion in the following section on a regression model based on field measurements).

As with carbon sequestration, the effect of elevated atmospheric nitrogen deposition on N_2O emissions from European forest soils was assessed by using the empirical approach outlined above. The calculated average N_2O emission during 1960–2000 was compared to the emission in the reference year 1960 (reference nitrogen deposition rates) using available data for all level I plots, representing a total area of 162 million hectares. In this way, we estimated the effect of the cumulative additional nitrogen deposition during 1960–2000, compared with that in 1960, on the cumulative additional N_2O emissions. The clay content and carbon pool in the mineral soil and the pH-KCl were based on measurements up to a depth of 20 cm. The mean annual precipitation and the mean annual temperature were based on an interpolation of 30-year average values for a high-resolution grid in Europe during 1970–2000 (Mitchell *et al.*, 2004). N_2O-N emission for 1960 was estimated at 66,000 t, corresponding to an average of 0.41 kg N_2O-N/ha/year. The difference between this value and the average European estimate with the PnET-N-DNDC model (0.58 kg N_2O-N/ha/year) is mainly due to a different schematization: the use of generic soil data for a geographic resolution of 50 × 50 km by Kesik *et al.* (2005) compared to the measured data at level I plots in this study. Using an average additional nitrogen deposition of 2.8 kg/ha/year during 1960–2000 leads to an average increase of 0.05 kg N_2O-N/ha/year. For 162 million hectares of forests, the average additional N_2O emission can be estimated at 8100 t N_2O-N/year. Comparing this value with the emission in 1960, it follows that the impact of nitrogen deposition on N_2O emissions in the last 40 years is ~12%.

Regression models based on field measurements

In the IPCC methodology for accounting N_2O emissions from agriculture (IPCC,

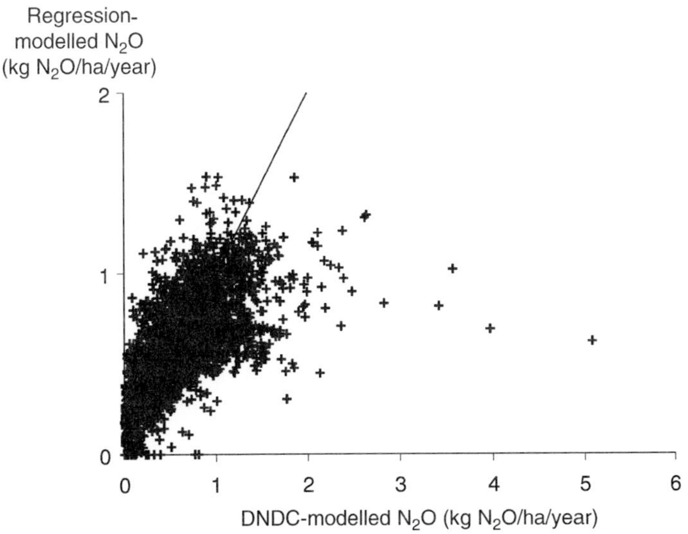

Fig. 17.4. Comparison of N_2O emission estimates with 2527 grids of 50 × 50 km with an empirical regression model and the PnET-N-DNDC model.

Table 17.5. Estimated nitrous oxide (N_2O) emission factors for deciduous forests and coniferous forests. (From Denier van der Gon and Bleeker, 2005.)

	N_2O emission factor (%)			
Type of forests	Mean[a]	Weighted mean[a]	Median	Regression
Coniferous forests	3.7	2.6	2.0	1.4
Deciduous forests	5.7	5.6	4.4	6.4

[a]In calculating mean emission each location and/or study is weighted equally, whereas the weighted mean weighs the average by the number of observations.

1996), the N_2O released from atmospheric nitrogen following its deposition on, for example, forest soils is simply calculated as a fraction of the amount of NH_3-N lost from agriculture. The emission factor (1% of nitrogen lost) for these indirect N_2O emissions from agriculture, multiplied with the total amount of NH_3-N lost from agriculture, provides a rough estimate of the indirect N_2O emissions. The empirical data in Table 17.2 show, however, that the current IPCC default value of 1% for indirect emissions is underestimating the N_2O emissions from forests. Results indicate that the N_2O emission fraction (derived as N_2O emissions divided by nitrogen deposition) is generally higher than 1%, specifically in deciduous forests.

Derivation of an average N_2O emission factor from field measurements data has some methodological artefacts. For example, our data presented in Tables 17.2 and 17.5 clearly indicate that emission factors depend on the way average emissions are calculated. Median values or average values, either weighted by the number of observations or unweighted, assume no N_2O emissions when nitrogen deposition is negligible. The average emission factors thus calculated are ~2.5% for coniferous forests and ~5% for deciduous forests. An averaging approach accounting for a certain N_2O emission when nitrogen deposition is negligible is linear regression analysis. Results from such an analysis indicate average emission factors of 1.4% for coniferous forests and 5.4% for deciduous forests as shown in Eq. 17.8 (with N_2O-N emissions and nitrogen deposition both in kg N/ha/year):

$$N_2O\text{-N emission} = 0.088\,\text{kg} + 0.014\,\text{N deposition } (r^2 = 0.28) \text{ for coniferous forests} \qquad (17.8a)$$

$$N_2O\text{-N emission} = 0.054\,\text{N deposition } (r^2 = 0.29) \text{ for deciduous forests} \qquad (17.8b)$$

These results are slightly different from those obtained by Denier van der Gon and Bleeker (2005), by removing several outliers, such as an nitrogen-fixing red alder stand with an emission of ~8 N_2O-N/ha/year (see Fig. 17.5). The relatively low N_2O emission factor for coniferous forests (1.4%) is due to a relatively large number of low emission observations (Table 17.2) that may not be properly weighted in the regression as it is biased by the sites with many observations. However, this emission factor is closer to the value of 1.8% derived by the PnET-DNDC model (Eq. 17.7).

For the Höglwald spruce site a simple regression model has been derived, based on weekly measured NH_4 input via throughfall and weekly measured N_2O emissions (Butterbach-Bahl *et al.*, 1998):

$$N_2O\ (\mu g\ N/m^2/h) = 4.7 + 1.4\ NH_4\text{-N input } (mmol/m^2/week) \qquad (17.9a)$$

Reworking this to the units that are relevant for our calculations on an annual basis, we get:

$$N_2O \text{ emission (kg N/ha/year)} = 0.41 + 0.0167\ NH_4\text{-N input (kg N/ha/year)} \qquad (17.9b)$$

The formula is derived for NH_4 deposition in the range of 0–8 mmol/m²/week, and the regression coefficient of the formula is

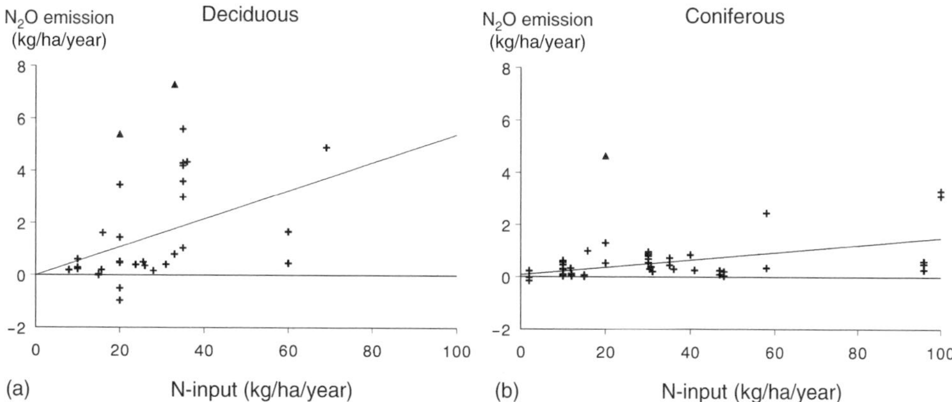

Fig. 17.5. N_2O-N emission as a function of nitrogen input for (a) deciduous forests and (b) coniferous forests based on the literature. (From Denier van der Gon and Bleeker, 2005.)

$r^2 = 0.38$. According to Eq. 17.9b, 1.67% of the nitrogen deposited is lost as N_2O. This estimate again is in good agreement with the independent estimate of 1.4%, derived from literature data for coniferous forests (using linear regression) and the 1.8% derived from the PnET-N-DNDC application (see Eq. 17.7). A much greater impact of nitrogen deposition was found in a pine forest in north-east Germany, where both nitrogen inputs and emissions of N_2O, NO and CH_4 were measured (Butterbach-Bahl *et al.*, 2002a). Here 14% of atmospheric wet nitrogen input was lost as N_2O. Considering a wet/dry deposition ratio of 2 would mean 7% of the total nitrogen input. This high percentage was, however, found at a site with a very high nitrogen input (>20 kg N/ha/year as wet deposition), largely exceeding nitrogen uptake.

Applying the nitrogen deposition–N_2O emission relationships in Eq. 17.8a and Eq. 17.8b leads to an estimated average annual N_2O-N emission of 0.3 kg/ha/year for 1960, corresponding to 48600 t N_2O-N for the whole of Europe. Using an average additional nitrogen deposition of 2.8 kg/ha/year during 1960–2000, while applying Eq. 17.8a and 17.8b, leads to an average increase of 0.098 kg N_2O-N/ha/year, which is equal to 15900 t N_2O-N/year.

Comparing this additional emission to the estimated emission in the reference year 1960 suggests that the impact of nitrogen deposition on N_2O emissions in the last 40 years has been ~33%. This is three times larger than the estimated contribution of nitrogen deposition on N_2O emission increase using Eq. 17.7. Although the discrepancies in the N_2O emission estimates by the two approaches are not very large (1960 N_2O emissions of 66,000 and 486,00 t N_2O-N, respectively and a nitrogen dose–effect curve leading to 8000 and 15,900 t N_2O-N/year, respectively), a calculated difference of 12% and 33% additional impact of nitrogen deposition on N_2O emissions in the last 40 years is quite significant. Considering both approaches, which are equally relevant, a general conclusion could be that the overall impact of nitrogen deposition on N_2O emissions in the last 40 years is ~22% ± 10%. However, the impact of 33% is likely to be an overestimation since the regression for deciduous forests in Eq. 17.8b implies that the background emission is zero, which is not true. In general, the difference in estimates on effects of nitrogen deposition on N_2O exchange is largely due to missing information on N_2O fluxes under unperturbed, pre-industrial and natural conditions, which would allow us to estimate the magnitude of background emissions.

17.4.3 Impact of nitrogen deposition on CH_4 exchange

Increased atmospheric deposition of nitrogen increases [NH_4^+] in the soil and usually decreases CH_4 uptake by well-drained soils although reverse effects can also be observed in nitrogen-deficient forests as discussed in Section 17.2. For the Höglwald (spruce) in Germany, Butterbach-Bahl *et al.* (1998, 2002a) found the following relationship:

$$CH_4 \text{ uptake } (\mu g\ CH_4/m^2/h) = 34.7 - 3.4\ NH_4\text{-N input } (mmol/m^2/week) \quad (17.10a)$$

Reworking this on an annual basis leads to

$$CH_4 \text{ uptake } (kg\ CH_4/ha/year) = 3.04 - 0.041\ NH_4\text{-N input } (kg\ NH_4\text{-N}/ha/year) \quad (17.10b)$$

Equation (17.10b) indicates that the uptake rate of atmospheric CH_4 is decreased by 4.1% for each kilogram of NH_4-N entering the system. As with N_2O emissions, a rather extreme impact of nitrogen deposition on CH_4 uptake was found for pine forests in north-east Germany (Butterbach-Bahl *et al.*, 2002a), but this result cannot be considered a representative. Assuming the Höglwald forest as a representative for Europe and considering an additional average NH_4 deposition of 1.2 kg N/ha/year during 1960–2000, there would have been an average decrease in atmospheric CH_4 uptake of 0.05 kg CH_4/ha/year at this site. This amount equals 1.6% of the total uptake of 3.04 kg CH_4/ha/year at zero NH_4-N input (see Eq. 17.10b). Clearly, the effect of the increased atmospheric nitrogen deposition on CH_4 uptake by forest soils is small. Furthermore, this result is rather uncertain, given the small number of studies and the complexities involved. Therefore, no further attempt has been made to quantify the effects of atmospheric deposition on net CH_4 uptake by European forest soils.

17.5 Discussion and Conclusions

17.5.1 Net emissions of CO_2, N_2O and CH_4 from forests compared to agriculture

The various estimates of carbon sequestration by European forests indicate a value of ~0.1 Pg C/year. This is comparable to European grasslands that are also thought to be a sink for atmospheric CO_2 of 0.1 Pg C/year (CarboEurope GHG, 2004). Inversely, Janssens *et al.* (2003) estimated that arable land in Europe loses 0.3 Pg C/year. Thereby, arable land is the largest annual biospheric source of CO_2 to the atmosphere in Europe. However, it must be acknowledged that the uncertainty in estimates for forests, grasslands and arable land is large.

Boeckx and Van Cleemput (2001) evaluated the emission of N_2O from agricultural soils in Europe using the IPCC methodology (Mosier *et al.*, 1998). They estimated that the average direct N_2O emission from agricultural soils across Europe is 5.6 kg N/ha/year. Lower numbers were calculated by Freibauer and Kaltschmitt (2003), who considered climate, soil and management factors in addition to the amount of nitrogen fertilization. For agricultural, minerotrophic soils for the temperate and sub-boreal climate regions in Europe, they estimated a mean N_2O emission of 2.0 kg/ha/year. For forest soils, we estimated an average emission of 0.5–0.6 kg N/ha/year. This implies that forest soils are approximately four- to tenfold smaller as a source for N_2O than agricultural soils are. With the data derived by Freibauer and Kaltschmitt (2003), Freibauer (2003) estimated the annual total N_2O emissions from agriculture in the EU-15 at 535,000 t N_2O-N, which included the N_2O emissions from animal husbandry (emissions from animal houses and manure storages). The percentage of soil-borne emissions was estimated at 82% or 440,000 t N_2O-N. Assuming estimated N_2O emissions by forest soils of 80,000 t N_2O-N (this study) would imply a contribution of 15% when neglecting the N_2O emissions from agricultural soils outside the EU-15. The agriculturally utilized area in EU-15 is 128 million hectares and that in EU-25 is 166 million hectares. The area used for agricultural in the non-member states of the EU (e.g. Norway, Switzerland, Belarus, Ukraine, Bulgaria, Russia) is relatively large but agricultural practices are less intensive, and hence N_2O emissions per unit surface area are probably less in these countries than in EU-15. All in all, this would suggest that the emission strength of forest soils in Europe most likely ranges between 5% and 15% of the emission strength of agricultural soils. As agricultural soils are the major source of N_2O in Europe (e.g. Freibauer, 2003), we consider

European forests a significant source of atmospheric N_2O in Europe.

For CH_4 the situation is quite different. Although the uncertainties are large, there is enough evidence to conclude that exchange rates of CH_4 between forest soils and atmosphere are very low compared to the release rates from wetlands. For forests, literature suggests a range in mean net CH_4 uptake of 0.2–3.0 kg CH_4/ha/year (Brumme *et al.*, 2004; Lindner *et al.*, 2004). Using an average value of 1.5 kg CH_4/ha/year and an area of 162 million hectares for European forests leads to a CH_4 sink of ~240,000 t. Freibauer (2003) estimated the CH_4 emissions from agriculture at 7,980,000 t CH_4 for the EU-15 countries of which ~75% is due to enteric fermentation and the remaining 25% due to animal housing and manure storage. Uptake of CH_4 by agricultural soils was not taken into account in the study of Freibauer (2003), but is probably in the same order of magnitude as that by forest soils. Comparision of those numbers shows that the contribution of forests to the net CH_4 emissions from the biosphere in Europe is small (<3%), specifically considering the fact that the estimates for agriculture presented by Freibauer (2003) are for the EU-15 only. Clearly, mitigation measures for CH_4 emissions should focus on animal agriculture and not on forests.

17.5.2 Impact of nitrogen deposition on net emissions of CO_2, N_2O and CH_4 by European forests

A comparison of the net exchange of the three investigated GHGs by European forests can best be done in terms of their GWP. The GWP is an index defined as the cumulative radiative forcing between the present and a chosen future time horizon (by convention 100 years), caused by a unit mass of gas emitted at present (by convention CO_2). Using this approach, N_2O and CH_4 emissions are expressed in terms of CO_2 equivalents. In this study it is assumed that 1 kg N_2O equals 296 kg CO_2 equivalents and 1 kg CH_4 equals 23 kg CO_2 equivalents (Ramaswamy, 2001). Other values frequently used in the literature are 310 and 21 kg CO_2 equivalents for N_2O and CH_4, respectively (IPCC, 1996).

An overview of the net exchange of CO_2, N_2O and CH_4 by European forests in terms of CO_2 equivalents is presented in Table 17.6. Data on the carbon sequestration are based on results by Kauppi *et al.* (1992), Nabuurs *et al.* (1997), Liski *et al.* (2002) and De Vries *et al.* (2005b) for forests (trees), and by Nabuurs and Schelhaas (2002), Liski *et al.* (2002) and De Vries *et al.* (2005b) for forest soils. These data are quite consistent and a reliable estimate can be made of 0.08–0.12 Pg C/year for trees and 0.015–0.04 Pg C/year for soil. Divided by a forested area of 162 million hectares, this implies an average carbon sequestration of 500–650 kg C/ha/year by stem wood and ~100–250 kg C/ha/year by soil. This represents an average total sink of ~600–900 kg C/ha/year or ~2200–3300 kg CO_2 equivalents/ha/year.

For N_2O the average source estimate for Europe ranges between 0.3 and 0.7 kg N_2O-N/ha/year (Schulte-Bisping *et al.*, 2003). Multiplying these values with 44/28 (N_2O versus N) and the GWP of 296. gives an average total source of 140–325 kg CO_2 equivalents/ha/year. For CH_4 the average sink estimate for Europe ranges from 0.2 to 3.0 kg CH_4/ha/year (Table 17.3: Schulte-Bisping *et al.*, 2001; Brumme *et al.*, 2004; Lindner *et al.*, 2004). This represents an average total sink of ~5–70 kg CO_2 equivalents/ha/year.

Using the average values given above, it is clear that CO_2 sequestration by forests is much larger than their N_2O emissions and CH_4 uptake. The CO_2 sequestration by forests is approximately ten times larger than the N_2O emissions, in terms of CO_2 equivalents (N_2O emission is ~10% of the CO_2 sequestration with a variation of 4–18%; see Table 17.6), whereas the CH_4 sink is almost negligible. The results imply that carbon sequestration in forests is clearly outweighing the N_2O emissions, but a more reliable N_2O estimate is important to estimate the counter-effect. Especially during the re-establishment of a forest plantation several years of high N_2O emissions (>3 kg N/ha/year) can be expected due to the missing plant nitrogen sink, which is not accounted for at present.

Table 17.6. Estimated ranges in long-term annual average carbon dioxide (CO_2), nitrous oxide (N_2O) and methane (CH_4) emissions and the impact of nitrogen deposition on those emissions including a comparison of their global warming potential (GWP) in CO_2 equivalents.

	Long-term average emissions (kg/ha/year)		GWP (kg CO_2 equivalents/ha/year)	
Greenhouse gas	Total estimates	Nitrogen deposition impacts[a]	Total estimates	Nitrogen deposition impacts[a]
CO_2-C	–600–900 (–750)	–60–90 (–75)	–2200–3300 (–2750)	–220–330 (–275)
N_2O-N	+0.3–0.7 (0.5)	+0.05–0.11 (0.08)	+140–325 (230)	+20–50 (35)
CH_4	–0.2–3.0 (1.6)	+0.003–0.05 (0.026)	–5–70 (–40)	+0.1–1.1 (0.6)

[a]The nitrogen deposition impacts are given for an estimated increase in total nitrogen deposition of 2.8 kg N/ha/year and 1.2 kg NH_4-N/ha/year for the period 1960–2000.

The impact of nitrogen deposition on the net exchange of CO_2, N_2O and CH_4 by European forests has also been assessed in terms of GWP, as shown in Table 17.6. The average estimated contribution of nitrogen deposition to the increase in CO_2 exchange (sink) is ~26 kg C/ha/year per kilogram of nitrogen of which 11.1 kg C/ha/year is sequestered in stem wood and 14.8 kg C/ha/year in soil. During 1960–2000, the average additional nitrogen deposition was 2.8 kg N/ha/year, causing an additional overall carbon sequestration of ~75 kg C/ha/year. The uncertainty is likely to be 20% (De Vries *et al.*, 2005b) leading to a range of 60–90 kg C/ha/year, implying a nitrogen deposition impact of ~10% (see Table 17.6). The average contribution of nitrogen deposition to the increase in N_2O emissions is estimated at 18–40 g/ha/year per kilogram of nitrogen, which is the range found in DNDC model estimates for Europe (first estimate) and in empirical data for coniferous and deciduous forest (second estimate). Using an average additional nitrogen deposition of 2.8 kg N/ha/year during 1960–2000 has led to an average increase of 0.05–0.11 kg N_2O-N/ha/year. The average contribution of nitrogen deposition to the reduction of the CH_4 sink is 0.041 kg CH_4/ha/year per kilogram of NH_4 for a spruce forest in Germany (Höglwald). This forest has an estimated CH_4 sink of ~3.0 kg NH_4-N/ha/year at zero NH_4-N input, which is the upper value of the CH_4 sink estimate for Europe. At an additional NH_4-N input of 1.2 kg during 1960–2000, this becomes 0.05 kg CH_4/ha/year, 1.6% that of forests. Assuming a constant percentage of nitrogen deposition impact, the range in reduction of the CH_4 sink becomes 0.003–0.05 kg CH_4/ha/year.

In summary, Table 17.6 shows that the impact of nitrogen deposition on the exchange in GHGs in terms of GWP is on average –275 kg CO_2 equivalents due to carbon sequestration by forests which is offset by ~35 kg CO_2 equivalents due to N_2O emissions (13%), whereas the contribution of a reduced CH_4 sink is negligible. This means that the increase in carbon sequestration in response to nitrogen deposition clearly outweighs the increased N_2O emissions, but again a more reliable N_2O estimate in response to nitrogen deposition is important to estimate the counter-effect more precisely.

17.5.3 Uncertainties in nitrogen deposition impact on CO_2, N_2O and CH_4 emissions

There are large uncertainties in the GHG emission estimates and in the estimated effects of nitrogen deposition on those emissions. The range in values presented in Table 17.6 may in reality be even larger due to other aspects not included in the quantification, such as the occurrence of forest disturbances, the neglect of off-site carbon sequestration in hardwood products, the neglect of indirect N_2O emissions and the occurrence of lag times between changes in nitrogen deposition and GHG emissions.

The neglect of off-site carbon sequestration causes an underestimation of the real nitrogen deposition impact, whereas the other aspects cause an overestimation.

Forest disturbances

Information on effects of disturbances on the productivity of forest ecosystems, either natural as a consequence of fire, wind, pest or disease, or managed such as forest logging, has been achieved by the use of chronosequences of forest stands at different ages. Results show that disturbances decrease photosynthetic capacity, completely in the case of logging, and typically convert a forest from a carbon sink into a carbon source (see also Hymus and Valentini, Chapter 2, this volume). The time required for a stand to become carbon-neutral and ultimately sequester carbon after the initial disturbance depends on the type and intensity of the disturbance and on post-disturbance management for evergreen forests (Thornton *et al.*, 2002). Periods generally vary between 10 and 12 years but periods of ~25 years have also been found (Hymus and Valentini, Chapter 2, this volume). In this context, it is also important to mention that N_2O emissions can substantially increase in the period following clear-cutting. In an overview article, Bowden (1986) estimated an increase of up to 0.5 kg N_2O-N/ha/year, but this might be much higher in nitrogen-saturated forests. First results for such forests indicate huge emissions of even more than 3 kg N_2O-N/ha/year in the first 4–7 years after clear-cutting (Butterbach-Bahl, 2006, personal communication). Including these estimates over the rotation time of a forest could increase the average N_2O emission by ~0.2–0.5 kg N_2O-N/ha/year, which is very substantial compared to the estimates presented earlier (see also Table 17.6). Similarly, nitrogen leaching is substantially increased in this period. This was already shown in an overview of numerous studies from the 1960s and 1970s in the USA (Vitousek and Melillo, 1979), and was recently substantiated by an additional overview of numerous studies since the 1980s in Europe and partly in the USA (Gundersen *et al.*, 2005). The results show that, in general, the nitrate concentration in soil and stream waters increases directly after clear-cutting with peak concentrations within 2–3 years. Gundersen *et al.* (2005) found that the highest responses (the difference in nitrate concentration between cut and reference stands) were observed in central Europe (5 mg N/l in stream or seepage water as a mean over the region). Assuming an average precipitation excess of 100–300 mm/year would imply an increase in the nitrogen leaching rate of 5–15 kg/ha/year. Using the standard fraction of 2.5% for indirect N_2O emission for each kilogram of nitrogen leached (Mosier *et al.*, 1998) would cause an estimated increase in indirect N_2O emissions of ~0.1–0.3 kg N_2O-N/ha/year. The nitrate concentration, however, often returns to pre-cutting levels within 5 years.

Incomplete carbon accounting

In this study we assume that wood which is harvested and removed from a site is ultimately released as CO_2 into the atmosphere. Thus we only account for the carbon sequestered in standing biomass. However, harvested wood can often reside in solid wood products, recycled products or landfills for centuries. Sometimes a large fraction of harvested wood is also used for energy production. This type of full accounting is often used for carbon sequestration, and results show that increases in off-site carbon can be sizeable, perhaps matching increases in on-site carbon (e.g. Pacala *et al.*, 2001). If nitrogen deposition accelerates forest growth, the potential for the off-site carbon sequestration (storage in products or in landfills, bioenergy offsets of fossil fuel emissions) is thus increased.

Indirect N_2O emissions

The impact of additional nitrogen deposition on indirect N_2O emissions from surface water, induced by additional nitrogen leaching or runoff, has not been included in the calculation because of its extreme uncertainties. In general, nitrogen leaching is negligible below an atmospheric input of 10 kg N/ha/year (Dise *et al.*, 1998; Gundersen *et al.*, 1998; De Vries *et al.*, 2003a). At higher nitrogen inputs, leaching is clearly higher in 'nitrogen-enriched' sites (carbon/nitrogen ratio in the organic layer below 23) than in 'carbon-enriched' sites (carbon/nitrogen ratio in the organic layer above 23). In the first case, a linear relationship has been derived according to nitrogen leaching = −4.3 + 0.67 × nitrogen deposition in kg N/ha/year (final report CNTER project 2005; Gundersen, 2006, personal communication). Using an average nitrogen deposition of 12.3 kg N/ha/year for forests in Europe (De Vries *et al.*, 2005b) would imply an average nitrogen leaching rate of 5.3 kg N/ha/year, which is close to 40% of the nitrogen deposition. However, in most cases the carbon/nitrogen ratio of the organic layer is above 23 and nitrogen leaching is generally lower. On average, the nitrogen leaching rate at 121 intensively monitored plots was only ~1 kg N/ha/year, which is ~7% of the nitrogen input (~13 kg N/ha/year; De Vries *et al.*, 2001). In the IPCC approach, nitrogen leaching is estimated at 30% of the nitrogen input (Mosier *et al.*, 1998). Using the standard fraction of 2.5% for indirect N_2O emissions for each kilogram of nitrogen leached (Mosier *et al.*, 1998) and a leaching fraction varying between 0.1 and 0.4 implies a net N_2O emission ranging from ~0.25% to 1.0% for each additional kilogram of nitrogen deposited. Compared to the average estimated value of 1.8–4.0% for direct N_2O emissions, this implies an average increase of ~10–20%, but the uncertainty in this value is high and the contribution may also be high at certain plots.

Lag times in nitrogen deposition effects

Considering the various factors affecting N_2O emissions, it is in general not correct to assume that N_2O emissions will decrease proportionally with reduced nitrogen deposition. Because the soil is a large reservoir of nitrogen, especially in nitrogen-saturated forest soils, decreases or increases in atmospheric nitrogen deposition may not cause direct changes in N_2O emissions. This is illustrated by a key experiment, published by Borken *et al.* (2002), describing the application of normal 'polluted' versus cleaned (to produce natural, unpolluted precipitation) throughfall to soil under roofed plots of a 70-year-old Norway spruce plantation in Germany. Although the average N_2O emission at the end of the experiment was slightly lower in the cleaned sites, no significant differences in N_2O emission rates were found after 7 years of treatment (Borken *et al.*, 2002). The results suggest that soil acidification and nitrogen eutrophication had a negligible effect on N_2O emissions of this temperate spruce forest. It should be noted that the N_2O emissions from the spruce forest were low (~0.3 kg N_2O-N/ha/year) and thus the mitigation potential was limited. Clearly the results cannot be extrapolated either to deciduous forests or 'high N_2O out' forest systems or to forests that are largely nitrogen-limited like many boreal forests. Nitrogen-input manipulations, i.e. N_2O response studies in various systems, would be an important addition to quantify the mitigation potential of reduced nitrogen inputs. Borken *et al.* (2002) also monitored the CH_4 oxidation rate and, like for N_2O emissions, no changes in CH_4 oxidation were observed upon a 7-year reduction of nitrogen deposition. The mechanism behind these observations remains obscured. It may be the result of a lag phase in the response of a forest that has received high nitrogen inputs over a long period, or because the original nitrogen inputs were below a level where strong effects are seen. Clearly further study is necessary to elucidate the potential to reverse N_2O emission levels and CH_4 oxidation capacities of forests exposed to elevated nitrogen deposition levels.

17.5.4 Research needs to improve the quantification

In summary, results of this study show that the average reduction in GWP due to CO_2 sequestration by European forests is set off by N_2O emissions for ~10%, whereas the net uptake of CH_4 is negligible compared to CO_2 sequestration. On average, the effect of nitrogen deposition on increasing N_2O emissions is estimated at ~10–15% of the increased CO_2 sequestration, implying that the positive effect on increasing the vegetation CO_2 sink is also much larger than the effect on increasing the N_2O emissions. The effect of nitrogen deposition on reduced CH_4 uptake is very small and highly uncertain. In this approach, the impacts of indirect N_2O emissions and additional N_2O emissions following forest disturbances are not accounted for, but the same holds for off-site carbon sequestration. Both aspects might nearly double the real emissions or sequestration, leading to a similar percentage in which N_2O emissions set off CO_2 sequestration.

Although the general conclusions are robust, the complexity of the processes involved and the large scale and diversity of the forests indicate that the impact of environmental factors on the emissions and sinks of CO_2, especially on N_2O and CH_4, remains highly uncertain. The research to improve the estimates of GHG emissions and the nitrogen deposition impacts on those emissions is an integrated interdisciplinary approach consisting of: (i) field measurements (including satellite observations above large forest areas) in extensive continental networks including measurements of N_2O and CH_4; (ii) process studies on the interaction of carbon and nitrogen in the laboratory; (iii) further development (using results of process studies) and testing (using field measurements) of detailed process-oriented biogeochemical models at plot scale; and (iv) upscaling of results by developing and using empirical models and simplified process-based models, in connection with results from field measurements and detailed model studies. Such an approach is foreseen in the Nitro Europe Project, which will be carried out during 2006–2011. Using such an integrated approach enables identification of knowledge gaps and/or ecosystems that are underrepresented, and the approach towards these gaps should be prioritized to maximize the effectiveness of the research investments.

Regarding field measurements, it is important to perform year-round measurements, preferably for a longer time period. By increasing the number of measurements to cover at least 1 year, and with data on weather and environmental conditions, the empirical models can be significantly improved. A better understanding of the basics of N_2O production as a function of soil moisture, temperature and nitrogen availability is needed to improve plot-scale models and estimates of the European greenhouse budget, and subsequently to devise strategies to minimize emissions. Such information can be derived from process studies using modern stable isotope techniques and advanced soil incubation techniques.

Apart from field and laboratory studies, the uncertainty in GHG emissions and nitrogen deposition impacts on them may be significantly reduced by the further development and testing of process-oriented models at a plot level. Recently several biogeochemical models such as the DNDC (Li *et al.*, 1992) and CENTURY (Parton *et al.*, 1996), which are able to mimic the complexity of processes observed in field and laboratory studies, have been developed not only to simulate GHG exchange but also for nitrate leaching or NH_3 volatilization. They are already capable of simulating a wide variety of ecosystems, and such models can be used to extrapolate results to other combinations of weather and climate, land use, soil type and (forest) management practices by running them for longer time periods (e.g. 10–30 years). However, it should be realized that these simulation models are able to 'explain' at best only 50% of the measured daily variation in N_2O fluxes over time.

One of the most challenging aspects, also attempted in this chapter, is to extrapolate results from the plot scale to the regional, continental and even global scale. Application of a detailed model, such as DNDC, implies the use of many assumptions and generic estimates regarding inputs, management and model parameters. This may completely offset the advantage of using validated detailed plot-scale models on a large regional scale. The data uncertainties (model inputs and model parameters) involved in this process can be ascertained by applying such a model on a plot with detailed input and output data. Comparison of the model results using these input and output data on a national or continental scale gives information on the loss of reliability caused by upscaling because of less detailed data information.

Considering the problem of data availability in detailed models, it might be just as reliable to use results of simulations of process models to generate artificial datasets from which simple 'empirical' relationships are derived, such as Eq. 17.7. These simple relationships can then be used to derive large-scale soil emissions, as done for N_2O. In this context, a relatively simple process-based model (e.g. INITIATOR2; De Vries *et al.*, 2005a) can also be a compromise between a simple empirical model and a detailed process-oriented model approach (e.g. DNDC). This implies the need for further development and application of both detailed and simple process-oriented models assessing the loss of reliability in emissions estimated caused by upscaling because of less detailed data information by comparing the results of both types of models using the approach mentioned earlier.

Finally, it is important to realize that for IPCC estimates only the anthropogenic CO_2, N_2O and CH_4 emissions are relevant. So the natural CO_2, N_2O and CH_4 source strength itself is not relevant, except for the deviation from the natural base line. IPCC estimates of biogenic GHG emissions thus focus on sources that are directly influenced by human activities, such as fertilizer-induced N_2O emissions from agricultural soils while N_2O emissions from forests are accounted for as indirect emissions (currently only related to effects of agricultural NH_3 emissions and not to NO_x emissions). In this context, it is important to establish an improved linkage between anthropogenic nitrogen deposition and atmospheric nitrogen emissions by using improved large-scale atmospheric dispersion models. This aspect is also foreseen in the Nitro Europe project. Only when the information about the impact per unit of nitrogen deposited on the CO_2, N_2O and CH_4 emissions – positive or negative (sink) – is properly linked to nitrogen emitted by anthropogenic sources can the potential impact of changes in nitrogen emissions on GHG emissions be quantified.

Acknowledgements

We thank Dr Dave Reay (University of Edinburgh) and Dr Graham Hymus (University of Tuscia) for their constructive comments on this chapter.

References

Adamsen, A. and King, G. (1993) Methane consumption in temperate and subarctic forest soils: rates, vertical zonation and responses to water and nitrogen. *Applied and Environmental Microbiology* 59, 485–490.

Amann, M., Johansson, M., Lükewille, A., Schöpp, W., Apsimon, H., Warren, R., Gonzales, T., Tarrason, L. and Tsyro, S. (2001) An integrated assessment model for fine particulate matter in Europe. *Water, Air, and Soil Pollution* 130, 223–228.

Ambus, P., Jensen, J.M., Priemé, A., Pilegaard, K. and Kjøller, A. (2001) Assessment of CH_4 and N_2O fluxes in a Danish beech (*Fagus sylvatica*) forest and an adjacent N-fertilised barley (*Hordeum vulgare*) field: effects of sewage sludge amendments. *Nutrient Cycling in Agroecosystems* 60, 15–21.

Asner, G.P., Townsend, A.R., Riley, W.J., Matson, P.A., Neff, J.C. and Cleveland, C.C. (2001) Physical and biogeochemical controls over terrestrial ecosystem responses to nitrogen deposition. *Biogeochemistry* 54, 1–39.

Baldocchi, D.D. and Wilson, K.B. (2001) Modeling CO_2 and water vapor exchange of a temperate broad-leaved forest across hourly to decadal time scales. *Ecological Modelling* 142, 155–184.

Bartlett, K., Crill, P., Bonassi, J., Richey, J. and Harriss, R. (1990) Methane flux from the Amazon river floodplain: emissions during rising water. *Journal of Geophysical Research* 95, 16773–16788.

Bauer, G.A., Bazzaz, F.A., Minocha, R., Long, S., Magill, A., Aber, J. and Berntson, G.M. (2004) Effects of chronic N additions on tissue chemistry, photosynthetic capacity, and carbon sequestration potential of a red pine (*Pinus resinosa* Ait.) stand in the NE United States. *Forest Ecology and Management* 196, 173–186.

Beier, C., Rasmussen, L., Pilegaard, K., Ambus, P., Mikkelsen, T., Jensen, N.O., Kjøller, A., Priemé, A. and Ladekarl, U.L. (2001) Fluxes of NO_3^-, NH_4^+, NO, NO_2, and N_2O in an old Danish beech forest. *Water, Air, and Soil Pollution: Focus* 1, 187–195.

Bender, M. and Conrad, R. (1992) Kinetics of CH_4 oxidation in oxic soils exposed to ambient air or high CH_4 mixing ratios. *FEMS Microbiology Letters* 101, 261–270.

Berg, B. and Matzner, E. (1997) Effect of N deposition on decomposition of plant litter and soil organic matter in forest systems. *Environmental Reviews* 5, 1–25.

Bleeker, A. and Draaijers, G.P.J. (2001) *Literatuurstudie naar de invloed van kroonstructuur en bosranden op de atmosferische depositie in bossen* (in Dutch). TNO Milieu en Procesinnovatie, Apeldoorn, The Netherlands. TNO-MEP R2001/580.

Bleeker, A., Reinds, G.J., Vermeulen, A.T., de Vries, W. and Erisman, J.W. (2004) *Critical Loads and Present Deposition Thresholds of Nitrogen and Acidity and Their Exceedances at Level II and Level I Monitoring Plots in Europe*. Energy Research Centre of The Netherlands. ECN Report ECN-C-04-117.

Bobbink, R., Hornung, M. and Roelofs, J.G.M. (1998) The effects of air-borne nitrogen pollutants on species diversity in natural and semi-natural European vegetation. *Journal of Ecology* 86, 717–738.

Bodelier, P.L.E. and Laanbroek, H.J. (2004) Nitrogen as a regulatory factor of methane oxidation in soils and sediments. *FEMS Microbiology Ecology* 47, 265–277.

Boeckx, P. and Van Cleemput, O. (2001) Estimates of N_2O and CH_4 fluxes from agricultural lands in various regions in Europe. *Nutrient Cycling in Agroecosystems* 60, 35–47.

Börjesson, G. and Nohrsted, H.Ö. (2000) Fast recovery of atmospheric methane consumption in a Swedish forest soil after single-shot N-fertilization. *Forest Ecology and Management* 134, 83–88.

Borken, W. and Brumme, R. (1997) Liming practice in temperate forest ecosystems and the effects on CO_2, N_2O and CH_4 fluxes. *Soil Use and Management* 13, 251–257.

Borken, W., Beese, F., Brumme, R. and Lamersdorf, N. (2002) Long-term reduction in nitrogen and proton inputs did not affect atmospheric methane uptake and nitrous oxide emission from a German spruce forest soil. *Soil Biology and Biochemistry* 34, 1815–1819.

Bousquet, P., Ciais, P., Peylin, P., Ramonet, M. and Monfray, P. (1999) Inverse modeling of annual atmospheric CO2 sources and sinks: 1. Method and control inversion. *Journal of Geophysical Research* 104, 26161–26178.

Bowden, W.B. (1986) Gaseous nitrogen emissions from undisturbed terrestrial ecosystems: an assessment of their impacts on local and global nitrogen budgets. *Biogeochemistry* 2, 249–279.

Bowden, R.D., Melillo, J.M., Steudler, P.A. and Aber, J.D. (1991) Effects of nitrogen additions on annual nitrous oxide fluxes from temperate forest soils in the Northeastern United States. *Journal of Geophysical Research* 96, 9321–9328.

Bowden, R.D., Rullo, G., Stevens, G.R. and Steudler, P.A. (2000) Soil fluxes of carbon dioxide, nitrous oxide, and methane at a productive temperate deciduous forest. *Journal of Environmental Quality* 29, 268–276.

Bowden, R.D., Davidson, E., Savage, K., Arabia, C. and Steudler, P. (2004) Chronic nitrogen additions reduce total soil respiration and microbial respiration in temperate forest soils at the Harvard Forest. *Forest Ecology and Management* 196, 43–56.

Bradford, M.A., Wookey, P.A., Ineson, P. and Lappin-Scott, H.M. (2001) Controlling factors and effects of chronic nitrogen and sulphur deposition on methane oxidation in a temperate forest soil. *Soil Biology and Biochemistry* 33, 93–102.

Brumme, R. and Beese, F. (1992) Effects of liming and nitrogen fertilization on emissions of CO_2 and N_2O from a temperate forest. *Journal of Geophysical Research* 97, 851–858.

Brumme, R. and Borken, W. (1999) Site variation in methane oxidation as affected by atmospheric deposition and type of temperate forest ecosystem. *Global Biogeochemical Cycles* 13, 493–501.

Brumme, R., Borken, W. and Finke, S. (1999) Hierarchical control on nitrous oxide emission in forest ecosystems. *Global Biogeochemical Cycles* 13, 1137–1148.

Brumme, R., Verchot, L.V., Martikainen, P.J. and Potter, C.S. (2004) Contribution of trace gases nitrous oxide (N_2O) and methane (CH_4) to the atmospheric warming balance of forest biomes. In: Griffiths, H. and Jarvis, P.G. (eds) *The Carbon Balance of Forest Biomes*. BIOS Scientific Publishers, Southampton, pp. 291–315.

Butterbach-Bahl, K. and Papen, H. (2002) Four years continuous record of CH_4-exchange between the atmosphere and untreated and limed soil of a N-saturated spruce and beech forest ecosystem in Germany. *Plant and Soil* 240, 77–90.

Butterbach-Bahl, K., Gasche, R., Breuer, L. and Papen, H. (1997) Fluxes of NO and N_2O from temperate forest soils: impact of forest type, N deposition and of liming on the NO and N_2O emissions. *Nutrient Cycling in Agroecosystems* 48, 79–90.

Butterbach-Bahl, K., Gasche, R., Huber, C., Kreutzer, K. and Papen, H. (1998) Impact of N-input by wet deposition on N-trace gas fluxes and CH_4-oxidation in spruce forest ecosystems of the temperate zone in Europe. *Atmospheric Environment* 32, 559–564.

Butterbach-Bahl, K., Stange, F., Papen, H. and Li, C. (2001) Regional inventory of nitric oxide and nitrous oxide emissions for forest soils of southeast Germany using the biogeochemical model PnET-N-DNDC. *Journal of Geophysical Research* 106, 34155–34166.

Butterbach-Bahl, K., Breuer, L., Gasche, R., Willibald, G. and Papen, H. (2002a) Exchange of trace gases between soils and the atmosphere in Scots pine forest ecosystems of the northeastern German Lowlands: 1. Fluxes of N_2O, NO/NO_2 and CH_4 at forest sites with different N-deposition. *Forest Ecology and Management* 167, 123–134.

Butterbach-Bahl, K., Gasche, R., Willibald, G. and Papen, H. (2002b) Exchange of N-gases at the Höglwald Forest: a summary. *Plant and Soil* 240, 117–123.

Butterbach-Bahl, K., Rothe, A. and Papen, H. (2002c) Effect of tree distance on N_2O- and CH_4-fluxes from soils in temperate forest ecosystems. *Plant and Soil* 240, 91–103.

Butterbach-Bahl, K., Kesik, M., Miehle, P., Papen, H. and Li, C. (2004) Quantifying the regional source strength of N-trace gases across agricultural and forest ecosystems with process based models. *Plant and Soil* 260, 311–329.

CarboEurope GHG (2004) *Greenhouse Gas Emissions from European Grasslands*. European Commission. Report 4, Specific Study 3. Viterbo, Italy.

Castro, M.S., Peterjohn, W.T., Melillo, J.M., Steudler, P.A., Gholz, H.L. and Lewis, D. (1994) Effects of nitrogen-fertilization on the fluxes of N_2O, CH_4, and CO_2 from soils in a Florida slash pine plantation. *Canadian Journal of Forest Research* 24, 9–13.

Castro, M., Steudler, P., Melillo, J., Aber, J. and Bowden, R. (1995) Factors controlling atmospheric methane consumption by temperate forest soils. *Global Biogeochemical Cycles* 9, 1–10.

Chan, A.S.K., Steudler, P.A., Bowden, R.D., Gulledge, J. and Cavanaugh, C.M. (2005) Consequences of nitrogen fertilization on soil methane consumption in a productive temperate deciduous forest. *Biology and Fertility of Soils* 41, 182–189.

Conrad, R. (1996) Soil microorganisms as controllers of atmospheric trace gases (H_2, CO, CH_4, OCS, N_2O, and NO). *Microbiological Reviews* 60, 609–640.

Corre, M.D., Pennock, D.J., van Kessel, C. and Elliott, D.K. (1999) Estimation of annual nitrous oxide emissions from a transitional grassland-forest region in Saskatchewan, Canada. *Biogeochemistry* 44, 29–49.

Crill, P., Bartlett, K., Harriss, R., Gorham, E., Verry, E., Sebacher, D., Madzar, L. and Sanner, W. (1988) Methane flux from Minnesota peatlands. *Global Biogeochemical Cycles* 2, 371–384.

Crutzen, P.J. (1995) Ozone in the troposphere. In: Sing, H.B. (ed.) *Composition, Chemistry, and Climate of the Atmosphere*. Van Nostrand-Reinold, New York, pp. 349–393.

Del Grosso, S.J., Parton, W.J., Mosier, A.R., Ojima, D.S., Potter, C.S., Borken, W., Brumme, R., Butterbach-Bahl, K., Crill, P.M., Dobbie, K. and Smith, K.A. (2000) General CH_4 oxidation model and comparisons of CH_4 oxidation in natural and managed systems. *Global Biogeochemical Cycles* 14, 999–1019.

Denier van der Gon, H.A.C. and Bleeker, A. (2005) Indirect N_2O emission due to atmospheric N deposition for the Netherlands. *Atmospheric Environment* 39, 5827–5838.

Denier van der Gon, H.A.C., Bleeker, A., Ligthart, T., Duyzer, J.H., Kuikman, P.J., Van Groenigen, J.W., Hamminga, W., Kroeze, C., De Wilde, H.P.J. and Hensen, A. (2004) *Indirect Nitrous Oxide Emissions from The Netherlands: Source Strength, Methodologies, Uncertainties and Potential for Mitigation*. TNO-Environment and Process Innovation, Apeldoorn, The Netherlands. TNO Report No. 2004/275.

De Vries, W., van Grinsven, J.J.M., van Breemen, N., Leeters, E.E.J.M. and Jansen, P.C. (1995) Impacts of acid deposition on concentrations and fluxes of solutes in acid sandy forest soils in The Netherlands. *Geoderma* 67, 17–43.

De Vries, W., Reinds, G.J., van Kerkvoorde, M.A., Hendriks, C.M.A., Leeters, E.E.J.M., Gross, C.P., Voogd, J.C.H. and Vel, E.M. (2000) *Intensive Monitoring of Forest Ecosystems in Europe. Technical Report 2000*. UN/ECE, EC, Forest Intensive Monitoring Coordinating Institute, Geneva.

De Vries, W., Reinds, G.J., van der Salm, C., Draaijers, G.P.J., Bleeker, A., Erisman, J.W., Auee, J., Gundersen, P., Kristensen, H.L., van Dobben, H., de Zwart, D., Derome, J., Voogd, J.C.H. and Vel, E.M. (2001) *Intensive Monitoring of Forest Ecosystems in Europe. Technical Report 2001.* UN/ECE, EC, Forest Intensive Monitoring Coordinating Institute, Geneva.

De Vries, W., Reinds, G.J., Van der Salm, C., van Dobben, H., Erisman, J.W., de Zwart, D., Bleeker, A., Draaijers, G.-P.J., Gundersen, P., Vel, E.M. and Haussmann, T. (2003a) Results on nitrogen impacts in the EC and UN/ECE ICP Forests programme. In: Achermann, B. and Bobbink, R. (eds) *Empirical Critical Loads for Nitrogen.* Proceedings of an Expert Workshop in Berne, Switzerland, Novemeber 11–13. Swiss Agency for the Environment, Forests and Landscape (SAEFL) Environmental Documentation 164, pp. 199–207.

De Vries, W., Reinds, G.J. and Vel, E. (2003b) Intensive monitoring of forest ecosystems in Europe: 2. Atmospheric deposition and its impacts on soil solution chemistry. *Forest Ecology and Management* 174, 97–115.

De Vries, W., van der Salm, C., Reinds, G.J., Dise, N.B., Gundersen, P., Erisman, J.W. and Posch, M. (2003c) *Assessment of the Dynamics in Nitrogen and Carbon Sequestration of European Forest Soils.* Alterra, Wageningen, The Netherlands. Alterra-rapport 818.

De Vries, W., Vel, E., Reinds, G.J., Deelstra, H., Klap, J.M., Leeters, E.E.J.M., Hendriks, C.M.A., Kerkvoorden, M., Landmann, G. and Herkendell, J. (2003d) Intensive monitoring of forest ecosystems in Europe: 1. Objectives, set-up and evaluation strategy. *Forest Ecology and Management* 174, 77–95.

De Vries, W., Kros, J. and Velthof, G. (2005a) Integrated Evaluation of Agricultural Management on Environmental Quality with a Decision Support System. In: Zhu, Z., Minami, K. and Xing, G. (eds) *3rd International Nitrogen Conference,* 12–16 October 2004, Nanjing China, Science Press, pp. 859–870.

De Vries, W., Reinds, G.J. and Gundersen, P. (2006a) Impacts of nitrogen deposition on carbon sequestration by forests in Europe. *Global Change Biology* 12, 1151–1173.

De Vries, W., Bahl, K.B., Denier van der Gon, H. and Oenema, O. (2006b) Quantification of the impact of nitrogen deposition on the exchange of greenhouse gases from European forests. *Global Change Biology* (submitted).

Dise, N.B., Matzner, E. and Gundersen, P. (1998) Synthesis of nitrogen pools and fluxes from European forest ecosystems. *Water, Air, and Soil Pollution* 105, 143–154.

Freibauer, A. (2003) Regionalised inventory of biogenic greenhouse gas emissions from European agriculture. *European Journal of Agronomy* 19, 135–160.

Freibauer, A. and Kaltschmitt, M. (2003) Nitrous oxide emissions from agricultural mineral soils in Europe – controls and models. *Biogeochemistry* 63(1), 93–115.

Friedlingstein, P., Fung, I., Holland, E., John, J., Brasseur, G., Erickson, D. and Schimel, D. (1995) On the contribution of CO_2 fertilization to the missing biospheric sink. *Global Biogeochemical Cycles* 9, 541–556.

Galloway, J.N. (1998) The global nitrogen cycle: changes and consequences. *Environmental Pollution* 102, 15–24.

Galloway, J.N., Dentener, F.J., Capone, D.G., Boyer, E.W., Howarth, R.W., Seitzinger, S.P., Asner, G.P., Cleveland, C.C., Green, P.A., Holland, E.A., Karl, D.M., Michaels, A.F., Porter, J.H., Townsend, A.R. and Vörösmarty, C.J. (2004) Nitrogen cycles: past, present and future. *Biogeochemistry* 70, 153–226.

Gasche, R. and Papen, H. (1999) A 3-year continuous record of nitrogen trace gas fluxes from untreated and limed soil of a N-saturated spruce and beech forest ecosystem in Germany: 2. NO and NO_2 fluxes. *Journal of Geophysical Research* 104, 18505–18520.

Goldman, M.B., Groffman, P.M., Pouyat, R.V., McDonnell, M.J. and Pickett, S. (1995) CH_4 uptake and N availability in forest soils along an urban to rural gradient. *Soil Biology and Biochemistry* 27, 281–286.

Granli, T. and Bøckman, O.C. (1994) Nitrous oxide from agriculture. *Norwegian Journal of Agricultural Sciences Supplement* 12, 7–128.

Groffman, P.M., Brumme, R., Butterbach-Bahl, K., Dobbie, K.E., Mosier, A.R., Ojima, D., Papen, H., Parton, W.J., Smith, K.A. and Wagner-Riddle, C. (2000) Evaluating annual nitrous oxide fluxes at the ecosystem scale. *Global Biogeochemical Cycles* 14, 1061–1070.

Gulledge, J. and Schimel, J.P. (1998) Low-concentration kinetics of atmospheric CH_4 oxidation in soil and mechanism of NH_4^+ inhibition. *Applied and Environmental Microbiology* 64, 4291–4298.

Gundersen, P., Callesen, I. and de Vries, W. (1998) Nitrate leaching in forest ecosystems is related to forest floor C/N ratios. *Environmental Pollution* 102, 403–407.

Gundersen, P., Schmidt, I.K. and Raulund-Rasmussen, K. (2006) Leaching of nitrate from temperate forests: effects of air pollution and forest management. *Environmental Reviews* 14, 1–57.

Hagedoorn, F., Spinnler, D. and Siegwolf, R. (2003) Increased N deposition retards mineralisation of old soil organic matter. *Soil Biology and Biochemistry* 35, 1683–1692.

Harrison, A.F., Harkness, D.D., Rowland, A.P., Garnett, J.S. and Bacon, P.J. (2000) Annual carbon and nitrogen fluxes in soils along the European forest transect determined using ^{14}C-bomb. In: Schulze, E.D. (ed.) *Carbon and Nitrogen Cycling in European Forest Ecosystems. Ecological Studies 142*. Springer, Berlin, pp. 237–256.

Hasenauer, H., Nemani, R.R., Schadauer, K. and Running, S.W. (1999) Forest growth response to changing climate between 1961 and 1990 in Austria. *Forest Ecology and Management* 122, 209–219.

Heinen, M. (2005) Simplified denitrification models: overview and properties. *Geoderma* 133, 444–463.

Holland, E.A., Braswell, B.H., Lamarque, J.F., Townsend, A.R., Sulzman, J., Müller, J.F., Dentener, F., Brasseur, G., Levy, H. II, Penner, J.E. and Roelofs, G.J. (1997) Variations in the predicted spatial distribution of atmospheric nitrogen deposition and their impact on carbon uptake by terrestial ecosystems. *Journal of Geophysical Research* 102, 15849–15866.

Houweling, S., Kaminski, T., Dentener, F., Lelieveld, J. and Heimann, M. (1999) Inverse modelling of methane sources and sinks using the adjoint of a global transport model. *Journal of Geophysical Research* 104, 26137–26160.

Hudson, R.J.M., Gherini, S.A. and Goldstein, R.A. (1994) Modeling the global carbon-cycle: nitrogen-fertilization of the terrestrial biosphere and the missing CO_2 sink. *Global Biogeochemical Cycles* 8, 307–333.

Hungate, B.A., Dukes, J.S., Shaw, M.R., Luo, Y. and Field, C.B. (2003) Nitrogen and climate change. *Science* 302, 1512–1513.

Hunter, I.R. and Schuck, A. (2002) Increasing forest growth in Europe: possible causes and implications for sustainable forest management. *Plant Biosystems* 136, 133–141.

Ineson, P., Coward, P.A., Benham, D.G. and Robertson, S.M.C. (1998) Coniferous forests as 'secondary agricultural' sources of nitrous oxide. *Atmospheric Environment* 32, 3321–3330.

IPCC (1996) *Climate Change 1995: The Science of Climate Change*. Contribution of Working Group I to the Second Assessment Report of the Intergovernmental Panel on Climate Change. Cambridge University Press, Cambridge.

Janssens, I.A., Freibauer, A., Ciais, P., Smith, P., Nabuurs, G.-J., Folberth, G., Schlamadinger, B., Hutjes, R.W.A., Ceulemans, R., Schulze, E.D., Valentini, R. and Dolman, A.J. (2003) Europe's terrestrial biosphere absorbs 7 to 12% of European anthropogenic CO_2 emissions. *Science* 300, 1538–1542.

Jungkunst, H.F., Fiedler, S. and Stahr, K. (2004) N_2O emissions of a mature Norway spruce (*Picea abies*) stand in the Black Forest (southwest Germany) as differentiated by the soil pattern. *Journal of Geophysical Research* 109, D07302.

Kauppi, P.E., Mielikäinen, K. and Kuusela, K. (1992) Biomass and carbon budget of European forests, 1971 to 1990. *Science* 256, 70–74.

Keller, M., Goreau, T., Kaplan, W., Wofsy, S. and McElroy, M. (1983) Production of nitrous oxide and consumption of methane by forest soils. *Geophysical Research Letters* 10, 1156–1159.

Kesik, M., Ambus, P., Baritz, R., Brüggemann, N., Butterbach-Bahl, K., Damm, M., Duyzer, J., Horvarth, L., Kiese, R., Kitzler, B., Leip, A., Li, C., Pihlatie, M., Pilegaard, K., Seufert, G., Simpson, D., Smiatek, G., Skiba, U., Vesala, T. and Zechmeister-Boltenstern, S. (2005) Inventories of N_2O and NO emissions from European forest soils. *Biogeosciences* 2, 353–375.

Klap, J.M., de Vries, W., Erisman, J.W. and van Leeuwen, E.P. (1997) *Relationships Between Forest Condition and Natural and Anthropogenic Stress Factors on the European Scale: Pilot Study*. DLO Winand Staring Centre for Integrated Land, Soil and Water Research, Wageningen, The Netherlands. Report 150.

Klemedtsson, L., Kasimir Klemedtsson, Å., Moldan, F. and Weslien, P. (1997) Nitrous oxide emission from Swedish forest soils in relation to liming and simulated increased N-deposition. *Biology and Fertility of Soils* 25, 290–295.

Lamarque, J.F., Kiehl, J.T., Brasseur, G.P., Butler, T., Cameron-Smith, P., Collins, W.D., Collins, W.J., Granier, C., Hauglustaine, D., Hess, P.G., Holland, E.A., Horowitz, L., Lawrence, M.G., McKenna, D., Merilees, P., Prather, M.J., Rach, P.J., Rotman, D., Shindell, D. and Thornton, P. (2005) Assessing future nitrogen deposition and carbon cycle feedback using a multimodel approach: analysis of nitrogen deposition. *Journal of Geophysical Research* 110, D19303.

Le Mer, J. and Roger, P. (2001) Production, oxidation, emission and consumption of methane by soils: a review. *European Journal of Soil Biology* 37, 25–50.

Li, C., Frolking, S. and Frolking, A. (1992) A model of nitrous oxide evolution from soil driven by rainfall events: 1. Model structure and sensitivity. *Journal of Geophysical Research* 97, 9759–9776.

Li, C., Aber, J., Stange, F., Butterbach-Bahl, K. and Papen, H. (2000) A process-oriented model of N_2O and NO emissions from forest soils: 1. Model development. *Journal of Geophysical Research* 105, 4369–4384.

Li, C., Frolking, S. and Butterbach-Bahl, K. (2005) Carbon sequestration in arable soils is likely to increase nitrous oxide emissions, offsetting reductions in climate radiative forcing. *Climatic Change* 72, 321–328.

Lindner, M., Lucht, W., Bouriaud, O., Green, T. and Janssens, I.A. (2004) *Specific Study on Forest Greenhouse Gas Budget*. CarboEurope-GHG concerted action synthesis of the European greenhouse gas budget. Report 8/2004, specific study 1.

Liski, J., Perruchoud, D. and Karjalainen, T. (2002) Increasing carbon stocks in the forest soils of western Europe. *Forest Ecology and Management* 169, 159–175.

Martin, P.H., Valentini, R., Jacques, M., Fabbri, K., Galati, D., Quaratino, R., Moncrieff, J.B., Jarvis, P., Jensen, N.O., Lindroth, A., Grelle, A., Aubinet, M., Ceulemans, R., Kowalski, A.S., Vesala, T., Keronen, P., Matteucci, G., Grainer, A., Berbingier, P., Loustau, D., Schulze, E.D., Tenhunen, J., Rebman, C., Dolman, A.J., Elbers, J.E., Bernhofer, C., Grunwald, T., Thorgeirsson, H., Kennedy, P. and Folving, S. (1998) New estimate of the carbon sink strength of EU forests integrating flux measurements, field surveys, and space observations: 0.17–0.35 Gt (C). *Ambio* 27, 582–584.

Matson, P.A., Gower, S.T., Volkmann, C., Billow, C. and Grier, C.C. (1992) Soil nitrogen cycling and nitrous oxide fluxes in fertilized Rocky Mountain Douglas-fir forests. *Biogeochemistry* 18, 101–117.

Melillo, J.M., McGuire, A.D., Kicklighter, D.W., Moore, B. III, Vorosmarty, C.J. and Schloss, A.L. (1993) Global climate change and terrestrial net primary production. *Nature* 363, 234–240.

Mitchell, T., Carter, T.R., Jones, P.D., Hulme, M. and New, M. (2004) *A Comprehensive Set of High-Resolution Grids of Monthly Climate for Europe and the Globe: The Observed Record (1901–2000) and 16 Scenarios (2001–2100)*. Working paper 55. Tyndall Centre, Norwich, UK.

Mogge, B., Kaiser, E.A. and Munch, J.C. (1998) Nitrous oxide emissions and denitrification N-losses from forest soils in the bornhöved lake region (northern Germany). *Soil Biology and Biochemistry* 30, 703–710.

Mosier, A., Kroeze, C., Nevison, C., Oenema, O., Seitzinger, S. and van Cleemput, O. (1998) Closing the global N_2O budget: nitrous oxide emissions through the agricultural nitrogen cycle. (OECD/IPCC/IEA phase II development of IPCC guidelines for national greenhouse gas inventory methodology.) *Nutrient Cycling in Agroecosystems* 52, 225–248.

Mosier, A.R., Duxbury, J.M., Freney, J.R., Heinemeyer, O., Minami, K. and Johnson, D.E. (1994) Mitigating agricultural emissions of methane. *Climatic Change* 40, 39–40.

Myneni, R.B., Keeling, C.D., Tucker, C.J., Asrar, G. and Nemani, R.R. (1997) Increased plant growth in the northern high latitudes from 1981 to 1991. *Nature* 386, 698–702.

Nabuurs, G.J. and Schelhaas, M.J. (2002) Carbon profiles of typical forest types across Europe assessed with CO_2FIX. *Ecological Indicators* 1, 213–223.

Nabuurs, G.J. and Schelhaas, M.J. (2003) Spatial distribution of whole-tree carbon stocks and fluxes across the forests of Europe: where are the options for bio-energy? *Biomass and Bioenergy* 24, 311–320.

Nabuurs, G.J., Päivinen, R., Sikkema, R. and Mohren, G.M.J. (1997) The role of European forests in the global carbon cycle: a review. *Biomass and Bioenergy* 13, 345–358.

Nabuurs, G.J., Schelhaas, M.J., Mohren, G.M.J. and Field, C.B. (2003) Temporal evolution of the European forest sector carbon sink from 1950 to 1999. *Global Change Biology* 9, 152–160.

Nadelhoffer, K.J., Emmett, B.A., Gundersen, P., Kjønaas, O.J., Koopmans, C.J., Schleppi, P., Tietema, A. and Wright, R.F. (1999) Nitrogen deposition makes a minor contribution to carbon sequestration in temperate forests. *Nature* 398, 145–148.

Osborne, C.P., Mitchell, P.L., Sheehy, J.E. and Woodward, F.I. (2000) Modelling the recent historical impacts of atmospheric CO_2 and climate change on Mediterranean vegetation. *Global Change Biology* 6, 445–458.

Oura, N., Shindo, J., Fumoto, T., Toda, H. and Kawashima, H. (2001) Effects of nitrogen deposition on nitrous oxide emissions from the forest floor. *Water, Air, and Soil Pollution* 130, 673–678.

Pacala, S.W., Hurtt, G.C., Baker, D., Peylin, P., Houghton, R.A., Birdsey, R.A., Heath, L., Sundquist, E.T., Stallard, R.F., Ciais, P., Moorcroft, P., Caspersen, J.P., Shevliakova, E., Moore, B., Kohlmaier, G., Holland, E., Gloor, M., Harmon, M.E., Fan, S.-M., Sarmiento, J.L., Goodale, C.L., Schimel, D. and Field, C.B. (2001) Consistent land- and atmosphere-based U.S. Carbon sink estimates. *Science* 292, 2316–2320.

Papale, D. and Valentini, R. (2003) A new assessment of European forests carbon exchanges by eddy fluxes and artificial neural network spatialization. *Global Change Biology* 9, 525–535.

Papen, H. and Butterbach-Bahl, K. (1999) A 3-year continuous record of nitrogen trace gas fluxes from untreated and limed soil of a N-saturated spruce and beech forest ecosystem in Germany: 1. N_2O emissions. *Journal of Geophysical Research* 104, 18487–18503.

Parton, W.J., Mosier, A.R., Ojima, D.S., Valentine, D.W., Schimel, D.S., Weier, K. and Kulmala, A.E. (1996) Generalized model for N_2 and N_2O production from nitrification and denitrification. *Global Biogeochemical Cycles* 10, 401–412.

Pilegaard, K., Hummelshøj, P. and and Jensen, N.O. (1999) Nitric oxide emission from a Norway spruce forest floor. *Journal of Geophysical Research* 104, 3433–3445.

Poth, M. (1986) Dinitrogen production from nitrite by a Nitrosomonas isolate. *Applied and Environmental Microbiology* 52, 957–959.

Poth, M. and Focht, D.D. (1985) ^{15}N kinetic analysis of N_2O production by Nitrosomonas europaea: an examination of nitrifier denitrification. *Applied and Environmental Microbiology* 49, 1134–1141.

Priemé, A. and Christensen, S. (1997) Seasonal and spatial variation of methane oxidation in a Danish spruce forest. *Soil Biology and Biochemistry* 29, 1165–1172.

Ramaswamy, V. (2001) Radiative forcing of climate change. In: Houghton, J.T. (ed.). *Climate Change 2001: The Scientific Basis*. IPCC Third Assessment Report. Cambridge University Press, Cambridge, pp. 350–416.

Reay, D.S. and Nedwell, D.B. (2004) Methane oxidation in temperate soils: effects of inorganic N. *Soil Biology and Biochemistry* 36, 2059–2065.

Reay, D.S., Nedwell, D.B., McNamara, N. and Ineson, P. (2005) Effect of tree species on methane and ammonium oxidation capacity in forest soils. *Soil Biology and Biochemistry* 37, 719–730.

Regina, K., Nykanen, H., Maljanene, M., Silvola, J. and Martikainen, P.J. (1998) Emissions of N_2O and NO and net nitrogen mineralization in a boreal forested peatland treated with different nitrogen compounds. *Canadian Journal of Forest Research* 28, 132–140.

Rehfuess, K.-E., Ågren, G.I., Andersson, F., Cannell, M.G.R., Friend, A., Hunter, I., Kahle, H.-P., Prietzel, J. and Spiecker, H. (1999) *Relationships Between Recent Changes of Growth and Nutrition of Norway Spruce, Scots Pine and European Beech Forests in Europe-Recognition*. Working paper 19. European Forest Institute, Joensuu, Finland.

Rennenberg, H., Kreutzer, K., Papen, H. and Weber, P. (1998) Consequences of high loads of nitrogen for spruce (*Picea abies*) and beech (*Fagus sylvatica*) forests. *New Phytologist* 139, 71–86.

Roulet, N., Ash, R. and Moore, T. (1992) Low boreal wetlands as a source of atmospheric methane. *Journal of Geophysical Research* 97, 3739–3749.

Schnell, S. and King, G.M. (1994) Mechanistic analysis of ammonium inhibition of atmospheric methane consumption in forest soil. *Applied and Environmental Microbiology* 60, 3514–3521.

Schuck, A., van Brusselen, J., Päivinen, R., Häme, T., Kennedy, P. and Folving, S. (2002) *Compilation of a Calibrated European Forest Map Derived from NOAA-AVHRR data*. European Forest Institute, Joensuu. EFI Internal Report 13.

Schulte-Bisping, H., Brumme, R. and Priesack, E. (2001) *Freisetzung und Konsumption von N_2O und CH_4 aus den Wäldern Deutschlands – Ein Beitrag zur Abschätzung globaler Bilanzen*. Institut für Bodenkunde Waldernährung, Göttingen, Germany. Abschlu_bericht zum BMBF-Projekt 01 LA 9807.

Schulte-Bisping, H., Brumme, R. and Priesack, E. (2003) Nitrous oxide emission inventory of German forest soils. *Journal of Geophysical Research* 108, 4132.

Schulze, E.D., Högberg, L., van Oene, H., Persson, T., Harrison, A.F., Read, D., Kjøller, A. and Matteucci, G. (2000) Interactions between the carbon and nitrogen cycle and the role of biodiversity: a synopsis of a study along a north–south transect through Europe. In: Schulze, E.D. (ed.) *Carbon and Nitrogen Cycling in European Forest Ecosystems: Ecological Studies 142*. Springer, Berlin, pp. 468–492.

Shaffer, M.J., Ma, L. and Hansen, S. (eds) (2001) *Modeling Carbon and Nitrogen Dynamics for Soil Management*. Lewis Publishers, Boca Raton, Florida.

Sitaula, B.K., Bakken, L.R. and Abrahamsen, G. (1995) CH_4 uptake by temperate forest soil: effect of N input and soil acidification. *Soil Biology and Biochemistry* 27, 871–880.

Six, J., Ogle, S.M., Breidt, F.J., Conant, R.T., Mosier, A.R. and Paustian, K. (2004) The potential to mitigate global warming with no-tillage management is only realized when practiced in the long term. *Global Change Biology* 10, 155–160.

Skiba, U., Sheppard, L., Pitcairn, C.E.R., Leith, I., Crossley, A., van Dijk, S., Kennedy, V.H., Fowler, D., van der Hoek, K.W., Erisman, J.W., Smeulders, S. and Wisniewski, J.R. (1998) Soil nitrous oxide and nitric oxide emissions as indicators of elevated atmospheric N deposition rates in seminatural ecosystems. *Environmental Pollution* 102, 457–461.

Spiecker, H., Mielikäinen, K., Kölh, M. and Skovsgaard, J.P. (eds) (1996) *Growth Trends in European Forests*. Springer, Berlin.

Stange, F., Butterbach-Bahl, K., Papen, H., Zechmeister-Boltenstern, S., Li, C. and Aber, J. (2000) A process oriented model of N_2O and NO emission from forest soils: 2. Sensitivity analysis and validation. *Journal of Geophysical Research* 105, 4385–4398.

Steinkamp, R., Butterbach-Bahl, K. and Papen, H. (2000) Methane oxidation by soils of an N-limited and N fertilized spruce forest in the Black Forest, Germany. *Soil Biology and Biochemistry* 33, 145–153.

Steudler, P.A., Bowden, R.D., Melillo, J.M. and Aber, J.D. (1989) Influence of nitrogen fertilization on methane uptake in temperate forest soils. *Nature* 341, 314–315.

Striegl, R.G. (1993) Diffusional limits to the consumption of atmospheric methane by soils. *Chemosphere* 26, 715–720.

Teepe, R., Brumme, R. and Beese, F. (2000) Nitrous oxide emissions from frozen soils under agricultural, fallow and forest land. *Soil Biology and Biochemistry* 32, 1807–1810.

Thornton, P.E., Law, B.E., Gholz, H.L., Clark, K.L., Falge, E., Ellsworth, D.S., Goldstein, A.H., Monson, R.K., Hollinger, D., Falk, M., Chen, J. and Sparks, J.P. (2002) Modeling and measuring the effects of disturbance history and climate on carbon and water budgets in evergreen needleleaf forests. *Agricultural and Forest Meteorology* 113, 185–222.

Tietema, A., Emmett, B.A., Gundersen, P., Kjønaas, O.J. and Koopmans, C. (1998) The fate of 15N-labelled nitrogen deposition in coniferous forest ecosystems. *Forest Ecology and Management* 101, 19–27.

Tilman, D., Fargione, J., Wolff, B., D'Antonio, C., Dobson, A., Howarth, R., Schindler, D., Schlesinger, W.H., Simberloff, D. and Schwackhamer, D. (2001) Forecasting agriculturally driven global environmental change. *Science* 292, 281–284.

Tognetti, R., Cherubini, P. and Innes, J.L. (2000) Comparative stem-growth rates of Mediterranean trees under background and naturally enhanced ambient CO_2 concentrations. *New Phytologist* 146, 59–74.

Townsend, A.R., Braswell, B.H., Holland, E.A. and Penner, J.E. (1996) Spatial and temporal patterns in terrestrial carbon storage due to deposition of fossil fuel nitrogen. *Ecological Applications* 6, 806–814.

Townsend, A.R., Howarth, R.W., Bazzaz, F.A., Booth, M.S., Cleveland, C.C., Collinge, S.K., Dobson, A.P., Epstein, P.R., Holland, E.A., Keeney, D.R., Mallin, M.A., Rogers, C.A., Wayne, P. and Wolfe, A.H. (2003) Human health effects of a changing global nitrogen cycle. *Frontiers in Ecology and the Environment* 1, 240–246.

UN/ECE and EC (2004) *Forest Condition in Europe. Results of the 2003 Survey*. Brussels, Geneva.

Van den Pol-van Dasselaar, A., van Beusichem, M.L. and Oenema, O. (1999) Effects of nitrogen input and grazing on methane fluxes of extensively and intensively managed grasslands in the Netherlands. *Biology and Fertility of Soils* 29, 24–30.

Van Dijk, S.M. and Duyzer, J.H. (1999) Nitric oxide emissions from forest soils. *Journal of Geophysical Research* 104, 15955–15961.

Vitousek, P.M. and Melillo, J.M. (1979) Nitrate losses from disturbed forests: patterns and mechanisms. *Forest Science* 25, 605–619.

Whalen, S.C. and Reeburgh, W.S. (2000) Effect of nitrogen fertilization on atmospheric methane oxidation in boreal forest soils. *Chemosphere – Global Change Science* 2, 151–155.

Wolfe, A.H. and Patz, J.A. (2002) Reactive nitrogen and human health: acute and long-term implications. *Ambio* 31, 120–125.

Wrage, N., Velthof, G.L., van Beusichem, M.L. and Oenema, O. (2001) Role of nitrifier denitrification in the production of nitrous oxide. *Soil Biology and Biochemistry* 33, 1723–1732.

Yavitt, J.B., Downey, D., Lang, G. and Sexstone, A. (1990a) Methane consumption in two temperate forest soils. *Biogeochemistry* 9, 39–52.

Yavitt, J.B., Lang, G.E. and Sexstone, A.J. (1990b) Methane fluxes in wetland and forest soils, beaver ponds and low-order streams of a temperate forest ecosystem. *Journal of Geophysical Research* 95, 22463–22474.

Zechmeister-Boltenstern, S. and Meger, S. (1997) Nitrous oxide emissions from two beech forests near Vienna, Austria. *7th International Workshop on Nitrous Oxide Emissions*. Cologne, Germany, pp. 429–432.

Zechmeister-Boltenstern, S., Hahn, M., Meger, S. and Jandl, R. (2002) Nitrous oxide emissions and nitrate leaching in relation to microbial biomass dynamics in a beech forest soil. *Soil Biology and Biochemistry* 34, 823–832.

Index

(Bold numbers denote figures or tables)